Fundamentals of Electromagnetics with MATLAB®
Second Edition

To our wives: Vicki, Rossi, and Vickie

Fundamentals of Electromagnetics with MATLAB®
Second Edition

Karl E. Lonngren
Department of Electrical and Computer Engineering
The University of Iowa
Iowa City, Iowa

Sava V. Savov
Department of Electronic Engineering
Technical University of Varna
Varna, Bulgaria

Randy J. Jost
Space Dynamics Laboratory
Department of Electrical and Computer Engineering
Department of Physics
Utah State University
Logan, Utah

SCITECH PUBLISHING, INC.

Raleigh, NC
www.scitechpub.com

SciTech
PUBLISHING, INC.

©2007 by SciTech Publishing, Inc.
All rights reserved. No part of this book may be reproduced or used in any form whatsoever without written permission except in the case of brief quotations embodied in critical articles and reviews. For information, contact SciTech Publishing.

Printed in Taiwan
10 9 8 7 6 5 4 3

Hardcover ISBN: 978-1-891121-58-6
Paperback ISBN: 978-1-61353-000-9

Publisher and Editor: Dudley R. Kay
Production Director: Susan Manning
Production Assistant: Robert Lawless
Production and Design Services: Aptara
Copyeditor: Deborah Stirling
Cover Design: Kathy Gagne
Cover Illustration: NASA
Illustration Reviewer: Michael Georgiev

SciTech Publishing, Inc.
911 Paverstone Drive, Suite B
Raleigh, NC 27615
Phone: (919) 847-2434 Fax: (919) 847-2568
Email: info@scitechpub.com
Updates: www.scitechpub.com/lonngren2e.htm

Information contained in this work has been obtained from sources believed to be reliable. However, neither SciTech Publishing nor its authors guarantee the accuracy or completeness of any information published herein, and neither SciTech Publishing nor its authors shall be responsible for any errors, omissions, or damages arising out of use of this information. This work is published with the understanding that SciTech Publishing and its authors are supplying information but are not attempting to render engineering or other professional services. If such services are required the assistance of an appropriate professional should be sought.

Library of Congress Cataloging-in-Publication Data

Lonngren, Karl E. (Karl Erik), 1938-
  Fundamentals of electromagnetics with MATLAB / Karl E. Lonngren, Sava V. Savov, Randy J. Jost.
    p. cm.
  Includes bibliographical references and index.
  ISBN-13: 978-1-891121-58-6 (hardback : alk. paper)
  ISBN-10: 1-891121-58-8 (hardback : alk. paper)
 1. Electromagnetic theory. 2. Electric engineering. 3. MATLAB. I. Savov, Sava Vasilev. II. Jost, Randy J. III. Title.

QC670.6.L56 2007
537–dc22

2006101377

Publisher's Note to the Second Printing

The diligent efforts of many dedicated individuals to help eliminate errata and improve fine points of this text bear special commendation. In barely six months from the first printing, the authors, SciTech advisors, adopting instructors, their students, and very talented teaching assistants went through every page, figure, example, equation and problem to bring subsequent students and users this remarkably clean printing. Among the improvements that will benefits students and readers are the following:

- Chapter One – *MATLAB, Vectors, and Phasors* was rewritten to provide an even better review of vector analysis and a more detailed introduction to phasors and phasor notation.

- Sixteen new problems were added to Chapter Two – *Electrostatic Fields*

- All problems were checked for clarity and 100% accurate answers in Appendix G

- Highly detailed Solutions were derived for instructors in editable Word files

- Short sections were added on topics of EMF, power flow and energy deposition, and complex vectors

We continue to encourage submission of errata and suggestions for improvements.

We welcome **Dr. Jonathan Bagby** of Florida Atlantic University to the author team. Though Dr. Bagby is primarily concerned with the Optional Topics on the Student CD that will evolve into the *Intermediate Electromagnetics with MATLAB* text, his strong background in Mathematics was ideal in strengthening Chapter 1 and overseeing numerous improvements in Chapter 7's *Transmission Lines* notation.

Prof. Sven Bilen and his colleague **Prof. Svetla Jivkova** of Pennsylvania State University used their teaching experience from the text and invited student feedback to submit numerous corrections and suggested improvements to factual and conceptual aspects of the book. In addition, their student **Mickey Rhoades** had a keen eye for errors that went undetected by others. Taken all together, a profound debt of gratitude is owed to the PSU spring 2007 EM course team. Similarly, **Prof. Doran Baker** and **Prof. Donald Cripps** at Utah State gathered student feedback and submitted a summary for our reprint consideration.

The accuracy checks of problems and their selected answers in Appendix G, plus the improvements to the detailed steps in the Solutions Manual for instructors, are the work of three talented PhD honors students: **David Padgett** of North Carolina State University, **Zhihong Hu** of the University of Iowa, and **Avinash Uppuluri** of

Utah State University. The feedback of a remarkable undergraduate student who provided a voice to the eyes and mind of the typical neophyte reader must also be acknowledged: **David Ristov** of South Dakota State University.

We like to call our textbook and its supplements "organic" because they are continually growing and being nurtured by those who are passionate about the subject and teaching undergraduates. Using book reprintings and the electronic opportunities of CDs and the internet, our authors and contributors are dedicated to fixing flaws and adding helpful resources whenever they present themselves. We gratefully thank all who share our dedication to the subject, the process, and the delight in learning.

Dudley R. Kay – President and Founder
SciTech Publishing, Inc.
Raleigh, NC
July, 2007

Contents

Preface x

Editorial Advisory Board in Electromagnetics xx

Notation Table xxi

Chapter 1 — MATLAB, Vectors, and Phasors — 3

1.1 Understanding Vectors Using MATLAB 4
1.2 Coordinate Systems 17
1.3 Integral Relations for Vectors 29
1.4 Differential Relations for Vectors 37
1.5 Phasors 52
1.6 Conclusion 56
1.7 Problems 56

Chapter 2 — Electrostatic Fields — 61

2.1 Coulomb's Law 61
2.2 Electric Field 67
2.3 Superposition Principles 69
2.4 Gauss's Law 77
2.5 Potential Energy and Electric Potential 85
2.6 Numerical Integration 100
2.7 Dielectric Materials 109
2.8 Capacitance 114
2.9 Conclusion 118
2.10 Problems 119

Chapter 3 — Magnetostatic Fields — 123

3.1 Electrical Currents 123
3.2 Fundamentals of Magnetic Fields 128
3.3 Magnetic Vector Potential and the Biot-Savart Law 138
3.4 Magnetic Forces 146
3.5 Magnetic Materials 157
3.6 Magnetic Circuits 162

　　　　　3.7　Inductance　166
　　　　　3.8　Conclusion　171
　　　　　3.9　Problems　172

Chapter 4　Boundary Value Problems Using MATLAB　　177

　　　　　4.1　Boundary Conditions for Electric and Magnetic Fields　178
　　　　　4.2　Poisson's and Laplace's Equations　186
　　　　　4.3　Analytical Solution in One Dimension—Direct Integration Method　191
　　　　　4.4　Numerical Solution of a One-Dimensional Equation—Finite Difference Method　201
　　　　　4.5　Analytical Solution of a Two-Dimensional Equation—Separation of variables　211
　　　　　4.6　Finite Difference Method Using MATLAB　220
　　　　　4.7　Finite Element Method Using MATLAB　226
　　　　　4.8　Method of Moments Using MATLAB　241
　　　　　4.9　Conclusion　251
　　　　　4.10　Problems　252

Chapter 5　Time-Varying Electromagnetic Fields　　257

　　　　　5.1　Faraday's Law of Induction　257
　　　　　5.2　Equation of Continuity　270
　　　　　5.3　Displacement Current　274
　　　　　5.4　Maxwell's Equations　280
　　　　　5.5　Poynting's Theorem　285
　　　　　5.6　Time-Harmonic Electromagnetic Fields　290
　　　　　5.7　Conclusion　293
　　　　　5.8　Problems　294

Chapter 6　Electromagnetic Wave Propagation　　297

　　　　　6.1　Wave Equation　297
　　　　　6.2　One-Dimensional Wave Equation　302
　　　　　6.3　Time-Harmonic Plane Waves　318
　　　　　6.4　Plane Wave Propagation in a Dielectric Medium　325
　　　　　6.5　Reflection and Transmission of an Electromagnetic Wave　335
　　　　　6.6　Conclusion　349
　　　　　6.7　Problems　349

Chapter 7　Transmission Lines　　353

　　　　　7.1　Equivalent Electrical Circuits　354
　　　　　7.2　Transmission Line Equations　357

Contents

- 7.3 Sinusoidal Waves 362
- 7.4 Terminations 367
- 7.5 Impedance and Matching of a Transmission Line 373
- 7.6 Smith Chart 381
- 7.7 Transient Effects and the Bounce Diagram 390
- 7.8 Pulse Propagation 397
- 7.9 Lossy Transmission Lines 402
- 7.10 Dispersion and Group Velocity 406
- 7.11 Conclusion 414
- 7.12 Problems 414

Chapter 8 Radiation of Electromagnetic Waves 419

- 8.1 Radiation Fundamentals 419
- 8.2 Infinitesimal Electric Dipole Antenna 427
- 8.3 Finite Electric Dipole Antenna 434
- 8.4 Loop Antennas 440
- 8.5 Antenna Parameters 443
- 8.6 Antenna Arrays 455
- 8.7 Conclusion 466
- 8.8 Problems 467

Appendix A Mathematical Formulas 469

Appendix B Material Parameters 474

Appendix C Mathematical Foundation of the Finite Element Method 477

Appendix D Transmission Line Parameters of Two Parallel Wires 483

Appendix E Plasma Evolution Adjacent to a Metallic Surface 487

Appendix F Bibliography 490

Appendix G Selected Answers 493

Appendix H Greek Alphabet 523

Index 524

Preface

Overview

Professors ask, "Why another textbook (edition)?" while students ask, "Why do I need to study electromagnetics?" The concise answers are that today's instructor needs more flexible options in topic selection, and students will better understand a difficult subject in their world of microelectronics and wireless if offered the opportunity to apply their considerable computer skills to problems and applications. We see many good textbooks but none with the built-in flexibility for instructors and computer-augmented orientation that we have found successful with our own students.

Virtually every four-year electrical and computer engineering program requires a course in electromagnetic fields and waves encompassing Maxwell's equations. Understanding and appreciating the laws of Nature that govern the speed of even the smallest computer chip or largest power line is fundamental for every electrical and computer engineer. Practicing engineers review these principles constantly, many regretting either their inattention as undergrads or the condensed, rushed nature of the single course. What used to be two or more terms of required study has been whittled down to one very intense term, with variations of emphasis and order. Recently, there has been a resurgence of the two-term course, or at least an elective second term, that is gathering momentum as a desirable, career-enhancing option in a wireless world. Students today have grown up with computers; they employ sophisticated simulation and calculation programs quite literally as child's play. When one considers the difficult challenges of this field of study, the variation among schools and individual instructors in course structure and emphasis, and the diverse backgrounds and abilities of students, you have the reason for another textbook in electromagnetics: learning by doing on the computer, using the premier software tool available in electrical engineering education today: MATLAB.

Textbook and Supplements on CD

Actually, this is much more than a mere textbook. The book itself offers a structural framework of principles, key equations, illustrations, and problems. With that crucial supporting structure, each instructor, student, or reader can turn to the supplemental files provided with this book or available online to customize and decorate each topic room. The entire learning package is "organic" as we the authors, contributing EM instructors, and SciTech Publishing strive to bring you an array of supporting material through the CD, the Internet, and files stored on your computer. It is very important, therefore, that you **register your book** and bookmark the URL that will always be available as a starting and reference point for ever-changing supplementary materials: **www.scitechpub.com/lonngren2e.htm**.

Approach Using MATLAB®[1]

Our underlying philosophy is that you can learn and apply this subject's difficult principles much more easily, and possibly even enjoyably, using MATLAB. Numerical computations are readily solved using MATLAB. Also, abstract theory of unobservable waves can be strikingly visualized using MATLAB. Perhaps you are either familiar with MATLAB through personal use or through a previous course and can immediately apply it to your study of electromagnetics. However, if you are unfamiliar with MATLAB, you can learn to use it on your own very quickly. A *MATLAB Tutorial* is supplied on the book's enclosed student CD. The extensive Lesson 0 is all you really need to establish a solid starting point and build on it. If you do not have the CD or a computer handy, Chapter One provides a brief overview of MATLAB operations and a review of vector analysis. For more information and instruction on using MATLAB, SciTech Publishing provides a list of MATLAB books and CDs in Appendix F, available at special discount prices to registered users of this book. If your book came without the CD (a used book purchase, perhaps), or even if you want to be sure of obtaining the latest files, you can purchase an electronic license for a year's access to all files at the URL shown above.

Within the book, MATLAB is used numerous ways. You will always be able to see where MATLAB is either applied or has the potential to be applied by the universal icon furnished by MATLAB's parent company MathWorks. Each time this icon appears, you will know that either MATLAB's M-files of program code are supplied on your student CD or else your instructor or TA has them, most typically on Problem solutions. Use and distribution of these solution M-files are at the discretion of individual instructors and are not, therefore, furnished to students. Non-student readers can contact the publisher for selected solutions and M-files if registered.

---

[1] MATLAB is a registered trademark of The MathWorks, Inc. For MATLAB product information and cool user code contributions, go to www.mathworks.com, write The MathWorks, Inc., 3 Apple Hill Dr., Natick, MA 01760-2098 or call (508) 647-7101.

- *Examples* – Worked-out examples run throughout the text to show how a proof can be derived or a problem solved in steps. Each is clearly marked with a heading and also appears in a tinted blue box. When the MATLAB icon appears, it means the worked solution also has an equivalent M-file. Here is how Example sections appear:

EXAMPLE 7.3

The voltage wave that propagates along a transmission line is detected at the indicated points. From this data, write an expression for the wave. Note that there is a propagation of the sinusoidal signal to increasing values of the coordinate z.

- *Figures* – Numerous figures within the text were generated using MATLAB. Not only can you obtain and manipulate the program code with its corresponding M-file, but you can also view this figure in full color from the CD. Here is how a MATLAB-generated figure will appear:

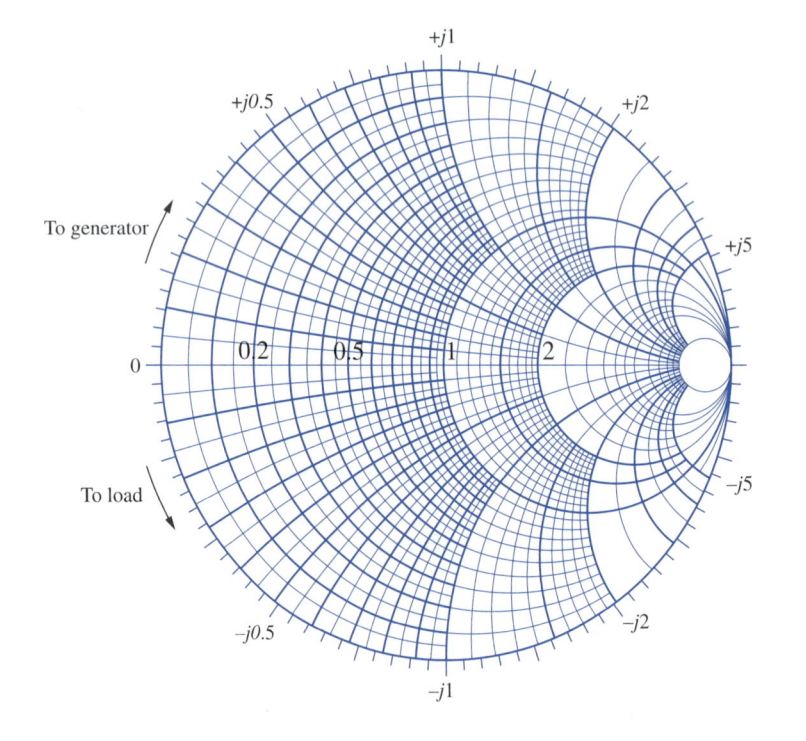

FIGURE 7-9

A Smith chart created with MATLAB.

Preface   xiii

- *Problems* – Each chapter contains numerous problems of varying complexity. Problem numbers correspond to the text sections so you can review any problems that prove difficult to handle at first pass. When an icon appears, you will know this problem can be solved using your MATLAB skills. Discuss with your instructor if the M-files will be made available for checking your work. Answers to selected problems are provided in Appendix G

   7.6.1 Using a Smith chart, find the impedance Z_{in} of a 50-Ω coaxial cable that is terminated in a load $Z_L = (25 + j25)\ \Omega$. The coaxial cable has a length of $3\lambda/8$.

   7.6.2 Using a Smith chart, find the admittance Y_{in} of a 50-Ω coaxial cable that is terminated in a load $Z_L = (25 + j25)\ \Omega$. The coaxial cable has a length of $\lambda/8$.

   7.6.3 Using a Smith chart, find the distance from a load impedance $Z_L = (25 + j25)\ \Omega$ that is connected to a 50-Ω coaxial cable where the normalized input

   7.7.3. Sketch the current profile at $z = \mathcal{L}/2$ as a function of time $0 < t < 4(\mathcal{L}/v)$ for the transmission line stated in Problem 7.7.2.

   7.7.4. Two transmission lines are joined with a resistor R_L.

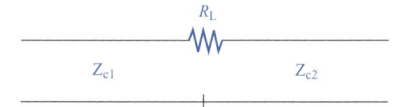

- *Animations* – It is possible to portray electromagnetic principles in animation, also sometimes called "movies." Authors, instructors, and even students have contributed a number of such animations showing principles at work. However, the real fun may be in manipulating the variables to produce different results. Thus the M-files become a starting point for that and also an instructive demo for leading you to your own creations. Animations at the time of this book's printing are indicated by a "flying disk" in the margin, close to the most pertinent text discussion of the underlying principle. You can create and submit animations to the publisher for posting and credit to you, using the submission form on the website. Also, check back for new animations.

Student CD

The Student CD enclosed with your textbook is a powerful resource. Not only does it contain files that are immediately and directly related to your course study, but it also offers a wealth of supplementary and advanced material for a 2nd term of study or your personal explorations. By registering your book, any new material produced for future CDs will be offered to you as web downloads. Here is what the Student CD contains:

MATLAB Tutorial

Readers will come to this book with widely varying exposures to MATLAB. A self-paced tutorial has been included on the CD. Divided into lessons, MATLAB operations and tools are introduced within the context of Electromagnetics extensive notation, subject areas, examples, and problems. That is, the MATLAB tutorial gets you started with basics first and then develops text topics incrementally. You will eventually learn to perform relatively complex operations, problem-solving, and visualizations. Your instructor may later choose to assign projects involving multi-step problem-solving in MATLAB. Independent readers seeking these Projects should contact SciTech after registering the book. Also, an array of helpful books and tutorials about MATLAB for engineers is kept up to date on the SciTech website, always at discount prices.

Optional Topics

Some schools require a second term of Electromagnetics and most at least offer a second term as an elective Advanced Electromagnetics course. Most of today's textbooks contain somewhat more material than can be covered in one term but not enough material for a full and flexible second term. Therefore, a second textbook is often required for the follow-up course, one that is costly, probably does not match the notation of the first textbook, and may even contradict the first book in places because of the difference in notation. These inconveniences are overcome with the CD's extended topics. *Optional Topics* are provided in PDF files that match the two-color design of the text, integrate MATLAB throughout, and contain the same array of problems. The book and CD, therefore, satisfy the needs of most two-term courses and many elective second-term courses. Additional optional topics are being added continually, as they are suggested and contributed by instructors with course-specific needs, such as biomedical engineering, wireless communications, materials science, military applications, and so forth. Contact SciTech Publishing if you wish to suggest or contribute a new topic.

Applications

While brief references are made throughout the text to real-world applications of electromagnetic principles, we have chosen not to interrupt text flow with lengthy application discussions. Instead, applications are done proper justice in three–five page descriptions with graphics, in most cases, on the CD in PDF format. These applications point the way toward the utility of later courses in Microwave and RF, Wireless Communications, Antennas, High-speed Electronics, and many other career and research interests. They answer the age-old student question: "Why do I have to know this stuff?" Instructors, their TAs, and students are encouraged to submit additional applications for inclusion in the web-based Shareware Community.

M-Files

The Student CD contains M-files for selected examples, figures, and animations. Any additional MATLAB items will usually be accompanied by M-files, so check the website and register your book for notifications of new items. New submissions and suggested improvements to existing M code will always be gratefully received and acknowledged.

Note to Students: Equation Importance and Notational Schemes May Vary

You will encounter scads of equations in your study of electromagnetics. They are not all of equal importance. We have tried to make them clear by setting off most of them from the text and numbering them for reference. Note that we have taken the additional step of drawing a box around the most important equations, so look for those when you review.

"Mathematical notation" is how physical quantities, unit dimensions, and concepts are put into mathematical expression. While some notation is simple and common to the physical sciences, others can be arbitrary and a matter of choice to the author. The only firm rule is that the author, or other user of notation (such as an instructor writing on the board, in PowerPoint, or on a test), be clear about what the symbols represent and be consistent in their usage. Students may sometimes be unduly concerned about their textbook's notation if it is different from a previous text or what their instructor uses. Adapting to varying notational schemes is simply part of the learning process that "comes with the territory" in physics and engineering. In choosing the notation for this text we called upon our Editorial Advisory Board to determine preferences and precedents in the most widely referenced electromagnetics books. Our notational scheme is shown on page xix, and we define our symbols when first used in the text. We are acutely aware of the confusion and worry that poor notation causes students. However, be forewarned that your instructor or TA may choose to use their own notational schemes. There is no 'right' or "wrong" method. If you find any apparent inconsistencies or absence of clear definition within the textbook, please report them to us.

Instructor Resources

Apart from the text and Student CD, numerous resources are available to instructors. Our Shareware Community of invited contributions is intended to grow the nature and number of teaching and evaluation tools. Instructors using this book as the required text are entitled to the following materials:

- *Solutions* to all chapter exercises: step-by-step Word files and MATLAB M-files for MATLAB-solvable problems designed by the MATLAB icon

- *Exam Sets* comprised of three exams in each set, including answers and solutions. Additional exams are solicited to add to the Shareware database

- *PowerPoint Slides* of all figures in the text, organized by chapter, including the full color version of MATLAB-generated figures. Additional supplementary figures may be added over time as part of the Shareware Program.

- *Projects* – the "starter set" of complex, multi-step problems involving use of MATLAB, including recommended student evaluation scoring sheets that break down the credit to be given for every step and aspect. Projects are an exciting component that makes excellent use of MATLAB skills to test full understanding and application of principles. They are a prime element of the SciTech Shareware Program.

Shareware in Electromagnetics (SWEM) Program

As the name implies, Shareware in electromagnetics has been set up by SciTech Publishing as a means for instructors, teaching assistants, and even students to share their ideas and methods for learning, appreciating, and applying electromagnetic principles. Some items are accessible only by our textbook adopters and buyers (exam sets and their solutions, for example), while others are set up to be shared publicly, the only stipulation being registration and a contribution, however modest, to SWEM. Think of it as a neighborhood block party for the electromagnetics teaching and learning community. Bring your "covered dish, salad, or dessert" and share in the overall goodies, fun, and collegiality. For major SWEM items such as Projects, Applications, and Optional Topics, custom submission forms have been created on the website for ease of a contribution. For general ideas and items, such as a cool web link or figures, a generic form may be used. In any case, one can simply submit an email to SWEM@scitechpub.com. Acknowledgements will always be given for submissions unless anonymity is requested. See the available shareware and submission forms under "Instructors" at www.scitechpub.com/lonngren2e.htm.

We recognize that there are several different approaches to teaching electromagnetics in the usual engineering curriculum. We have tried to make our Second Edition adaptable to every approach. Every course will quickly review mathematical techniques and background material from previous math and physics courses. After that, curriculum sequences move down different paths.

Historical Approach – One traditional sequence is to follow a historical approach, where topics are covered in a sequence similar to the historical development of the subject matter, paralleling the experiments that revealed electromagnetic phenomena from ancient times. Thus, the usual course starts with electrostatics, then covers magnetostatics, introduces Maxwell's equations, wave phenomena, and follows up with applications such as transmission lines, waveguides, antennas, etc. This has probably been the most common curriculum sequence used until recently, and our book maintains this traditional approach.

Maxwell's Equations Approach – Another approach is to present Maxwell's equations early, develop wave phenomenon from them and then cover electrostatics, magnetostatics, and applications. A common variation on this with physics departments is to show how Maxwell's equations follow from relativity. This approach tends to require more mathematical sophistication from students, but it is popular with some instructors because of its independence from experimentally derived laws.

Transmission Lines Approach – Because of the perceived higher level of mathematics associated with electromagnetics, an increasing number of instructors today prefer to build upon the subject matter with which the student is already familiar. Thus, they prefer to introduce transmission lines early as a logical extension of the circuit theory that students

Fundamental Physical Constants

Quantity	Symbol	Value	Unit
Avogadro constant	N_A	6.0221415×10^{23}	mol^{-1}
Boltzmann constant	k	$1.3806505 \times 10^{-23}$	J/K
Planck constant	h	$6.6260693 \times 10^{-34}$	J sec
electron charge	e	$1.60217653 \times 10^{-19}$	C
electron mass	m_e	$9.1093826 \times 10^{-31}$	kg
proton mass	m_p	$1.67262171 \times 10^{-27}$	kg
proton-electron mass ratio	m_p/m_e	1836.15267261	
Gravitation constant	G	6.6742×10^{-11}	$m^3/kg\ sec^2$
standard acceleration of gravity	g_n	9.80665	m/sec^2
permittivity in vacuum	ε_0	$8.854187817 \times 10^{-12}$	F/m
Approx. value ε_0	ε_0	$(1/36\pi) \times 10^{-9}$	F/m
permeability in vacuum	μ_0	$4\pi \times 10^{-7}$	H/m
Approx. value μ_0	μ_0	$12.566370614 \times 10^{-7}$	H/m
speed of light in vacuum	c, c_0	$299,792,458$	m/sec
characteristic impedance in vacuum	Z_0	376.730313461	Ω

Additional physical constants can be found on the National Institute of Standards and Technology (NIST) website at: http://physics.nist.gov/cuu/Constants/

SI Prefixes

10^n	Prefix	Symbol	10^n	Prefix	Symbol
10^{24}	yotta	Y	10^{-1}	deci	d
10^{21}	zetta	Z	10^{-2}	centi	c
10^{18}	exa	E	10^{-3}	milli	m
10^{15}	peta	P	10^{-6}	micro	μ
10^{12}	tera	T	10^{-9}	nano	n
10^{9}	giga	G	10^{-12}	pico	p
10^{6}	mega	M	10^{-15}	femto	f
10^{3}	kilo	k	10^{-18}	atto	a
10^{2}	hecto	h	10^{-21}	zepto	z
10^{1}	deca, deka	da	10^{-24}	yocto	y

have studied prior to arriving in their electromagnetics courses. This approach has several advantages, beyond the obvious one of using familiar circuit analogies to help the student develop their physical insight. Using transmission lines, students are exposed to applications of the electromagnetics theory early in the class, providing them with a rationale for the need of this subject matter. Secondly, transmission lines naturally incorporate many of the concepts that students sometimes find difficult to visualize when talking about fields and waves, such as time delay, dispersion, attenuation, etc.

Students should appreciate that there is no one best way to learn (or teach) this material. Each of these approaches is equally valid, and chances are that the student can learn this material, whatever the approach, if a sustained effort is made. To support both the student and the instructor in this educational effort, we have tried to make this text flexible enough to be used with a variety of curriculum sequences with varying degrees of emphasis. For instance, we made Chapter 7 on transmission lines, independent of other chapters, so those instructors wishing to cover this material first can do so, with a minimum of backtracking required. On the other hand, the instructor and student can start at the beginning of the text and work forward through the chapter material in a more conventional sequencing of chapters. Additionally, instructors can choose to skip more advanced sections of the chapters, so they can cover more topics at the expense of depth of topic coverage. Thus, this text can be used for a one or two quarter format, or in a one or two semester format, especially when supplemented by the topics on the accompanying CD. Following are some suggestions for course syllabi, depending on what the instructor wishes to emphasize and how much time he or she has available.

- A traditional one-quarter course primarily emphasizing static fields could be covered using the first five chapters.

- A traditional one-semester course with reduced emphasis on static fields, but including transmission line applications would include Chapters 1, 2, 3, and portions of Chapters 4 and 5 (Sections 4-1 through 4-6 and Sections 5-1 through 5-5), followed by Chapter 6 and portions of Chapter 7 (Sections 7-1 through 7-8).

- A one-semester "transmission lines first" approach, consisting of Chapter 1, followed by Chapter 7 (Sections 7-1 through 7-8), then Chapters 2 and 3, followed by portions of Chapter 4 (Sections 4-1 through 4-6), Chapter 5 (Sections 5-1 through 5-5) and Chapter 6.

A second semester course can be developed from the remainder of the text, as well as selected supplemental topics from the student CD. For instance, a second semester course emphasizing EM waves and their applications would consist of a review of the material previously covered in Chapters 4–7, and additional selections from Chapter 4 (Sections 4-6 through 4-8), as well as Chapter 8, and CD selections on transmission lines, waveguides and antennas. The Instructor's Resource CD offers additional suggestions, and students and instructors should check the appropriate sections of the website for updates.

Of course, these are just suggestions; the actual course content will reflect the interests of the instructors and the programs for which they are preparing their students. It is for this reason that we have tried to enhance the flexibility of this text with supplemental material covering a variety of electromagnetic topics. Your suggestions and contributions of additional topics are invited and welcomed. Let us know your thoughts.

Acknowledgments

We are extremely proud of this new edition and the improvements made to virtually every aspect of the book, its supplements, and the web support. In very large part these upgrades are due to our adopting professors, their students, and the editorial advisory board members who volunteered to assist. All share our passion for electromagnetics. Adopters who taught from the first edition and passed along many helpful corrections and suggestions were:

Jonathon Wu – Windsor University
David Heckmann – University of North Dakota
Perry Wheless, Jr. – University of Alabama
Fran Harackiewicz – Southern Illinois University at Carbondale

Advisors who have given conceptual suggestions, read chapters closely, and critiqued the supplements in detail are:

Sven Bilén – Pennsylvania State University
Don Dudley – University of Arizona
Chuck Bunting – Oklahoma State University
Jim West – Oklahoma State University
Kent Chamberlin – University of New Hampshire
Christos Christodoulou – University of New Mexico
Larry Cohen – Naval Research Laboratory
Atef Elsherbeni – The University of Mississippi
Cindy Furse – University of Utah
Mike Havrilla – Air Force Institute of Technology
Anthony Martin – Clemson University
Wilson Pearson – Clemson University

The publication of an undergraduate textbook represents a close collaboration among authors, technical advisors, and the publisher's staff. We consider ourselves extremely fortunate to be working with a highly personal, committed, and passionate publisher like SciTech Publishing. President and editor Dudley Kay has believed in our approach and the book's potential for wide acceptance through two editions. Susan Manning is our production director, and we have learned to appreciate her with awe and astonishment for all the bits and pieces of a textbook that must be managed. Production Assistant Robert Lawless, an engineering graduate student at North Carolina State University, provided incredible support, understanding, and suggestions for reader clarity. Bob Doran's unfailing good humor and optimism is just what every author wants in a sales director.

Special thanks to Michael Georgiev, Sava Savov's graduate student, for his early work on the illustrations program, its technical accuracy, and checks on its conversion to two-color renderings. Graphic designer Kathy Gagne's cover art, chapter openings, web pages, and advertising pages are all a remarkably cohesive effort that helps support our vision brilliantly. Dr. Bob Roth and daughter Anne Roth have extended Kathy's design to beautifully matching student support and instructor feedback web pages.

The authors would like to acknowledge Professor Louis A. Frank, Dr. John B. Sigwarth, and NASA for permission to use the picture on the cover of this book, Professor Er-Wei Bai for discussions concerning MATLAB, and Professor Jon Kuhl for his support of this project.

Errors and Suggestions

Our publisher assures us that the perfect book has yet to be written. We can reasonably expect some errors, but we need not tolerate them. We also strive for increased clarity on a challenging subject. We invite instructors, students, and individual readers to contribute to the errata list and recommendations via the feedback forms on the website. SciTech will maintain a continually updated errata sheet and corrected page PDFs on the website, and the corrections and approved suggestions will be incorporated into each new printing of the book. We vow to make this project "organic" in the sense that it will be continually growing, shedding dead and damaged leaves, and becoming continuously richer and more palatable. If we are successful and realize our goal, every reader and user of the book and its supplements will have a hand in its evolving refinements. We accept all responsibility for any errors and shortcomings and invite direct comments at any time.

Karl E. Lonngren
University of Iowa
Iowa City, IA
lonngren@engineering.uiowa.edu

Sava V. Savov
Technical University of Varna
Varna, Bulgaria
svsavov@ms.ieee.bg

Randy Jost
Utah State University
Logan, UT
rjost@engineering.usu.edu

Editorial Advisory Board in Electromagnetics

Dr. Randy Jost–Chair
Space Dynamics Lab and
ECE Department Utah State University

Dr. Jon Bagby
Department of Electrical Engineering
Florida Atlantic University

Dr. Sven G. Bilen
Penn State University
University Park, PA

Dr. Chuck Bunting
ECE Department Oklahoma State University

Mr. Kernan Chaisson
Electronic Warfare Consultant
Rockville, MD

Dr. Kent Chamberlin
ECE Department University
of New Hampshire

Dr. Christos Christodoulou
ECE Department University
of New Mexico

Mr. Larry Cohen
Naval Research Lab Washington, D.C.

Dr. Atef Elsherbeni
ECE Department University of Mississippi

Dr. Thomas Xinzhang Wu
School of Electrical Engineering
and Computer Science University
of Central Florida

Dr. Cynthia Furse
ECE Department University of Utah

Dr. Hugh Griffiths
Principal, Defence College
of Management & Technology
Cranfield University Defence Academy
of the United Kingdom

Dr. William Jemison
ECE Department Lafayette College

Mr. Jay Kralovec
Harris Corporation Melbourne, FL

Mr. David Lynch, Jr.
DL Sciences, Inc.

Dr. Anthony Martin
ECE Department Clemson University

Dr. Wilson Pearson
ECE Department Clemson University

Mr. Robert C. Tauber
SAIC–Las Vegas EW University–Edwards AFB

Notation Table

Coordinate System	Coordinates
Cartesian	(x, y, z)
Cylindrical	(ρ, ϕ, z)
Spherical	(r, θ, ϕ)

Quantity	Symbol	SI Unit	Abbreviations	Dimensions
Amount of substance		mole	mol	
Angle	ϕ, θ	radian	rad	
Electric current	$I; i$	ampere	A	I
Length	$\mathscr{L}, \ell, \ldots$	meter	m	L
Luminous intensity		candela	cd	
Mass	$M; m$	kilogram	kg	M
Thermodynamic temperature		Kelvin	K	K
Time	t, T	second	s	T
Admittance	Y	Siemens	S	I^2T^3/ML^2
Capacitance	C	farad	F	I^2T^4/ML^2
Charge; point charge	$Q; q$	Coulomb	C	IT
Charge density, line	ρ_ℓ	coulomb/meter	C/m	IT/L
Charge density, surface	ρ_s	coulomb/meter^2	C/m^2	IT/L^2
Charge density, volume	ρ_v	coulomb/meter^3	C/m^3	IT/L^3
Conductance	G	Siemens	S	I^2T^3/ML^2
Conductivity	σ	Siemens/meter	S/m	I^2T^3/ML^3
Current density	$\mathbf{J}$	ampere/meter^2	A/m^2	I/L^2
Current density, surface	$\mathbf{J}_\ell$	ampere/meter	A/m	I/L
Electric flux density	$\mathbf{D}$	coulomb/meter^2	C/m^2	IT/L^2
Electric dipole moment	$\mathbf{p}$	coulomb $\cdot$ meter	C $\cdot$ m	ITL
Electrical field intensity	$\mathbf{E}$	Volt/meter	V/m	ML/IT^3
Electric potential; Voltage	V	Volt	V	ML^2/IT^3

Quantity	Symbol	SI Unit	Abbreviations	Dimensions
Energy, Work	W	joule	J	ML^2/T^2
Energy density, volume	w	joule/meter^3	J/m^3	M/LT^2
Force	**F**	Newton	N	ML/T^2
Force density, volume	**f**	Newton/meter^3	N/m^3	$M/L^2/T^2$
Frequency	f	Hertz	Hz	$1/T$
Impedance	Z	Ohm	Ω	$ML^2/I^2/T^3$
Inductance	$L; M$	Henry	H	$ML^2/I^2/T^2$
Magnetic dipole moment	**m**	ampere · meter^2	A · m^2	IL^2
Magnetic field intensity	**H**	ampere/meter	A/m	I/L
Magnetic flux	Φ	Weber	Wb	$ML^2/I/T^2$
Magnetic flux density	**B**	Tesla	T	$M/I/T^2$
Magnetic vector potential	**A**	Tesla · meter	T · m	$ML/I/T^2$
Magnetic voltage	V_m	ampere	A	I
Magnetization	**M**	ampere/meter	A/m	I/L
Period	T	second	s	T
Permeability	μ	Henry/meter	H/m	$ML/I^2/T^2$
Permittivity	ε	farad/meter	F/m	$I^2T^4/M/L^3$
Polarization	**P**	coulomb/meter^2	C/m^2	IT/L^2
Power	P	watt	W	ML^2/T^3
Power density, volume	p	watt/meter^3	W/m^3	$M/L/T^3$
Power density, surface	S	watt/meter^2	W/m^2	$M/L/T^2$
Propagation constant	γ	1/meter	1/m	$1/L$
Reluctance	Γ_m	1/Henry	1/H	$I^2T^2/M/L^2$
Resistance	R	Ohm	Ω	$ML^2/I^2/T^3$
Torque	**T**	Newton · meter	N · m	ML^2/T^2
Velocity	**v**	meter/second	m/s	L/T
Wavelength	λ	meter	m	L

CHAPTER 1

MATLAB, Vectors, and Phasors

1.1	Understanding Vectors Using MATLAB	4
1.2	Coordinate Systems	17
1.3	Integral Relations for Vectors	29
1.4	Differential Relations for Vectors	37
1.5	Phasors	52
1.6	Conclusion	56
1.7	Problems	56

This chapter introduces readers to MATLAB, either as a gentle beginning or as a refresher of specific properties and operations. Your book should include a CD containing a self-study MATLAB tutorial with more depth and practice than can be provided here. If you did not receive a CD with your book, if you need a replacement, or if you want to be sure you have the most recent files, go to www.scitechpub.com/lonngren2e.htm. Meanwhile, this chapter provides a convenient application of MATLAB and review of vectors to our study of electromagnetics.

MATLAB, widely used in engineering studies and industry, is a software tool that allows you to obtain visual images easily of various electromagnetic phenomena that we will soon encounter. Also, vectors, which are so crucial in describing electromagnetic phenomena, can be easily manipulated using MATLAB. Numerous figures in this text have been created using MATLAB, as can be seen by the MATLAB icon beside them. For simplicity, we will emphasize Cartesian coordinates in this review. The vector operations in other coordinate systems are included in Appendix A. Our motivation in employing vectors is that electromagnetic fields are vector quantities and their use will permit us to use a fairly compact notation to represent sets of partial differential equations. This review will include a presentation of the vector differential operations of the gradient, divergence, and curl. The transformation of a vector from one coordinate system to another will be discussed. It is often helpful to be able to transform a problem from one coordinate system to another in order to see additional symmetry and gain further insight. Readers who feel comfortable with vector terminology can easily skip this section of the chapter and move on with no loss of continuity. Just note that in this text, bold-face type will be used to denote a vector **A**, and the symbol **u**$_A$ will be used to indicate the unit vector corresponding to this vector. This chapter concludes with some brief comments on phasors.

1.1 Understanding Vectors Using MATLAB

MATLAB software is widely available at most schools and departments of electrical and computer engineering. That is why we make extensive use of it, and no doubt your instructor understands its capabilities and advantages. MATLAB is a very useful tool in assisting understanding of engineering electromagnetics, because it combines both the capabilities of a traditional calculator as well as a programming language. The two- and three-dimensional plotting capabilities are exploited throughout this text since a picture or a graph aids in the physical interpretation of an equation. Here we present several important features of this program that will be useful in comprehending electromagnetic theory.

Various functions such as trig functions appear in a MATLAB library that can be easily called up and used to visualize an abstract equation. The user can customize and add to this list by writing a program in an M-file. Numerous MATLAB figures and problem statements are included throughout this textbook, and the files that have been used to create them are available on the enclosed CD. These programs are characterized with file names such as "example_103" and "figure_103" to indicate the third example and the third figure in Chapter 1, respectively. The example and figure captions are clearly identified in the book with the MATLAB icon to let you know an M-file exists. We assume that the reader is able to call up MATLAB with the familiar MATLAB prompt ">>" appearing on the screen. Typing the words "help *topic*" after the prompt brings on-screen help to the user when needed.

Let's start with a simple example. Type the following command after the prompt, press the Enter key, and note the following statements that appear on the screen:

$$>>x = 3$$
$$x =$$
$$3$$
$$>> \tag{1.1}$$

The computer has assigned a value for the variable *x* that it remembers until changed or until we exit the program. This is an example of using MATLAB in the calculator mode, with the result available immediately after entering the value for the variable. MATLAB is now ready for the next input. Let us

1.1 Understanding Vectors Using MATLAB

choose the value $y = 4$ but indicate that it not print back this number immediately. In MATLAB, certain functions are implemented through the use of special characters. Many of these special characters mean different things, depending on the context in which they are used. One of these special characters is the semicolon ";". When placed at the end of a line, it suppresses the printing of the output, as we see in the following example:

$$>>y = 4; \tag{1.2}$$
$$>>$$

This may not seem important at this stage, but a simple statement in a program could lead to screen clutter or waste of printer paper as numbers spew forth.

Mathematical operations with these two numbers will follow, and we will write a mathematical operation at the prompt. In the table given below, the following three lines will appear after we push the Enter key.

Addition	Subtraction	Multiplication	Division
$>>z = x + y$	$>>z = x - y$	$>>z = x * y$	$>>z = x / y$
$z =$	$z =$	$z =$	$z =$
7	−1	12	0.7500
>>	>>	>>	>>

Note the four-place accuracy in the last column. The user can control the accuracy.

The special character semicolon, ";" also has several other uses. Using the semicolon, it is possible to write several commands on one line. For example, the addition program can also be written in one line as

$$>>x = 3; \quad y = 4; \quad z = x + y; \tag{1.3}$$

In order to obtain the solution using this operation, just type "z" at the MATLAB prompt and the program responds

$$>>z$$
$$z =$$
$$7 \tag{1.4}$$
$$>>$$

Another useful special character is the percent sign, "%". The percent sign signals the logical end of a line, thus, anything that comes after the percent sign is ignored. It is a convenient way to add comments to document a program or subroutine.

In electromagnetics, we frequently use the concept of a ***field***. A field is an assignment of a physical quantity to a point in space. Typically a field encompasses a physical quantity that extends over a large, quantifiable region of space. The physical quantities that make up the fields we most often encounter in day-to-day life are usually defined by either a magnitude or a magnitude and a direction. Quantities that can be described by a magnitude only are called *scalars*. Energy, temperature, weight, and speed are all examples of scalar quantities. Other quantities, called *vectors* require both a magnitude and a direction to fully characterize them. Examples of vector quantities include force, velocity, and acceleration. Thus, a car traveling at 30 miles per hour (mph) can be described by the scalar quantity speed. However, a car traveling 30 mph in a northwest direction can be described by the vector quantity velocity, which has both a magnitude—the 30 mph speed—and a direction—northwest.

The ability to employ vector notation allows us the convenience of visualizing problems with or without the specification of a coordinate system. After choosing the coordinate system that most concisely describes the distribution of the field, we then specify the field with the components determined with regard to that coordinate system (i.e., Cartesian, cylindrical, spherical). Detailed exposition of vector operations will be given in Cartesian coordinates with the equivalent results just stated in the other systems. There are a large number of orthogonal coordinate systems, and there is a generalized orthogonal coordinate system. The term *orthogonal* implies that every point in a particular coordinate system can be defined as the intersection of three mutually-perpendicular surfaces in that coordinate system. This will be further examined later.

A vector can be specified in MATLAB by stating its three components. We use a capital letter to identify a vector in MATLAB notation, while lowercase letters are reserved for scalar quantities. This is not required, but it adds clarity to the work. The unit vector corresponding to a given vector is defined as a vector whose magnitude is equal to one and oriented in the same direction as the vector. For example, in a Cartesian coordinate system the vector $\mathbf{A} = A_x \mathbf{u_x} + A_y \mathbf{u_y} + A_z \mathbf{u_z}$, where A_x is the magnitude of the x component of the vector $\mathbf{A}$, and $\mathbf{u_x}$ is a unit vector directed along the x axis. This is written as

$$>> A = [A_x \ A_y]; \\ >> A = [A_x \ A_y \ A_z]; \quad (1.5)$$

in two and three dimensions, respectively. In MATLAB notation, this vector can be displayed by just typing 'A' at the prompt.

$$>> A \\ A = \\ \quad A_x \ A_y \ A_z \\ >> \quad (1.6)$$

We must insert a space or a comma between components of the vectors that are numbers.

1.1 Understanding Vectors Using MATLAB

Let us now specify numerical values for the three components: $A = [1\ 2\ 3]$. A second vector $\mathbf{B} = 2\mathbf{u}_x + 3\mathbf{u}_y + 4\mathbf{u}_z$ is written as

$$>> B = [2\ 3\ 4]; \tag{1.7}$$

where we again employed the semicolon in order to save space.

Having "stored" the two vectors $\mathbf{A}$ and $\mathbf{B}$ into computer memory, we can then perform various mathematical operations. The vectors can be added as $\mathbf{C} = \mathbf{A} + \mathbf{B}$ by typing

$$\begin{aligned}&>> C = A + B\\&C =\\&\quad 3\ 5\ 7\\&>>\end{aligned} \tag{1.8}$$

The vector is interpreted as $\mathbf{C} = 3\mathbf{u}_x + 5\mathbf{u}_y + 7\mathbf{u}_z$.

The two vectors can also be subtracted as $\mathbf{D} = \mathbf{A} - \mathbf{B}$ with the command

$$\begin{aligned}&>> D = A - B\\&D =\\&\quad -1\ -1\ -1\\&>>\end{aligned} \tag{1.9}$$

The vector is interpreted to be $\mathbf{D} = -1\mathbf{u}_x - 1\mathbf{u}_y - 1\mathbf{u}_z$. We will incorporate other vector operations using MATLAB—such as the scalar (or "dot") product, the vector (or "cross") product, and various differential operations—when they are introduced within the text.

Finally, the magnitude of a vector $\mathbf{A}$, denoted $|\mathbf{A}|$ or A, can also been computed using the MATLAB command *norm (A)*. The unit vector $\mathbf{u}_A$ can be found using the norm function. It is equal to the vector divided by the magnitude of the vector. This is illustrated in Example 1.1.

EXAMPLE 1.1

Given the vectors $\mathbf{A} = 3\mathbf{u}_x$ and $\mathbf{B} = 4\mathbf{u}_y$, use MATLAB to compute the sum $\mathbf{C} = \mathbf{A} + \mathbf{B}$. Find the magnitude of $\mathbf{C}$ and the unit vector $\mathbf{u}_C$. Plot and label these vectors and the unit vectors $\mathbf{u}_x$ and $\mathbf{u}_y$ to illustrate the "tip-to-tail" addition method.

Answer. The sum is $\mathbf{C} = 3\mathbf{u}_x + 4\mathbf{u}_y$. The magnitude is $|\mathbf{C}| = \sqrt{3^2 + 4^2} = 5$ and the unit vector is $\mathbf{u}_C = (3\mathbf{u}_x + 4\mathbf{u}_y)/\sqrt{3^2 + 4^2} = 0.6\mathbf{u}_x + 0.8\mathbf{u}_y$. MATLAB gives this result to the (user-controllable) default accuracy of four decimal places. A title and captions have been added to the plot using MATLAB plot options.

At the present time, MATLAB does not have a feature to create a vector directly by drawing with arrows. However, thanks to Jeff Chang and Tom Davis, there exists a user-contributed file entitled *arrow3* at *http://www.mathworks.com/matlabcentral/fileexchange/loadFile.do?objectId=1430*

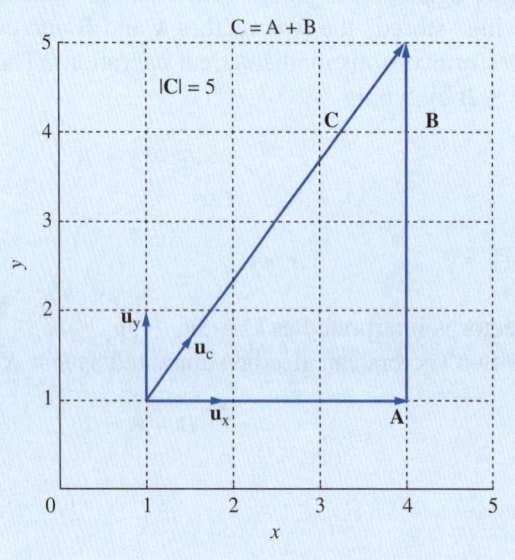

The operation of multiplication on vectors can be carried out in two different ways, yielding two very different results. The first multiplication operation is called either the scalar product or the dot product. One definition of the scalar product of two vectors is given by

$$\mathbf{A} \cdot \mathbf{B} \equiv |\mathbf{A}||\mathbf{B}|\cos\theta = AB\cos\theta \qquad (1.10)$$

This multiplication results in a scalar product that is equal to the product of the magnitude of vector **A** times the magnitude of vector **B** times the cosine of the smaller of the two angles between the two vectors. An equivalent definition of the dot product is given by

$$\mathbf{A} \cdot \mathbf{B} \equiv A_x B_x + A_y B_y + A_z B_z \qquad (1.11)$$

where the first definition could be considered a geometric definition of the dot product while the second definition could be considered an algebraic definition. With the use of the dot product, we can determine several useful quantities or properties associated with the combination of these two vectors. For instance, we can determine if two vectors are perpendicular or parallel to each other with the use of the dot product. In examining equation 1.10, we note that if **A** and **B** are perpendicular to each other, then the angle between them is 90°, and cos(90°) = 0, which means the dot product is equal to zero. In similar fashion, we note that if two vectors are parallel, then the magnitude of the dot product equals the product of the magnitudes of the two vectors. Finally, if we take the dot product of a vector with itself, we obtain the square of the magnitude of the vector, or

$$\mathbf{A} \cdot \mathbf{A} = |\mathbf{A}|^2 = A^2 \tag{1.12}$$

Another quantity we can obtain from the dot product is called the scalar projection of one vector onto another. For instance, if we want to obtain the scalar projection of the vector **A** onto the vector **B**, we can compute this as follows:

$$proj_\mathbf{B} \mathbf{A} = \frac{\mathbf{A} \cdot \mathbf{B}}{|\mathbf{B}|} \tag{1.13}$$

Note that this is a scalar quantity, and that we can also define the projection of the vector **B** onto the vector **A** in a similar fashion. To see the geometrical illustration of this, see Figure 1–1. We can obtain a more familiar form of the scalar projection by re-writing (1.13) using (1.10) to obtain

$$proj_\mathbf{B} \mathbf{A} = |\mathbf{A}| \cos \theta \tag{1.14}$$

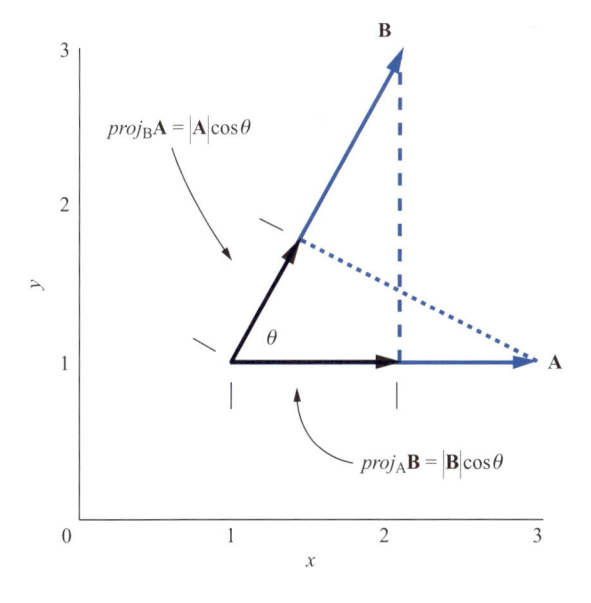

FIGURE 1-1

Illustration of the scalar product of the two vectors $\mathbf{A} = 2\mathbf{u}_x + 0\mathbf{u}_y + 0\mathbf{u}_z$ and $\mathbf{B} = 1\mathbf{u}_x + 2\mathbf{u}_y + 0\mathbf{u}_z$. The scalar product of the two vectors is equal to 2.

Finally, we can simplify this even further, if the vector **B** is a unit vector. In that case, the projection is simply the dot product:

$$proj_{u_B} \mathbf{A} = |\mathbf{A}|\cos\theta = \mathbf{A} \bullet \mathbf{u}_B \qquad (1.15)$$

One very important physical application of the scalar or dot product is the calculation of work. We can use the dot product to calculate the amount of work done when impressing a force on an object. For example, if we are to move an object a distance Δx in the prescribed direction, x, we must apply a force, **F**, in the same direction. The total amount of work expended, ΔW, is given by the expression

$$\Delta W = \mathbf{F} \bullet (\Delta x)\mathbf{u}_x = |\mathbf{F}|\Delta x \cos\theta \qquad (1.16)$$

This operation will be very useful later, when we start moving charges around in an electric field and we want to know how much work is required to do so. We will also use the dot product to help us find the amount of flux crossing a surface. Other useful things that can be done using the dot product and its variations include finding the components of a vector if the other vector is a unit vector, or finding the direction cosines of a vector in three-space. The MATLAB command that permits taking a scalar product of the two vectors **A** and **B** is either ***dot (A, B)*** or ***dot (B, A)***, since these are equal.

The second vector multiplication of two vectors is called the vector product or the cross-product and it is defined as

$$\mathbf{A} \times \mathbf{B} \equiv AB\sin\theta\, \mathbf{u}_{\mathbf{A}\times\mathbf{B}} \qquad (1.17)$$

This multiplication yields a vector whose direction is determined by the "right hand rule." This rule states that if you take the fingers of your right hand (vector **A**) and close them in order to make a fist (vector **B**), the unit vector $\mathbf{u}_{\mathbf{A}\times\mathbf{B}}$ will be in the direction of your thumb. Therefore, we find that the cross product is "anti-commutative": $\mathbf{B} \times \mathbf{A} = -\mathbf{A} \times \mathbf{B}$.

In Cartesian coordinates, we can easily calculate the vector product by remembering the expansion routine of the following determinant.

$$\begin{aligned}\mathbf{A} \times \mathbf{B} &= \begin{vmatrix} \mathbf{u}_x & \mathbf{u}_y & \mathbf{u}_z \\ A_x & A_y & A_z \\ B_x & B_y & B_z \end{vmatrix} \\ &= (A_y B_z - A_z B_y)\mathbf{u}_x + (A_z B_x - A_x B_z)\mathbf{u}_y + (A_x B_y - A_y B_x)\mathbf{u}_z\end{aligned} \qquad (1.18)$$

1.1 Understanding Vectors Using MATLAB

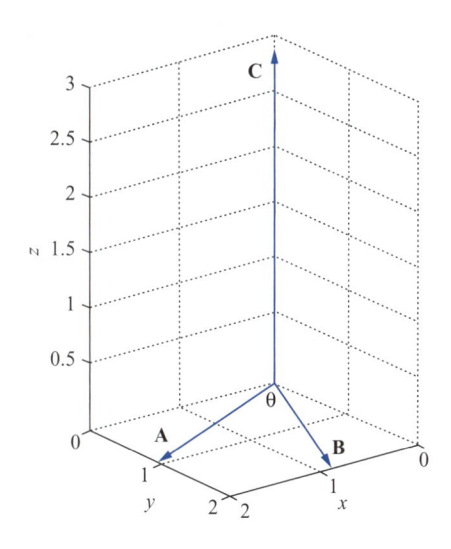

FIGURE 1–2

The cross product of the two vectors $\mathbf{A} = 2\mathbf{u}_x + 1\mathbf{u}_y + 0\mathbf{u}_z$ and $\mathbf{B} = 1\mathbf{u}_x + 2\mathbf{u}_y + 0\mathbf{u}_z$ is shown. The vector product of the two vectors $\mathbf{A}$ and $\mathbf{B}$ is equal to $\mathbf{C} = 0\mathbf{u}_x + 0\mathbf{u}_y + 3\mathbf{u}_z$.

The MATLAB command that summons the vector product of two vectors $\mathbf{A}$ and $\mathbf{B}$ is *cross (A, B)*. In Figure 1–2, we illustrate the cross-product of two vectors.

EXAMPLE 1.2

Show that the magnitude of the cross product of two vectors may be interpreted as the area of the parallelogram defined by the vectors.

Answer. It is possible to give a geometric interpretation for the magnitude of the vector product. The magnitude $|\mathbf{A} \times \mathbf{B}|$ is the area of the parallelogram whose sides are specified by the vectors $\mathbf{A}$ and $\mathbf{B}$ as shown in the figure. In this case $\mathbf{A} = 3\mathbf{u}_x$ and $\mathbf{B} = 3\mathbf{u}_x + 3\mathbf{u}_y$.

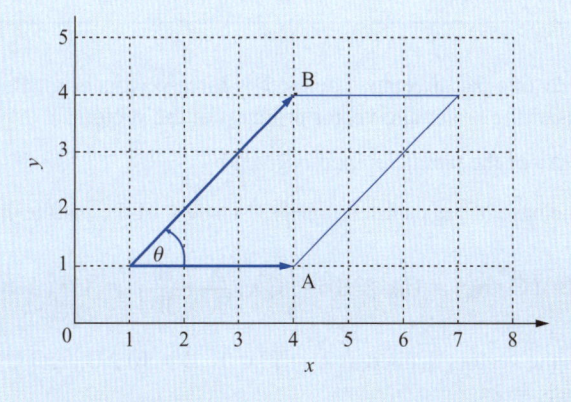

A convenient method of stating that two non-zero vectors **A** and **B** are perpendicular or orthogonal ($\theta = 90°$) is to use the scalar product. If $\mathbf{A} \cdot \mathbf{B} = 0$ and neither vector is zero, then the two vectors are perpendicular, since $\cos 90° = 0$. To state that the two non-zero vectors are parallel ($\theta = 0°$) or antiparallel ($\theta = 180°$), we may use the vector product. If $\mathbf{A} \times \mathbf{B} = 0$, then the two vectors are parallel or antiparallel, since $\sin 0° = \sin 180° = 0$.

Two triple products encountered in electromagnetic theory are included here. The first is called the ***scalar triple product***. It is defined as

$$\mathbf{A} \cdot (\mathbf{B} \times \mathbf{C}) = \mathbf{B} \cdot (\mathbf{C} \times \mathbf{A}) = \mathbf{C} \cdot (\mathbf{A} \times \mathbf{B}) \tag{1.19}$$

Note the cyclical permutation of the vectors in (1.19). There are several additional possible permutations to this product because the change in vector product gives us a minus sign:

$$\mathbf{B} \times \mathbf{C} = -\mathbf{C} \times \mathbf{B} \tag{1.20}$$

The second triple product is called the ***vector triple product***.

$$\mathbf{A} \times (\mathbf{B} \times \mathbf{C}) = \mathbf{B}(\mathbf{A} \cdot \mathbf{C}) - \mathbf{C}(\mathbf{A} \cdot \mathbf{B}) \tag{1.21}$$

The inclusion of the parentheses in this triple product is critical. This triple product is sometimes called the "bac-cab" rule, since this is an easy way to remember how the vectors are ordered. We will use this particular vector identity extensively when the topic of Poynting's vector is introduced.

EXAMPLE 1.3

Show that the volume Δv of a parallelepiped defined by three vectors originating at a point can be defined in terms of the scalar and vector products of the vectors.

Answer. The volume Δv of the parallelepiped is given by

$$\Delta v = (\text{area of the base of the parallelepiped}) \times (\text{height of the parallelepiped})$$

$$\Delta v = (|\mathbf{A} \times \mathbf{B}|)(\mathbf{C} \cdot \mathbf{u_n}) = (|\mathbf{A} \times \mathbf{B}|) \cdot \left(\mathbf{C} \cdot \left(\frac{\mathbf{A} \times \mathbf{B}}{|\mathbf{A} \times \mathbf{B}|}\right)\right) = \mathbf{C} \cdot (\mathbf{A} \times \mathbf{B})$$

This is illustrated with the vectors defined as $\mathbf{A} = [3\ 0\ 0]$; $\mathbf{B} = [0\ 2\ 0]$; and $\mathbf{C} = [0\ 2\ 4]$. The calculated volume is = 24.

1.1 Understanding Vectors Using MATLAB

Note that the height of the parallelepiped is given by the scalar product of the vector **C** with the unit vector $(\mathbf{A} \times \mathbf{B})/|\mathbf{A} \times \mathbf{B}|$ that is perpendicular to the base.

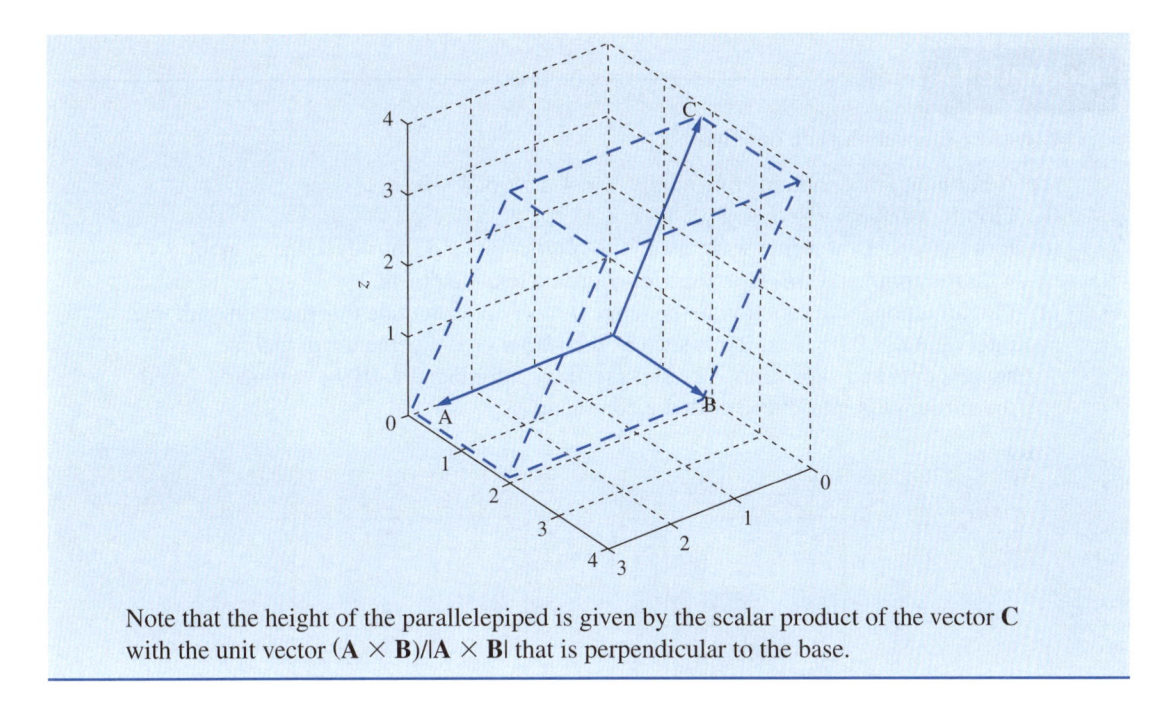

MATLAB provides extensive two- and three-dimensional graphical plotting routines. The data to be plotted can be generated from an internal program, or it can be imported from an external program. Use the command *fplot* to specify and plot using functions that are included in the MATLAB library. Labels and titles using different fonts and font sizes and styles can be positioned on the graphs, and the plots can be distinguished with different symbols. The MATLAB notation for a superscript and a subscript require the additional statement ^ and _, respectively, in the text command. We will present several examples here to illustrate the variety of two-dimensional plots that are available. Additional graphs can be located on one plot with the *hold* command. Either axis can have a logarithmic scale. In addition, the command *subplot* permits us to put more than one graph on a page—either vertically or horizontally displaced. The command *subplot (1, 2, 1)* states that there are to be two graphs next to one another, and this command will be used to select the left one. The command *subplot (2, 1, 2)* states that there are two graphs, one on top of the other, and the command selects the bottom one. Other commands that follow then detail the characteristics of that particular graph. This is best illustrated in Example 1.4.

It is possible to customize a graph by changing the characteristics of the line. This is illustrated by plotting the same function, say $\sin\theta$, versus the independent variable θ. In addition, the text item "(a)" is sequenced in the program using the command *s(2) = setstr(s(2) + 1)* after the initial inclusion of the statement *s = '(a)'* in the program. This is illustrated in Figure 1–3.

EXAMPLE 1.4

Construct four subplots on one figure.

(a) A bar graph that contains five numbers $x = 2, 4, 6, 8, 10$.
(b) Plot the numbers $y = 5, 4, 3, 2, 1$ vs. x.
(c) Plot two cycles of a sine wave using the *fplot* command. Introduce the symbol θ with the command \theta in the *xlabel* or in a text statement.
(d) Plot an exponential function in the range $0 < x < 3$. Calculate this function with the interval $\Delta x = 0.01$. Text items such as the *ylabel* or a statement can include superscripts and subscripts. The superscript is introduced with the command ^ and the subscript is introduced with the command _.

Answer.

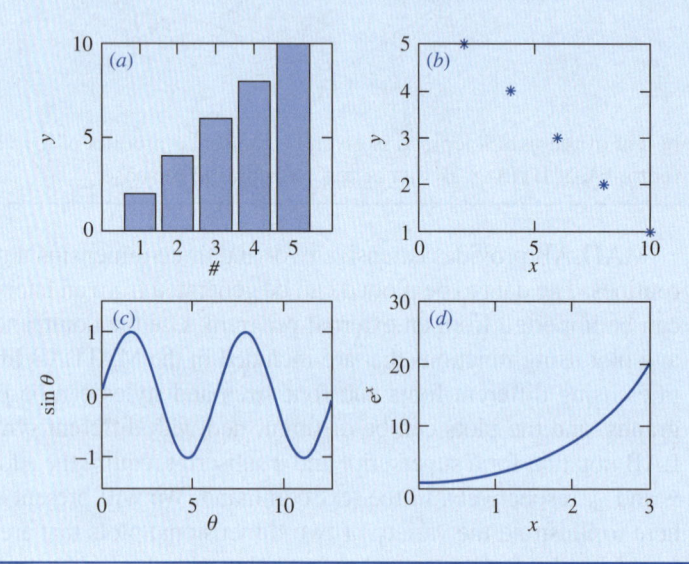

MATLAB can also display three-dimensional graphical representations. The horizontal space is subdivided into a large number of points (x_j, y_k), and the function $z = z(x_j, y_k)$ has to be evaluated at each of these points. In order to accomplish this, a "." (period) must follow each of the independent variables in a program. The results of a three-dimensional picture is illustrated in Figure 1–4. There are two distinct plot commands, *mesh* and *surf*. In addition, there are also commands that allow the user to change the viewing angle, both in rotation and in elevation.

In addition to plotting the figure in Cartesian coordinates, it is also possible to plot it in polar coordinates. This will be useful when examin-

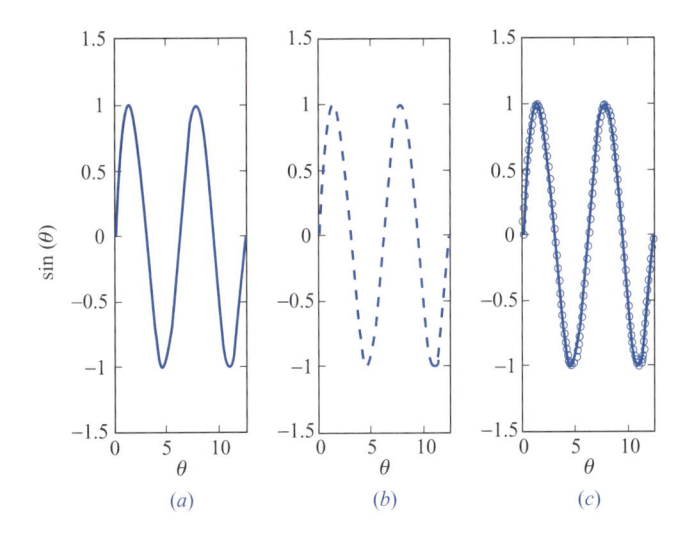

FIGURE 1-3

Illustration of a variety of different styles for the lines in a graph.
(*a*) Solid line.
(*b*) Dashed line.
(*c*) Alternate 'o –'.

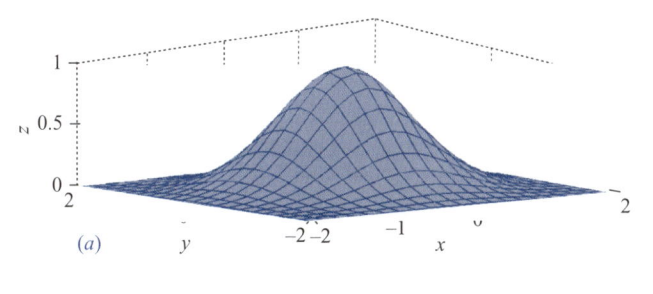

FIGURE 1-4

Three-dimensional plots of a Gaussian function.
(*a*) Mesh plot.
(*b*) Surf plot.

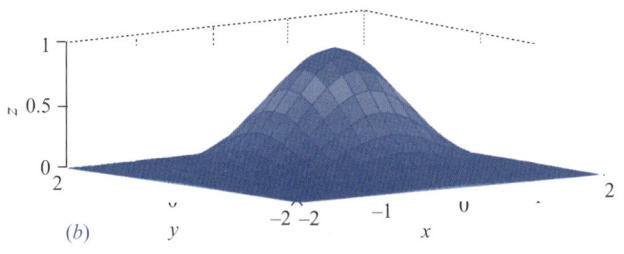

ing the radiation pattern of antennas. In addition, one can plot graphs in a semilog or a log-log format. These graphs are useful when ascertaining the variation of a function $y = x^n$—say in interpreting data collected in a laboratory experiment. Examples of these figures are shown in Figure 1–5.

All the functions cited above are included in the MATLAB library of functions. This library can be expanded to include user-created functions. This is done by creating a text file with extension ".m", known as an M-file, with a unique name, say *custom.m*. Once the M-file is created,

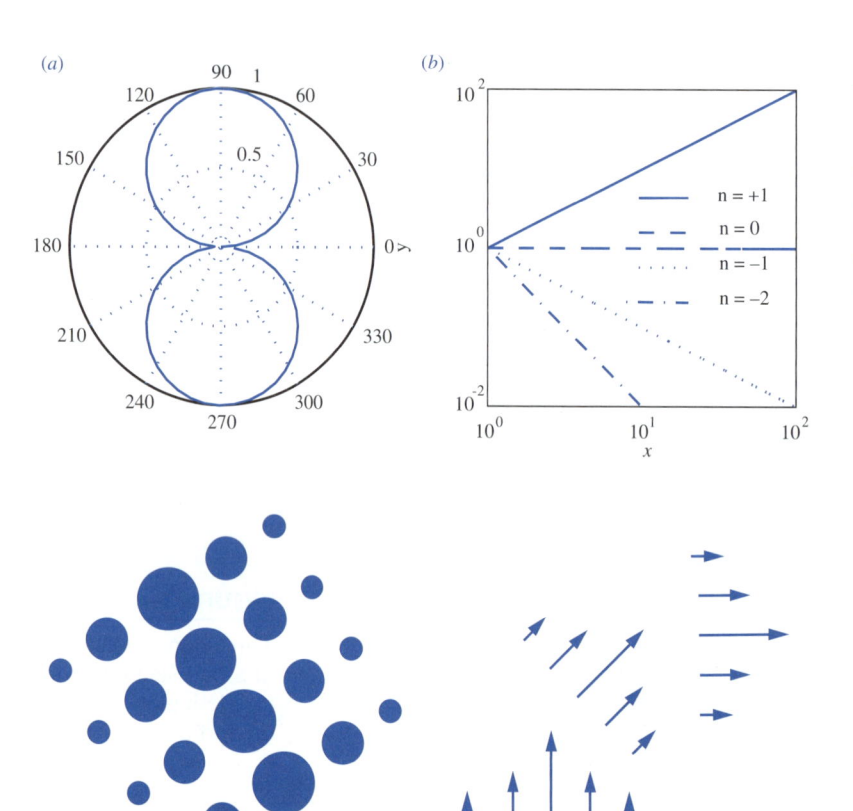

FIGURE 1–5

Polar plot and a log–log graph. (*a*) |sin θ| versus the angle θ where $0 < \theta < 2\pi$. (*b*) $y = x^n$ for different values of n where $1 < x < 100$.

FIGURE 1–6

Scalar and vector fields. (*a*) The magnitude of a scalar is specified by the size of the circle. (*b*) The magnitude and the direction of the vector at any point are indicated with the length and the orientation of the arrows.

it becomes a part of the user's personal library. In order to use this file, simply type the word *custom* after the prompt ">>" and this particular file is activated. These files are frequently shared over the internet and, in fact, all of the files that have been used to create the figures and the examples in this book are available on the enclosed CD or the web site www.scitechpub.com/lonngren2e.htm.

In the discussion above, we have focused on single vectors. Later in this text we will encounter a distribution of vectors called a ***vector field***. An example of a vector field is the wind distribution in a region, where the wind at any point has a magnitude and a direction associated with it. We typically would characterize a vector field made up of different-length vectors as representing a nonuniform distribution of wind. Scalar fields also exist. An example is the distribution of temperature across the nation, characterized by numbers at each location. Examples of the two types of fields are shown in Figure 1–6.

1.2 Coordinate Systems

In this text, we will frequently encounter problems where there is a source of an electromagnetic field. To be able to specify the field at a point in space caused by a source, we have to refer to a coordinate system. In three dimensions, the coordinate system can be specified by the intersection of three surfaces. An orthogonal coordinate system is defined when these three surfaces are mutually orthogonal at every point. Coordinate surfaces may be planar or curved. A general orthogonal coordinate system is illustrated in Figure 1–7.

In Cartesian coordinates, all of the surfaces are planes, and they are specified by each of the independent variables x, y, and z separately having prescribed values. In cylindrical coordinates, the surfaces are two planes and a cylinder. In spherical coordinates, the surfaces are a sphere, a plane, and a cone. We will examine each of these in detail in the following discussion. There are many other coordinate systems that can be employed for particular problems, and there are formulas that allow one to easily transform vectors from one system to another.

The three coordinate systems used in this text are pictured in Figure 1–8 as (*a*), (*b*), and (*c*). The directions along the axes of the coordinate systems

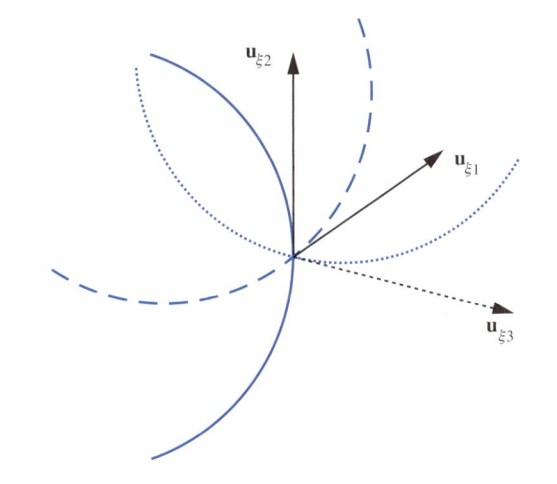

FIGURE 1–7

A general orthogonal coordinate system. Three surfaces intersect at a point, and the unit vectors are mutually orthogonal at that point.

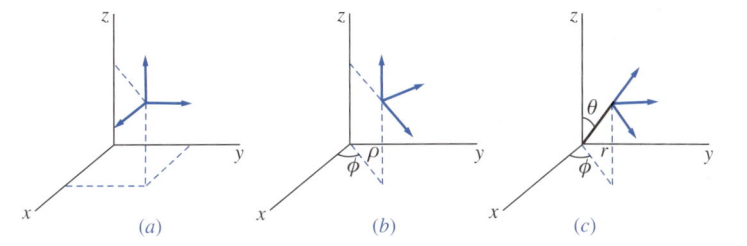

FIGURE 1–8

The three coordinate systems that will be employed in this text. The unit vectors are indicated. (*a*) Cartesian coordinates. (*b*) Cylindrical coordinates. (*c*) Spherical coordinates.

are given by the sets of unit vectors $(\mathbf{u}_x, \mathbf{u}_y, \mathbf{u}_z)$, $(\mathbf{u}_\rho, \mathbf{u}_\phi, \mathbf{u}_z)$, and $(\mathbf{u}_r, \mathbf{u}_\theta, \mathbf{u}_\phi)$ for Cartesian, cylindrical, and spherical coordinates, respectively. In each of the coordinate systems, the unit vectors are mutually orthogonal at every point.

In each coordinate system, the unit vectors point in the direction of increasing coordinate value. In Cartesian coordinates, the direction of the unit vectors is independent of position, whereas in cylindrical and spherical coordinates, unit vector directions at a point in space depend on the location of that point. For example, in spherical coordinates, the unit vector $\mathbf{u}_r$ is directed radially away from the origin at every point in space; it will be directed in the $+z$ direction if $\theta = 0$, and it will be directed in the $-z$ direction if $\theta = \pi$. Since we will employ these three coordinate systems extensively in the following chapters, it is useful to summarize the important properties of each one.

1.2.1 Cartesian Coordinates

The unit vectors in Cartesian coordinates depicted in Figure 1–8a are normal to the intersection of three planes as shown in Figure 1–9. Each of the surfaces depicted in this figure is a plane that is individually normal to a coordinate axis.

For the unit vectors that are in the directions of the x, y, and z axes, we have $\mathbf{u}_x \cdot \mathbf{u}_x = \mathbf{u}_y \cdot \mathbf{u}_y = \mathbf{u}_z \cdot \mathbf{u}_z = 1$ and $\mathbf{u}_x \cdot \mathbf{u}_y = \mathbf{u}_x \cdot \mathbf{u}_z = \mathbf{u}_y \cdot \mathbf{u}_z = 0$. The following rules also apply to the cross products of the unit vectors, since this is a right-handed system:

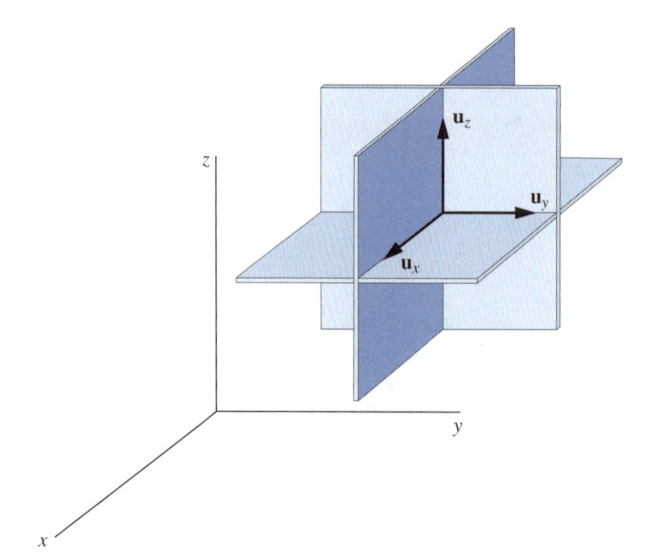

FIGURE 1–9

A point in Cartesian coordinates is defined by the intersection of the three planes: x = constant; y = constant; z = constant. The three unit vectors are normal to each of the three surfaces.

1.2 Coordinate Systems

$$\left.\begin{array}{c} \mathbf{u}_x \times \mathbf{u}_y = \mathbf{u}_z \\ \mathbf{u}_y \times \mathbf{u}_z = \mathbf{u}_x \\ \mathbf{u}_z \times \mathbf{u}_x = \mathbf{u}_y \end{array}\right\} \quad (1.22)$$

All other cross products of unit vectors follow from the facts that the cross product is anti-symmetric ($\mathbf{u}_x \times \mathbf{u}_y = -\mathbf{u}_y \times \mathbf{u}_x$, etc.), and the cross product of any vector with itself is zero ($\mathbf{u}_x \times \mathbf{u}_x = 0$, etc.).

A vector $\mathbf{r}$ from the origin to the point (x, y, z) in Cartesian coordinates is $\mathbf{r} = x\mathbf{u}_x + y\mathbf{u}_y + z\mathbf{u}_z$. Given two vectors, $\mathbf{A} = A_x\mathbf{u}_x + A_y\mathbf{u}_y + A_z\mathbf{u}_z$ and $\mathbf{B} = B_x\mathbf{u}_x + B_y\mathbf{u}_y + B_z\mathbf{u}_z$, their dot product is given by

$$\mathbf{A} \cdot \mathbf{B} = A_x B_x + A_y B_y + A_z B_z \quad (1.23)$$

and their cross product is given by equation (1.18).

EXAMPLE 1.5

State the MATLAB commands that identify the three unit vectors in Cartesian coordinates. A general feature of any orthogonal coordinate systems is that the unit vectors at any point defined by the intersection of the three surfaces are mutually orthogonal at that point.

Answer. The MATLAB commands for the unit vectors are written as

$\mathbf{u_x}$: $ux = [1\ 0\ 0]$, $\mathbf{u_y}$: $uy = [0\ 1\ 0]$, and $\mathbf{u_z}$: $uz = [0\ 0\ 1]$

The unit vectors are depicted in the figure below.

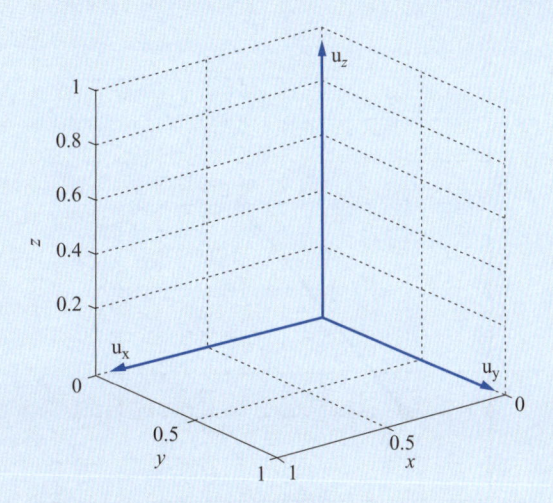

We will perform line, surface, and volume integrals in the following chapters. Figure 1–10 depicts the differential line element, surface elements, and volume elements in Cartesian coordinates. Note that there are six possible differential surface elements, each corresponding to one of the six faces of the differential volume. In each case, the vector direction is the outward normal direction.

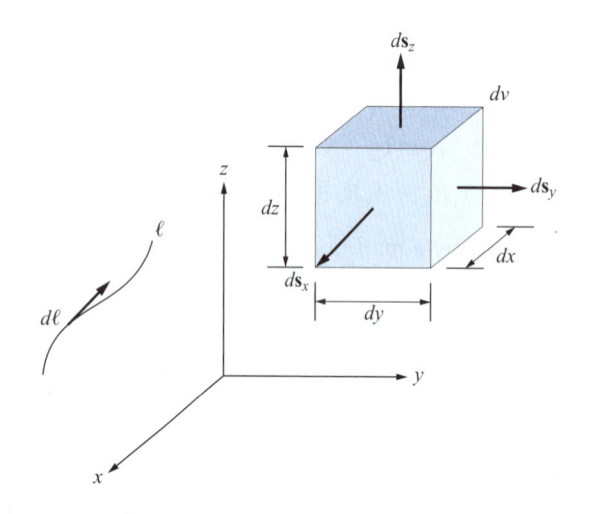

FIGURE 1–10

A differential line element $d\ell = \mathbf{u}_x dx + \mathbf{u}_y dy + \mathbf{u}_z dz$ is shown. Three of six differential surface elements, $d\mathbf{s}_x = \mathbf{u}_x dydz$, $d\mathbf{s}_y = \mathbf{u}_y dxdz$, and $d\mathbf{s}_z = \mathbf{u}_z dxdy$ are shown along with the differential volume element $dv = dxdydz$.

EXAMPLE 1.6

Find the expression for the vector $\mathbf{G}_{AB}$ that extends from the point $P_A = (2, 2, 0)$ to the point $P_B = (6, 7, 0)$ in Cartesian coordinates and plot it on a graph. Also, find the corresponding unit vector $\mathbf{u_G}$.

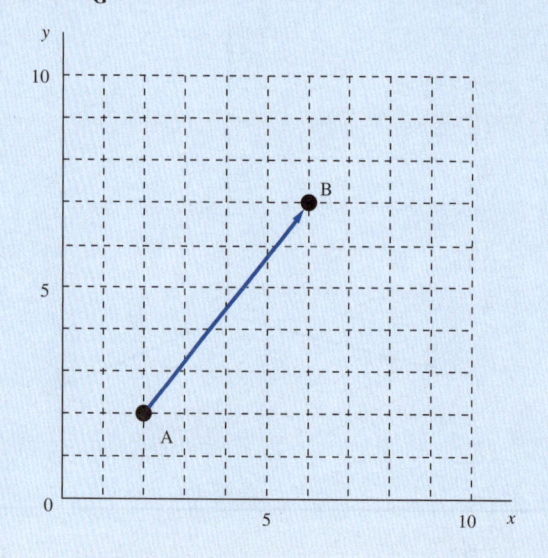

1.2 Coordinate Systems

Answer. The vector is given by

$$G_{AB} = (x_B - x_A)\mathbf{u}_x + (y_B - y_A)\mathbf{u}_y + (z_B - z_A)\mathbf{u}_z$$
$$= (6 - 2)\mathbf{u}_x + (7 - 2)\mathbf{u}_y + (0 - 0)\mathbf{u}_z = 4\mathbf{u}_x + 5\mathbf{u}_y + 0\mathbf{u}_z$$

The corresponding unit vector is

$$\mathbf{u}_G = \frac{4\mathbf{u}_x + 5\mathbf{u}_y + 0\mathbf{u}_z}{\sqrt{4^2 + 5^2 + 0^2}}$$

EXAMPLE 1.7

Given the two vectors $\mathbf{A} = 3\mathbf{u}_x + 4\mathbf{u}_y$ and $\mathbf{B} = 12\mathbf{u}_x + 5\mathbf{u}_y$ in Cartesian coordinates, evaluate the following quantities. In addition, state the MATLAB commands that can be used to check your answers. The vectors are written in MATLAB notation as *A = [3 4 0]* and *B = [12 5 0]*.

(a) the scalar product $\mathbf{A} \cdot \mathbf{B}$
(b) the angle between the two vectors
(c) the scalar product $\mathbf{A} \cdot \mathbf{A}$
(d) the vector product $\mathbf{A} \times \mathbf{B}$

Answer.

(a) The scalar product $\mathbf{A} \cdot \mathbf{B}$ is given by $\mathbf{A} \cdot \mathbf{B} = 36\mathbf{u}_x \cdot \mathbf{u}_x + 15\mathbf{u}_x \cdot \mathbf{u}_y + 48\mathbf{u}_y \cdot \mathbf{u}_x + 20\mathbf{u}_y \cdot \mathbf{u}_y$. Note: the scalar product of two orthogonal unit vectors is equal to zero and two collinear unit vectors is equal to one. This leads to $\mathbf{A} \cdot \mathbf{B} = 36 + 0 + 0 + 20 = 56$. In MATLAB, use *dot(A, B)*.

(b) The angle between the two vectors is computed from the definition of the scalar product.

$$\cos\theta = \frac{\mathbf{A} \cdot \mathbf{B}}{|\mathbf{A}||\mathbf{B}|} = \frac{56}{\sqrt{3^2 + 4^2}\sqrt{12^2 + 5^2}} = \frac{56}{65} \text{ or } \theta = 30.5°$$

In MATLAB, *theta = acos(dot(A, B)/(norm(A)*norm(B)))*(180/pi)*

(c) The scalar product $\mathbf{A} \cdot \mathbf{A}$ is given by $\mathbf{A} \cdot \mathbf{A} = 9\mathbf{u}_x \cdot \mathbf{u}_x + 16\mathbf{u}_y \cdot \mathbf{u}_y = 25$. The scalar product $\mathbf{A} \cdot \mathbf{A}$ is a convenient method to determine the magnitude of the vector $\mathbf{A}$ since $A = |\mathbf{A}| = \sqrt{\mathbf{A} \cdot \mathbf{A}}$. The MATLAB command is *dot(A,A)*.

(d) The vector product $\mathbf{A} \times \mathbf{B}$ is given by

$$\mathbf{A} \times \mathbf{B} = \begin{vmatrix} \mathbf{u}_x & \mathbf{u}_y & \mathbf{u}_z \\ 3 & 4 & 0 \\ 12 & 5 & 0 \end{vmatrix} = -33\mathbf{u}_z$$

The MATLAB command is *cross(A, B)*.

1.2.2 Cylindrical Coordinates

The unit vectors in cylindrical coordinates depicted in Figure 1–8b are normal to the intersection of three surfaces as shown in Figure 1–11. Two of the surfaces depicted in this figure are planes, and the third surface is a cylinder that is centered on the z axis. A point (ρ, ϕ, z) in cylindrical coordinates is located at the intersection of the two planes and the cylinder. The value of ρ is the distance away from the z axis and the value of ϕ is the angle between the projection onto the $x-y$ plane and the x axis. The mutually-perpendicular unit vectors $\mathbf{u}_\rho$, $\mathbf{u}_\phi$, and $\mathbf{u}_z$ are in the direction of increasing coordinate value; note that unlike Cartesian unit vectors, the directions of $\mathbf{u}_\rho$ and $\mathbf{u}_\phi$ vary with location.

As usual, the dot product of a unit vector with itself is equal to one, and the dot product of one unit vector with another is equal to zero. Also, since this is a right-handed system, the cross products are given by

$$\left. \begin{array}{l} \mathbf{u}_\rho \times \mathbf{u}_\phi = \mathbf{u}_z \\ \mathbf{u}_\phi \times \mathbf{u}_z = \mathbf{u}_\rho \\ \mathbf{u}_z \times \mathbf{u}_\rho = \mathbf{u}_\phi \end{array} \right\} \quad (1.24)$$

The negative of these results holds when the terms are interchanged, and the cross product of any unit vector with itself is zero. A vector $\mathbf{A}$ in cylindrical coordinates is given in terms of its components as

$$\mathbf{A} = A_\rho \mathbf{u}_\rho + A_\phi \mathbf{u}_\phi + A_z \mathbf{u}_z \quad (1.25)$$

The differential line element $d\ell$, the differential surface elements $d\mathbf{s}_\rho$, $d\mathbf{s}_\phi$, and $d\mathbf{s}_z$, and the differential volume element dv are given in cylindrical coordinates as

$$d\ell = \mathbf{u}_\rho d\rho + \mathbf{u}_\phi \rho d\phi + \mathbf{u}_z dz \quad (1.26)$$

$$\begin{array}{l} d\mathbf{s}_\rho = \mathbf{u}_\rho \rho d\phi dz \\ d\mathbf{s}_\phi = \mathbf{u}_\phi d\rho dz \\ d\mathbf{s}_z = \mathbf{u}_z \rho d\rho d\phi \end{array} \quad (1.27)$$

$$dv = \rho d\rho d\phi dz \quad (1.28)$$

A vector in cylindrical coordinates can be transformed to a vector in Cartesian coordinates or vice versa. A vector given in Cartesian coordinates, such as $\mathbf{A} = A_x \mathbf{u}_x + A_y \mathbf{u}_y + A_z \mathbf{u}_z$, may also be expressed in cylindrical coordinates, as $\mathbf{A} = A_\rho \mathbf{u}_\rho + A_\phi \mathbf{u}_\phi + A_z \mathbf{u}_z$. The question is how to relate the components and unit vectors in the two forms.

1.2 Coordinate Systems

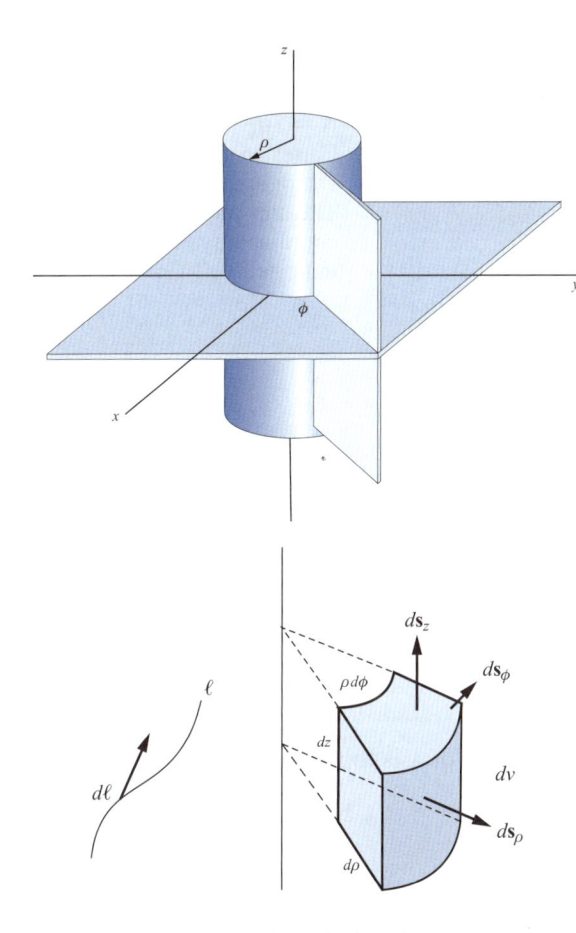

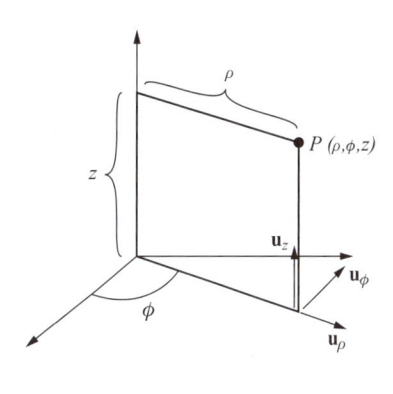

FIGURE 1-11

The cylindrical coordinate system. A point is located at the intersection of a cylinder and two planes. The variables ρ, ϕ, and z are shown. A differential line element $d\ell$, differential surface elements $d\mathbf{s}_\rho$, $d\mathbf{s}_\phi$, and $d\mathbf{s}_z$, and a differential volume element dV are depicted.

The relation between components in the two coordinate systems may be found using the scalar product and recalling that the scalar product of a vector with a unit vector may be interpreted as the amount of the vector in the direction of the unit vector. First note that A_z is identical in both coordinate systems. Now, the x component of a vector may be found as the dot product of the vector with the unit vector $\mathbf{u}_x$. Given $\mathbf{A}$ in cylindrical coordinates, this means

$$A_x = \mathbf{A} \cdot \mathbf{u}_x = (A_\rho \mathbf{u}_\rho + A_\phi \mathbf{u}_\phi + A_z \mathbf{u}_z) \cdot \mathbf{u}_x$$
$$= A_\rho \mathbf{u}_\rho \cdot \mathbf{u}_x + A_\phi \mathbf{u}_\phi \cdot \mathbf{u}_x$$

From Figure 1-12 or Appendix A, we note that

$$\mathbf{u}_\rho \cdot \mathbf{u}_x = \cos\phi$$
$$\mathbf{u}_\phi \cdot \mathbf{u}_x = -\sin\phi$$

Therefore

$$A_x = A_\rho \cos\phi - A_\phi \sin\phi \tag{1.29}$$

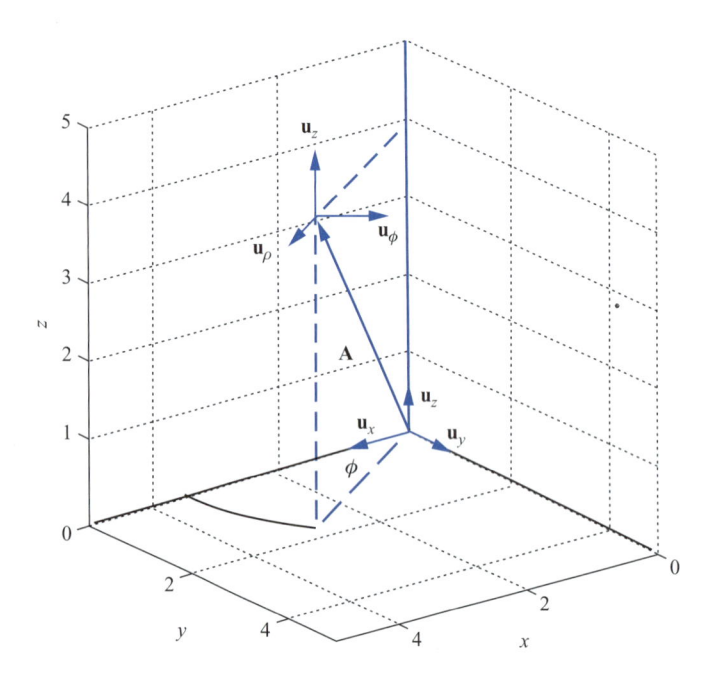

FIGURE 1–12

The transformation of a vector $\mathbf{A} = 3\mathbf{u}_x + 2\mathbf{u}_y + 4\mathbf{u}_z$ in Cartesian coordinates into a vector in cylindrical coordinates. The unit vectors of the two coordinate systems are indicated.

In a similar fashion, the y component of $\mathbf{A}$ is found by the dot product with $\mathbf{u}_y$:

$$A_y = \mathbf{A} \cdot \mathbf{u}_y = (A_\rho \mathbf{u}_\rho + A_\phi \mathbf{u}_\phi + A_z \mathbf{u}_z) \cdot \mathbf{u}_y$$

$$= A_\rho \mathbf{u}_\rho \cdot \mathbf{u}_y + A_\phi \mathbf{u}_\phi \cdot \mathbf{u}_y$$

From the figure or Appendix A,

$$\mathbf{u}_\rho \cdot \mathbf{u}_y = \sin\phi$$

$$\mathbf{u}_\phi \cdot \mathbf{u}_y = \cos\phi$$

Hence

$$A_y = A_\rho \sin\phi + A_\phi \cos\phi \tag{1.30}$$

So far, we have assumed that the vector $\mathbf{A}$ is a constant. However, in many cases the vector may be a function of the cylindrical variables ρ, ϕ, and z. In this case, the variables must also be transformed from cylindrical to Cartesian coordinates. The transformation is found in Appendix A as

$$x = \rho \cos\phi$$

$$y = \rho \sin\phi \tag{1.31}$$

$$z = z$$

1.2 Coordinate Systems

Care must be used in choosing the correct quadrant for the arc-tangent function. The inverse transformation from Cartesian to cylindrical coordinates from Appendix A is

$$\rho = \sqrt{x^2 + y^2}$$
$$\phi = \tan^{-1}\left(\frac{y}{x}\right) \quad (1.32)$$
$$z = z$$

There are commands in MATLAB that will effect this transformation between cylindrical and Cartesian coordinates. In addition, the command *cylinder* (that includes additional parameters) will create a picture of a cylinder.

1.2.3 Spherical Coordinates

The unit vectors in spherical coordinates depicted in Figure 1–8c are normal to the intersection of three surfaces as shown in Figure 1–13. One of the surfaces depicted in this figure is a plane, another surface is a sphere, and the third surface is a cone. The latter two surfaces are centered on the z axis. A point in spherical coordinates is specified by the intersection of the three surfaces. The unit vectors $\mathbf{u}_r$, $\mathbf{u}_\theta$, and $\mathbf{u}_\phi$ are perpendicular to the sphere, the cone, and the plane. The variables and unit vectors in spherical coordinates are also shown in the figure.

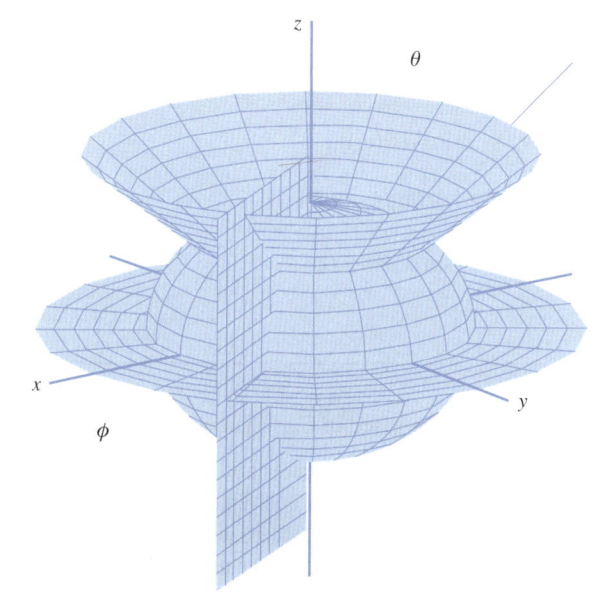

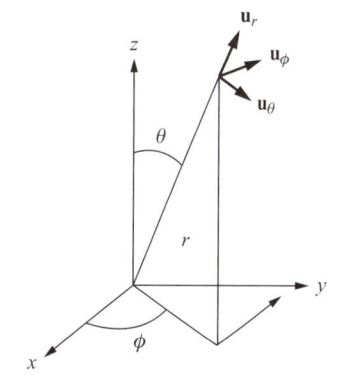

FIGURE 1–13

Spherical coordinates. A point is defined by the intersection of a sphere whose radius is r, a plane that makes an angle ϕ with respect to the x axis, and a cone that makes an angle θ with respect to the z axis.

A point (r, θ, ϕ) in spherical coordinates is located at the intersection of the sphere, cone and plane. The value of r is the distance away from the origin, θ is the angle from the z axis, and ϕ is the same angle as in cylindrical coordinates. The mutually-perpendicular unit vectors $\mathbf{u}_r$, $\mathbf{u}_\theta$, and $\mathbf{u}_\phi$ are in the direction of increasing coordinate value; note that unlike Cartesian unit vectors, the directions of the unit vectors vary with location.

As usual, the dot product of a unit vector with itself is equal to one, and the dot product of one unit vector with another is equal to zero. Also, since this is a right-handed system, the cross products are given by

$$\mathbf{u}_r \times \mathbf{u}_\theta = \mathbf{u}_\phi$$
$$\mathbf{u}_\theta \times \mathbf{u}_\phi = \mathbf{u}_r \quad (1.33)$$
$$\mathbf{u}_\phi \times \mathbf{u}_r = \mathbf{u}_\theta$$

The negative of these results holds when the terms are interchanged, and the cross product of any unit vector with itself is zero. A vector $\mathbf{A}$ in spherical coordinates is given in terms of its components as

$$\mathbf{A} = A_r \mathbf{u}_r + A_\theta \mathbf{u}_\theta + A_\phi \mathbf{u}_\phi \quad (1.34)$$

The differential line element $d\ell$, the differential surface elements $d\mathbf{s}_r$, $d\mathbf{s}_\theta$, and $d\mathbf{s}_\phi$, and the differential volume element dv are given in spherical coordinates as

$$d\ell = \mathbf{u}_r dr + \mathbf{u}_\theta r d\theta + \mathbf{u}_\phi r \sin\theta d\phi \quad (1.35)$$

$$d\mathbf{s}_r = r^2 \sin\theta d\theta d\phi \, \mathbf{u}_r$$
$$d\mathbf{s}_\theta = r\sin\theta dr d\phi \, \mathbf{u}_\theta \quad (1.36)$$
$$d\mathbf{s}_\phi = r dr d\theta \, \mathbf{u}_\phi$$

$$dv = r^2 \sin\theta \, dr d\theta d\phi \quad (1.37)$$

EXAMPLE 1.8

Show that a vector given in spherical coordinates can be expressed in Cartesian coordinates.

Answer. The vector is $\mathbf{A} = A_x \mathbf{u}_x + A_y \mathbf{u}_y + A_z \mathbf{u}_z$ in Cartesian coordinates and $\mathbf{A} = A_r \mathbf{u}_r + A_\theta \mathbf{u}_\theta + A_\phi \mathbf{u}_\phi$ in spherical coordinates. The transformation between the two coordinate systems is found by taking the scalar product of the unit vector in the new coordinate system with the vector in the other coordinate system and correctly interpreting the scalar products of the unit vectors.

$$A_x = \mathbf{A} \cdot \mathbf{u}_x$$
$$= A_r \mathbf{u}_r \cdot \mathbf{u}_x + A_\theta \mathbf{u}_\theta \cdot \mathbf{u}_x + A_\phi \mathbf{u}_\phi \cdot \mathbf{u}_x$$

1.2 Coordinate Systems

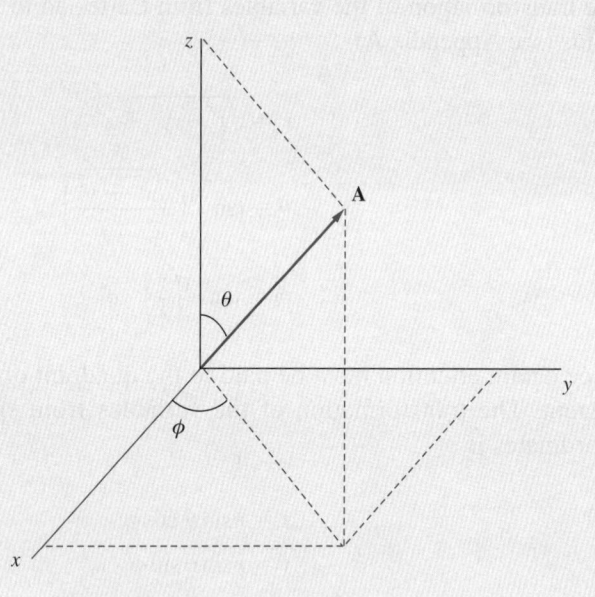

From the figure and from Appendix A we find that

$$\mathbf{u_r} \cdot \mathbf{u_x} = \sin\theta \cos\phi = \frac{x}{\sqrt{x^2 + y^2 + z^2}}$$

$$\mathbf{u_\theta} \cdot \mathbf{u_x} = \cos\theta \cos\phi = \frac{xz}{\sqrt{x^2 + y^2}\sqrt{x^2 + y^2 + z^2}}$$

$$\mathbf{u_\phi} \cdot \mathbf{u_x} = -\sin\phi = -\frac{y}{\sqrt{x^2 + y^2}}$$

In a similar fashion, we write

$$\mathbf{u_r} \cdot \mathbf{u_y} = \sin\theta \sin\phi = \frac{y}{\sqrt{x^2 + y^2 + z^2}}$$

$$\mathbf{u_\theta} \cdot \mathbf{u_y} = \cos\theta \sin\phi = \frac{yz}{\sqrt{x^2 + y^2}\sqrt{x^2 + y^2 + z^2}}$$

$$\mathbf{u_\phi} \cdot \mathbf{u_y} = \cos\phi = \frac{x}{\sqrt{x^2 + y^2}}$$

and

$$\mathbf{u_r} \cdot \mathbf{u_z} = \cos\theta = \frac{z}{\sqrt{x^2 + y^2 + z^2}}$$

$$\mathbf{u_\theta} \cdot \mathbf{u_z} = -\sin\theta = -\frac{\sqrt{x^2 + y^2}}{\sqrt{x^2 + y^2 + z^2}}$$

$$\mathbf{u_\phi} \cdot \mathbf{u_z} = 0$$

The transformation of the variables from Cartesian to spherical coordinates yields (see Appendix A)

$$r = \sqrt{x^2 + y^2 + z^2}$$

$$\theta = \tan^{-1}\left(\frac{\sqrt{x^2 + y^2}}{z}\right) \tag{1.38}$$

$$\phi = \tan^{-1}\left(\frac{y}{x}\right)$$

Once again, attention must be paid to the quadrant of the result of the arctangent. The transformation of the variables from spherical to Cartesian coordinates is

$$x = r \sin\theta \cos\phi$$
$$y = r \sin\theta \sin\phi \tag{1.39}$$
$$z = r \cos\theta$$

A summary of the unit vectors, the differential lengths, the differential surfaces, and the differential volumes for the three coordinate systems is given in Table 1–1. (See also Appendix A—Mathematical Formulas.)

TABLE 1–1 Three Orthogonal Coordinate Systems

Coordinate system	Cartesian (x, y, z)	Cylindrical (ρ, ϕ, z)	Spherical (r, θ, ϕ)
Unit vectors	$\mathbf{u}_x, \mathbf{u}_y, \mathbf{u}_z$	$\mathbf{u}_\rho, \mathbf{u}_\phi, \mathbf{u}_z$	$\mathbf{u}_r, \mathbf{u}_\theta, \mathbf{u}_\phi$
Differential length **dl**	$dx\,\mathbf{u}_x$ $dy\,\mathbf{u}_y$ $dz\,\mathbf{u}_z$	$d\rho\,\mathbf{u}_\rho$ $\rho d\phi\,\mathbf{u}_\phi$ $dz\,\mathbf{u}_z$	$dr\,\mathbf{u}_r$ $r d\theta\,\mathbf{u}_\theta$ $r\sin\theta\,d\phi\,\mathbf{u}_\phi$
Differential surface area $d\mathbf{s}$	$dy\,dz\,\mathbf{u}_x$ $dx\,dz\,\mathbf{u}_y$ $dx\,dy\,\mathbf{u}_z$	$\rho d\phi\,dz\,\mathbf{u}_\rho$ $d\rho dz\,\mathbf{u}_\phi$ $\rho d\rho d\phi\,\mathbf{u}_z$	$r^2\sin\theta\,d\theta d\phi\,\mathbf{u}_r$ $r\sin\theta\,dr d\phi\,\mathbf{u}_\theta$ $r dr\,d\theta\,\mathbf{u}_\phi$
Differential volume dv	$dx\,dy\,dz$	$\rho d\rho d\phi\,dz$	$r^2\sin\theta\,dr d\theta d\phi$

TABLE 1–2 Summary of the Transformation between Coordinate Systems

Transformation			MATLAB Command
Cartesian to cylindrical			
$\rho = \sqrt{x^2 + y^2}$	$\phi = \tan^{-1}\left(\dfrac{y}{x}\right)$	$z = z$	$[\phi, \rho, z] =$ cart2pol(x,y,z)
Cartesian to spherical			
$r = \sqrt{x^2 + y^2 + z^2}$	$\theta = \tan^{-1}\left(\dfrac{\sqrt{x^2 + y^2}}{z}\right)$	$\phi = \tan^{-1}\left(\dfrac{y}{x}\right)$	$[\phi, \psi, \rho] =$ cart2sph(x,y,z) note $\psi = \dfrac{\pi}{2} - \theta$
Cylindrical to Cartesian			
$x = \rho\cos\phi$	$y = \rho\sin\phi$	$z = z$	$[x,y,z] =$ pol2cart(ϕ, ρ, z)
Spherical to Cartesian			
$x = r\sin\theta\cos\phi$	$y = r\sin\theta\sin\phi$	$z = r\cos\theta$	$[x,y,z] =$ sph2cart (ϕ, ψ, ρ) note $\psi = \dfrac{\pi}{2} - \theta$

A summary of the transformations of the variables between coordinate systems is given in Table 1–2. In addition, the MATLAB commands that will perform these operations are also presented. Appendix A provides a summary of the vector operations in these three coordinate systems.

The simple vector mathematics of addition and subtraction requires a little more care when it comes to the other coordinate systems. All of the vectors must be defined with reference to the same point.

1.3 Integral Relations for Vectors

We will find that certain integrals involving vector quantities are important when describing the material presented later in this text. These integrals will be useful initially in deriving vector operations and later in gaining an understanding of electromagnetic fields. For simplicity, most presentations will utilize Cartesian coordinates. The fact that a field's behavior could depend on its local position should not be too surprising. Recall the effects of a change in the gravitational field while watching the astronauts walking into a ground-based spacecraft, then floating within the spacecraft in atmosphere. The integrals we will focus on are listed in Table 1–3.

TABLE 1-3 Integrals of Vector Fields and Densities

Line integral of a vector field **F** along a prescribed path from the location a to the location b.	$\int_a^b \mathbf{F} \cdot d\mathbf{l}$
Surface integral of a vector field **F** through a surface $\Delta \mathbf{s}$.	$\int_{\Delta s} \mathbf{F} \cdot d\mathbf{s}$
Volume integral of a density ρ_v over the volume Δv.	$\int_{\Delta v} \rho_v dv$

1.3.1 Line Integral

Let us first examine a line integral. One possible application of this integral would be to compute the work W that would be required to push the cart with a force **F** from point a to point b along a prescribed path as shown in Figure 1–14. This path could be dictated by metallic rails underneath the cart. The line integral is written as

$$\int_a^b \mathbf{F} \cdot d\mathbf{l} \tag{1.40}$$

The differential length element **dl** can be written in the three orthogonal coordinate systems, and these were included in Table 1–1. The limits a and b determine the sign of the integration, i.e., $+$ or $-$. This integral states that no work is expended in moving the cart if the direction of the force applied to the object is perpendicular to the path of the motion. If we were to push the cart completely around the path so it returned to the original point, we would call this a ***closed line integral*** and indicate it with a circle at the center of the integral sign as in (1.41) below:

$$\oint \mathbf{F} \cdot d\mathbf{l} \tag{1.41}$$

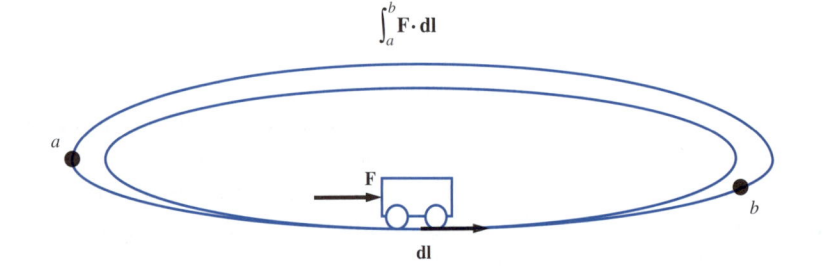

FIGURE 1-14

The cart is constrained to move along the prescribed path from points a to b.

1.3 Integral Relations for Vectors

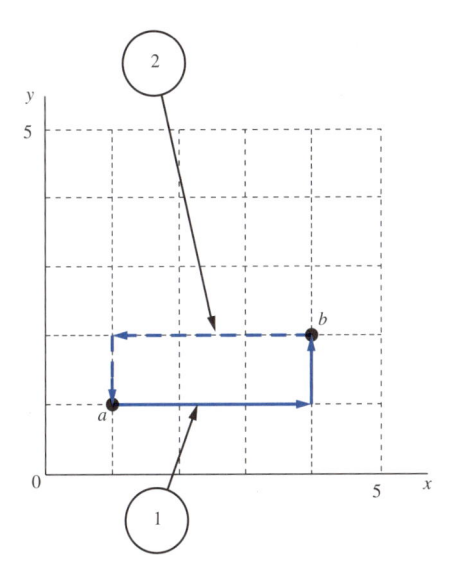

FIGURE 1–15

Two of the many possible paths along which the line integral could be evaluated.

To illustrate this point, let us calculate the work required to move the cart along path 1 as indicated in Figure 1–15 against a force field **F** where

$$\mathbf{F} = 3xy\mathbf{u_x} + 4xy\mathbf{u_y} \quad (1.42)$$

In this example, we are able to specify a numerical value for one of the variables along each segment of the total path, since each path is chosen to be parallel to an axis of the Cartesian coordinate system. This is not always possible, and one of the variables may have to be specified in terms of the other variable, or these dependent variables may be a function of another independent parameter, for example, time. In this example, the work is found using the line integral. This integral will consist of two terms, since the path of integration is initially parallel to the x axis and then parallel to the y axis. In the first term, the incremental change in y is zero, hence $dy = 0$ and the differential length becomes $\mathbf{dl} = dx\,\mathbf{u_x}$. Similarly, $\mathbf{dl} = dy\,\mathbf{u_y}$ in the second integration since $dx = 0$. Therefore, we write

$$\Delta W = \int_{(1,1)}^{(4,2)} \mathbf{F} \cdot \mathbf{dl} = 3\frac{x^2}{2}\bigg|_1^4 + 16\frac{y^2}{2}\bigg|_1^2 = \frac{93}{2} \quad (1.43)$$

We could return from point b back to point a along the same path that we followed earlier or along a different path—say, path 2 in Figure 1–15. We calculate the work along this new path. The differential length **dl** remains the same even though there is a change of direction in the integration. The limits of the integration will specify the final sign to be encountered from the integration.

$$\Delta W = \int_{(4,2)}^{(1,1)} \mathbf{F} \cdot \mathbf{dl}$$

$$= \int_4^1 (3xy\mathbf{u_x} + 4xy\mathbf{u_y})\big|_{y=2} \cdot dx\, \mathbf{u_x} + \int_2^1 (3xy\mathbf{u_x} + 4xy\mathbf{u_y})\big|_{x=1} \cdot dy\, \mathbf{u_y} \qquad (1.44)$$

$$= 6\frac{x^2}{2}\bigg|_4^1 + 4\frac{y^2}{2}\bigg|_2^1 = -51$$

The total work required to move the cart completely around this closed path is not equal to zero! A closed path is defined as any path that returns us to the original point. In Figure 1–15, the cart could have been pushed completely around the loop. There may or may not be something enclosed within the closed path. In order to emphasize this point, think of walking completely around the perimeter of a green on a golf course. This would be an example of a closed path. The entity that would be enclosed within this path and rising above the ground would be the flag. If the closed line integral over all possible paths were equal to zero, then the vector **F** would belong to a class of fields that are called *conservative fields*.[1] The example that we have just encountered would correspond to the class of *nonconservative fields*. Both conservative and nonconservative fields will be encountered in electromagnetics.

As electrical and computer engineers, you have already encountered this integral in the first course in electrical circuits without even knowing it. If we sum up the voltage drops around a closed loop, we find that they are equal to zero. This is, of course, just one of Kirchhoff's laws. For the cases that we have encountered in that early circuit's course, this would be an example of a conservative field.

EXAMPLE 1.9

Calculate the work ΔW required to move the cart along the closed path if the force field is $\mathbf{F} = 3\mathbf{u_x} + 4\mathbf{u_y}$.

Answer. The closed line integral is given by the sum of four integrals.

$$\Delta W = \oint \mathbf{F} \cdot \mathbf{dl} = \int_{(1,1)}^{(4,1)} (3\mathbf{u_x} + 4\mathbf{u_y}) \cdot dx\mathbf{u_x} + \int_{(4,1)}^{(4,2)} (3\mathbf{u_x} + 4\mathbf{u_y}) \cdot dy\, \mathbf{u_y}$$

$$+ \int_{(4,2)}^{(1,2)} (3\mathbf{u_x} + 4\mathbf{u_y}) \cdot dx\, \mathbf{u_x} + \int_{(1,2)}^{(1,1)} (3\mathbf{u_x} + 4\mathbf{u_y}) \cdot dy\, \mathbf{u_y}$$

$$= [3 \times 3] + [4 \times 1] + [3 \times (-3)] + [4 \times (-1)] = 0$$

In this case, the force field **F** is a conservative field.

---

[1] The field $\mathbf{F} = 3\mathbf{u_x} + 4\mathbf{u_y}$ is a conservative field that is demonstrated in Example 1–9; we will also find that non-constant fields may be conservative (think of work in the gravitational field).

EXAMPLE 1.10

Calculate the work ΔW required to move the cart along the circular path from point A to point B if the force field is $\mathbf{F} = 3xy\mathbf{u_x} + 4x\mathbf{u_y}$.

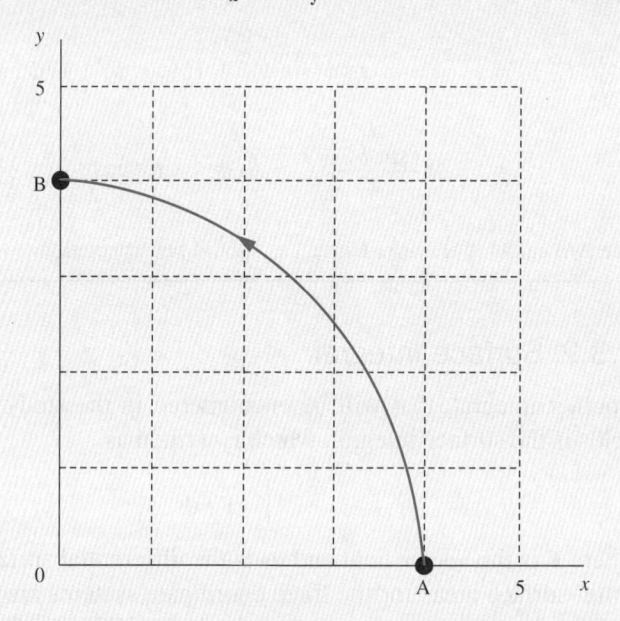

Answer. The integral can be performed in Cartesian coordinates or in cylindrical coordinates. In Cartesian coordinates, we write

$$\mathbf{F} \cdot \mathbf{dl} = (3xy\mathbf{u_x} + 4x\mathbf{u_y}) \cdot (dx\,\mathbf{u_x} + dy\,\mathbf{u_y})$$

$$= 3xy\,dx + 4x\,dy$$

The equation of the circle is $x^2 + y^2 = 4^2$. Hence

$$\int_A^B \mathbf{F} \cdot \mathbf{dl} = \int_4^0 3x\sqrt{16 - x^2}\,dx + \int_0^4 4\sqrt{16 - y^2}\,dy$$

$$= -(16 - x^2)^{3/2}\Big|_4^0 + 4\left(\frac{y}{2}\sqrt{16 - y^2} + 8\sin^{-1}\left(\frac{y}{4}\right)\right)\Big|_0^4 = -64 + 16\pi$$

In cylindrical coordinates, we write

$$\mathbf{F} \cdot \mathbf{dl} = (3xy\mathbf{u_x} + 4x\mathbf{u_y}) \cdot (d\rho\,\mathbf{u_\rho} + \rho d\phi\,\mathbf{u_\phi} + dz\,\mathbf{u_z})$$

Since the integral is to be performed along the indicated path where only the angle ϕ is changing, we have $d\rho = 0$ and $dz = 0$. Also $\rho = 4$. Therefore

$$\mathbf{F} \cdot \mathbf{dl} = (3xy\mathbf{u_x} + 4x\mathbf{u_y}) \cdot (4d\phi\,\mathbf{u_\phi})$$

From Appendix A, we write the scalar products as

$$\mathbf{u}_x \cdot \mathbf{u}_\phi = -\sin\phi \text{ and } \mathbf{u}_y \cdot \mathbf{u}_\phi = \cos\phi$$

The integral becomes

$$\int_A^B F \cdot dl = \int_0^{\pi/2} (-48\cos\phi(\sin\phi)^2 + 16(\cos\phi)^2)\, 4d\phi$$

$$= \left(-48\frac{(\sin\phi)^3}{3}\bigg|_0^{\pi/2} + 8(\phi + \sin\phi\cos\phi)\big|_0^{\pi/2}\right) 4 = -64 + 16\pi$$

The results of the two calculations are identical as should be expected.

1.3.2 Surface Integral

Another integral, that will be encountered in the study of electromagnetic fields is the surface integral, which is written as

$$\int_{\Delta s} \mathbf{F} \cdot \mathbf{ds} \qquad (1.45)$$

where **F** is the vector field and **ds** is the differential surface area. The differential surface areas for the three coordinate systems are given in Table 1–1. A field flowing through a surface is shown in Figure 1–16 for an arbitrary surface. The vector **F** at this stage could represent a fluid flow. In some sense, the loop monitors the flow of the field.

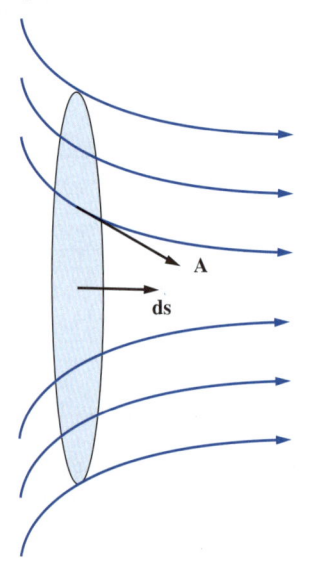

FIGURE 1-16

A surface integral for an arbitrary surface. At the particular location of the loop, the component of **A** that is tangent to the loop does not pass through the loop. The scalar product **A** • *ds* eliminates this contribution.

1.3 Integral Relations for Vectors

The differential surface element is, by definition, a vector since a direction is associated with it. The vector orientation of **ds** is in the direction that is normal to the surface outward. For a closed surface, this is the obvious direction. However, for a nonclosed surface such as a plane or our golfing green, the user must specify it in the absence of any obvious outward direction. Using the "right hand rule" convention, we rely on which way the thumb points if the fingers of the right hand follow the perimeter of the surface counterclockwise. A person standing atop the golfing green would observe a different direction than an individual underneath the green.

The surface integral allows us to ascertain the amount of the vector field **A** that is passing through a surface Δ**s**, which has a differential surface element **ds**. This vector field is frequently called a flux. A vector **A** that is directed tangential to the surface will have the scalar product **A** • **ds** = 0—i.e., the vector **A** does not pass through the surface.

If we integrated the vector field over an entire closed surface, the notation

$$\oint \mathbf{A} \cdot \mathbf{ds} \tag{1.46}$$

is employed. As we will see later, this *closed surface integral* can be either: greater than zero, equal to zero, or less than zero depending on what is contained within the enclosed volume.

For the cubical surface shown in Figure 1–17, there are six vectors **ds** associated with the six differential surfaces. The vectors

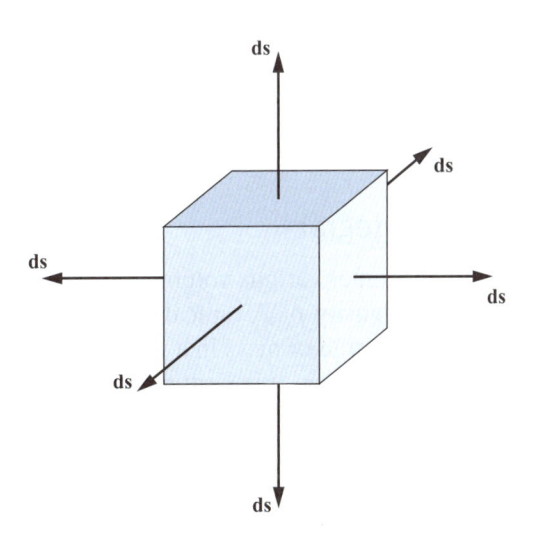

$$\mathbf{ds} = dxdy\mathbf{u}_z \quad \text{and} \quad \mathbf{ds} = dxdy(-\mathbf{u}_z)$$

FIGURE 1–17

There are six differential surface vectors associated with a cube. The vectors are directed outwards.

EXAMPLE 1.11

Assume that vector field $\mathbf{A} = A_0/r^2 \mathbf{u_r}$ exists in a region surrounding the origin of a spherical coordinate system. Find the value of the closed-surface integral $\oint \mathbf{A} \cdot \mathbf{ds}$ over the unit sphere.

Answer. The closed-surface integral is given by

$$\oint \mathbf{A} \cdot \mathbf{ds} = \int_{\phi=0}^{\phi=2\pi} \int_{\theta=0}^{\theta=\pi} \left(\frac{A_0}{r^2}\mathbf{u_r}\right) \cdot (r^2 \sin\theta \, d\theta d\phi \, \mathbf{u_r}) = 4\pi A_0$$

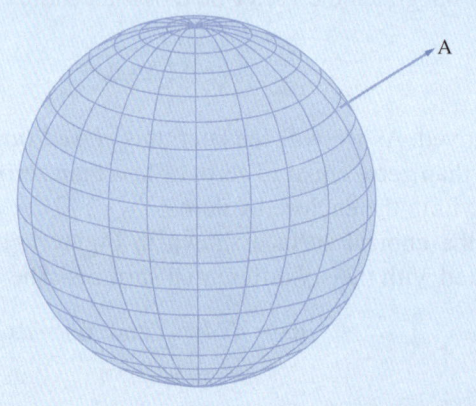

In this integral, we have used the differential surface area in spherical coordinates that has a unit vector $\mathbf{u_r}$. If the vector $\mathbf{A}$ had any additional components directed in the $\mathbf{u}_\theta$ or $\mathbf{u}_\phi$ directions, their contribution to this surface integral would be zero, since the scalar product of these terms will be equal to zero. The MATLAB command *sphere* produces this sphere.

1.3.3 Volume Integral

Finally, we will encounter various volume integrals of scalar quantities, such as a volume charge density ρ_v. A typical integration would involve the computation of the total charge or mass in a volume if the volume charge density or mass density were known. It is written as

$$Q = \int_{\Delta v} \rho_v \, dv \tag{1.47}$$

The differential volumes for the three coordinate systems are given in Table 1–1. This will be demonstrated with an example.

1.4 Differential Relations for Vectors

EXAMPLE 1.12

Find the volume of a cylinder that has a radius a and a length $\mathscr{L}$.

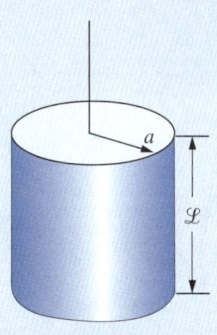

Answer. The volume of a cylinder is calculated to be

$$\Delta v = \int_{\Delta v} dv = \int_{z=0}^{z=\mathscr{L}} \int_{\phi=0}^{\phi=2\pi} \int_{\rho=0}^{\rho=a} \rho \, d\rho \, d\phi \, dz = \pi a^2 \mathscr{L}$$

1.4 Differential Relations for Vectors

In addition to the integral relations for vectors, there are also differential operations that will be encountered frequently in our journey through electromagnetic theory. Each of these differential operators can be interpreted in terms of physical phenomena. We will concentrate on vector operations in Cartesian coordinates. In addition, the operations in cylindrical and spherical coordinates are included in this discussion. The three vector operations are given in Table 1–4. A summary of these vector operations in the three coordinate systems is found in Appendix A.

1.4.1 Gradient

It is possible to methodically measure scalar quantities (such as a temperature) at various locations in space. From these data, it is possible to connect

TABLE 1–4 Three important vector operations expressed with the Del operator

Gradient of a scalar function	∇a
Divergence of a vector field	$\nabla \cdot \mathbf{A}$
Curl of a vector field	$\nabla \times \mathbf{A}$

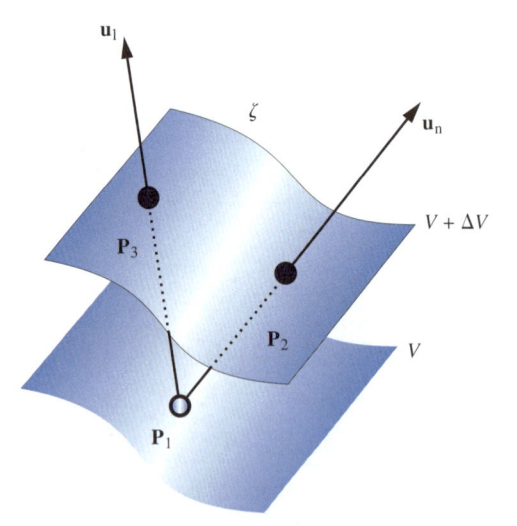

FIGURE 1–18

Equipotential surfaces in space.

the locations where the temperatures are the same. When placed on a graph in a two dimensional plot, these equitemperature contours are useful in interpreting various effects. This could include determining the magnitude and direction where the most rapid changes occur or ascertaining the amount of heat that will flow in a particular direction. This could be useful in planning a ski trip or beach vacation. The gradient of the scalar quantity (which in this case is the temperature) allows us to compute the magnitude and the direction we should follow to find the maximum spatial rate of change in the scalar quantity to attain the desired conditions.

In Figure 1–18, we sketch two equipotential surfaces in space; the potential of one surface is arbitrarily chosen to have the value V, and the potential of the other surface is $V + \Delta V$. Point 1 is located on the first surface. The unit vector $\mathbf{u_n}$ that is normal to this surface at P_1 intersects the second surface at point P_2. The magnitude of the distance between these two points is Δn. Point P_3 is another point on the second surface, and the vector distance between P_1 and P_3 is Δl. The unit vector from P_1 to P_3 is $\mathbf{u_l}$. The angle between the two vectors is ζ. The distance Δl is greater than Δn. Therefore,

$$\frac{\Delta V}{\Delta n} \geq \frac{\Delta V}{\Delta l}$$

This allows us to define the first differential operation.

The gradient is defined as the vector that represents both the magnitude and the direction of the maximum spatial rate of increase of a scalar function. It depends on the position where the gradient is to be evaluated, and it

1.4 Differential Relations for Vectors

may have different magnitudes and directions at different locations in space. Referring to Figure 1–18, we write the gradient as

$$\text{grad } V = \nabla V \equiv \lim_{\Delta n \to 0} \frac{\Delta V}{\Delta n}\mathbf{u}_n = \frac{dV}{dn}\mathbf{u}_n \tag{1.48}$$

In writing (1.48), we have used the common notation of replacing grad with the symbol ∇ ("del"). In addition, we have assumed that the separation distance between the two surfaces is small, and let $\Delta n \to dn$, which is indicative of a derivative.

The definition of a directional derivative is self-explanatory. In Figure 1–18 we want it in the $\mathbf{u}_l$ direction, and we write

$$\frac{\Delta V}{\Delta l}\mathbf{u}_l \to \frac{dV}{dl}\mathbf{u}_l$$

where we have again let $\Delta l \to dl$. Using the chain rule, we find that

$$\frac{dV}{dl} = \frac{dV}{dn}\frac{dn}{dl} = \frac{dV}{dn}\cos\zeta = \frac{dV}{dn}\mathbf{u}_n \cdot \mathbf{u}_l = \nabla V \cdot \mathbf{u}_l \tag{1.49}$$

We realize that the directional derivative in the $\mathbf{u}_l$ direction is the projection of the gradient in that particular direction. Equation (1.49) can be written as

$$dV = \nabla V \cdot dl\, \mathbf{u}_l = \nabla V \cdot \mathbf{dl} \tag{1.50}$$

The gradient of the scalar function $a(x,y,z)$ is given in Cartesian coordinates as

$$\nabla a = \frac{\partial a}{\partial x}\mathbf{u}_x + \frac{\partial a}{\partial y}\mathbf{u}_y + \frac{\partial a}{\partial z}\mathbf{u}_z \tag{1.51}$$

This equation prompts us to define the "del operator" in Cartesian coordinates (only) as $\nabla = \mathbf{u}_x \partial/\partial x + \mathbf{u}_y \partial/\partial x + \mathbf{u}_z \partial/\partial x$.

The gradient of the scalar function $a(\rho,\phi,z)$ in cylindrical coordinates is

$$\nabla a = \frac{\partial a}{\partial \rho}\mathbf{u}_\rho + \frac{1}{\rho}\frac{\partial a}{\partial \phi}\mathbf{u}_\phi + \frac{\partial a}{\partial z}\mathbf{u}_z \tag{1.52}$$

The gradient of the scalar function $a(r,\theta,\phi)$ in spherical coordinates is

$$\nabla a = \frac{\partial a}{\partial r}\mathbf{u}_r + \frac{1}{r}\frac{\partial a}{\partial \theta}\mathbf{u}_\theta + \frac{1}{r\sin\theta}\frac{\partial a}{\partial \phi}\mathbf{u}_\phi \tag{1.53}$$

MATLAB also provides the capability of performing the gradient operation. In order to use this command, we must first calculate the contours that connect the points that have the same value.

EXAMPLE 1.13

Assume that there exists a surface that can be modeled with the equation $z = e^{-(x^2 + y^2)}$. Calculate ∇z at the point $(x = 0, y = 0)$. In addition, use MATLAB to illustrate the profile and to calculate and plot this field.

Answer. $\nabla z = -2x e^{-(x^2 + y^2)} \mathbf{u_x} - 2y e^{-(x^2 + y^2)} \mathbf{u_y}$. At the point $(x = 0, y = 0)$, $\nabla z = 0$. Using MATLAB, the function is illustrated in Figure (a). The contours with the same value are connected together, and the resulting field is indicated in (b). The length of the vectors and their orientation clearly indicate the distribution of the field in space. The commands *gradient*, *contour*, and *quiver* have been employed to create the figure.

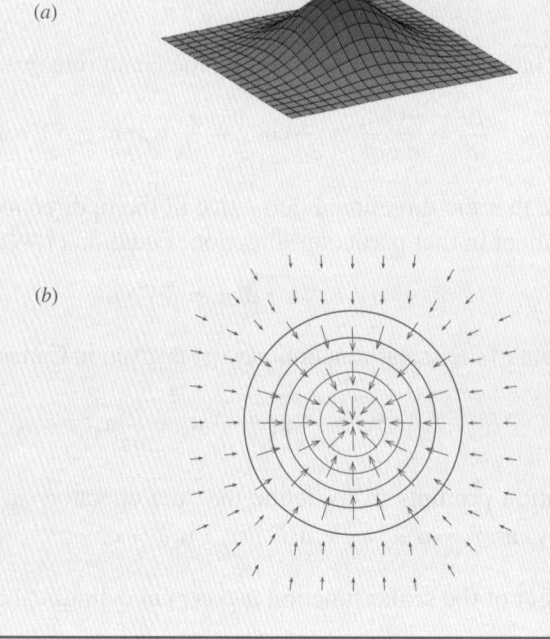

(a)

(b)

1.4.2 Divergence

The second vector derivative that should be reviewed is the divergence operation. The divergence operator is useful in determining if there is a source or a sink at locations in space where a vector field exists. For electromagnetic fields, these sources and sinks will turn out to be positive and negative electrical charges. This region could also be situated in a river where water would be flowing, as shown in Figure 1–19. This could be a very porous box

1.4 Differential Relations for Vectors

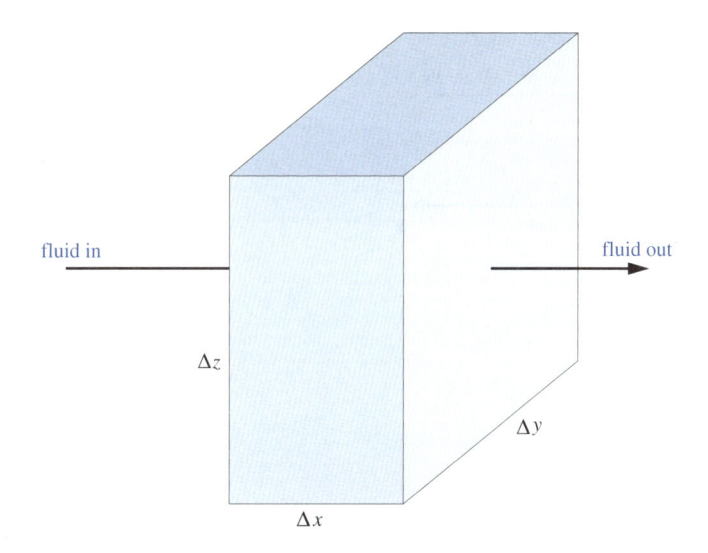

FIGURE 1–19

Schematic of a source or a sink in a region where a fluid is flowing.

that contained either a drain or faucet that was connected with an invisible hose to the shore, where the fluid could either be absorbed or from which it could be extracted.

The divergence of a vector that applies at a point is defined by the expression

$$\text{div } \mathbf{A} = \nabla \cdot \mathbf{A} \equiv \lim_{\Delta v \to 0} \frac{\oint \mathbf{A} \cdot d\mathbf{s}}{\Delta v} \tag{1.54}$$

The symbol $\oint \mathbf{A} \cdot d\mathbf{s}$ indicates an integral over the entire closed surface that encloses the volume Δv. The point where the divergence is evaluated is within the volume Δv, and the surface for the closed-surface integral is the surface that surrounds this volume. As we will see, the application of the "$\nabla \cdot$" notation (where ∇ is the *del operator*) will help us in remembering the terms that appear in the operation.

We evaluate the surface integral over the pair of constant x surfaces only; the remaining surface integrals are similar and need not be repeated. The only component of the vector $\mathbf{A}$ involved is the component A_x, due to the dot products with $d\mathbf{s}$. We use a Taylor series expansion of A_x about the midpoint of the volume. Recall that an arbitrary function $f(x)$ has the Taylor series

$$f(x) = f(x_0) + \left.\frac{\partial f(x)}{\partial x}\right|_{x_0} (x - x_0) + \left.\frac{\partial^2 f(x)}{\partial x^2}\right|_{x_0} (x - x_0)^2 + \cdots \tag{1.55}$$

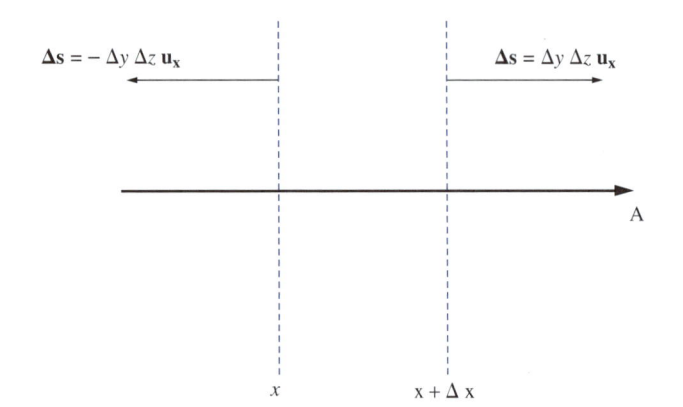

FIGURE 1-20

Two surfaces that are located at x and at $x + \Delta x$.

$\Delta \mathbf{s} = -\Delta y \Delta z \, \mathbf{u}_x$ $\Delta \mathbf{s} = \Delta y \Delta z \, \mathbf{u}_x$

x $x + \Delta x$

We will retain only the first two terms of the series, since in our application x will approach x_0 in the limit, and hence higher powers in $(x - x_0)$ will be very small. Furthermore, we will assume that the functions to be expanded are sufficiently smooth that indicated derivatives pose no problems.

Figure 1–20 depicts the two surfaces over which we integrate. Since the volume is very small, the quantity A_x may be assumed constant over the surfaces. The surface integral in this case yields A_x multiplied by the surface area $\Delta y \Delta z$. Note that the normal directions for the two surfaces differ in sign, as shown in the figure. Thus, the surface integral over the two surfaces is given by

$$\oint (A_x \mathbf{u}_x) \cdot d\mathbf{s} \approx (A_x \mathbf{u}_x|_{x + \Delta x}) \cdot (\Delta y \Delta z \mathbf{u}_x) + (A_x \mathbf{u}_x|_x) \cdot (-\Delta y \Delta z \mathbf{u}_x)$$

$$\approx \left\{ \left[\left(A_x + \Delta x \frac{\partial A_x}{\partial x} \right) \mathbf{u}_x \right] \cdot (\Delta y \Delta z \mathbf{u}_x) + (A_x \mathbf{u}_x) \cdot (-\Delta y \Delta z \mathbf{u}_x) \right\} \quad (1.56)$$

$$= \left(\frac{\partial A_x}{\partial x} \right) (\Delta x \Delta y \Delta z)$$

Note that if A_x doesn't change with x, then the integral is zero. This corresponds to the situation where all fluid flowing in through one surface exits the other; there is no source nor sink in the intermediate region.

Using the result of equation (1.56) in the definition (1.54) of the divergence gives the result

$$\nabla \cdot (A_x \mathbf{u}_x) = \lim_{\Delta v \to 0} \frac{\left(\frac{\partial A_x}{\partial x} \right)(\Delta x \Delta y \Delta z)}{(\Delta x \Delta y \Delta z)} = \frac{\partial A_x}{\partial x} \quad (1.57)$$

1.4 Differential Relations for Vectors

where the volume in question is $\Delta v = \Delta x \, \Delta y \, \Delta z$. We may repeat this analysis for the remaining pairs of faces of the cube and sum the results to obtain the well-known result

$$\nabla \cdot \mathbf{A} = \frac{\partial A_x}{\partial x} + \frac{\partial A_y}{\partial y} + \frac{\partial A_z}{\partial z} \tag{1.58}$$

Once again, in Cartesian coordinates this suggests the definition of the del operator, this time in a scalar product with a vector. However, be cautious: the del operator assumes this simple form only in Cartesian coordinates, and it is an *operator*, not a vector. In other words, $\nabla \cdot \mathbf{A} \neq \mathbf{A} \cdot \nabla$!

EXAMPLE 1.14

Find the divergence of the vector $\mathbf{A} = 3x\mathbf{u}_x + xy^2\mathbf{u}_y - 2xye^{-z}\mathbf{u}_z$ at the point $(1, -1, 2)$.

Answer. Using equation (1.58), we find

$$\nabla \cdot \mathbf{A} = \frac{\partial}{\partial x}(3x) + \frac{\partial}{\partial y}(xy^2) - \frac{\partial}{\partial z}(2xye^{-z})$$

$$= 3 + 2xy + 2xye^{-z}$$

At the point $(1, -1, 2)$ this gives $\nabla \cdot \mathbf{A} = 3 - 2 - 2e^{-2}$.

We have found the divergence of a vector and we can suggest a physical interpretation of it. If the divergence of a vector $\mathbf{A}$ is equal to zero, then there are neither sources to create the vector $\mathbf{A}$ nor sinks to absorb it at that location. In this case, everything that enters the volume will leave it unscathed. If the divergence of a vector is greater than or less than zero, then there is either a source or a sink for the vector at that location; there is a net divergence or convergence of flux. In the context of electromagnetic theory, the integral of the divergence of a vector field determines whether positive or negative electric charges exist within a volume (in the case of electric fields), or whether magnetic charges or "magnetic monopoles" exist in the region (for the case of magnetic fields, no such charges have been detected in nature).

We derived the divergence in Cartesian coordinates. The extension to cylindrical and spherical coordinates is similar. In cylindrical coordinates, we find (Appendix A)

$$\nabla \cdot \mathbf{A} = \frac{1}{\rho}\frac{\partial (\rho A_\rho)}{\partial \rho} + \frac{1}{\rho}\frac{\partial A_\phi}{\partial \phi} + \frac{\partial A_z}{\partial z} \tag{1.59}$$

In spherical coordinates, we have

$$\nabla \cdot \mathbf{A} = \frac{1}{r^2}\frac{\partial (r^2 A_r)}{\partial r} + \frac{1}{r \sin \theta}\frac{\partial (A_\theta \sin \theta)}{\partial \theta} + \frac{1}{r \sin \theta}\frac{\partial A_\phi}{\partial \phi} \tag{1.60}$$

EXAMPLE 1.15

Find the divergence of the vector field $\mathbf{A} = e^{-(\rho/\alpha)^2} \mathbf{u}_\rho$, where α is a constant, both analytically and by application of the MATLAB *divergence* function.

Answer. From equation (1.59), we have

$$\nabla \cdot \mathbf{A} = \frac{1}{\rho}\frac{\partial}{\partial \rho}(\rho A_\rho) = \frac{1}{\rho}\frac{\partial}{\partial \rho}(\rho e^{-(\rho/\alpha)^2}) = \frac{1}{\rho}\left[e^{-(\rho/\alpha)^2} - 2\rho\left(\frac{\rho}{\alpha}\right)e^{-(\rho/\alpha)^2}\left(\frac{1}{\alpha}\right)\right]$$

$$= \frac{e^{-(\rho/\alpha)^2}}{\rho}\left[1 - 2\left(\frac{\rho}{\alpha}\right)^2\right]$$

The plot of the vector field $\mathbf{A}$ using the *quiver* function is shown in Figure (*a*), and the contours of the divergence, $\nabla \cdot \mathbf{A}$, are presented in Figure (*b*).

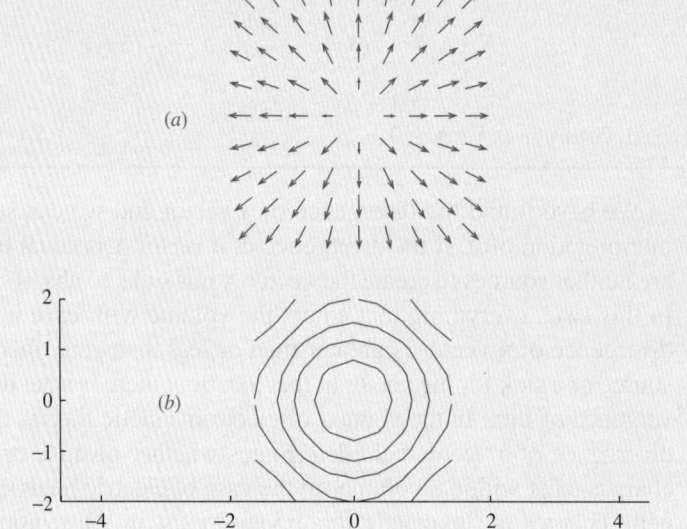

From the definition of the divergence (1.54), we can also find a useful relation between a volume integral of the vector's divergence and the integral of the vector field over the closed surface enclosing the volume Δv. This can be obtained from the following "hand-waving" argument. From (1.54), we write

$$\oint \mathbf{A} \cdot \mathbf{ds} \approx (\nabla \cdot \mathbf{A})\Delta v \approx \int_{\Delta v}(\nabla \cdot \mathbf{A})dv \tag{1.61}$$

In passing from the second term that appears in the definition of the divergence to the integral in the third term, we have let the volume Δv be so small

1.4 Differential Relations for Vectors

that the volume integral of the vector's divergence is approximately equal to the product of the volume and the divergence.

Equating the two terms involving the integrals and replacing the approximately equal symbol with the equal sign, we obtain the ***divergence theorem***.

$$\oint \mathbf{A} \cdot \mathbf{ds} = \int_{\Delta v} (\nabla \cdot \mathbf{A}) dv \qquad (1.62)$$

The integral on the right-hand side is over the surface enclosing the volume. This theorem will prove very useful in later developments regarding electromagnetic fields, since it allows us to move easily between a volume integral and a closed-surface integral. It is also known as ***Gauss's theorem***.

EXAMPLE 1.16

Evaluate both sides of the divergence theorem for a vector field $\mathbf{A} = x\mathbf{u}_x$ within the unit cube centered about the origin.

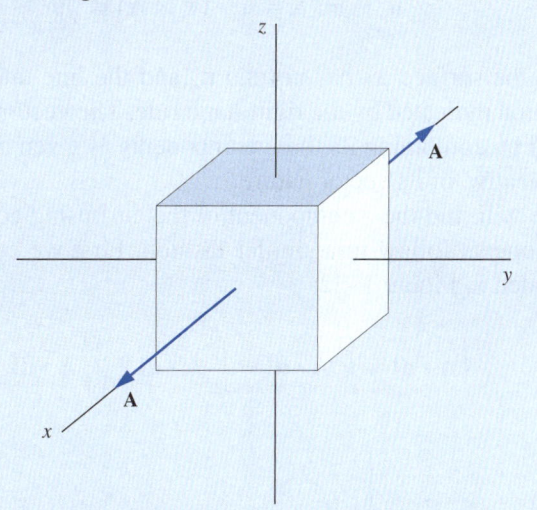

Answer. The volume integral is given by

$$\int_{\Delta v} \nabla \cdot \mathbf{A} \, dv = \int_{x=-1/2}^{x=1/2} \int_{y=-1/2}^{y=1/2} \int_{z=-1/2}^{z=1/2} dz \, dy \, dx = 1$$

The closed-surface integral consists of two terms that are evaluated at $x = -1/2$ and at $x = +1/2$. We write

$$\oint \mathbf{A} \cdot \mathbf{ds} = \int_{y=-1/2}^{y=1/2} \int_{z=-1/2}^{z=1/2} (x|_{x=-1/2}) \mathbf{u}_x \cdot (-dz\,dy\,\mathbf{u}_x)$$
$$+ \int_{y=-1/2}^{y=1/2} \int_{z=-1/2}^{z=1/2} (x|_{x=+1/2}) \mathbf{u}_x \cdot (dz\,dy\,\mathbf{u}_x) = 1$$

As we expected, the two answers are the same.

1.4.3 Curl

The curl is a vector operation that can be used to determine whether there is a rotation associated with a vector field. This is visualized most easily by considering the experiment of inserting a small paddle wheel in a flowing river as shown in Figure 1–21. If the paddle wheel is inserted in the center of the river, it will not rotate since the velocity of the water a small distance on either side of the center will be the same. However, if the paddle wheel was situated near the edge of the river, it would rotate since the velocity near the edge will be less than in a region further from the edge. Note that the rotation will be in the opposite directions at the two edges of the river. The curl operation determines both the sense and the magnitude of the rotation.

The curl of a vector $\mathbf{A}$ gives a vector result. The $\mathbf{u}_n$ component of the curl is defined by

$$\mathbf{u}_n \cdot \text{curl } \mathbf{A} = \mathbf{u}_n \cdot (\nabla \times \mathbf{A}) \equiv \lim_{\Delta s \to 0} \frac{\oint \mathbf{A} \cdot d\mathbf{l}}{\Delta s} \tag{1.63}$$

Here, the surface Δs has normal $\mathbf{u}_n$ and the line integral is traversed in the direction indicated by the right-hand rule. The vector curl, $\nabla \times \mathbf{A}$, is then obtained by combining its three components as given above. The notation "$\times$" is indicative of its vector nature.

We will find the z component of the curl using equation (1.63); the other components follow in a similar fashion. First we compute the line integral indicated in Figure 1–22:

$$\oint \mathbf{A} \cdot d\mathbf{l} = \int_1^2 \mathbf{A} \cdot d\mathbf{l} + \int_2^3 \mathbf{A} \cdot d\mathbf{l} + \int_3^4 \mathbf{A} \cdot d\mathbf{l} + \int_4^1 \mathbf{A} \cdot d\mathbf{l} \tag{1.64}$$

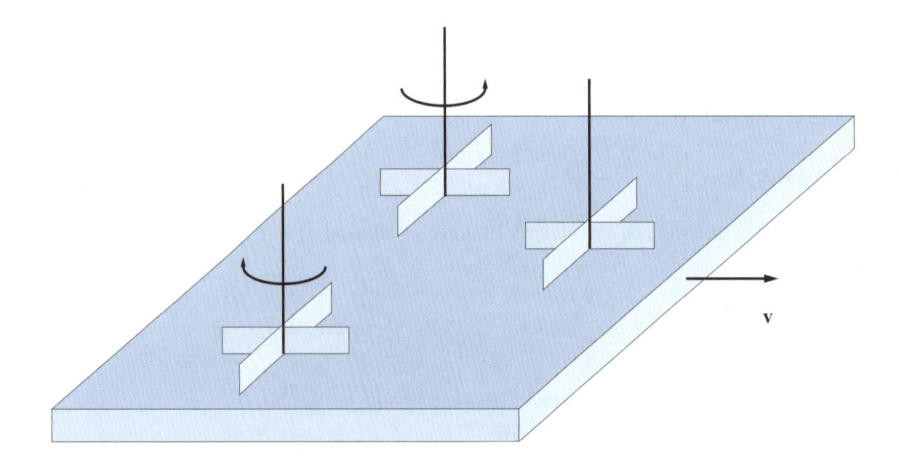

FIGURE 1–21

The paddle wheels inserted in a river will rotate if they are near the edges, since the river velocity just at the edge is zero. The wheel at the center of the river will not rotate.

1.4 Differential Relations for Vectors

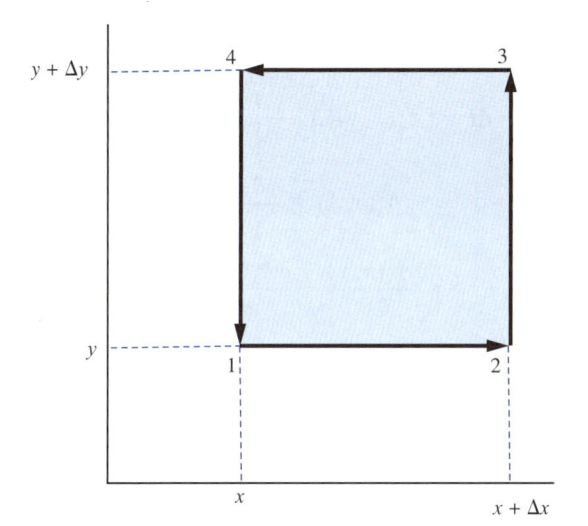

FIGURE 1-22

Orientation of the loop required to find the $\mathbf{u}_z$ component of curl $\mathbf{A}$ at the point (x, y).

The right-hand rule indicates that we follow the counterclockwise path shown. If the surface is sufficiently small, $\mathbf{A}$ is approximately constant on each segment, so we have

$$\oint \mathbf{A} \cdot d\mathbf{l} \approx \mathbf{A}(x, y) \cdot \int_1^2 d\mathbf{l} + \mathbf{A}(x + \Delta x, y) \cdot \int_2^3 d\mathbf{l}$$
$$+ \mathbf{A}(x, y + \Delta y) \cdot \int_3^4 d\mathbf{l} + \mathbf{A}(x, y) \cdot \int_4^1 d\mathbf{l} \qquad (1.65)$$

The line integrals along the segments are

$$\int_1^2 d\mathbf{l} = \mathbf{u}_x \Delta x \qquad \int_2^3 d\mathbf{l} = \mathbf{u}_y \Delta y \qquad (1.66)$$

$$\int_3^4 d\mathbf{l} = -\mathbf{u}_x \Delta x \qquad \int_4^1 d\mathbf{l} = -\mathbf{u}_y \Delta y$$

We now use the Taylor series expansion of a function $f(x, y)$ of two variables, and retain only the low-order terms since Δs will tend to zero:

$$f(x, y) \approx f(x_0, y_0) + \left.\frac{\partial f(x, y)}{\partial x}\right|_{x_0, y_0} \Delta x + \left.\frac{\partial f(x, y)}{\partial y}\right|_{x_0, y_0} \Delta y \qquad (1.67)$$

Application to each term in (1.65) gives

$$\oint \mathbf{A} \cdot \mathbf{dl} \approx [A_x(x, y)]\Delta x + \left[A_y(x, y) + \left.\frac{\partial A_y(x, y)}{\partial x}\right|_{x, y} \Delta x\right]\Delta y$$

$$- \left[A_x(x, y) + \left.\frac{\partial A_x(x, y)}{\partial y}\right|_{x, y} \Delta y\right]\Delta x - [A_y(x, y)]\Delta y \quad (1.68)$$

$$= \left[\frac{\partial A_y(x, y)}{\partial x} - \frac{\partial A_x(x, y)}{\partial y}\right]\Delta x \Delta y$$

We have found the z component of the curl; the other components follow in a similar fashion. Collecting all terms in Cartesian coordinates yields the result for $\Delta \times \mathbf{A}$. It is most easily expressed and remembered as the determinant of a matrix:

$$\text{curl } \mathbf{A} = \begin{vmatrix} \mathbf{u}_x & \mathbf{u}_y & \mathbf{u}_z \\ \frac{\partial}{\partial x} & \frac{\partial}{\partial y} & \frac{\partial}{\partial z} \\ A_x & A_y & A_z \end{vmatrix} \quad (1.69)$$

$$= \left(\frac{\partial A_z}{\partial y} - \frac{\partial A_y}{\partial z}\right)\mathbf{u}_x + \left(\frac{\partial A_x}{\partial z} - \frac{\partial A_z}{\partial x}\right)\mathbf{u}_y + \left(\frac{\partial A_y}{\partial x} - \frac{\partial A_x}{\partial y}\right)\mathbf{u}_z$$

We derived the curl in Cartesian coordinates. The extension to cylindrical and spherical coordinates follows. In cylindrical coordinates, we have

$$\text{curl } \mathbf{A} = \frac{1}{\rho}\begin{vmatrix} \mathbf{u}_\rho & \rho\mathbf{u}_\phi & \mathbf{u}_z \\ \frac{\partial}{\partial \rho} & \frac{\partial}{\partial \phi} & \frac{\partial}{\partial z} \\ A_\rho & \rho A_\phi & A_z \end{vmatrix} \quad (1.70)$$

In spherical coordinates, we have

$$\text{curl } \mathbf{A} = \frac{1}{r^2 \sin\theta}\begin{vmatrix} \mathbf{u}_r & r\mathbf{u}_\theta & r\sin\theta\,\mathbf{u}_\phi \\ \frac{\partial}{\partial r} & \frac{\partial}{\partial \theta} & \frac{\partial}{\partial \phi} \\ A_r & rA_\theta & r\sin\theta\,A_\phi \end{vmatrix} \quad (1.71)$$

EXAMPLE 1.17

Find the curl of the vector field $\mathbf{A} = \omega\rho e^{-(\rho/\alpha)^2}\mathbf{u}_\phi$, where α and ω are constant, both analytically and by use of the *curl* function.

1.4 Differential Relations for Vectors

Answer. From equation (1.70), the curl is given by

$$\nabla \times \mathbf{A} = \frac{1}{\rho} \begin{bmatrix} \mathbf{u}_\rho & \rho \mathbf{u}_\phi & \mathbf{u}_z \\ \frac{\partial}{\partial \rho} & \frac{\partial}{\partial \phi} & \frac{\partial}{\partial z} \\ 0 & \rho(\omega \rho e^{-(\rho/\alpha)^2}) & 0 \end{bmatrix}$$

$$= \frac{1}{\rho}\left[(0)\mathbf{u}_\rho + (0)\rho\mathbf{u}_\phi + \frac{\partial}{\partial \rho}(\omega \rho^2 e^{-(\rho/\alpha)^2})\mathbf{u}_z\right]$$

$$= 2\omega e^{-(\rho/\alpha)^2}\left[1 - \left(\frac{\rho}{\alpha}\right)^2\right]\mathbf{u}_z$$

A plot of the field is shown in (*a*) and contours of the *z*-component of the curl are shown in (*b*).

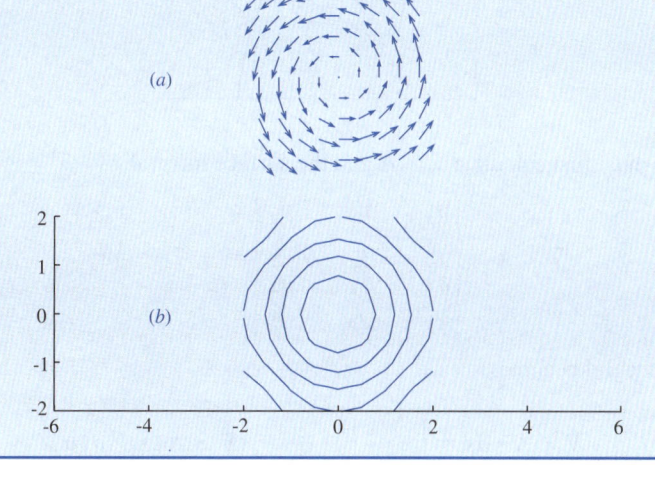

(*a*)

(*b*)

From the definition of the curl of a vector given in (1.63), we can obtain **Stokes's theorem** that relates a closed-line integral to a surface integral. Following the same "hand-waving" procedure that we used to derive the divergence theorem, we write

$$\oint \mathbf{A} \cdot d\mathbf{l} \approx (\nabla \times \mathbf{A})\Delta s \approx \int_{\Delta s} \nabla \times \mathbf{A} \cdot d\mathbf{s} \qquad (1.72)$$

This is finally written with the same caveats that we employed previously as **Stoke's theorem**.

$$\boxed{\oint \mathbf{A} \cdot d\mathbf{l} = \int_{\Delta s} \nabla \times \mathbf{A} \cdot d\mathbf{s}} \qquad (1.73)$$

The line integral on the left-hand side is along the perimeter of the surface in the direction indicated by the right-hand rule. Recall that in the right-hand convention that we are employing, the fingers of the right hand follow the path of the line integral **dl** and the thumb points in the direction of the vector surface element **ds**.

EXAMPLE 1.18

Given a vector field $\mathbf{A} = xy\mathbf{u}_x - 2x\mathbf{u}_y$, verify Stokes's theorem over one-quarter of a circle whose radius is 3.

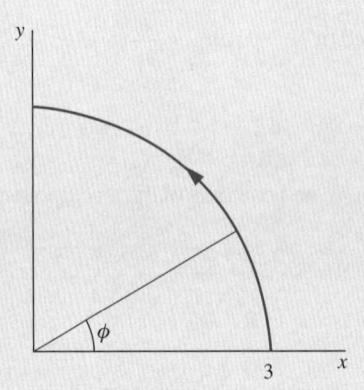

Answer. We must first calculate $\nabla \times \mathbf{A}$ and the surface integral.

$$\nabla \times \mathbf{A} = \begin{vmatrix} \mathbf{u}_x & \mathbf{u}_y & \mathbf{u}_z \\ \dfrac{\partial}{\partial x} & \dfrac{\partial}{\partial y} & \dfrac{\partial}{\partial z} \\ xy & -2x & 0 \end{vmatrix} = -(2 + x)\mathbf{u}_z$$

The surface integral becomes

$$\int_{\Delta s} \nabla \times \mathbf{A} \cdot d\mathbf{s} = \int_{y=0}^{y=3} \int_{x=0}^{x=\sqrt{9-y^2}} (\nabla \times \mathbf{A}) \cdot dxdy\mathbf{u}_z$$

$$= \int_{y=0}^{y=3} \int_{x=0}^{x=\sqrt{9-y^2}} -(2+x)\mathbf{u}_z \cdot dxdy\mathbf{u}_z$$

$$= -\int_{y=0}^{y=3} \left[2\sqrt{9-y^2} + \frac{(\sqrt{9-y^2})^2}{2} \right] dy$$

The integral $\int_{y=0}^{y=3} 2\sqrt{9-y^2}\, dy$ requires the substitution $y = 3\sin\phi$ and the identity $\cos^2\phi = \frac{1}{2}(1 + \cos 2\phi)$ to transform it to the integral $9\int_0^{\pi/2} (1 + \cos^2\phi)d\phi$ which can be evaluated. Therefore, we obtain

$$-\int_{y=0}^{y=3} \left[2\sqrt{9-y^2} + \frac{(\sqrt{9-y^2})^2}{2} \right] dy = -9\frac{\pi}{2} - \frac{27}{2} + \frac{27}{6} = -9\left(1 + \frac{\pi}{2}\right)$$

1.4 Differential Relations for Vectors

> The closed-line integral involves three terms, and using the right-hand convention for the integration sequence, we write
>
> $$\oint \mathbf{A} \cdot \mathbf{dl} = \int_{x=0,\,y=0}^{x=3,\,y=0} \mathbf{A} \cdot dx\,\mathbf{u_x} + \int_{\text{arc}} \mathbf{A} \cdot \mathbf{dl} + \int_{x=0,\,y=3}^{x=0,\,y=0} \mathbf{A} \cdot dy\,\mathbf{u_y}$$
>
> The two integrals that are along the two axes will contribute zero to the closed integral because the vector $\mathbf{A} = 0$ on the axis. The remaining integral becomes
>
> $$\oint \mathbf{A} \cdot \mathbf{dl} = \int_{\text{arc}} \mathbf{A} \cdot \mathbf{dl} = \int_{\text{arc}} (xy\,\mathbf{u_x} - 2x\,\mathbf{u_y}) \cdot (dx\,\mathbf{u_x} + dy\,\mathbf{u_y} + dz\,\mathbf{u_z})$$
>
> $$= \int (xy\,dx - 2x\,dy)$$
>
> $$= \int_3^0 x\sqrt{9-x^2}\,dx - 2\int_0^3 \sqrt{9-y^2}\,dy = -9\left(1 + \frac{\pi}{2}\right)$$
>
> As we should expect, the two answers are the same.

1.4.4 Repeated Vector Operations

Having defined the vector operations of the gradient, the divergence, and the curl, we may be curious about a repeated vector operation, such as the divergence of the curl of a vector. There are several methods of approaching this topic. A straightforward rigorous approach would be to perform the vector operations mechanically in order to find the answer. This approach is left for the problems section at the end of this chapter. A second approach that we will pursue here is based on intuitive arguments. The meaning of the various vector operations will likely become more clear as the discussion is presented.

The three vector operations that will be examined are

$$\nabla \cdot \nabla \times \mathbf{A} = 0 \quad (1.74)$$

$$\nabla \times \nabla a = 0 \quad (1.75)$$

$$\nabla \cdot \nabla a = \nabla^2 a \quad (1.76)$$

Other vector identities exist, and a list of useful vector identities is included in Appendix A.

Equation (1.74) can be interpreted in the following terms. The curl operation gives the magnitude and sense of vector rotation confined within a prescribed region. The quantity that this vector represents neither enters nor leaves the region. The divergence operation monitors a vector field's entry into or departure from a region due to a local source or sink within it. Therefore, a vector $\mathbf{A}$ that has a nonzero curl just rotates and neither enters nor leaves the region. One could think of a boat in a rotating whirlpool that cannot be paddled away from its impending doom as an example of this identity.

Equation (1.75) is understood from the following argument. The gradient of a scalar function expresses the direction and magnitude that an inertialess ball would take as it rolls down a mountain along the path of least resistance. This path would not be expected to close upon itself. The curl, however, would require that the ball return to the same point on the mountain to indicate rotation. This point could be back at the top, interrupting the ball's roll down the mountain under its own volition. Hence we can conclude that (1.75) is correct, since it is clear that the ball could not return unless there were suddenly some new laws of nature—such as anti-gravitational forces.

Equation (1.76) is the definition of the Laplacian operation. It states that there is a vector field ∇a where a is some scalar function. The divergence of this vector field determines whether a source or a sink exists at that point. In Cartesian coordinates, the **Laplacian operator** is written as

$$\nabla^2 a = \frac{\partial^2 a}{\partial x^2} + \frac{\partial^2 a}{\partial y^2} + \frac{\partial^2 a}{\partial z^2} \tag{1.77}$$

As will be seen later, this operation is important for finding the potential distribution caused by a charge distribution.

In cylindrical coordinates, the Laplacian operator is

$$\nabla^2 a = \frac{1}{\rho}\frac{\partial\left(\rho\frac{\partial a}{\partial \rho}\right)}{\partial \rho} + \frac{1}{\rho^2}\frac{\partial^2 a}{\partial \phi^2} + \frac{\partial^2 a}{\partial z^2} \tag{1.78}$$

In spherical coordinates, the Laplacian operator is

$$\nabla^2 a = \frac{1}{r^2}\frac{\partial\left(r^2\frac{\partial a}{\partial r}\right)}{\partial r} + \frac{1}{r^2 \sin\theta}\frac{\partial\left(\sin\theta\frac{\partial a}{\partial \theta}\right)}{\partial \theta} + \frac{1}{r^2 \sin^2\theta}\frac{\partial^2 a}{\partial \phi^2} \tag{1.79}$$

1.5 Phasors

In applications we often encounter situations where sources in a linear system exhibit harmonic time dependence at a single fixed frequency. For instance, in circuits we are often interested in the AC case where the voltage varies sinusoidally with time as

$$v(t) = V_0 \cos(\omega t + \phi) \tag{1.80}$$

Here, V_0 is the peak amplitude of the signal, ω is the angular frequency ($\omega = 2\pi f$, where f is the frequency in Hertz), and ϕ is the phase of the signal in radians. Such a signal is shown in Figure 1–23.

1.5 Phasors

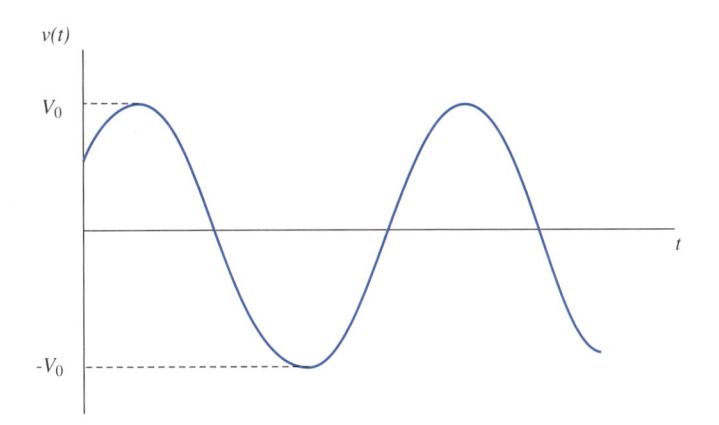

FIGURE 1-23

A time-harmonic signal $v(t)$.

In this case, all quantities of interest will vary sinusoidally with the same frequency ω (due to linearity), and it is convenient to work with the ***phasor*** representation of these quantities. The phasor V corresponding to the voltage (1.80) is defined by the relationship

$$v(t) = \text{Re}\{Ve^{j\omega t}\} \qquad (1.81)$$

Here the notation Re{ } indicates that we are to take the real part of the quantity in the brackets, and $j = \sqrt{-1}$. Recall Euler's identity, which states $e^{j\theta} = \cos\theta + j\sin\theta$; we see that in order for (1.80) to be equal to (1.81), we must have

$$V = V_0 e^{j\phi} \qquad (1.82)$$

Therefore, we see that $V = V_0 e^{j\phi}$ is the phasor corresponding to $v(t) = V_0\cos(\omega t + \phi)$. Then (1.81) gives $v(t) = \text{Re}\{Ve^{j\omega t}\} = \text{Re}\{V_0 e^{j\phi} e^{j\omega t}\} = \text{Re}\{V_0 e^{j(\omega t + \phi)}\} = V_0 \text{Re}\{\cos(\omega t + \phi) + j\sin(\omega t + \phi)\} = V_0 \cos(\omega t + \phi)$ as required. We will encounter in this text phasors that are functions of positions, such as $V(z) = V_0 e^{-j\beta z}$. Application of (1.81) reveals the time-varying voltage corresponding to such a phasor: $v(t,z) = \text{Re}\{Ve^{j\omega t}\} = \text{Re}\{V_0 e^{-j\beta z} e^{j\omega t}\} = \text{Re}\{V_0 e^{j(\omega t - \beta z)}\} = V_0 \cos(\omega t - \beta z)$. As we will see, this waveform represents a ***traveling wave*** of voltage, such as may be observed on a transmission line.

The primary advantage of using phasors in analysis of time-harmonic systems is the simplification that results in differentiation and integration with respect to time. If we denote the time-varying signal and its corresponding phasor using the "phasor transform" pair notation $v(t) \leftrightarrow V$, then we will demonstrate the following results: $d/dt[v(t)] \leftrightarrow j\omega V$ and $\int v(t)dt \leftrightarrow (1/j\omega)V$. In words, differentiation with respect to time becomes multiplication by a factor of $j\omega$ for the corresponding phasor, and integration with respect to time becomes divison by a factor of $j\omega$ for the corresponding phasor. These results follow easily from equation (1.81). Consider

$$\frac{dv}{dt} = \frac{d}{dt}\text{Re}\{Ve^{j\omega t}\} = \text{Re}\left\{V\frac{d}{dt}e^{j\omega t}\right\} = \text{Re}\{j\omega Ve^{j\omega t}\} \qquad (1.83)$$

We observe that the phasor corresponding to dv/dt (the factor multiplying $e^{j\omega t}$) is $j\omega V$, where V is the phasor corresponding to $v(t)$, as desired. In a similar fashion, we consider

$$\int v(t)dt = \int \text{Re}\{Ve^{j\omega t}\}dt = \text{Re}\left\{V\int e^{j\omega t}dt\right\} = \text{Re}\{(1/j\omega)Ve^{j\omega t}\} \qquad (1.84)$$

Here we observe that the phasor corresponding to $\int v(t)dt$ (the factor multiplying $e^{j\omega t}$) is $(1/j\omega)V$, where V is the phasor corresponding to $v(t)$, as desired.

It is important to remember that we are assuming that the system under consideration is linear. This implies that if all sources vary sinusoidally with a single fixed frequency ω, then all quantities of interest will vary sinusoidally with the same frequency. This is *not the case* when the system is non-linear. For instance, consider a non-linear system that produces the product of two signals of frequency ω. By the familiar trig identity, we have $\cos\omega t \cos\omega t = 1/2[1 + \cos 2\omega t]$. Thus the product contains a DC term plus a term varying at the 2ω.

EXAMPLE 1.19

Express $v(t) = 10\cos(120\pi t + 60°)$ volts in phasor notation.

Answer. This is written as

$$V = 10\, e^{j(\pi/3)} = 5 + j8.7 \text{ volts}$$

EXAMPLE 1.20

Express $v(t) = 3\cos\omega t - 4\sin\omega t$ as $A\cos(\omega t + \phi)$. Use phasor notation. Plot the function.

Answer. Let us use $\cos\omega t$ as the reference and add the two phasors.

$$3\cos\omega t \Rightarrow 3$$

$$-4\sin\omega t = -4\cos\left(\omega t - \frac{\pi}{2}\right) \Rightarrow -4e^{-j\pi/2} = j4$$

Therefore, we write

$$V = 3 + j4 = 5e^{j\tan^{-1}(4/3)} = 5e^{j53°}$$

The real part of the product of this phasor and $e^{j\omega t}$ yields

$$v(t) = 3\cos\omega t - 4\sin\omega t = \text{Re}(5e^{j(\omega t + 53°)}) = 5\cos(\omega t + 53°)$$

The plot of the time-harmonic function is shown in the figure below.

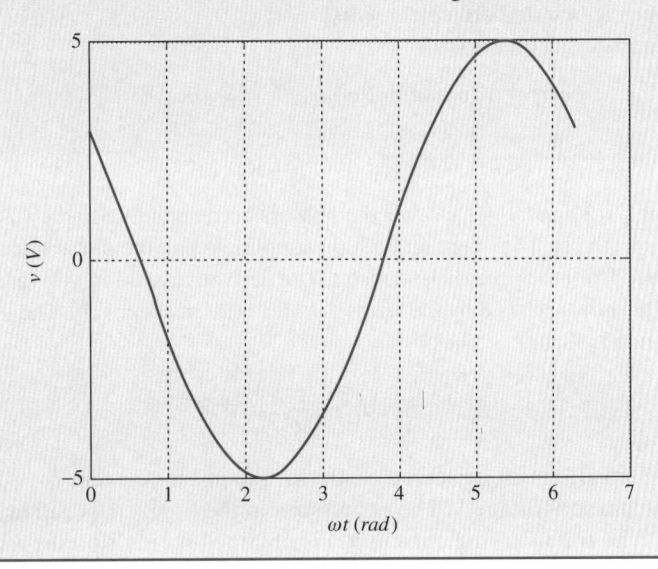

EXAMPLE 1.21

Express the loop equation for an *RLC* equation in phasor notation. The applied voltage is $v(t) = V_0 \cos \omega t$ and the loop equation is

$$L\frac{di}{dt} + Ri + \frac{1}{C}\int i\,dt = v$$

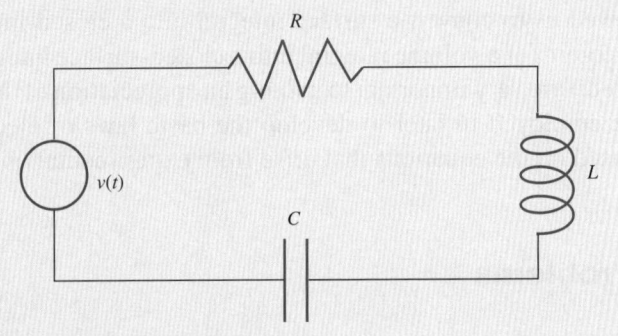

Answer. Write the current i as

$$i(t) = I_0 \cos(\omega t + \phi)$$

since we have chosen the cosine as the reference. Hence the differential equation becomes

$$I_0\left[-\omega L \sin(\omega t + \phi) + R\cos(\omega t + \phi) + \frac{1}{\omega C}\sin(\omega t + \phi)\right] = V_0 \cos \omega t$$

The mathematical manipulations beyond this point that would be required to determine i and ϕ are tedious at best and difficult at worst.

In contrast, using phasors,

$$v(t) = V_0 \cos \omega t = \text{Re}[V_0 e^{j\phi} e^{j\omega t}] = \text{Re}[V e^{j\omega t}]$$

and

$$i(t) = \text{Re}[I_0 e^{j\phi} e^{j\omega t}] = \text{Re}[I e^{j\omega t}]$$

Here, the terms $V = V_0$ and $I = I_0 e^{j\phi}$ are the phasors corresponding to the voltage $v(t)$ and current $i(t)$, respectively. They contain both the amplitude and the phase information that has been isolated from the time dependence t. The derivative and integral in the original loop equation are now replaced by the terms $j\omega$ and $1/j\omega$, respectively. The loop equation in phasor notation is then

$$\left[R + j\left(\omega L - \frac{1}{\omega C}\right)\right] I = V$$

This algebraic equation can be solved easily for the phasor current I in terms of the phasor voltage V. The expression in the brackets is called the **impedance** Z. The time-varying current $i(t)$ is obtained by multiplying I by $e^{j\omega t}$ and taking the real part of the product.

1.6 Conclusion

The study of electromagnetic fields as described in this book makes use of MATLAB, vectors, and the various integral and differential operations that have been introduced in this chapter. Moreover, the two theorems that allowed us to convert a surface integral into a closed-line integral (Stokes's theorem) or a volume integral into a closed-surface integral (divergence theorem) are very important to gaining an appreciation of these fields. They will be employed in later to develop the basic laws of electromagnetic theory based on the equations that arise from experimental observations.

1.7 Problems

1.1.1. Given two vectors $\mathbf{A} = 3\mathbf{u}_x + 4\mathbf{u}_y + 5\mathbf{u}_z$ and $\mathbf{B} = -5\mathbf{u}_x + 4\mathbf{u}_y - 3\mathbf{u}_z$, find $\mathbf{C} = \mathbf{A} + \mathbf{B}$ and $\mathbf{D} = \mathbf{A} - \mathbf{B}$. In addition, carefully illustrate these vectors using MATLAB.

1.1.2. Using the vectors defined in Problem 1.1.1, evaluate $\mathbf{A} \cdot \mathbf{B}$ and $\mathbf{A} \times \mathbf{B}$ using MATLAB.

1.1.3. Given two vectors $\mathbf{A} = \mathbf{u}_x + \mathbf{u}_y + \mathbf{u}_z$ and $\mathbf{B} = 2\mathbf{u}_x + 4\mathbf{u}_y + 6\mathbf{u}_z$, find $\mathbf{C} = \mathbf{A} + \mathbf{B}$ and $\mathbf{D} = \mathbf{A} - \mathbf{B}$. In addition, carefully illustrate these vectors using MATLAB.

1.1.4. Using the vectors defined in Problem 1.1.3, evaluate $\mathbf{A} \cdot \mathbf{B}$ and $\mathbf{B} \times \mathbf{A}$ using MATLAB.

1.7 Problems

1.1.5. Using MATLAB, write a program to convert degrees Celsius to degrees Fahrenheit. Plot the results.

1.1.6. Using MATLAB, write a program to convert a yardstick to a meterstick. Plot the results.

1.1.7. Using MATLAB, plot $y = e^{-x}$ on a linear and a semilog graph.

1.1.8. Using MATLAB, plot two cycles of $y = \cos(x)$ on a linear and a polar graph.

1.1.9. Using MATLAB, carefully plot a vector field defined by $\mathbf{A} = y^2\mathbf{u}_x - x\mathbf{u}_y$ in the region $-2 < x < +2$, $-2 < y < +2$. The length of the vectors in the field should be proportional to the field at that point. Find the magnitude of this vector at the point (3,2).

1.1.10. Using MATLAB, carefully plot a vector field defined by $\mathbf{A} = \sin x\, \mathbf{u}_x - \sin y\, \mathbf{u}_y$ in the region $0 < x < \pi$, $0 < y < \pi$. The length of the vectors in the field should be proportional to the field at that point. Find the magnitude of this vector at the point $(\pi/2, \pi/2)$.

1.2.1. Find the vector $\mathbf{A}$ that connects the two opposite corners of a cube whose volume is a^3. One corner of the cube is located at the center of a Cartesian coordinate system. Also write this vector in terms of the magnitude times a unit vector.

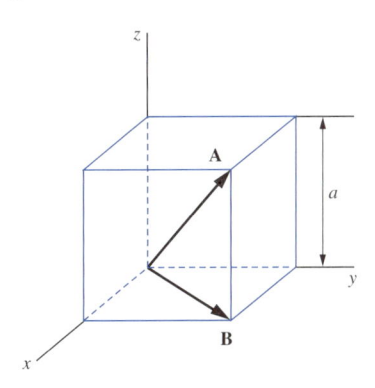

1.2.2. Find the vector $\mathbf{B}$ from the origin to the opposite corner that lies in the xy plane.

1.2.3. Find the scalar product of the two vectors defined by $\mathbf{A} = 3\mathbf{u}_x + 4\mathbf{u}_y + 5\mathbf{u}_z$ and $\mathbf{B} = -5\mathbf{u}_x + 4\mathbf{u}_y - 3\mathbf{u}_z$. Determine the angle between these two vectors. Check your answer using MATLAB.

1.2.4. Find the scalar product of the two vectors defined by $\mathbf{A} = \mathbf{u}_x + \mathbf{u}_y + \mathbf{u}_z$ and $\mathbf{B} = 2\mathbf{u}_x + 4\mathbf{u}_y + 6\mathbf{u}_z$. Determine the angle between these two vectors. Check your answer using MATLAB.

1.2.5. Find the projection of a vector from the origin to the point (1,2,3) on the vector from the origin to the point (2,1,6) in Cartesian coordinates. Find the angle between these two vectors. Check your answer using MATLAB.

1.2.6. Find the vector product of the two vectors defined by $\mathbf{A} = 3\mathbf{u}_x + 4\mathbf{u}_y + 5\mathbf{u}_z$ and $\mathbf{B} = -5\mathbf{u}_x + 4\mathbf{u}_y - 3\mathbf{u}_z$. Check your answer using MATLAB.

1.2.7. Find the vector product of the two vectors defined by $\mathbf{A} = \mathbf{u}_x + \mathbf{u}_y + \mathbf{u}_z$ and $\mathbf{B} = 2\mathbf{u}_x + 4\mathbf{u}_y + 6\mathbf{u}_z$. Check your answer using MATLAB.

1.2.8. Express the vector field $\mathbf{A} = 3\mathbf{u}_x + 4\mathbf{u}_y + 5\mathbf{u}_z$ in cylindrical coordinates. Check your answer using MATLAB.

1.2.9. Convert the vector $\mathbf{B} = 3\mathbf{u}_\rho + 4\mathbf{u}_\phi + 5\mathbf{u}_z$ that is in cylindrical coordinates into Cartesian coordinates. Check your answer using MATLAB.

1.2.10. Express the vector field $\mathbf{A} = 3\mathbf{u}_x + 4\mathbf{u}_y + 5\mathbf{u}_z$ in spherical coordinates. Check your answer using MATLAB.

1.2.11. Convert the vector $\mathbf{B} = 3\mathbf{u}_r + 4\mathbf{u}_\theta + 5\mathbf{u}_\phi$ that is in spherical coordinates into Cartesian coordinates. Check your answer using MATLAB.

1.2.12. For the vectors $\mathbf{A} = \mathbf{u}_x + \mathbf{u}_y + \mathbf{u}_z$, $\mathbf{B} = 2\mathbf{u}_x + 2\mathbf{u}_y + 2\mathbf{u}_z$, and $\mathbf{C} = 3\mathbf{u}_x + 3\mathbf{u}_y + 3\mathbf{u}_z$, show that $\mathbf{A} \times (\mathbf{B} \times \mathbf{C}) = \mathbf{B}(\mathbf{A} \cdot \mathbf{C}) - \mathbf{C}(\mathbf{A} \cdot \mathbf{B})$. Check your answer using MATLAB.

1.2.13. For the vectors $\mathbf{A} = \mathbf{u}_x + 3\mathbf{u}_y + 5\mathbf{u}_z$, $\mathbf{B} = 2\mathbf{u}_x + 4\mathbf{u}_y + 6\mathbf{u}_z$, and $\mathbf{C} = 3\mathbf{u}_x + 4\mathbf{u}_y + 5\mathbf{u}_z$, show that $\mathbf{A} \times (\mathbf{B} \times \mathbf{C}) = \mathbf{B}(\mathbf{A} \cdot \mathbf{C}) - \mathbf{C}(\mathbf{A} \cdot \mathbf{B})$. Check your answer using MATLAB.

1.2.14. Find the area of the parallelogram using vector notation. Compare your result with that found graphically.

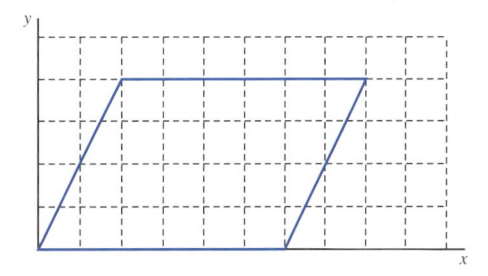

1.2.15. Show that we can use the vector definitions $\mathbf{A} \cdot \mathbf{B} = 0$ and $\mathbf{A} \times \mathbf{B} = 0$ to express that two vectors are perpendicular and parallel to each other, respectively.

1.2.16. Let $\mathbf{A} = -2\mathbf{u}_x + 3\mathbf{u}_y + 4\mathbf{u}_z$, $\mathbf{B} = 7\mathbf{u}_x + 1\mathbf{u}_y + 2\mathbf{u}_z$, and $\mathbf{C} = -1\mathbf{u}_x + 2\mathbf{u}_y + 4\mathbf{u}_z$. Find

a. $\mathbf{A} \times \mathbf{B}$

b. $(\mathbf{A} \times \mathbf{B}) \cdot \mathbf{C}$

c. $\mathbf{A} \cdot (\mathbf{B} \times \mathbf{C})$

1.3.1. Calculate the work required to move a mass m against a force field $\mathbf{F} = 5\mathbf{u}_x + 7\mathbf{u}_y$ along the indicated direct path from point a to point b.

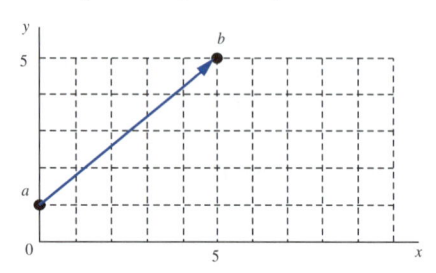

1.3.2. Calculate the work required to move a mass m against a force field $\mathbf{F} = y\mathbf{u}_x + x\mathbf{u}_y$ along the path abc and along the path adc. Is this field conservative?

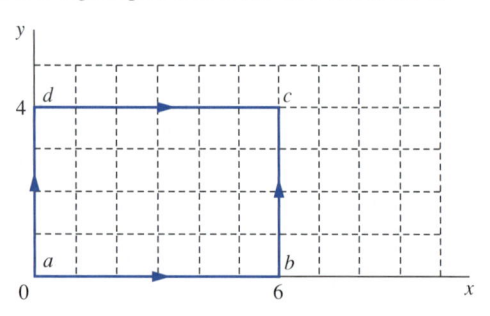

1.3.3. Calculate the work required to move a mass m against a force field $\mathbf{F} = \rho\mathbf{u}_\rho + \rho\phi\mathbf{u}_\phi$ along the path abc.

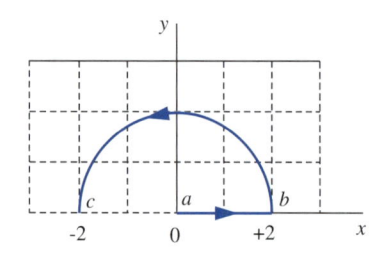

1.3.4. Calculate the work required to move a mass m against a force field $\mathbf{F} = \rho\phi\mathbf{u}_\phi$ if the radius of the circle is a and $0 \leq \phi \leq 2\pi$.

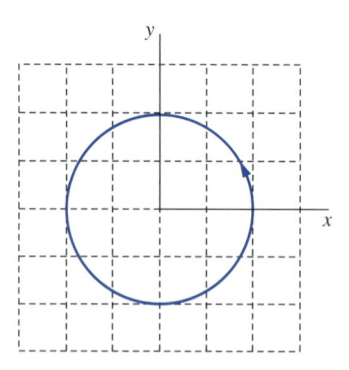

1.3.5. Calculate the closed-surface integral $\oint \mathbf{A} \cdot \mathbf{ds}$ if $\mathbf{A} = x\mathbf{u}_x + y\mathbf{u}_y$ and the surface is that of the cube shown below..

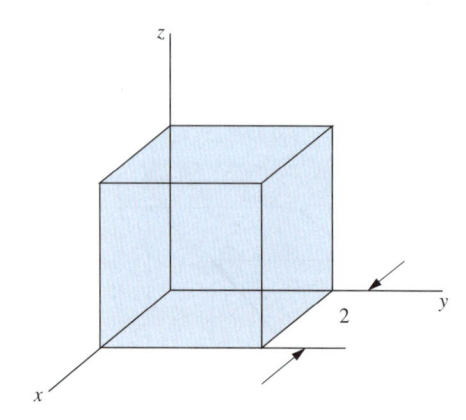

Apply the divergence theorem to solve the same integral.

1.3.6. Evaluate the closed-surface integral of the vector $\mathbf{A} = xyz\mathbf{u}_x + xyz\mathbf{u}_y + xyz\mathbf{u}_z$ over the cubical surface of Problem 1.3.5.

1.3.7. Evaluate the closed-surface integral of the vector $\mathbf{A} = 3\mathbf{u}_r$ over the spherical surface centered on the origin that has a radius a.

1.7 Problems

1.3.8. Find the surface area of a cylindrical surface shown below by setting up and evaluating the integral $A = \oint \mathbf{A} \cdot d\mathbf{s}$ where $\mathbf{A}$ is the surface normal $\mathbf{A} = \mathbf{u}_\rho$ on the side, and $\mathbf{A} = \pm \mathbf{u}_z$ on the top and bottom of the cylinder, respectively.

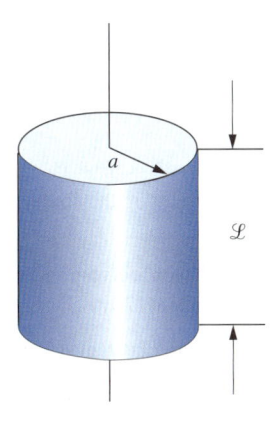

1.4.1. A hill can be modeled with the equation $H = 10 - x^2 - 3y^2$ where H is the elevation of the hill. Find the direction in which a frictionless ball would roll unimpeded if released from rest at location (x_0, y_0).

1.4.2. Find the gradient of the function $H = x^2yz$ and also the directional derivative of H in the direction specified by the unit vector $\mathbf{u} = a(\mathbf{u}_x + \mathbf{u}_y + \mathbf{u}_z)$ where a is a constant at the point (1,2,3). State the value for the constant a.

1.4.3. By direct differentiation show that

$$\nabla\left(\frac{1}{R}\right) = -\nabla'\left(\frac{1}{R}\right)$$

where

$$R = \sqrt{(x-x')^2 + (y-y')^2 + (z-z')^2}$$

and ∇' denotes differentiation with respect to the variables x', y', and z'.

1.4.4. Calculate the divergence of the vector $\mathbf{A} = x^3 y\sin(\pi z)\mathbf{u}_x + xy\sin(\pi z)\mathbf{u}_y + x^2 y^2 z^2 \mathbf{u}_z$ at the point (1,1,1).

1.4.5. Show that the divergence theorem is valid for the cube below, located at the center of a Cartesian coordinate system, for a vector $\mathbf{A} = x\mathbf{u}_x + 2\mathbf{u}_y$.

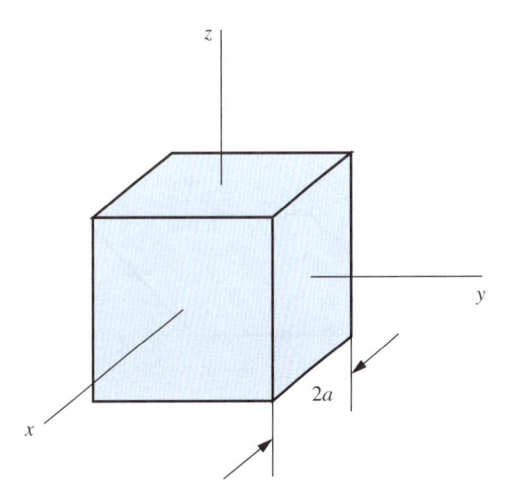

1.4.6. Show that the divergence theorem is valid for a sphere of radius a located at the center of a coordinate system for a vector $\mathbf{A} = r\mathbf{u}_r$.

1.4.7. The water that flows in a channel with sides at $x = 0$ and $x = a$ has a velocity distribution $\mathbf{v}(x, z) = [(a/2)^2 - (x - a/2)^2]z^2\mathbf{u}_y$. The bottom of the river is at $z = 0$. A small paddle wheel with its axis parallel to the z axis is inserted into the channel and is free to rotate. Find the relative rates of rotation at the points

$$\left(x = \frac{a}{4}, z = 1\right), \left(x = \frac{a}{2}, z = 1\right), \text{ and } \left(x = \frac{3a}{4}, z = 1\right)$$

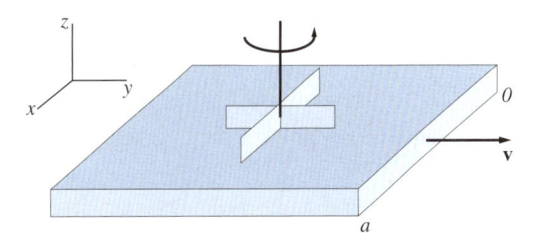

Will the paddle wheel rotate if its axis is parallel to the x axis or the y axis?

1.4.8. Evaluate the line integral of the vector function $\mathbf{A} = x\mathbf{u}_x + x^2\mathbf{u}_y + xyz\mathbf{u}_z$ around the square contour C. Integrate $\nabla \times \mathbf{A}$ over the surface bounded by C. Show that this example satisfies Stokes's theorem.

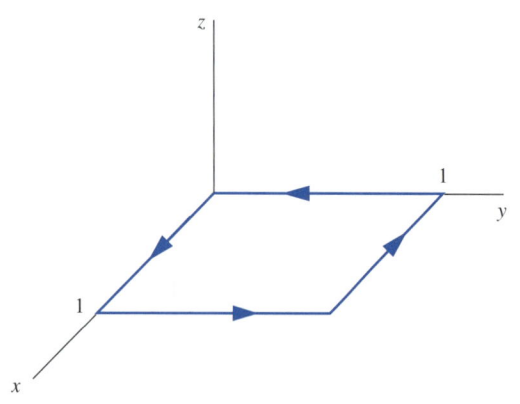

1.4.9. Show that $\nabla \times \mathbf{A} = 0$ if $\mathbf{A} = 1/\rho\, \mathbf{u}_\rho$ in cylindrical coordinates.

1.4.10. Show that $\nabla \times \mathbf{A} = 0$ if $\mathbf{A} = r^2\mathbf{u}_r$ in spherical coordinates.

1.4.11. In rectangular coordinates, verify that $\nabla \cdot \nabla \times \mathbf{A} = 0$ where $\mathbf{A} = x^2y^2z^2\,[\mathbf{u}_x + \mathbf{u}_y + \mathbf{u}_z]$ by carrying out the indicated derivatives.

1.4.12. In rectangular coordinates, verify that $\nabla \times \nabla a = 0$ where $a = 3x^2y + 4z^2x$ by carrying out the indicated derivatives.

1.4.13. In rectangular coordinates, verify that $\nabla \times (a\mathbf{A}) = (\nabla a) \times \mathbf{A} + a\nabla \times \mathbf{A}$ where $\mathbf{A} = xyz[\mathbf{u}_x + \mathbf{u}_y + \mathbf{u}_z]$ and $a = 3xy + 4zx$ by carrying out the indicated derivatives.

1.4.14. In rectangular coordinates, verify that $\nabla \cdot (a\mathbf{A}) = \mathbf{A} \cdot \nabla a + a\nabla \cdot \mathbf{A}$ where $\mathbf{A} = xyz\,[\mathbf{u}_x + \mathbf{u}_y + \mathbf{u}_z]$ and $a = 3xy + 4zx$ by carrying out the indicated derivatives.

1.4.15. By direct differentiation, show that $\nabla^2(1/R) = 0$ at all points where $R \neq 0$ where
$$R = \sqrt{(x-x')^2 + (y-y')^2 + (z-z')^2}$$

1.5.1. Express the signal $v(t) = 100\cos(120\pi t - 45°)$ in phasor notation.

1.5.2. Given a phasor $V = 10 + j5$, find the sinusoidal signal this represents if the frequency equals 60 Hz.

1.5.3. Find the phasor corresponding to $v(t) = \cos(120\pi t - 60°) - \sin(120\pi t)$.

1.5.4. Find and plot the current $i(t)$ in the circuit if $v(t) = 10\cos(120\pi t)$.

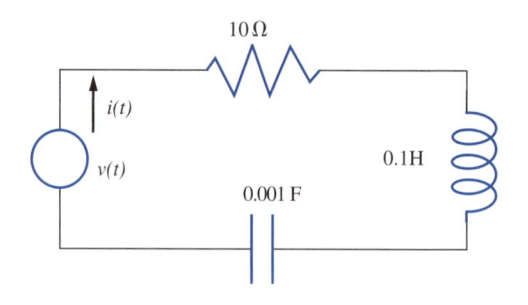

1.5.5. Repeat Problem 1.5.4 with $v(t) = 10\cos(120\pi t + 45°)$.

CHAPTER 2

Electrostatic Fields

2.1 Coulomb's Law 61
2.2 Electric Field 67
2.3 Superposition Principles 69
2.4 Gauss's Law 77
2.5 Potential Energy and Electric Potential 85
2.6 Numerical Integration 100
2.7 Dielectric Materials 109
2.8 Capacitance 114
2.9 Conclusion 118
2.10 Problems 119

The important properties of time-independent electrostatic fields will be reviewed in this chapter. This will include a review of the force between two stationary charges, the concept of an electric field, the energy stored in an electric field, and several procedures that are used to calculate the electric field. Capacitance will be defined in terms of electrical charges and electrical fields. The effects of dielectric materials that are introduced into the region containing an electric field will be described.

2.1 Coulomb's Law

The phenomenon that is the basis for the study of static electric fields has been known since ancient times. As early as 600 B.C., Thales of Miletus observed that the rubbing of amber against a cloth caused the amber rod to attract light objects to itself. The use of amber by this ancient Greek experimenter has had a dramatic influence on the discipline that we now call *electrical engineering*[1] and on the subject of electromagnetic fields. Indeed, these ancient observers have given us a word that is still in everyday use—the Greek word for amber is *élektron*. Materials other than amber also exhibit this process of electrification and today we can observe the same effect when we rub a glass rod on a silk cloth or take off a wool sweater too quickly. Both the rod and the cloth will attract small pieces of "fluff and stuff."

A new entity in nature that we call a ***charge*** was uncovered in those experiments. It is as fundamental a quantity as those that we have already encountered: mass, length, and time. The charge can be either positive or negative. In this text we will use the symbols Q, M, L, and T for the quantities charge, mass, length, and time when we are conducting a dimensional or unit analysis of an equation. We will call these "fundamental units". This nomenclature will be useful in "checking the dimensions" of an equation that we may have derived. We cannot claim that a lengthy derivation is correct if we end up with "apples" on one side of the equality sign and "oranges" on the other side.

As we pass through the middle of the eighteenth century, we find the names of many who have contributed to our understanding of this physical phenomenon: Benjamin Franklin, Joseph Priestley, Michael Faraday, Henry Cavendish, and Charles-Augustin de Coulomb. Through a series of experiments, these scientists uncovered the fact that there would be a force of attraction for unlike charges and a force of repulsion for like charges. This force is somewhat similar to the force of gravity. Both forces have the same geometrical dependence on the separation distance R between the two objects. Both forces also depend on the product of the magnitudes of fundamental quantities for each force—charges, Q_j, in the case of the electric force and masses, m_j, in the case of the gravitational force. After much experimentation, they concluded that the magnitude of this electrical force could be written as

$$F \propto \frac{Q_1 Q_2}{R^2} \tag{2.1}$$

In words, we say that the electric force between two charges is proportional to the product of the two charges and inversely proportional to the square of the distance between the two charges. This formulation is known as an inverse square law and is a very common one in the physical world. It applies not only to the gravitational force and the electric force, but to the

---

[1] This discipline is now frequently called "electrical and computer engineering."

2.1 Coulomb's Law

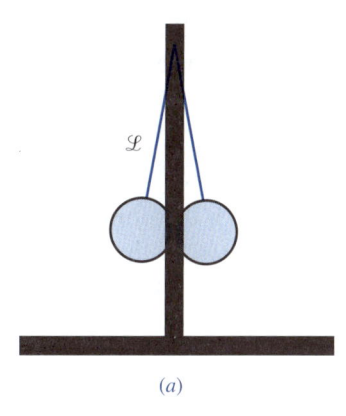

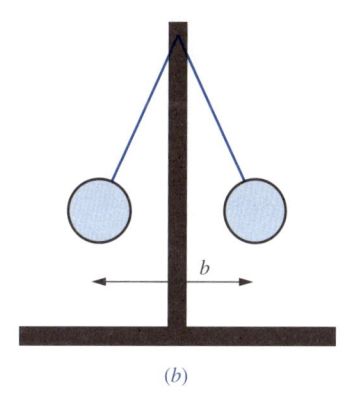

FIGURE 2–1

An experiment designed to demonstrate the electrostatic force. (*a*) Two uncharged pith balls are hanging from a vertical rod. The length of the string is $\mathscr{L}$. The only force the balls experience is in the vertical direction and it is due to gravity. (*b*) The pith balls repel each other due to the Coulomb force in the horizontal direction to a distance *b* after the same charge is distributed on each ball. The distance *b* will be determined by the amount of charge on the balls, as well as the gravitational force pulling the balls downward.

force between two electric currents, as well as describing the way light spreads from a flashlight or energy flows out of the antenna connected to your cell phone.

An ancient experimental system is depicted in Figure 2–1. A charge is induced on a glass rod by rubbing it on a cloth and then touching it simultaneously to two pith balls. In part (a), the "electrified" glass rod is just used to transfer the electrical charge to two stationary pith balls that are touching initially. The sign of the charge deposited on each ball is the same, and it was found that both balls experienced a force that caused the balls to separate. Let us assume that the experimenter performing the experiment shown in Figure 2–1 could accurately measure the following quantities:

(a) The magnitude and the sign of the charges
(b) The magnitude and the vector direction of the force
(c) The distance between the two pith balls
(d) The masses of the two pith balls

It was found in this experiment that the magnitude of a charge was an integer multiple of the magnitude of the charge of an electron. If charges with different signs had been placed individually on these two balls, the balls would not separate but would be attracted to each other.

From the experimental results, this force that we will call an *electrostatic force* or a *Coulomb force* can be written in MKS or SI units as

$$\mathbf{F} = \frac{Q_1 Q_2}{4\pi\varepsilon_0 R^2}\mathbf{u_R} \quad (\text{N}) \tag{2.2}$$

The unit of force is measured in terms of newtons (N), the unit of charge is measured in terms of coulombs (C), the distance between the charges is measured in terms of meters (m) and $\mathbf{u_R} = (\mathbf{r} - \mathbf{r}')/|\mathbf{r} - \mathbf{r}'|$ is the unit vector in the direction from $Q_1(\mathbf{r}')$ to $Q_2(\mathbf{r})$. The charge of an electron is

$$Q_e \approx -1.602 \times 10^{-19} \text{ C} \tag{2.3}$$

and the charge of a single proton is

$$Q_p \approx +1.602 \times 10^{-19} \text{ C} \tag{2.4}$$

Although the electron and proton have equal and opposite charges, they have very different masses, with the proton having at least 1836 times the mass of an electron. There are both positive and negative charges that exist in nature. All of the charges have values that are *integer multiples* of these values.

In a series of reports to the French Academy of Science from 1785 to 1791, Charles–Augustin de Coulomb described the results of a series of experiments involving a carefully constructed torsion balance in which he verified the equation of the electrostatic force (2.2). He also performed a series of experiments using small magnets and verified that the magnetic force between like and unlike magnetic poles would be either repulsive or attractive with the same geometrical dependence. We will discuss magnetic fields in the next chapter.

One coulomb is a very large amount of charge. For example, if we were to collect all the charge that is created in a single lightning strike, we would collect a total of only 10 to 20 coulombs. Considering the violent nature of such a stroke, this does not appear to be a very big number. There are, however, a very large number N of charged particles in a lightning strike. For example, we compute

$$N = \left(\frac{10}{1.602 \times 10^{-19}} \approx 6 \times 10^{19} \text{ particles}\right) \tag{2.5}$$

The constant ε_0 in (2.2) is called the ***permittivity of free space***. In SI units, it has the numerical value

$$\varepsilon_0 \approx \left(8.854 \times 10^{-12} \approx \frac{1}{36\pi} \times 10^{-9} \text{ F/m}\right) \tag{2.6}$$

2.1 Coulomb's Law

The approximate value $1/36\pi \times 10^{-9}$ is convenient to remember, although it is not the exact value. With this approximation, we frequently will be able to obtain a numerical result without having to resort to a calculator in a computation. As we will see later when we discuss electromagnetic waves, this approximation is useful and will yield the well-known numerical value of 3×10^8 m/s for the velocity of light instead of the more accurate value that is slightly less than this.

We now call equation (2.2) *Coulomb's law*. We can still experience the phenomenon of electrification every day when we stroll across a shag rug and receive a shock upon touching someone else or if we comb our hair and later pick up pieces of paper with the comb. But don't attempt to stand outside in a lightning storm! If you did this latter experiment, you might find that your hair would "stand on its end" due to your body conducting charges with the same sign from the ground to the tips of the strands of your hair. Not only would this cause your strands of hair to separate, but it would be an indication that you are about to become a conducting path from the sky above you to the ground beneath your feet. This is a warning that you are in the process of becoming a lightning rod! Since you are not made of metal, this would not be a good thing. At this point you should seek shelter immediately or, if none is available, lie down on the ground in a ditch or depression.

EXAMPLE 2.1

Using Coulomb's law, determine the units of the permittivity of free space.

Answer. The force equals the mass $\times$ acceleration or $M(L/T^2)$. Therefore, we write from Coulomb's law (2.2) that the units of ε_0 can be obtained from

$$\mathbf{F} = \frac{Q_1 Q_2}{4\pi\varepsilon_0 R^2}\mathbf{u_R} \Rightarrow M\frac{L}{T^2} = \frac{Q^2}{L^2}\frac{1}{\varepsilon_0} \Rightarrow \varepsilon_0 = \frac{Q^2 T^2}{ML^3}$$

Remember, 4 and π are just numbers that do not have any units associated with them. The unit vector is also dimensionless.

The vector direction of the force acting on charge 1 due to a charge 2 is directed along the line between the two charges. We indicate this direction with the unit vector $\mathbf{u_R}$. The vector direction of the Coulomb force between charges 1 and 2 is determined by the signs of the two charges. If the two charges have the same sign, either positive or negative, they will repel each other. If the two charges have different signs, they will attract each other. In the experimental system depicted in Figure 2–1, the charge with the same sign that was originally on the rod would subsequently be transferred to both pith balls. Therefore, the two balls will repel each other.

The Coulomb force equation is fundamental in the explanation of electromagnetic fields. It contains the new physical quantity—the charge—that makes electromagnetic theory unique. Charges that are in motion create currents that, in turn, create magnetic fields. The magnetic field will be described in the next chapter. The theory of relativity allows us to derive other laws of electromagnetic theory from the Coulomb force equation. Hence, this one simple equation will bear much fruit in our later discussions. Rather than invoke such an esoteric subject as relativity and considerable mathematical effort, we will examine these topics and follow in the footsteps of the giants who have walked ahead of us and who will guide us through the dark forest of seemingly unrelated experimental observations in order to obtain an explanation.

EXAMPLE 2.2

Find the magnitude of the Coulomb force that exists between an electron and a proton in a hydrogen atom. Compare the Coulomb force and the gravitational force between the two particles. The two particles are separated approximately by 1 Ångström = 1Å = 1 × 10^{-10} m.

Answer. The magnitude of the Coulomb force is computed from (2.2)

$$F_{\text{Coulomb}} = \frac{Q^2}{4\pi\varepsilon_0 R^2} \approx \frac{(1.602 \times 10^{-19})^2}{4\pi\left(\frac{1}{36\pi} \times 10^{-9}\right)(10^{-10})^2} \approx 2.3 \times 10^{-8} \text{ N}$$

The gravitational constant

$$G = 6.67 \times 10^{-11} \text{ N-m}^2/\text{kg}^2$$

$$F_{\text{gravitational}} = G\frac{m_{\text{electron}} M_{\text{proton}}}{R^2}$$

$$= (6.67 \times 10^{-11})\left(\frac{(9.11 \times 10^{-31})(1836 \times 9.11 \times 10^{-31})}{(1 \times 10^{-10})^2}\right)$$

$$= 1.02 \times 10^{-47} \text{ N}$$

2.2 Electric Field

> The ratio of the two forces is
>
> $$\frac{F_{\text{Coulomb}}}{F_{\text{gravitational}}} = 2.27 \times 10^{39}$$
>
> The fact that the Coulomb force is so large compared with the gravitational force helps explain why chemical bonds that hold atoms, molecules, and compounds together can be very strong.

Electric Field

When we step on the scale to determine our weight, we do not carry out a detailed calculation involving the Earth's mass, our mass, and the distance between the center of the Earth and our center of mass. We just assume that there is a gravitational field where we are standing and have the scale calibrated to indicate the multiplication of our mass times the gravitational field. The gravitational field is a vector quantity that is pointing toward the center of the Earth. We also encounter the same phenomenon with electric fields as will be described below.

The electric field **E** caused by a charge Q is a vector quantity that has the definition

$$\boxed{\mathbf{E} \equiv \frac{\mathbf{F}}{q}} \quad (\text{N}/\text{C}) \tag{2.7}$$

where **F** is the Coulomb force between the two charges Q and q. The charge q is a test charge that we use to determine the direction of the field due to the charge Q. By convention, we use a positive unit test charge. We place the test charge wherever we wish to determine the electric field, called the observation point, and observe the direction of the unit vector from Q to the observation point. By placing the test charge in a variety of locations, we can develop a map or plot of the electric field due to the charge Q. The standard symbol for the electric field is **E**. Since we're looking at static electric fields that do not depend upon time, this electric field is frequently called an ***electrostatic field***. The electric field in this region due to the charge Q is written as

$$\mathbf{E} = \frac{Q}{4\pi\varepsilon_0 R^2} \mathbf{u_R} \tag{2.8}$$

We will see later that the units for the electric field are also V/m, which will be used throughout the book.

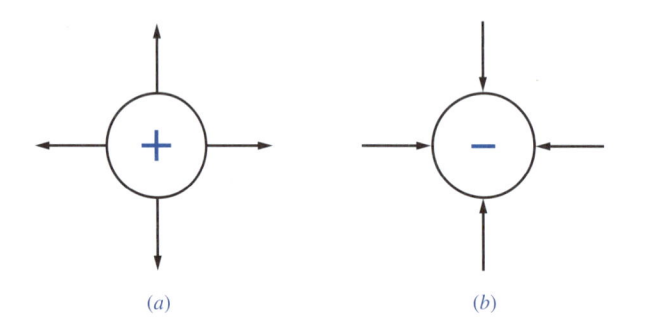

FIGURE 2-2

Electric fields:
(*a*) Emanate from a positive charge.
(*b*) Terminate on a negative charge.

Electric fields from a positive and a negative charge are depicted in Figure 2–2. We note that the direction of the electric field depends on the sign of the charge. Gravitational fields only cause two masses to be attracted to each other. In analogy with the relation between the gravitational field and the gravitational force, we can find the force on a charged particle that is brought into a region containing an electric field **E** by simply multiplying the electric field by the charge q, that is $\mathbf{F} = q\mathbf{E}$. This will be a particularly useful concept when we study the ballistic motion of charged particles in a region containing an electric field, say in a cathode ray tube. Knowing the spatial distribution of the electric field in a particular region will have important practical consequences.

It is worth discussing a conceptual point at this stage. We might be inclined to compare equations (2.2) and (2.8) and suggest that the electric field could be defined in terms of a derivative. We remember the definition of the derivative to be

$$\mathbf{E} \Rightarrow \frac{d\mathbf{F}}{dq} \equiv \lim_{\Delta q \to 0} \frac{\mathbf{F}(q + \Delta q) - \mathbf{F}(q)}{\Delta q} \tag{2.9}$$

where the definition of the derivative is also explicitly stated in (2.9). The operation of performing a differentiation certainly appears to give the correct mathematical result. However, this differentiation will *not* be correct since the smallest charge that has been observed in nature is that of an electron or a single proton whose charge has a magnitude of 1.602×10^{-19} C. The limiting procedure that is required in the definition of the derivative cannot be performed since the charge does not *continuously and smoothly* approach the value of zero. This is because charges have a total charge value that is an integer multiple of the value of the electron or the proton. Quantum electrodynamics has suggested that entities with a charge magnitude that is equal to one third of this value exist, but the mathematical limiting procedure in the derivative still fails. We must use the definition $\mathbf{E} \equiv \mathbf{F}/q$ for the electric field.

EXAMPLE 2.3

Calculate the electric field at a distance of 1 μm (1 μm = 10^{-6} m) from a proton. Calculate the Coulomb force on a second electron at this location.

Answer. From (2.8), we compute the electric field to be

$$\mathbf{E} = \frac{Q}{4\pi\varepsilon_0 R^2}\mathbf{u_R} \approx \frac{1.602 \times 10^{-19}}{4\pi\left(\frac{1}{36\pi} \times 10^{-9}\right)(10^{-6})^2}\mathbf{u_R} = 1440\,\mathbf{u_R}\ \text{V/m}$$

The Coulomb force is

$$\mathbf{F} = -q\mathbf{E} = (-1.602 \times 10^{-19}) \times (1440)\mathbf{u_R} = -2.3 \times 10^{-16}\mathbf{u_R}\ \text{N}$$

Note that we have employed the approximate value for the permittivity of free space.

2.3 Superposition Principles

If we had more than one charge and each charge were at a different location in a vacuum, the total electric field in the space external to the location of these charges would be the vector summation of the electric field originating from each individual charge. The vacuum is a linear medium. In fact, a vacuum has the greatest number of linear properties that can be found in nature. The principles of superposition apply in a vacuum. The only caveat that we will encounter will be that we must be careful to apply vector superposition principles and just not scalar superposition principles. Both the magnitude and the direction of the individual electric fields from each charge must be included in the addition. For N separate charges in the region of interest, this vector summation can be written as

$$\mathbf{E_T} = \mathbf{E}_1 + \mathbf{E}_2 + \mathbf{E}_3 + \cdots = \sum_{n=1}^{N} \mathbf{E}_n \tag{2.10}$$

The electric field created by each individual charge can be obtained from vector addition and this is illustrated in Figure 2–3 for the particular case of two charges, Q_1 and Q_2. Remember that we have to include the correct sign of the charge in the vector addition operation. For the case depicted in Figure 2–3, the total electric field $\mathbf{E_T}$ is calculated from the vector summation of the two individual components. Both charges are assumed to be positive; therefore, the electric fields will be directed away from the charges. The total electric field intensity is given by

$$\mathbf{E_T} = \mathbf{E}_1 + \mathbf{E}_2 = \frac{Q_1}{4\pi\varepsilon_0 R_1^2}\mathbf{u_{R_1}} + \frac{Q_2}{4\pi\varepsilon_0 R_2^2}\mathbf{u_{R_2}} \tag{2.11}$$

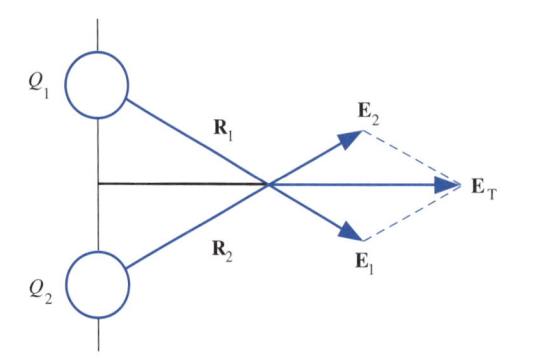

FIGURE 2-3

The total electric field is the vector sum of individual components.

where $\mathbf{u}_{R_j}$ indicates the unit vector associated with each individual charge Q_j to the point where the electric field is to be computed. The distance between the charge and this point is given by R_j. This vector addition is best illustrated with two examples.

EXAMPLE 2.4

Two charges $Q_1 = +4$ C and $Q_2 = -2$ C are located at the points indicated on the graph. The units of the graph are in meters. Find the electric field at the origin (0, 0) of the coordinate system.

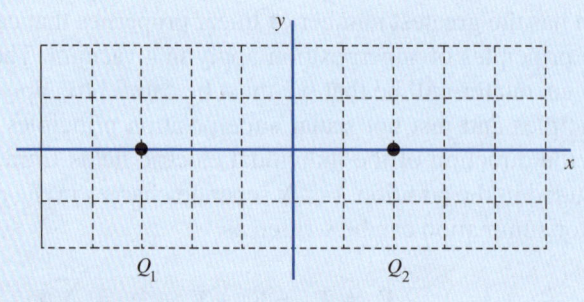

Answer. The electric field **E** is computed from (2.11)

$$\mathbf{E_T} = \mathbf{E}_1 + \mathbf{E}_2 = \frac{Q_1}{4\pi\varepsilon_0 R_1^2}\mathbf{u}_{R_1} + \frac{Q_2}{4\pi\varepsilon_0 R_2^2}\mathbf{u}_{R_2}$$

The vector direction of the electric field is directed from the charge 1 to the charge 2 at the origin. Using the numerical values specified for the charges and the distances as determined from the graph, we write

$$\mathbf{E_T} = \mathbf{E}_1 + \mathbf{E}_2 = \frac{4}{4\pi\varepsilon_0 3^2}\mathbf{u_x} + \frac{-2}{4\pi\varepsilon_0 2^2}(-\mathbf{u_x})$$

$$= \frac{1}{4\pi\varepsilon_0}\left(\frac{4}{9} + \frac{2}{4}\right)\mathbf{u_x} = \frac{1}{4\pi\varepsilon_0}\left(\frac{34}{36}\right)\mathbf{u_x} \text{ V/m}$$

Note that the electric fields from the two charges add up at the origin.

2.3 Superposition Principles

EXAMPLE 2.5

Three charges $Q_1 = +1$ C, $Q_2 = +2$ C, and $Q_3 = +3$ C are placed at the indicated points on the graph. Find the electric field at the point P.

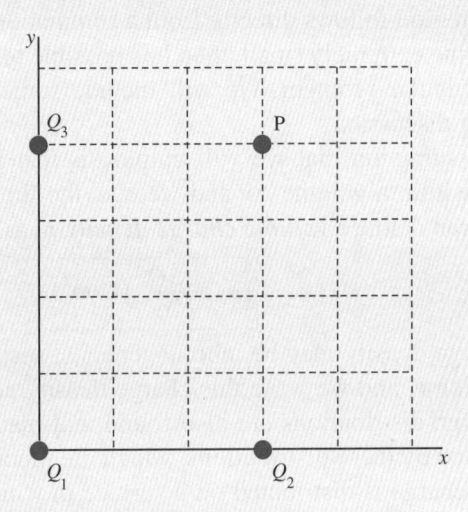

The electric field at the point P is computed from a linear superposition of the individual electric field components due to the individual charges. We write

$$\mathbf{E}(3,4) = \mathbf{E}_1(3,4) + \mathbf{E}_2(3,4) + \mathbf{E}_3(3,4)$$

$$= \frac{Q_1}{4\pi\varepsilon_0 R_1^2}\mathbf{u}_{R_1} + \frac{Q_2}{4\pi\varepsilon_0 R_2^2}\mathbf{u}_{R_2} + \frac{Q_3}{4\pi\varepsilon_0 R_3^2}\mathbf{u}_{R_3}$$

$$= \frac{1}{4\pi\varepsilon_0 5^2}\frac{3\mathbf{u}_x + 4\mathbf{u}_y}{5} + \frac{2}{4\pi\varepsilon_0 4^2}\mathbf{u}_y + \frac{3}{4\pi\varepsilon_0 3^2}\mathbf{u}_x$$

$$= \frac{1}{4\pi\varepsilon_0}\left[\left(\frac{3}{125} + \frac{3}{9}\right)\mathbf{u}_x + \left(\frac{4}{125} + \frac{2}{16}\right)\mathbf{u}_y\right] \text{ V/m}$$

The further evaluation of this electric field is straightforward, and yields the result $\mathbf{E}(3, 4) = (3.212\,\mathbf{u}_x + 1.411\,\mathbf{u}_y) \times 10^9$ V/m.

Up to this point, our discussion of electrostatic fields has assumed that it was possible to calculate the electric field by merely summing the vector contributions from each individual charge. In theory, this is the correct procedure that should always be followed. However, in practice, we would quickly run out of steam in following such a procedure when describing realistic situations where the number of charged particles in a confined volume may be expressed in powers of 10, say, 10^{15} particles. Numerical tools would soon be required to perform this summation.

If we can make certain assumptions concerning the distribution of the charges in a region and realize that an integration of the distributed charges over the region follows directly from a summation if we let a certain parameter become extremely small, then it is possible to obtain analytical solutions for a particular problem. We will include some of these solutions in the following discussion.

The assumption that we will employ is that if a total charge Q is distributed within a volume Δv and we take the limit as this volume $\Delta v \to 0$, then we can define a *volume charge density* ρ_v as depicted in Figure 2–4a.

$$\rho_v = \frac{\Delta Q}{\Delta v} \quad (\text{C/m}^3) \tag{2.12}$$

This charge density may be inhomogeneous, such that it depends on the local position $\mathbf{r}$, and we write this charge density as $\rho_v = \rho_v(\mathbf{r})$. Uniform volume charge distributions are also found and many important problems are represented by these distributions, which are much easier to compute.

If the charge is distributed on a surface in which the area is Δs, and it is independent of the distance normal to the surface, then we can define this as a *surface charge density* ρ_s as depicted in Figure 2–4b.

$$\rho_s = \frac{\Delta Q}{\Delta s} \quad (\text{C/m}^2) \tag{2.13}$$

We take the limit as $\Delta s \to 0$. This surface charge density could depend on its location on the surface $\mathbf{r}$ and we write $\rho_s = \rho_s(\mathbf{r})$. It could also be distributed uniformly on the surface, and the charge distribution would be a constant in this case.

The charge could also be distributed along a line whose length is $\Delta \ell$. The charge would have a uniform distribution in the two transverse coordinates

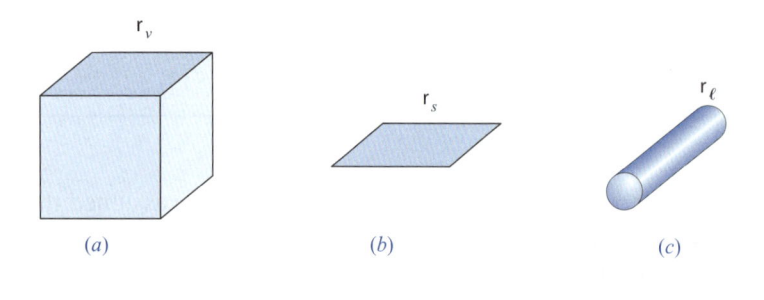

FIGURE 2–4

Distributed charge densities:
(a) The charge is distributed in a volume Δv, creating a volume charge density ρ_v.
(b) The charge is distributed on a surface Δs, creating a surface charge density ρ_s.
(c) The charge is distributed along a line $\Delta \ell$, creating a linear charge density ρ_ℓ.

2.3 Superposition Principles

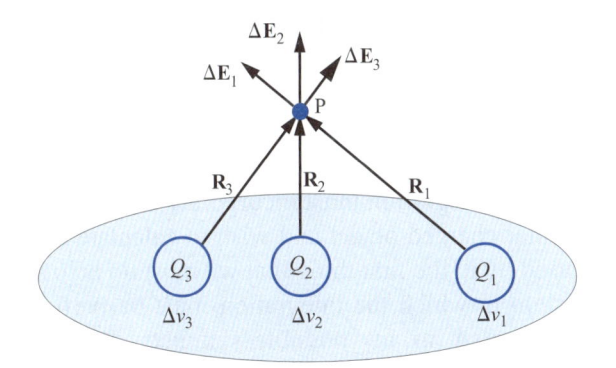

FIGURE 2–5

In order to calculate the electric field at the point P, the differential electric fields $\mathbf{E_j}$ caused by the charges in the differential volumes Δv_j are added together vectorially.

of the line. This would yield a *linear charge density* ρ_ℓ as depicted in Figure 2–4c.

$$\rho_\ell = \frac{\Delta Q}{\Delta \ell} \quad (\text{C/m}) \tag{2.14}$$

We take the limit $\Delta \ell \to 0$. Once again, the charge could be distributed non-uniformly or uniformly along the line.

We will find it advantageous to use all of the three definitions in later derivations. We will later encounter "infinite sheets" or "infinite lines" that have charge densities given by (2.13) and (2.14). This merely implies that an infinite amount of charge is distributed over these infinite surfaces or lines, but the ratios given in these two equations (2.13) and (2.14) are finite.

If we want to calculate the electric field that is created by any of the distributed charge density distributions, we will make use of the principle of superposition that was stated in (2.10) and is shown in Figure 2–5. A quick glimpse at Figure 2–5 should convince us that numerical techniques may have to be employed for most charge distributions in order to calculate the electric field. Fortunately for us, there are a few examples that can be treated analytically, and most of them will appear in this text.

If we let the differential volumes Δv_j become very small and the number of the small volumes become very large, then the summation of the distinct electric fields caused by the discrete charges within these volumes eventually will cause the summation to become an integration that must be performed over the entire volume Δv where the distributed charge density is located. This integration is written as

$$\mathbf{E} = \frac{1}{4\pi\varepsilon_0} \int_{\Delta v} \frac{\rho_v(r')}{R^2} \mathbf{u_R} \, dv' \tag{2.15}$$

Equation (2.15) implies that there exists a differential electric field that is directed radially from each of the differential charges that is enclosed within each of the differential volumes. The total electric field that will emanate from the entire volume Δv is calculated by integrating the charge density over the entire volume. Each of the incremental electric fields will have their individual unit vectors, and the integration must incorporate this fact.

If we are given a particular charged object and wish to calculate analytically the electric field caused by it, the first thing that we must do is to select the proper coordinate system in which the integration must be performed. This choice is usually predicated on any possible symmetry that can be found in the problem. For example, if the charged body were a sphere that was centered on the origin of a coordinate system, we should attempt the solution in spherical coordinates. If the charged body were a long cylindrical rod that was centered at the origin, we should use cylindrical coordinates.

The variables that appear in this integral are defined as follows. The variable R is the distance between the point of observation and the location of a particular charge element $\rho_v(\mathbf{r}')\,dv$ that is within the volume of integration Δv. In Cartesian coordinates, we write

$$R = \sqrt{(x-x')^2 + (y-y')^2 + (z-z')^2} \tag{2.16}$$

where x', y', and z' specify the location of the differential charge element and x, y, and z specify the location where the electric field is to be determined at the point P. The unit vector $\mathbf{u_R}$ is directed from this charge element to the point P. The unit vector $\mathbf{u_R}$ will change as the integration is performed. This will be noted when actually performing the integration. In the generalized equation (2.15), we have to be careful since vectors are present, and we would have to perform the integration separately over the three components in the differential volume dv'.

In order to illustrate the procedure involved in setting up the integral and identifying each term in the integral, we calculate the electric field from a finite amount of charge that is distributed uniformly along a finite line. The linear charge density on this line will be $\rho_\ell(z')$. This linear charge density is depicted in Figure 2–6. From this figure, we find that the unit vector from a differential charge that is localized on a section of the line whose length is dz to the point of observation is given by

$$\mathbf{u_R} = \frac{-z'\mathbf{u}_z + \rho\mathbf{u}_\rho}{\sqrt{(z')^2 + \rho^2}} \tag{2.17}$$

The variation of the unit vector with respect to the variable z' alluded to earlier is clearly displayed in (2.17). In the calculation, we will assume that there is symmetry in that the point of observation is taken to be at the mid-

2.3 Superposition Principles

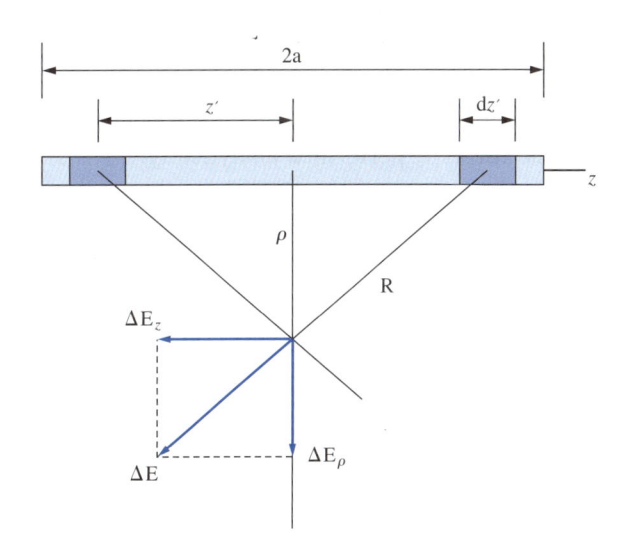

FIGURE 2-6

Calculating the electric field from a uniformly distributed finite line of charge. The radial axis is at the center of the charged line. Because of this symmetry, the tangential components $\Delta \mathbf{E}_z$ of the electric field cancel.

point of the line. Therefore, for every charge segment at a distance $+z'$, there will be an equivalent charge element located at $-z'$. This is an example of symmetry, and it is shown in Figure 2–6. Because of this symmetry, the components of the electric fields oriented in the $\pm z$ directions will cancel ($\Delta E_{z+} = \Delta E_{z-}$). If the line of charge were infinite in length, the center of symmetry could be placed anywhere along the line. The term *polarization* means that the electric field is oriented in a specific direction with respect to some reference direction. For instance, we speak of an electric field being vertically polarized with respect to the ground, when the electric field vector is oriented perpendicular to the ground. In the discussion to come, we will try to orient the fields so that the components of the electric field will cancel when possible. Put another way, we will try to make sure that field components have polarizations that cancel. We'll discuss polarization in more detail in the chapter on wave propagation.

The radial component of the electric field is given in terms of the differential electric field $\mathbf{dE}$ by

$$dE_\rho = dE \cos\theta = dE \frac{\rho}{R} = dE \frac{\rho}{\sqrt{\rho^2 + (z')^2}} \quad (2.18)$$

where the magnitude of differential electric field $\mathbf{dE}$ is calculated from the charge that is contained in the length dz'. This charge is equal to $\rho_\ell dz'$. Therefore, (2.15) becomes

$$dE = \frac{\rho_\ell dz'}{4\pi\varepsilon_0(\rho^2 + (z')^2)} \quad (2.19)$$

The total radial electric field is given by the summation of all of the infinitesimal components dE_ρ since this is a linear medium and superposition applies. This summation becomes an integration of the linear charge density over the length of the line, and it can be performed analytically:

$$E_\rho = \frac{1}{4\pi\varepsilon_0} \int_{-a}^{a} \frac{\rho_\ell \rho}{(\rho^2 + (z')^2)^{3/2}} dz' = \frac{\rho_\ell}{2\pi\varepsilon_0 \rho} \frac{a}{\sqrt{\rho^2 + a^2}} \quad (2.20)$$

This integral can be performed with the substitution of $z' = \rho \tan\theta$ or by using an integral table. As the length of the line is made extremely long ($2a \to \infty$), the radial component of the electric field decreases as the distance from the charge line increases.

$$E_\rho = \frac{\rho_\ell}{2\pi\varepsilon_0 \rho} \quad (2.21)$$

EXAMPLE 2.6

Calculate the electric field from an infinite charged plane. Assume that the plane consists of an infinite number of parallel charged lines as shown in the figure.

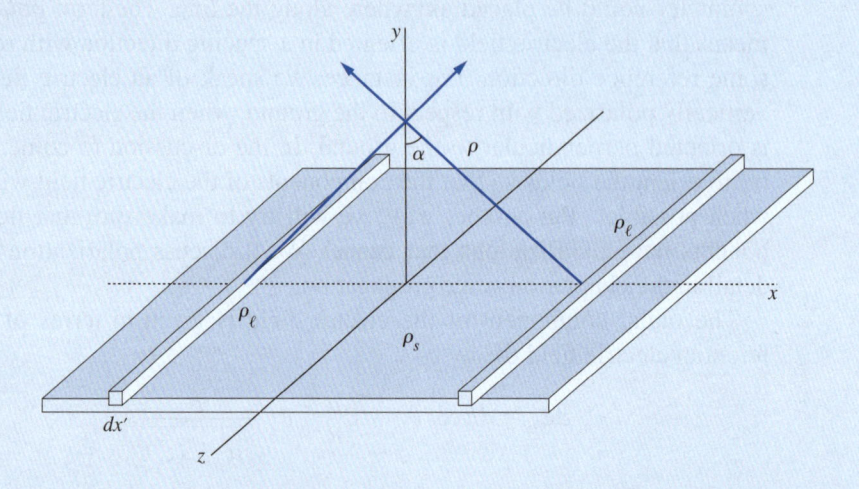

Answer. It is possible to consider the infinite plane as a parallel array of juxtaposed infinite charged lines. Hence, we can use (2.21) as our point of embarkation, where the distance $R = \sqrt{(x')^2 + y^2}$. The linear charge density ρ_ℓ of a particular line whose width is dx' is just equal to $\rho_\ell = \rho_s \, dx'$. Due to symmetry, the components of the electric field that

2.4 Gauss's Law

are tangent to the plane will cancel. Therefore, we only need find the sum of the components of the electric field that are normal to the plane.

$$E_y = \int_{-\infty}^{+\infty} dE \cos \alpha \, dx' = \int_{-\infty}^{+\infty} \frac{\rho_s}{2\pi\varepsilon_0 \sqrt{(x')^2 + y^2}} \frac{y}{\sqrt{(x')^2 + y^2}} dx'$$

$$= \frac{\rho_s}{2\pi\varepsilon_0} \int_{-\infty}^{+\infty} \frac{y}{\sqrt{(x')^2 + y^2}} dx' = \frac{\rho_s}{2\pi\varepsilon_0} \tan^{-1}\left(\frac{x'}{y}\right)\Bigg|_{-\infty}^{+\infty} = \frac{\rho_s}{2\varepsilon_0}$$

We find that the electric field is independent of the distance that it is above the infinite charged sheet. An alternative integration could be performed by assuming that the differential surface areas are concentric circular washers.

Due to the symmetry found in Figure 2–6 and Example 2.6, we have been able to obtain analytical solutions for the electric field from two different charge configurations using the integral given in (2.15). We already have the electric field due to a point charge in (2.8), which varies as $1/R^2$ from the location of the point charge. For the other two charge configurations we have studied, the field varies as $1/R^1$ in the case of the line charge and $1/R^0$, or is constant, in the case of an infinite surface charge. We should expect a different exponent for R since the infinite line charge and the infinite surface charge each contains an increasing order of infinities.

The assumption of symmetry has made these two examples solvable. There are, however, many more examples in which one cannot invoke these arguments of symmetry. The resulting integration may have to be performed numerically and we'll discuss this topic with reference to MATLAB later after we encounter the subject of the electric potential. The electric potential will permit us to neglect any vector notation, and this will simplify our discussion of that topic. In the material that immediately follows, we'll continue to make the symmetry assumption.

2.4 Gauss's Law

It is possible to find the electric field using the laws of vector calculus. We will present this technique by assuming symmetry since the resulting integrals will be easier to perform, but the principles hold even when there is no symmetry. The final result is more general than it appears at first. In order to introduce this procedure, we assume that there is a charge Q that is distributed

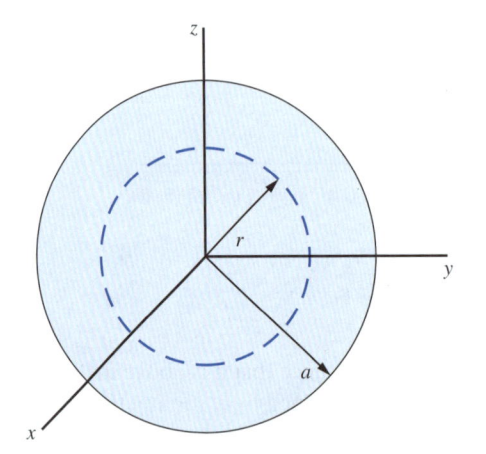

FIGURE 2–7

A charge Q is distributed uniformly within a sphere of radius a.

uniformly within a sphere whose radius is a as shown in Figure 2–7. There will be a uniform volume charge density $\rho_v = Q/(4\pi a^3/3)$ within the sphere.

The radial electric field at the surface of the sphere can be calculated by assuming that the distributed charge, which is distributed uniformly within the sphere, is localized at the center. Therefore, we write from (2.8) that

$$\mathbf{E} = \frac{Q}{4\pi\varepsilon_0 a^2}\mathbf{u_r} \qquad (2.22)$$

The next step is to integrate both sides of (2.22) over the entire spherical surface. This is a closed surface integral that is written as

$$\oint \mathbf{E}\cdot\mathbf{ds} = \oint \frac{Q}{4\pi\varepsilon_0 a^2}\mathbf{u_r}\cdot\mathbf{ds} \qquad (2.23)$$

At the surface of the sphere, the electric field is a constant and it is directed in the radial direction. This is in the same direction as the unit vector associated with the differential surface area $\mathbf{ds}$, which implies that the scalar product of the two unit vectors $\mathbf{u_r}\cdot\mathbf{u_r} = 1$. The closed surface integral yields the spherical surface area $4\pi a^2$. Hence (2.23) becomes

$$\boxed{\oint \mathbf{E}\cdot\mathbf{ds} = \frac{Q_{enc}}{\varepsilon_0}} \qquad (2.24)$$

where we have explicitly understood that $Q = Q_{enc}$ represents the charge that is enclosed within the closed surface. This is **Gauss's law**. It is common to refer to the closed surface as a **Gaussian surface**. In passing from (2.23) to (2.24), we have invoked symmetry arguments by stating that the electric field had a constant value on the spherical surface.

2.4 Gauss's Law

We can rewrite equation (2.24) using the ***divergence theorem***, which we previously encountered in Chapter 1, and express the enclosed charge Q_{enc} in terms of a charge density ρ_v as

$$\oint_{\Delta s} \mathbf{E} \cdot \mathbf{ds} = \int_{\Delta v} \nabla \cdot \mathbf{E} \, dv = \frac{\int_{\Delta v} \rho_v \, dv}{\varepsilon_0} \quad (2.25)$$

In order for the two volume integrals in (2.25) to be equal for any arbitrary volume Δv, the two integrands must be equal. This implies that

$$\boxed{\nabla \cdot \mathbf{E} = \frac{\rho_v}{\varepsilon_0}} \quad (2.26)$$

Equations (2.24) and (2.26) express one of the fundamental postulates of electrostatics. These equations are the integral form and the differential form of Gauss's law respectively. We will make extensive use of both forms.

As written, (2.24) is the integral form of Gauss's law, which allows us to ascertain the electric field in cases where there is significant symmetry inherent in the problem. This will be demonstrated for a charge Q that is distributed uniformly within a spherical volume and for a charge that is distributed uniformly on a surface.

Consider the spherical volume shown in Figure 2–7. A charge Q is distributed uniformly within the spherical volume $\Delta v = 4\pi a^3 / 3$. The volume charge density ρ_v is specified to be

$$\rho_v = \frac{Q}{\Delta v} = \frac{Q}{\left(\frac{4\pi a^3}{3}\right)} \quad (2.27)$$

Using (2.24) and (2.25), the total charge Q_{enc} that is enclosed within the spherical volume is calculated to be

$$Q_{\text{enc}} = \int_{\Delta v} \rho_v \, dv \quad (2.28)$$

This integral can be performed in this case and for several other cases. The charge that is enclosed within the volume is given by Q_{enc}, and this volume could have a radius r that is greater or less than the radius a of the sphere. This allows us to find the radial electric field both outside and within the spherical volume.

Let us first calculate the electric field outside of the spherical volume $r > a$. In this case, a spherical surface will enclose entirely the total charge Q. Therefore $Q_{\text{enc}} = Q$, and we have

$$Q = \int_{\Delta v} \rho_v \, dv \quad (2.29)$$

Since the unit vector associated with the differential surface area is in the radial direction, we will have only a radial component of the electric field. The closed surface integral can also be performed and this leads to

$$\oint_{\Delta s} \mathbf{E} \cdot \mathbf{ds} = 4\pi r^2 E_r \tag{2.30}$$

Therefore, we use Gauss's law in equation (2.24) and equate (2.29) and (2.30) to yield

$$4\pi r^2 E_r = \frac{Q}{\varepsilon_0} \tag{2.31}$$

Hence, the magnitude of the radial electric field E_r in the region $r > a$ is given by

$$E_r = \frac{Q}{4\pi\varepsilon_0 r^2} \tag{2.32}$$

This is the same result that was obtained in (2.8) as we should expect since the charge is enclosed entirely within this larger spherical surface.

Within the sphere, $r < a$ the total charge that is enclosed within this volume is given by

$$\frac{Q_{enc}}{\varepsilon_0} = \frac{\int_{\Delta v} \rho_v dv}{\varepsilon_0} = \frac{1}{\varepsilon_0} \frac{Q}{\left(\frac{4\pi a^3}{3}\right)} \left(\frac{4\pi r^3}{3}\right) = \frac{Q}{\varepsilon_0}\left(\frac{r}{a}\right)^3 \tag{2.33}$$

In (2.33), we have used the definition of the differential volume dv in spherical coordinates. The closed surface integral surrounding this charge is still given by (2.30). Hence the magnitude of the radial electric field within the charged sphere is found from equating these two expressions:

$$4\pi r^2 E_r = \frac{Q}{\varepsilon_0}\left(\frac{r}{a}\right)^3 \tag{2.34}$$

Solving for the radial electric field within the sphere, we compute

$$E_r = \frac{Qr}{4\pi\varepsilon_0 a^3} \tag{2.35}$$

We note that the electric field linearly increases with the radius r. This increase is due to the inclusion of more charge within this expanding spherical surface. A summary of the electric field as a function of radius is shown in Figure 2–8.

2.4 Gauss's Law

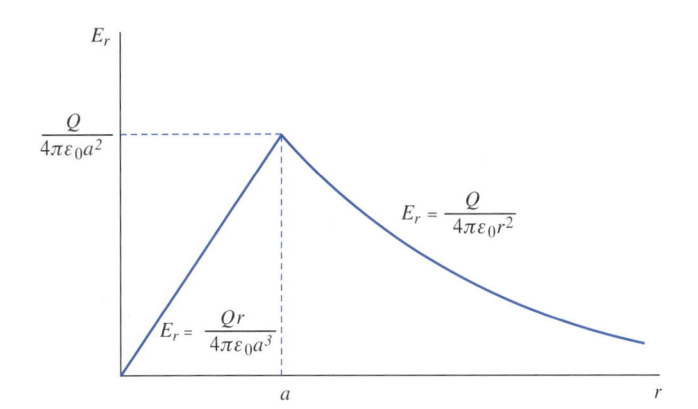

FIGURE 2–8

The variation of the electric field inside and outside of a uniformly charged sphere.

E_r

$\dfrac{Q}{4\pi\varepsilon_0 a^2}$

$E_r = \dfrac{Q}{4\pi\varepsilon_0 r^2}$

$E_r = \dfrac{Qr}{4\pi\varepsilon_0 a^3}$

EXAMPLE 2.7

Assume that charge is uniformly distributed in the axial direction of the concentric hollow cylinders yielding linear charge densities of $+\rho_\ell$ on the inner cylinder and $-\rho_\ell$ on the outer cylinder. There is no angular variation of these distributions. Find the electric field in all regions of space using Gauss's law.

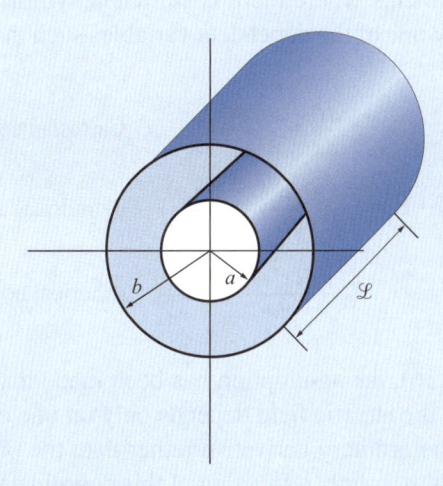

Answer. Because the structure is of infinite length, we can neglect the contributions of the electric field in the axial direction in our calculation. Due to the cylindrical symmetry that is found in this problem, the Gaussian surface will be a cylinder. In the region $\rho < a$, the enclosed charge is equal to zero. Hence the electric field within the inner cylinder is equal to zero. In the region $a < \rho < b$, we can assume that the inner cylinder can be replaced with a distributed line charge of density ρ_ℓ that is localized at the center of the cylinder. The total enclosed charge in a length $\mathscr{L}$ is equal to $+\rho_\ell\mathscr{L}$. Therefore, from Gauss's law in equation (2.24) in cylindrical coordinates, we write

$$\oint \mathbf{E}\cdot\mathbf{ds} = \int_{\phi=0}^{\phi=2\pi}\int_{z=0}^{z=\mathscr{L}} E_\rho dz d\phi = E_\rho(2\pi\rho\mathscr{L}) = \frac{Q_{enc}}{\varepsilon_0} = \frac{+\rho_\ell\mathscr{L}}{\varepsilon_0}$$

The first integral contributes a factor of 2π and the second integral contributes a factor of $\mathscr{L}$. Hence, the radial electric field in the region $a < \rho < b$ is equal to

$$E_\rho = \frac{+\rho_\ell}{2\pi\varepsilon_0\rho}$$

This is the same result that was obtained in (2.21). In the region $\rho > b$, the enclosed charge is equal to $\{+\rho_\ell\mathscr{L} - \rho_\ell\mathscr{L}\} = 0$ in the length $\mathscr{L}$. The electric field external to the outer cylinder will be equal to zero.

In order to further emphasize the physical meaning of Gauss's law, let us introduce a slightly different derivation. Its interpretation will allow us to see clearly the meaning of the term "enclosed charge Q_{enc}." This method will apply for problems where there is sufficient symmetry and there is a dependence on only one of the dependent variables such that (2.26) can be written as

$$\nabla\cdot\mathbf{E} = \begin{cases} \dfrac{dE}{dx} = \dfrac{\rho_v(x)}{\varepsilon_0} & \text{Cartesian coordinates} \\[6pt] \dfrac{1}{\rho}\dfrac{d(\rho E)}{d\rho} = \dfrac{\rho_v(\rho)}{\varepsilon_0} & \text{cylindrical coordinates} \\[6pt] \dfrac{1}{r^2}\dfrac{d(r^2 E)}{dr} = \dfrac{\rho_v(r)}{\varepsilon_0} & \text{spherical coordinates} \end{cases} \quad (2.36)$$

In writing (2.36), the assumption has been made that sufficient symmetry exists such that the electric field depends only on one of the coordinates. Hence, we can use the ordinary derivative rather than the partial derivative. Also, the charge density ρ_v is only a function of this coordinate. For example, the charge could be distributed within a spherical volume as shown in Figure 2–9.

The charge Q_{enc} that is enclosed within the spherical volume depicted in Figure 2–9 is given from (2.33), where the integrations over the transverse coordinates (θ and ϕ) yield a factor of only 4π. We are left with the following integral in the radial variable that has yet to be performed:

$$Q_{enc} = 4\pi\int_0^\rho \rho_v(\tilde{r})\tilde{r}^2 d\tilde{r} \quad (2.37)$$

2.4 Gauss's Law

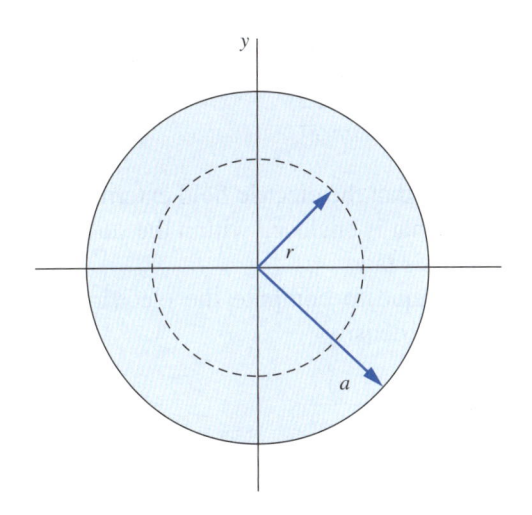

FIGURE 2-9

Charge is distributed within the spherical volume whose radius is a. The charge Q_{enc} refers to the charge that is enclosed within the sphere whose radius is r.

where $\tilde{r}$ is a dummy variable of integration. Using the chain rule for differentiation, we write the left-hand side of (2.36) as

$$\frac{1}{r^2}\frac{d(r^2 E_r)}{dr} = \frac{1}{r^2}\frac{d(r^2 E_r)}{dQ_{enc}}\frac{dQ_{enc}}{dr}$$

$$= \frac{1}{r^2}\frac{d(r^2 E_r)}{dQ_{enc}}[4\pi r^2 \rho_v(r)] \quad (2.38)$$

$$= \frac{d(r^2 E_r)}{dQ_{enc}}[4\pi \rho_v(r)]$$

where we have employed the definition for the differentiation of an integral using Leibnitz's rule.[2] The term $[4\pi r^2 \rho_v(r)]$ arises from the application of this definition. Since the charge density $\rho_v(r)$ is common to both sides of the equation, it cancels and (2.36) becomes

$$4\pi\frac{d(r^2 E)}{dQ_{enc}} = \frac{1}{\varepsilon_0} \quad (2.39)$$

---

[2] The differentiation of an integral is given by the expression that is known as Leibnitz's rule.

$$\frac{d\left(\int_{a(x)}^{b(x)} f(x,y)\,dy\right)}{dx} = \int_{a(x)}^{b(x)} \frac{df(x,y)}{dx}\,dy + f(x,b(x))\frac{db(x)}{dx} - f(x,a(x))\frac{da(x)}{dx}$$

This can be easily integrated to yield

$$E = \frac{Q_{enc}}{4\pi\varepsilon_0 r^2} \qquad (2.40)$$

Equation (2.40) explicitly states that the electric field external to a surface is determined by the charge Q_{enc} that is enclosed within the surface. This is the physical interpretation of Gauss's law.

We can also use this technique to compute the electric field within a sphere that has a uniform charge density

$$\rho_v = \frac{Q}{\left(\dfrac{4\pi a^3}{3}\right)} \qquad (2.41)$$

The charge Q_{enc} that is enclosed within the spherical volume $4\pi r^3/3$ is given by

$$Q_{enc} = \int_{\Delta v} \rho_v dv = \int_0^\pi \sin\theta\, d\theta \int_0^{2\pi} d\phi \int_0^r \frac{Q}{\left(\dfrac{4\pi a^3}{3}\right)} r'^2 dr' = Q\left(\frac{r}{a}\right)^3 \qquad (2.42)$$

Substitute this in (2.40) and we obtain

$$E = \frac{Q_{enc}}{4\pi\varepsilon_0 r^2} = \frac{Q\left(\dfrac{r}{a}\right)^3}{4\pi\varepsilon_0 r^2} = \frac{Qr}{4\pi\varepsilon_0 a^3} \qquad (2.43)$$

The result in (2.43) is the same result that was given in (2.35) but from a slightly different point of view. We should not expect to find and do not find a different result. What we have done is employ a modified form of **Lagrangian mass variables**. This change of variables has been used by our colleagues in fluid mechanics who let the independent variable of space r become the total mass m_{enc} that is enclosed within the volume that is defined by this spatial variable r. Using this technique, they have been able to advance the solutions for fairly difficult problems. We have merely borrowed and used their technique in order to further interpret the meaning of Gauss's law in electrostatics.

In using Gauss's law, we have made extensive use of various symmetry arguments. Because of this, we have been able to reduce the problem such that it depends on only one spatial variable. If the enclosing sphere is sufficiently larger than the container of the charge, then it may be a good approximation to assume that the enclosed charge is localized at a point that is at the center of the sphere in order to obtain an approximate solution for the electric field.

2.5 Potential Energy and Electric Potential

In conclusion, Gauss's law states that there must be a charge that exists within an enclosed surface in order to have an electric field emanate from or terminate within the enclosed surface. Otherwise, the electric field will just pass through this region unaffected.

A charged particle will gain a certain amount of potential energy as the particle is moved in a region against an electric field as shown in Figure 2–10. This is because work has to be done to overcome the force due to the electric field.

The energy ΔW_e in joules (J) that will be gained by the charged particle is calculated from the line integral

$$\Delta W_e = \int_a^b \mathbf{F} \cdot \mathbf{dl} = -Q \int_a^b \mathbf{E} \cdot \mathbf{dl} \ (J) \tag{2.44}$$

Note the appearance of a minus sign in this equation. This indicates that if the charge is positive, work must be done to overcome the electric field. Energy must be conserved in this process. Therefore, the positive charge will gain in energy as calculated from (2.44). The potential energy of the positive charge will be increased. A negatively charged particle will experience a decrease of potential energy if it followed the same path as indicated in Figure 2–10. The fact that the scalar product has been employed in this integral reflects the fact that no work is performed in regions where the force (or electric field) is perpendicular to the direction of the motion. This fact will be important in later work.

It is possible to define the total electrostatic energy stored in a volume using the following ***gedanken experiment***.[3] Let us assume that all charges initially are at $x = -\infty$ and none exist in the laboratory. When we say that all of the charges are at $x = -\infty$, we also imply that each of the charges is infinitely far from its neighbor and there are no Coulomb forces between them that will have to be included in our experiment. Any electric field originating at $x = -\infty$ will have decayed to have a value of zero in the laboratory.

Let us compute the total work required to bring the charges into the shaded region from infinity. This is illustrated in Figure 2–11. No work is required to bring the first charge Q_1 into the shaded region since no force is required to move this charge in our frictionless wagon, hence $W_1 = 0$. However, to bring the second charge Q_2 into the region, we will have to do some

---

[3] The word "gedanken" is German for the word "thought." Hence, we are to perform a thought experiment.

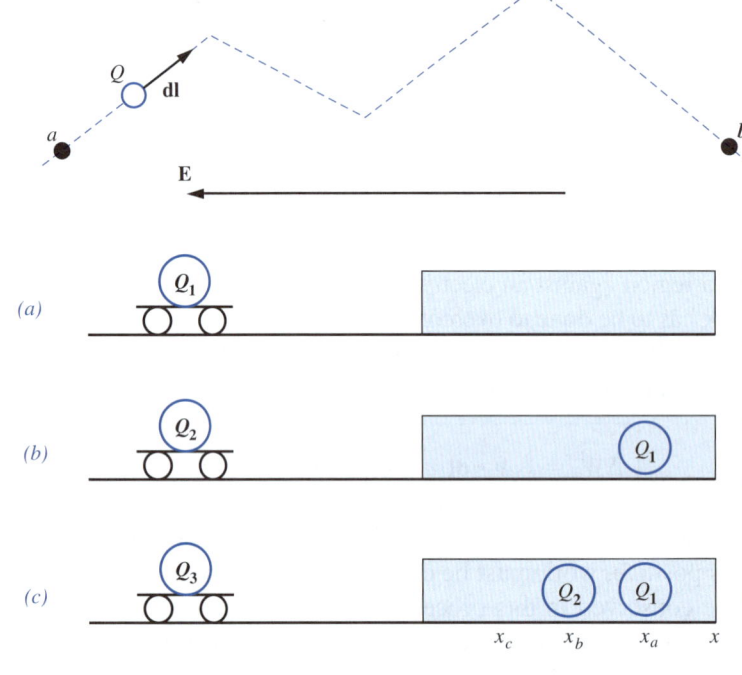

FIGURE 2-10

The transport of a charge Q against an electric field $\mathbf{E}$ from point a to point b causes the particle's potential energy to change.

FIGURE 2-11

Calculation of the work required to bring charges from $x = -\infty$ into the defined space.
(*a*) Moving the first charge Q_1 requires no work.
(*b*) Moving the second charge Q_2 requires work since the first charge Q_1 creates an electric field.
(*c*) Moving the third charge Q_3 requires work since there are two charges already present.

work since we have to overcome the Coulomb force of repulsion caused by the presence of the first charge in the laboratory. Hence, a minus sign will appear in this equation

$$W_2 = -\int_{-\infty}^{x_b} \frac{Q_1 Q_2}{4\pi\varepsilon_0 (x - x_a)^2} dx = \frac{Q_1 Q_2}{4\pi\varepsilon_0 |x_b - x_a|} = Q_2 V_1 \quad (2.45)$$

In (2.45), the energy depends on the magnitude of the distance separating the two charges. We factor the charge Q_2 from the remaining terms. The remaining terms are due to the presence of the first charge Q_1. We will call this collection of terms the potential difference between the two charges that is caused by charge 1 already residing in this region. The potential difference between this point and infinity is sometimes called the ***absolute potential***. The potential at infinity or a ground plane is taken to be equal to zero and it is called the ***ground potential***.

$$V_1 = -\int_{-\infty}^{x_b} \frac{Q_1}{4\pi\varepsilon_0 (x - x_a)^2} dx = \frac{Q_1}{4\pi\varepsilon_0 |x_b - x_a|} \quad (V) \quad (2.46)$$

The units of the voltage are volts (V). We will encounter the term potential again in a few lines where it will be given a physical interpretation. Initially, we will just use this integration as a mathematical entity.

2.5 Potential Energy and Electric Potential

Since the charges that are carried into the room in Figure 2–11 each have a label on them (1 or 2), it behooves us to ask the following question, "Would it have made any difference in calculating the total energy that had been expended if we had brought the charge labeled 2 in before the charge labeled 1?" The answer is no. Equation (2.45) could equivalently be written as $Q_1 V_2$ with no loss of generality. We'll encounter this point in a few minutes when we try to generalize this result.

Passing on to the next iteration of carrying charges into the room as depicted in Figure 2–11, we now bring charge Q_3 into the shaded region. The work that has to be performed, following the same procedure of calculating the work required to bring in Q_2, will be against the electric fields due to charges Q_1 and Q_2 already being in the laboratory.

$$W_3 = -\int_{-\infty}^{x_c} \frac{Q_1 Q_3}{4\pi\varepsilon_0 (x - x_a)^2} dx - \int_{-\infty}^{x_c} \frac{Q_2 Q_3}{4\pi\varepsilon_0 (x - x_b)^2} dx \qquad (2.47)$$

leads to the following expression:

$$W_3 = \frac{Q_1 Q_3}{4\pi\varepsilon_0 |x_c - x_a|} + \frac{Q_2 Q_3}{4\pi\varepsilon_0 |x_c - x_b|} = Q_3 V_1 + Q_3 V_2 \qquad (2.48)$$

The *total* work that has been expended in bringing the three charges into the shaded region is given by

$$W_{\text{total}} = W_1 + W_2 + W_3 = 0 + Q_2 V_{12} + Q_3 (V_{13} + V_{23}) \qquad (2.49)$$

The double subscript notation for the potential V_{ij} allows us to explicitly indicate that the potential due to charge i is to be evaluated at the location where charge j is eventually to be located.

The total energy that has been expended in order to bring the charges into the shaded region has to appear somewhere. None was converted into heat and subsequently lost since the charges were transported in frictionless vehicles. This energy is stored in this region as **electrostatic stored energy**. It can be recovered and used for other purposes at a later time. This energy has the potential to do work.

We could continue this process to include all N charges that were originally at the location $x = -\infty$. The procedure is straightforward and can be generalized to

$$W_{\text{total}} = \frac{1}{2} \sum_{i=1}^{N} \sum_{j=1}^{N(j \neq i)} \frac{Q_i Q_j}{4\pi\varepsilon_0 |x_{ij}|} \qquad (2.50)$$

where $|x_{ij}|$ is the magnitude of the distance between the charge Q_i and the charge Q_j. The factor of one half arises since the terms are counted twice in using the notation of this double summation. For example, a term with $i = 6$ and $j = 8$ will have the same value as the term with $i = 8$ and $j = 6$, and thus this value would be counted twice in the summation. The notation $N(j \neq i)$ indicates that this particular summation excludes the term $j = i$.

Let the potential at the ith charge due to all of the other charges be given by V_i, that is

$$V_i = \sum_{j=1}^{N(j \neq i)} \frac{Q_j}{4\pi\varepsilon_0 |x_{ij}|} \tag{2.51}$$

Hence, the total stored electrostatic energy can be written explicitly as

$$W_e = W_{\text{total}} = \frac{1}{2} \sum_{i=1}^{N} Q_i V_i \tag{2.52}$$

We have used the notation W_e to indicate the total stored electrostatic energy. Later, the symbol W_m will be used to indicate the total stored magnetic energy when we discuss magnetic fields. Although we have obtained an expression for the stored electrostatic energy using a one dimensional model, the result is also valid in three dimensions.

EXAMPLE 2.8

Explicitly demonstrate that the factor of 1/2 must be included in (2.50) for the case of $N = 3$ charges. Recall that there is no energy required to bring in the first charge.

Answer. Explicitly expand (2.50) for the case $N = 3$. We write

$$W_{\text{total}} = \frac{1}{2} \left(\frac{Q_1 Q_2}{4\pi\varepsilon_0 |x_{12}|} + \frac{Q_1 Q_3}{4\pi\varepsilon_0 |x_{13}|} + \frac{Q_2 Q_1}{4\pi\varepsilon_0 |x_{21}|} + \frac{Q_2 Q_3}{4\pi\varepsilon_0 |x_{23}|} \right.$$
$$\left. + \frac{Q_3 Q_1}{4\pi\varepsilon_0 |x_{31}|} + \frac{Q_3 Q_2}{4\pi\varepsilon_0 |x_{32}|} \right)$$

Since the distances satisfy $x_{jk} = x_{kj}$ and the products of the charges satisfy $Q_j Q_k = Q_k Q_j$, this can be written as

$$W_{\text{total}} = \frac{Q_1 Q_2}{4\pi\varepsilon_0 |x_{12}|} + \frac{Q_1 Q_3}{4\pi\varepsilon_0 |x_{13}|} + \frac{Q_2 Q_3}{4\pi\varepsilon_0 |x_{23}|} = Q_2 V_{12} + Q_3 V_{13} + Q_3 V_{23}$$

This is the same value for the total energy that was obtained in (2.49).

If the charge is distributed within a volume, we can further generalize the expression for the energy that is stored within a volume Δv. The charge Q_i in

2.5 Potential Energy and Electric Potential

(2.52) can be replaced with $\rho_v dv$ where ρ_v is the charge distribution and dv is the differential volume and the discrete potential V_i can be replaced with a continuous term V. The summation is replaced with an integration of the volume charge density over the volume dv where the distributed charge is located. Hence, the stored electrostatic energy is given by

$$W_e = \frac{1}{2} \int_{\Delta v} \rho_v V dv \; (\text{J}) \qquad (2.53)$$

where we have again defined the total electrostatic stored energy as W_e. There are alternative methods of writing this expression as will be noted later.

It is now possible to give a physical interpretation to the terms **electric potential** or the more common expression **voltage**. The ratio of the work required to move the charge against the electric field from point a to point b divided by the value of that charge is defined as the **electric potential difference** V_{ab} between the points a and b. This can be written with reference to the absolute potentials at the two points a and b as

$$V_a - V_b = \Delta V_{ab} = \frac{1}{Q}\left\{ \int_{-\infty}^{a} \mathbf{F} \cdot \mathbf{dl} - \int_{-\infty}^{b} \mathbf{F} \cdot \mathbf{dl} \right\}$$

$$= \frac{1}{Q}\left\{ Q\int_{-\infty}^{b} \mathbf{E} \cdot \mathbf{dl} - Q\int_{-\infty}^{a} \mathbf{E} \cdot \mathbf{dl} \right\}$$

or

$$\Delta V_{ab} = \int_a^b \mathbf{E} \cdot \mathbf{dl} \; (\text{V}) \qquad (2.54)$$

The units of the energy that have been expended to effect this action are given in SI units as joules (J). The units of energy when we consider the individual charge is a very large quantity and we frequently measure the energy in terms of the energy gained by an electron passing through a potential difference of 1 volt ≡ 1 joule/1 coulomb. This energy is given in terms of electron volts (eV).[4]

To illustrate this method of calculating the potential difference given in (2.54), let us calculate the work (work = charge × potential difference) required to move a charge q from a radius $r = b$ to a radius $r = a$ as shown in Figure 2–12. A charge Q is located at the center of the inner spherical

---

[4] One eV = 1.6×10^{-19} J.

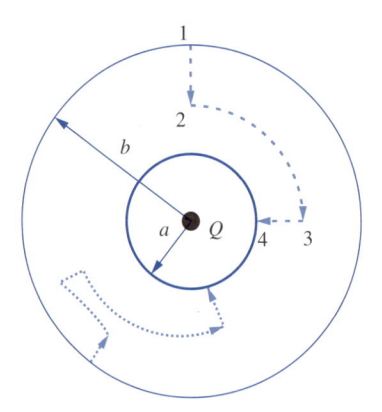

FIGURE 2-12

Two concentric spherical surfaces surrounding a charge Q. The dashed lines indicate a possible path that is to be followed in order to calculate the potential difference between the two spheres. The second set of lines indicates another possible path to evaluate the potential difference between the two spheres. We will find that using either path will yield the same result for the potential difference.

surface. The electric field between the concentric spherical surfaces shown in Figure 2–12 is calculated using (2.8)

$$\mathbf{E} = \frac{Q}{4\pi\varepsilon_0 r^2}\mathbf{u_r} \qquad (2.55)$$

Hence, the potential difference between the two spherical surfaces is computed to be

$$\Delta V_{ab} = \int_a^b \frac{Q}{4\pi\varepsilon_0 r^2}\mathbf{u_r} \cdot dr\mathbf{u_r} = -\frac{Q}{4\pi\varepsilon_0 r}\Big|_a^b = \frac{Q}{4\pi\varepsilon_0}\left(\frac{1}{a} - \frac{1}{b}\right) \qquad (2.56)$$

If the radius b of the outer sphere increased to a value of $b \to \infty$, this would then be the potential difference between $r = +\infty$ and $r = a$. The potential at $r = \infty$ is defined as being equal to zero. It is frequently designated as being the ***ground potential***. In a properly connected three-wire electrical cord, the third wire is connected to this far-off place. In practice this may actually be the copper tubing that brings the water into the room. In many student laboratories, copper tubing is frequently located above the pipe that contains the electrical power for the instruments. This tubing is supposed to be connected to ground for safety reasons and to have a well-defined ground potential in the laboratory. Imagine the red faces of teachers who might connect a light bulb to copper tubing in different parts of the laboratory or the building and have it shine brightly. The potential at the radius $r = a$ may be either positive or negative depending upon the sign of the charge. This electric potential with respect to the ground potential is defined as the ***absolute potential*** at that particular point.

We should note at this point that there is only a potential difference between the two spherical surfaces in Figure 2–12 (path $1 \to 2$ and path $3 \to 4$). If we

2.5 Potential Energy and Electric Potential

move along a circumferential line on one of the surfaces (path 2 → 3, which is at a constant radius), no work would be required since **E** and **dl** are then perpendicular to each other. From (2.54), the work required to effect this move is equal to zero. A surface that has the same potential is called an *equipotential surface*. Note also that moving along the second path in Figure 2–12 yields the same result, since it travels the same distance, when considering those portions of the path that are not along equipotential surfaces. This term will be encountered again in this text and we might expect to see it later in several practical situations. For example, a metal container surrounding an electronic device should be an equipotential surface, and that surface for safety reasons should be at the ground potential. The third wire in the electric cable connects this outer cover to the ground. If we reach for the container and the connection is faulty, a "zap" would quickly convince us that it is an unsafe implement to have in the home. The water pipes and the "third wire conductor" in the home are designed to be good connections to the ground potential.

EXAMPLE 2.9

Calculate the variation of the potential between two infinite concentric cylinders if the potential of the inner cylinder is V_0 and the potential of the outer cylinder is zero. A finite section of this structure is shown below.

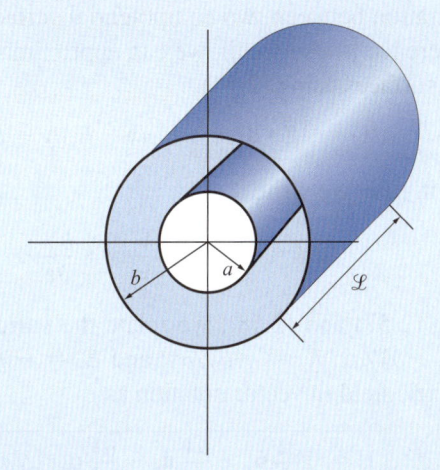

Answer. From Example 2.7, the radial electric field from an infinitely long cylinder that has a uniform linear charge density ρ_ℓ C/m on the external surface of the inner conductor is given by $E_\rho = \rho_\ell / 2\pi\varepsilon_0 \rho$. This charge density, however, is not known in this example, and it must be computed. Only the potential difference between the two cylinders was given. The spatial variation of the potential between the cylinders is computed from the electric field using (2.54) to yield

$$V(\rho) = -\int_b^\rho E_\rho d\tilde{\rho} = -\int_b^\rho \frac{\rho_\ell}{2\pi\varepsilon_0 \tilde{\rho}} d\tilde{\rho} = -\frac{\rho_\ell}{2\pi\varepsilon_0} \ln\left(\frac{\rho}{b}\right)$$

This satisfies the requirement that the potential at $\rho = b$ be equal to zero. In this case, the constant of integration is included in the charge density. In order to compute this constant, we evaluate the potential at $\rho = a$ to be V_0. This yields

$$V_0 = -\frac{\rho_\ell}{2\pi\varepsilon_0} \ln\left(\frac{a}{b}\right)$$

Eliminate the charge density ρ_ℓ between these two expressions to obtain the potential variation between the two cylinders.

$$V(\rho) = V_0 \frac{\ln\left(\frac{\rho}{b}\right)}{\ln\left(\frac{a}{b}\right)}$$

This electric potential satisfies the boundary conditions. Recall that when $\rho = b$, we have $\ln(1) = 0$. When $\rho = a$, the numerator and denominator cancel.

If the separation between two equipotential surfaces is very small and the potential difference is also small, we can approximate the voltage difference between two surfaces using (2.54).

$$dV \approx -\mathbf{E} \cdot \mathbf{dl} = -E_x dx - E_y dy - E_z dz \tag{2.57}$$

From the chain rule, we write

$$dV = \frac{\partial V}{\partial x} dx + \frac{\partial V}{\partial y} dy + \frac{\partial V}{\partial z} dz \tag{2.58}$$

In comparing (2.57) and (2.58), we relate the various terms of the electric field as $E_x = -\partial V/\partial x$; $E_y = -\partial V/\partial y$; and $E_z = -\partial V/\partial z$. Therefore, we can write the electric field in vector notation as

$$\boxed{\mathbf{E} = -\frac{\partial V}{\partial x}\mathbf{u_x} - \frac{\partial V}{\partial y}\mathbf{u_y} - \frac{\partial V}{\partial z}\mathbf{u_z} \quad (\text{V}/\text{m})} \tag{2.59}$$

If we knew the location and values of various equipotential surfaces, say, from a sequence of experimental measurements, then it would be possible to calculate both the magnitude and the vector direction associated with the electric field. In writing (2.59), we have made the statement

2.5 Potential Energy and Electric Potential

that the electric field is defined to be in the direction of the maximum rate of change of the potential. In addition, we see that the electric field can also have the units of

$$\text{volts / meter} = \text{V / m}$$

since the spatial derivative operation will introduce the unit of 1/(length). These units are probably the most commonly used in practice.

We identify (2.59) as being the three components of the *gradient operation* of the scalar electric potential in Cartesian coordinates. Hence the electric field can be found analytically by taking the negative gradient of the electrostatic potential. This is a written as

$$\boxed{\mathbf{E} = -\nabla V \ (\text{V / m})} \qquad (2.60)$$

This equation has important ramifications since it is usually easy to measure the electric potential at various points in the space. From these measurements, it is possible to connect all of the points in space that have the same potential. The resulting surfaces or lines are equipotential surfaces or lines. The application of (2.60) will then produce the magnitude and the direction of the resulting electric field in this space.

EXAMPLE 2.10

The potential is measured at several locations in space. Connecting the points that have the same value of the electric potential with a line produces equipotential contours that can be drawn on a graph. Find the electric field at the point P. The graph is 5 m × 5 m.

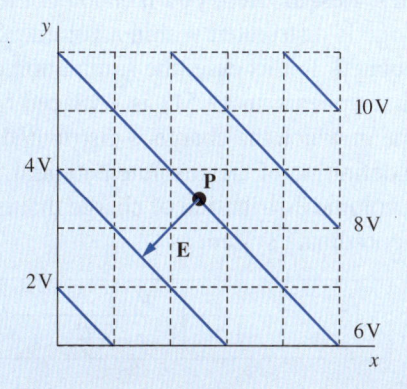

Answer. The electric field is computed from $\mathbf{E} = -\nabla V$ in two dimensions. The measured equipotential contours are indicated by the solid lines and are separated by a distance of

$\sqrt{1^2 + 1^2} = \sqrt{2}$ m. The electric field at the point P is

$$\mathbf{E} = -\nabla V = -\frac{\partial V}{\partial x}\mathbf{u_x} - \frac{\partial V}{\partial y}\mathbf{u_y} \approx -(\mathbf{u_x} + \mathbf{u_y}) \text{ (V/m)}$$

The electric field is a vector that is pointing from the higher potential to the lower potential. There are several cases in practice where this "graphical" procedure can be performed using MATLAB.

We can substitute the electric field that is given in (2.60) into (2.26) in order to obtain the dependence of the electrostatic potential upon the charge density. We write

$$\nabla \bullet (-\nabla V) = \frac{\rho_v}{\varepsilon_0}$$

or

$$\nabla^2 V = -\frac{\rho_v}{\varepsilon_0} \text{ (V/m}^2\text{)} \qquad (2.61)$$

Equation (2.61) is called **Poisson's equation**. If the charge density ρ_v is equal to zero, this is called **Laplace's equation**. These two equations are extremely important in obtaining solutions for the electric potential and the electric field in terms of the charge density. They will be studied in further detail since the methods of solution will involve either analytical or numerical techniques. Rest assured, you'll encounter this equation again.

If the charge is distributed within a volume Δv, the absolute potential can also be calculated. In this case, the summation of the individual charge contributions that appears in (2.51) is replaced with an integration over the entire volume in which the charge is distributed.

From the definition of the absolute potential, we would write the potential caused by a volume distribution of charge that is not centered at the origin of a Cartesian coordinate system as

$$V(x, y, z) = \frac{1}{4\pi\varepsilon_0}\int_{\Delta v}\frac{\rho_v(x', y', z')}{R}dx'\,dy'\,dz' \text{ (V)} \qquad (2.62)$$

where the distance R is given by $R = \sqrt{(x - x')^2 + (y - y')^2 + (z - z')^2}$. The expressions for the distance in other coordinate systems are given in Appendix A. This distance is equal to the magnitude of the difference

2.5 Potential Energy and Electric Potential 95

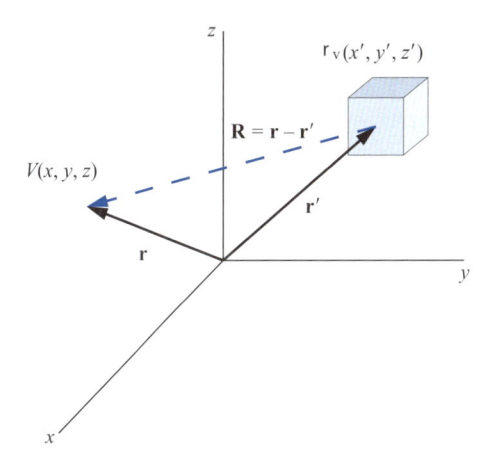

FIGURE 2–13

The voltage at a location in space (x, y, z) that is caused by a volume charge distribution that is located at a different point (x', y', z') in space. The distance between the volume charge distribution and the point where the voltage is to be determined is given by the magnitude of the difference of the two vectors **r** and **r'**.

between the two vectors **r'** and **r** shown in Figure 2–13. In MATLAB notation, we define the distances using the command *norm*.

The gradient operation is carried forth at the observer's position, and we can assume that the two vectors **r** and **r'** are independent from each other. Hence, the gradient operation ∇ that is in the unprimed system can be brought freely inside the integral since the integration is performed in the primed system. It is left as a problem to verify that

$$\nabla\left(\frac{1}{R}\right) = -\frac{\mathbf{u_R}}{R^2} \quad (2.63)$$

where $\mathbf{u_R} = (\mathbf{r} - \mathbf{r'})/|\mathbf{r} - \mathbf{r'}|$ is a unit vector. Therefore, we can compute the electric field once the potential field is known from (2.60). There may be certain advantages in finding the electrostatic potential first using (2.62), which contains the integration operation. The reason is that there is only one integration that is involved in (2.62) and it is a scalar integral. The direct calculation of the electric field from a distributed volume charge distribution given in (2.15) was a vector integral. This implies that the integration must be performed over the three coordinates in order to obtain the three components of the electric field.

In (2.53), we calculated the electrostatic energy stored in the volume Δv after bringing in additional charges. We are now prepared to obtain other expressions for this energy by replacing the volume charge density ρ_v using (2.26).

$$W_e = \frac{1}{2}\int_{\Delta v} \rho_v V \, dv = \frac{1}{2}\int_{\Delta v} (\varepsilon_0 \nabla \cdot \mathbf{E}) V \, dv \quad (2.64)$$

This equation contains the product of the divergence of the electric field and the scalar electric potential. Using the vector identity from Appendix A,

$$\nabla \cdot (a\mathbf{B}) = \mathbf{B} \cdot \nabla a + a\nabla \cdot \mathbf{B}$$

we rewrite (2.64).

$$W_e = \frac{\varepsilon_0}{2} \int_{\Delta v} [\nabla \cdot (V\mathbf{E}) - \mathbf{E} \cdot \nabla V] \, dv \qquad (2.65)$$

The first term on the right-hand side of (2.65) can be converted to a closed surface integral using the divergence theorem:

$$\int_{\Delta v} \nabla \cdot (V\mathbf{E}) \, dv = \oint_{\Delta s} V\mathbf{E} \cdot d\mathbf{s} \qquad (2.66)$$

In this case, the surface Δs encloses the volume Δv. Let us assume that this volume has a spherical shape and the volume charge density is localized near the center of the sphere. The electric field and the electric potential on the spherical surface will depend upon the radius of the sphere as $1/R^2$ and $1/R$ as given in (2.55) and (2.47) respectively. The surface area of the spherical surface will increase with increasing radius as R^2. Therefore, (2.66) will approach zero as the radius $R \to \infty$. The conclusion is that (2.66) is equal to zero for an infinite volume.

The total electrostatic energy can be written as

$$W_e = \frac{\varepsilon_0}{2} \int_{\Delta v} (-\mathbf{E} \cdot \nabla V) \, dv = \frac{\varepsilon_0}{2} \int_{\Delta v} (\mathbf{E} \cdot \mathbf{E}) \, dv$$

or

$$\boxed{W_e = \frac{\varepsilon_0}{2} \int_{\Delta v} E^2 \, dv \ (\text{J})} \qquad (2.67)$$

where we have incorporated the relation that the electric field can be derived from the gradient of the scalar electric potential stated in (2.60). Note that the electrostatic energy depends upon the scalar quantity of the magnitude of the square of the electric field. We will encounter (2.67) later.

The following example, which makes use of the superposition principles that were described above, will introduce the reader to the subject of dielectric materials. A material consists of a very large number of atoms. The Bohr model of a hydrogen atom assumes that there is a positive charge at the center of the atom and there is a negative charge that is located at a distance of approximately 10^{-10} meters from the center. This is a very small separation and our colleagues in physics have defined a new unit called the Ångström, named after the Swedish scientist, where $1\text{Å} = 10^{-10}$ meters.

2.5 Potential Energy and Electric Potential

EXAMPLE 2.11

Find the potential V due to two equal charges that have the opposite sign and are separated by a distance d in a vacuum. If the point where the voltage is to be determined is much greater than the separation distance d, this configuration is known as an ***electric dipole***. Using MATLAB, sketch the electric potential distribution surrounding the two charges.

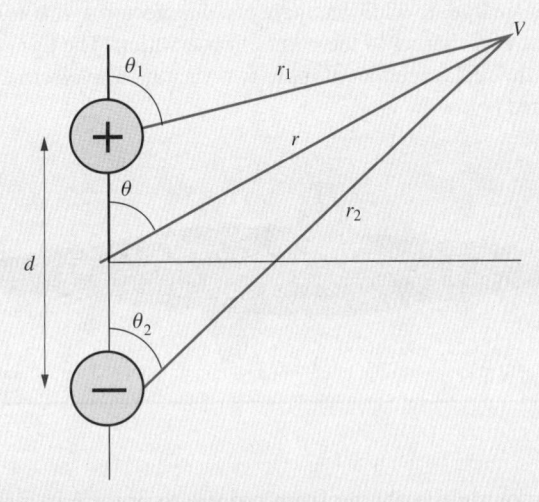

Answer. Superposition will apply in that the total electric potential is computed by adding the individual contributions together. We find

$$V = \frac{Q}{4\pi\varepsilon_0 r_1} - \frac{Q}{4\pi\varepsilon_0 r_2}$$

We have chosen that $r \gg d$. Therefore, we can assume that the three lines r, r_1, and r_2 are almost parallel and the three angles θ, θ_1, and θ_2 are approximately equal. With these assumptions, we write that

$$r_1 \approx r - \frac{d}{2}\cos\theta \text{ and } r_2 \approx r + \frac{d}{2}\cos\theta$$

Using these approximations, the voltage is computed to be

$$V \approx \frac{Q}{4\pi\varepsilon_0\left(r - \frac{d}{2}\cos\theta\right)} - \frac{Q}{4\pi\varepsilon_0\left(r + \frac{d}{2}\cos\theta\right)}$$

$$= \frac{Q}{4\pi\varepsilon_0 r\left(1 - \frac{d}{2r}\cos\theta\right)} - \frac{Q}{4\pi\varepsilon_0 r\left(1 + \frac{d}{2r}\cos\theta\right)}$$

The ratio of the distances $d/2r$ is a small quantity, and we can use it as a small parameter to expand the terms in the denominator and obtain

$$V \approx \frac{Q}{4\pi\varepsilon_0 r}\left(1 + \frac{d}{2r}\cos\theta - 1 + \frac{d}{2r}\cos\theta\right) = \frac{Qd}{4\pi\varepsilon_0 r^2}\cos\theta$$

Let us define an *electric dipole moment* vector $\mathbf{p} = Q\mathbf{d}$ that is directed from the negative charge to the positive charge. In addition, there is a unit vector $\mathbf{u_r}$ that is directed from the midpoint between the two charges to the point of observation. The term $Qd\cos\theta$ can be interpreted as being the scalar product of these two vectors. The electric potential distribution is depicted below.

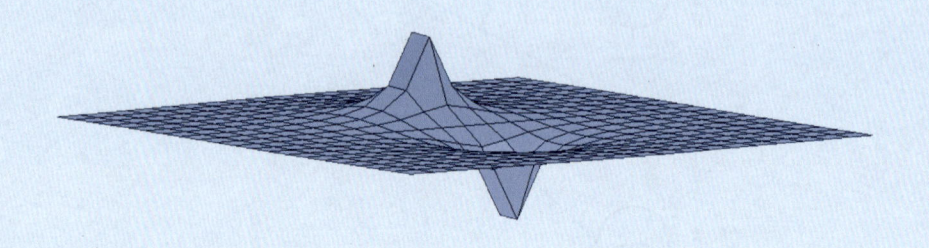

In moving the charge from point a to point b in a region that contained an electric field, we found from (2.45) that work was required. If we move it back to point a along a slightly different path, as shown in Figure 2–14 in an electrostatic field, we will find that the expended energy is returned to us. In this case, we write (2.45) as

$$\frac{W_e}{Q} = \oint \mathbf{E} \cdot \mathbf{dl} = 0 \qquad (2.68)$$

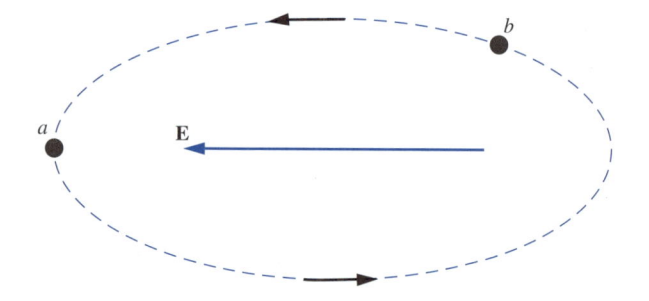

FIGURE 2-14

The closed path of integration from a to b and then back to a. There is an electric field in the region surrounding the two points a and b.

2.5 Potential Energy and Electric Potential

where an integral over a closed contour is indicated. Equation (2.68) states that no energy is either expended or acquired in this process. In this case, the electrostatic field belongs to a class of fields that are called *conservative fields*. You may recognize that (2.68) is almost identical to the Kirchhoff's voltage law that states "The sum of the voltage drops around a closed loop is equal to zero."

Equation (2.68) is the second postulate of electrostatics in integral form. This equation can be converted into a surface integral via Stokes's theorem. We write

$$0 = \oint \mathbf{E} \cdot \mathbf{dl} = \int_{\Delta s} \nabla \times \mathbf{E} \cdot \mathbf{ds} \tag{2.69}$$

In order for this integral to be zero for any arbitrary surface, the integrand must be equal to zero. This allows us to obtain the second postulate of electrostatic fields in differential form.

$$\boxed{\nabla \times \mathbf{E} = 0} \tag{2.70}$$

Equation (2.70) states that an electrostatic field is *irrotational*. We will encounter these postulates of electrostatics later when time-varying fields are described.

EXAMPLE 2.12

Calculate the work required to move a charge $Q = 2C$ around the closed path if there is an electric field $\mathbf{E} = 3\mathbf{u_x}$ in the region.

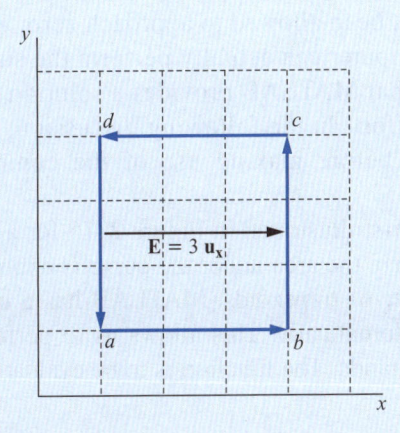

The total work is computed by evaluating the closed line integral

$$W_e = Q \oint \mathbf{E} \cdot \mathbf{dl}$$

$$= Q\left(\int_a^b \mathbf{E} \cdot \mathbf{dl} + \int_b^c \mathbf{E} \cdot \mathbf{dl} + \int_c^d \mathbf{E} \cdot \mathbf{dl} + \int_d^a \mathbf{E} \cdot \mathbf{dl} \right)$$

$$= 2\left(\int_{(1,1)}^{(4,1)} 3\mathbf{u_x} \cdot dx\mathbf{u_x} + \int_{(4,1)}^{(4,4)} 3\mathbf{u_x} \cdot dy\mathbf{u_y} \right.$$

$$\left. + \int_{(4,4)}^{(1,4)} 3\mathbf{u_x} \cdot dx\mathbf{u_x} + \int_{(1,4)}^{(1,1)} 3\mathbf{u_x} \cdot dy\mathbf{u_y} \right)$$

$$= 2(9 + 0 - 9 + 0) = 0$$

The limits that are included in each integral will determine the sign of a particular integration in this closed loop. This electric field is a conservative field.

2.6 Numerical Integration

In the previous section, we were able to calculate the electric potential that results from a distributed charge density. The examples that were presented all require that there is sufficient symmetry in order to find the electric potential. However, in practice, we usually encounter situations that do not have the required symmetry, and we are forced to embark on a slightly different path. One of these paths requires the numerical solution of an electrostatics problem. Remembering that the integration is just a summation in which the number of distinct differential volumes, differential areas, or differential lengths has been allowed to approach zero, we now investigate whether a digital computer can actually perform the summation operation for us. It turns out that MATLAB provides a solution to this problem that requires minimal effort. In the following discussion, we will initially develop the procedure before making use of the commands that are available in MATLAB.

This is first illustrated in Figure 2–15 for a simple one-dimensional integration where the area under the curve is approximated with the summation of a number of trapezoids. MATLAB has a command that creates the trapezoids automatically. This allows us to perform this calculation with just three commands. The function is to be evaluated at every point separated by Dx.

$$x\ min:\ Dx = b - a:\ x\ max;\ y(x);\ z = trapz\ (x,\ y) \qquad (2.71)$$

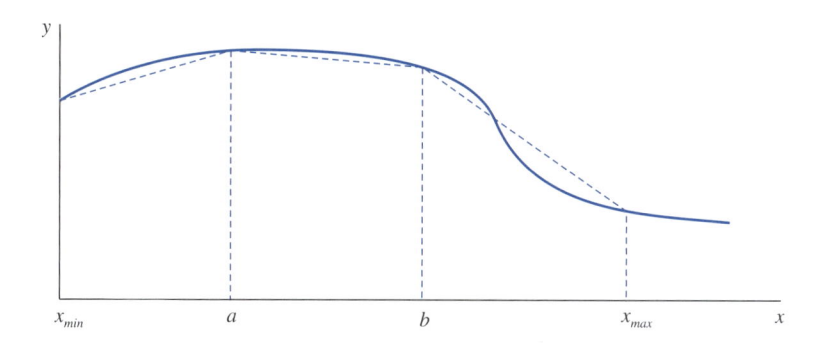

2.6 Numerical Integration

FIGURE 2-15
The area under the curve $y = y(x)$ is obtained numerically by subdividing the area into small trapezoidal subareas and adding the areas of the individual trapezoids.

The choice of the value of the increment Dx is somewhat arbitrary and it depends upon the conflict between the desired accuracy and computational time. There are additional numerical integration programs that can be employed and the evaluation of the truncation errors has received mathematical attention that is beyond the scope of this book.

In addition, this integration can be performed in MATLAB using Simpson's rule with the commands

$$quad(func, x\ min, x\ max)$$

$$dblquad(func, x\ min, x\ max, y\ min, y\ max) \qquad (2.72)$$

$$triplequad(func, x\ min, x\ max, y\ min, y\ max, z\ min, z\ max)$$

where the function *func* is defined by the user. The numerical values for the end points of the integration are also stated in these commands. The default tolerance for the integration is 10^{-6}. For example, this function could be the product of the three variables xyz where we would write

$$func = inline('x.*y.*z') \qquad (2.73)$$

In the definition for the function, we must remember to include the "." after each of the first two variables.

EXAMPLE 2.13

Compare the analytical and the numerical evaluation of the area under the curve defined by the function $y = x^2$ in the interval $0 \leq x \leq 4$. Use both the *trapz* command and the *quad* command.

Answer. The solution that is obtained from an analytical integration is

$$\text{area} = \int_0^4 x^2 dx = \left.\frac{x^3}{3}\right|_0^4 = \frac{64}{3} = 21.3333$$

Using the *trapz* command, we write

$$x = 0: 0.001: 4; y = x.\wedge 2;$$
$$\text{ztrapezoidal} = \textit{trapz(x, y)};$$

Typing ztrapezoidal yields the numerical value of 21.3333.
The *quad* command requires the definition of the function

$$\text{func} = \textit{inline('x.} \wedge \textit{2')};$$
$$\text{zquadrature} = \textit{quad(func, 0, 4)};$$

Typing zquadrature yields the numerical value of 21.3333.

In this example, it was possible to obtain an analytical solution for this problem. The reader can explore the convergence of a numerical calculation for different values of the increment Δx using the *trapz* command.

We find the electric potential due to an object that has a finite size as shown in Figure 2–16. The potential from an arbitrary body of charge was obtained in (2.62) and, for convenience, we rewrite it below.

$$V(x, y, z) = \frac{1}{4\pi\varepsilon_0} \int_{\Delta v} \frac{\rho_v(x', y', z')}{R} dx' dy' dz' \text{ (V)} \qquad (2.74)$$

In our later development of a MATLAB program, we'll replace this distance with the command *norm(r - r')*. The primed variables refer to the location of the charge and the unprimed variables refer to the location at which the potential is to be computed, often called the observation or field point. This permits us to assume that neither the charged object nor the calculated voltage have to be at the origin of the coordinate system.

The procedure to perform the integration numerically will be developed in full detail. After this development, we will use one of the commands that is presented in (2.72). We assume that a finite charge Q is uniformly distributed on a thin finite sheet that is located at $z' = 0$. This results in a uniform surface charge distribution ρ_s at every point on the sheet. Let us also assume that the sheet has a rectangular shape that is centered on a Cartesian coordinate system as shown in Figure 2–17. The procedure that we will follow is to subdivide this large sheet into a number of small subareas and assume that the charge in each small subarea is localized at its individual center. Therefore, we have reduced the uniformly distributed charge to a large number of discrete individual charges. This reduction allows us to convert the integration (2.74) into a summation.

2.6 Numerical Integration

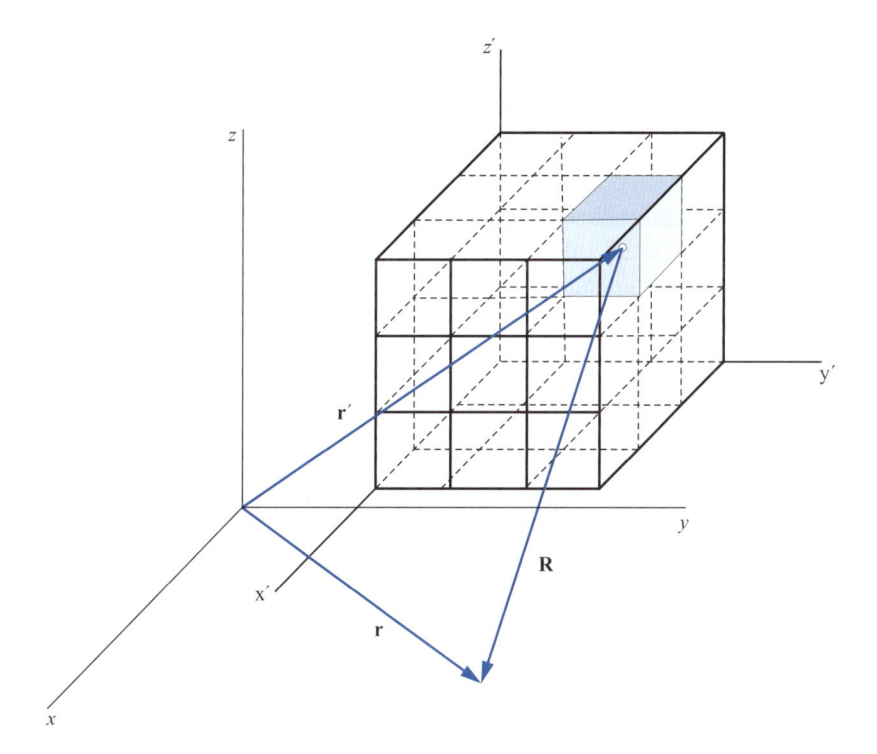

FIGURE 2–16

Electric charge is distributed within a volume identified with the prime. The voltage is to be determined in the un-primed location using the vector notation.

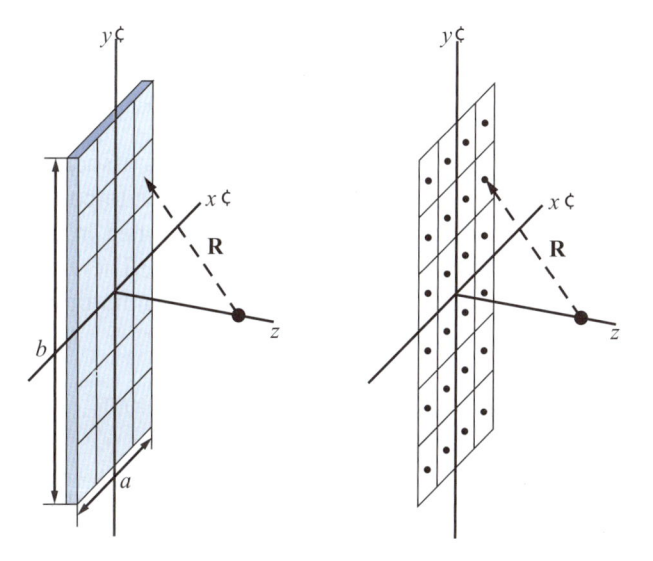

FIGURE 2–17

The area containing the charge is subdivided into a number of small subareas. Each subarea is replaced with an individual charge whose value is equal to the charge contained in the individual subarea.

Therefore, (2.74) can be written as

$$V(x, y, z) = \sum_{j=1}^{N-1} \sum_{k=1}^{M-1} \frac{\rho_s(j, k)\Delta x' \Delta y'}{4\pi\varepsilon_0 R} \qquad (2.75)$$

where the rectangular sheet has been subdivided into $(N - 1) \times (M - 1)$ subareas. The area of each subarea is given by $\Delta x' \Delta y'$. The charge that is at the center of an individual subarea is $\rho_s(j, k) \Delta x' \Delta y'$. The charge density is not required to be uniform on the entire sheet. However, it is assumed to be uniform within each subarea.

It is convenient to assume that the sheet is centered upon a Cartesian coordinate system since later we will be able to invoke certain symmetry arguments in order to simplify the calculation. In addition, we assume that it is located at $z' = 0$. With these assumptions, (2.74) becomes the following double integral:

$$V(x, y, z) = \int_{-b/2}^{b/2} \int_{-a/2}^{a/2} \frac{\rho_s}{4\pi\varepsilon_0 R} dx' dy' \qquad (2.76)$$

where the uniformly distributed charge density is ρ_s. Using the definition for the distance R, we write.

$$V(x, y, z) = \int_{-b/2}^{b/2} \int_{-a/2}^{a/2} \frac{\rho_s}{4\pi\varepsilon_0 \sqrt{(x - x')^2 + (y - y')^2 + (z)^2}} dx' dy' \qquad (2.77)$$

Since the potential is to be determined along the z axis, this simplifies to

$$V(0, 0, z) = \int_{-b/2}^{b/2} \int_{-a/2}^{a/2} \frac{\rho_s}{4\pi\varepsilon_0 \sqrt{(x')^2 + (y')^2 + (z)^2}} dx' dy' \qquad (2.78)$$

There are several cases that have a certain degree of symmetry associated with them. Symmetry may reduce the computational time required to perform the resulting calculation. The integration will have to be performed in only one quadrant of the surface, say $0 \le x' \le a/2, 0 \le y' \le b/2$. The computed value resulting from the integration will then just have to be multiplied by a factor of four. Equation (2.78) becomes

$$V(0, 0, z) = 4 \int_0^{b/2} \int_0^{a/2} \frac{\rho_s}{4\pi\varepsilon_0 \sqrt{(x')^2 + (y')^2 + (z)^2}} dx' dy' \qquad (2.79)$$

In order to evaluate the integral numerically (2.79), we subdivide the entire plane in the quadrant into small rectangles. We will be required to identify the edges of each of the subareas in a methodical manner. One such

2.6 Numerical Integration

procedure assumes that the point at the bottom-left corner is identified as being ($j = 1$, $k = 1$) and the point at the upper-right corner is identified as being ($j = N$, $k = M$). Therefore, there will be $(N - 1) \times (M - 1)$ small subareas in the subdivision process of the large area. The area of each individual subarea is equal to

$$\Delta A = \frac{a}{(N-1)} \frac{b}{(M-1)} = h_x h_y \qquad (2.80)$$

The total charge ΔQ within each subarea is

$$\Delta Q = \rho_s \Delta A = \rho_s h_x h_y \qquad (2.81)$$

and we assume that this charge is localized at the center of the subarea. If the charge has a nonuniform distribution, then the charge distribution ρ_s will have different values at each individual subarea. The incremental potential $V_{j,k}$ due to the localized charge that is identified with the label "j,k" is given by

$$\Delta V_{j,k} = \frac{1}{4\pi\varepsilon_0} \left(\frac{\Delta Q}{\sqrt{\left(jh_x - \frac{h_x}{2}\right)^2 + \left(kh_y - \frac{h_y}{2}\right)^2 + z^2}} \right) \qquad (2.82)$$

The center of a particular subarea is identified as

$$\left[\left(jh_x - \frac{h_x}{2}\right), \left(kh_y - \frac{h_y}{2}\right)\right] \qquad (2.83)$$

All that we need do now is use superposition and sum the incremental potentials due to each incremental charge.

$$V = 4 \sum_{j=1}^{N-1} \sum_{k=1}^{M-1} \Delta V_{j,k} \qquad (2.84)$$

or

$$V = 4 \frac{1}{4\pi\varepsilon_0} \sum_{j=1}^{N-1} \sum_{k=1}^{M-1} \left(\frac{\Delta Q}{\sqrt{\left(jh_x - \frac{h_x}{2}\right)^2 + \left(kh_y - \frac{h_y}{2}\right)^2 + z^2}} \right) \qquad (2.85)$$

Hence the double integral in (2.79) has been converted into a double summation (2.85). The number of small subareas is determined by the compromise that must be made between accuracy and computational time that is required to effect this calculation.

EXAMPLE 2.14

Evaluate the potential at the point $z = a$ due to a charge Q being distributed uniformly upon a square surface whose area is equal to a^2 if the number of subareas is equal to one and four. The center of the square is the z axis, which creates significant symmetry in the problem.

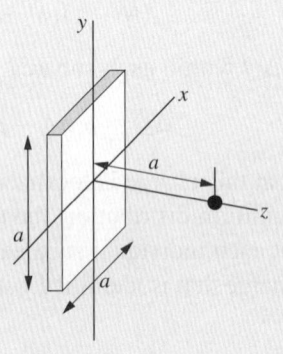

Answer. The first iteration assumes that the charge is localized at the center of the square. Therefore, we compute from (2.46) that

$$V = \frac{Q}{4\pi\varepsilon_0 a}$$

The second iteration is evaluated using (2.85) since there are now four subareas. We calculate the potential to be

$$V = 4 \frac{1}{4\pi\varepsilon_0} \left(\frac{\left(\frac{Q}{a^2}\right)\left(\frac{a^2}{4}\right)}{\sqrt{\left(\frac{a}{2} - \frac{a}{2}\right)^2 + \left(\frac{a}{2} - \frac{a}{2}\right)^2 + a^2}} \right)$$

$$= \frac{Q}{4\pi\varepsilon_0 a} \left(\frac{1}{\sqrt{\frac{1}{16} + \frac{1}{16} + 1}} \right) = (0.9428) \frac{Q}{4\pi\varepsilon_0 a}$$

We could continue with this analytical procedure. However, we find that this is better left for the computer. Using MATLAB, we obtain the following numerical coefficients for the voltage versus the number of subareas n. The results are presented using the above procedure along with the *dblquad* command that is included in MATLAB.

2.6 Numerical Integration

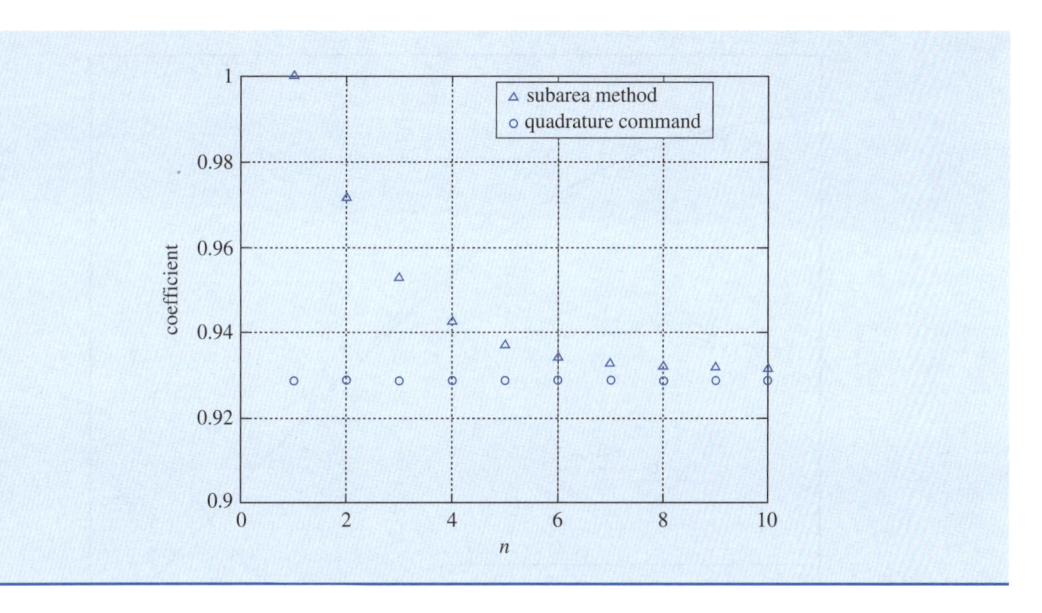

The procedure of subdividing an area into a large number of subareas or a volume into a large number of subvolumes with the incremental charges located at the center of the subarea or the subvolume could be continued. As noted in Example 2.13, there are significant inaccuracies in the resulting numerical computation results. Therefore, we will exploit the MATLAB commands given in (2.72) in the following computation.

In particular, we focus on the following question: "If one moves far away from the object, should it not appear that the charges are localized at a point?" Remember, the sun is bigger than the earth although it appears to be a small ball in the sky. We will answer this question by examining the dependence of the voltage as we move far away from the sheet of uniformly distributed charge. This is demonstrated with an example.

EXAMPLE 2.15

Plot the spatial variation that will be called the coefficient for the electric field as a function of distance z from a square that contains a uniform charge distribution of ρ_s. The z axis is at the center of the square.

Answer. The electric field is computed with the quadrature command that is given in (2.72). In this case, there is an additional variable z.

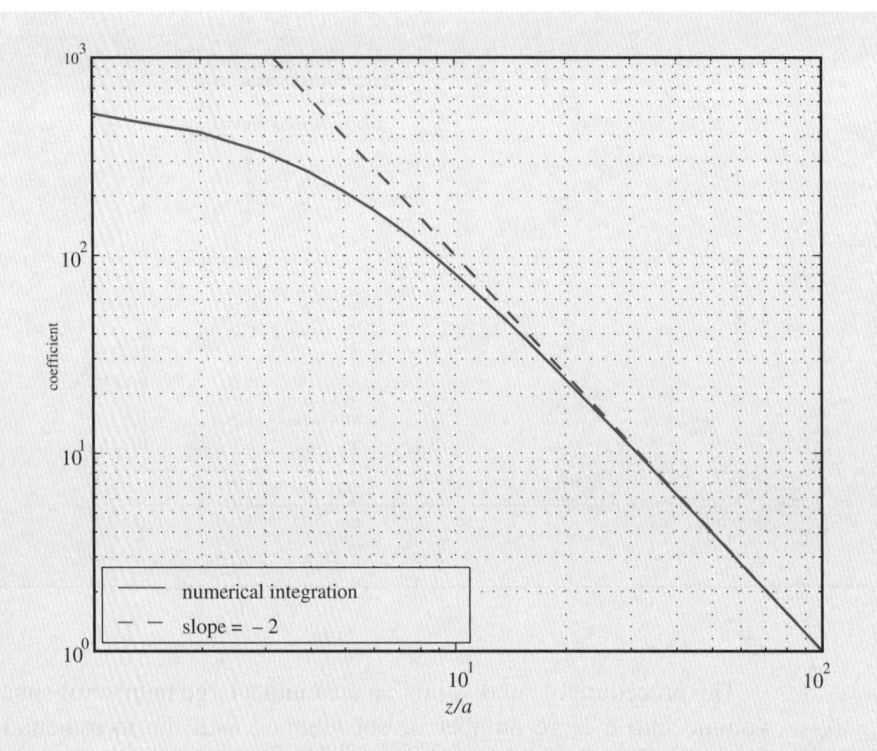

It is convenient to understand the asymptotic values on the spatial dependence of the electric field coefficient by plotting the calculated result using a log–log graph. The slope of the dashed line is equal to -2. This is the same dependence that was obtained for the electric field from a point charge in (2.8).

This example clearly illustrates that the potential will decrease at large distances and the unit square of uniform charge density will appear almost as a point charge.

EXAMPLE 2.16

Plot the potential in the x–y plane in the region $x > 0$ due to a uniformly charged line that is 10 units long that is located at $y = 0$. Perform the integration with the *quad* function.

Answer. In MATLAB, a function is defined using the command *inline*. Since the calculation is to be performed as part of a '*for x = 1: 20*' loop, the numerical value of x must be converted to a string variable that can be incorporated into the *inline* command. This is accomplished with a command *num2str*. The result of the calculation is shown below.

2.7 Dielectric Materials

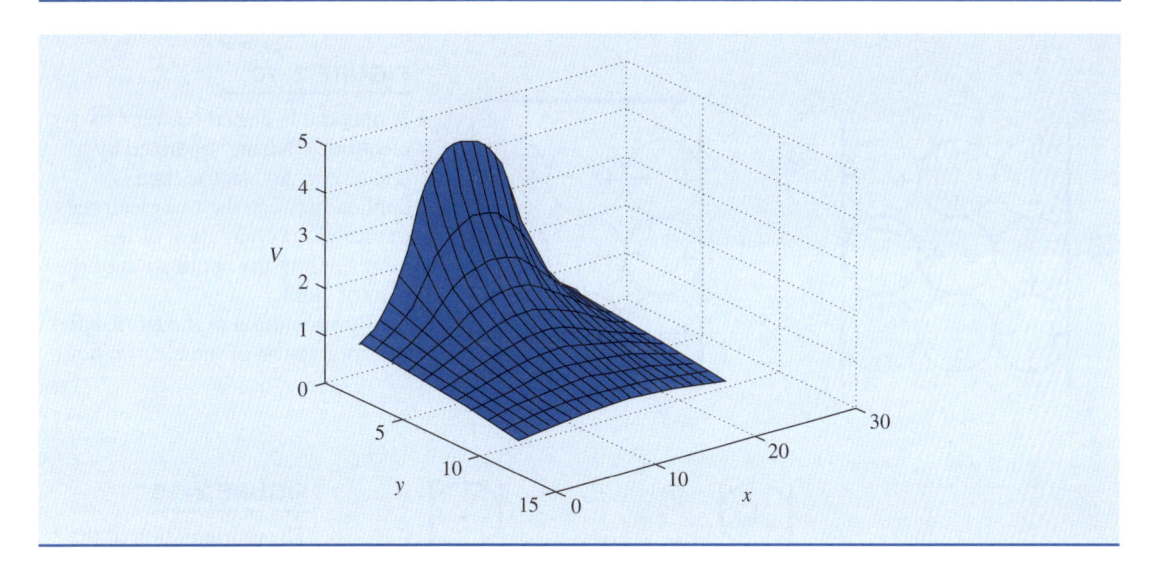

There are other important problems in electromagnetics that will require numerical techniques that are far beyond the brief introduction that is presented here.

2.7 Dielectric Materials

Up to this point, we have examined the behavior of electric fields in a vacuum. The results were correct but we may now be wondering what would be the effects of applying the electric field in a material. The wearing of rubber gloves seems to have some desirable protective features when one is close to touching a high voltage line. Manufacturers of capacitors or integrated circuits usually insert an oxide layer between the two metal surfaces in order to keep the top conductor from falling down and touching the bottom conductor. How do these materials affect the electric field? Some answers will be provided here.

As noted earlier, materials consist of atoms and, in a simple model, these atoms can be considered to be a large collection of randomly oriented small electric dipoles as shown in Figure 2–18. Certain molecules, called *polar molecules*, normally have a permanent displacement between the positively charged nucleus at the center of the atom and the negatively charged electron at the edge. This distance is of the order of 10^{-10} meters. Each pair of charges acts as an electric dipole. If an electric field is applied externally to this material, then the dipoles may reorient themselves. If the field is strong enough, there will actually be an additional displacement of the positive and

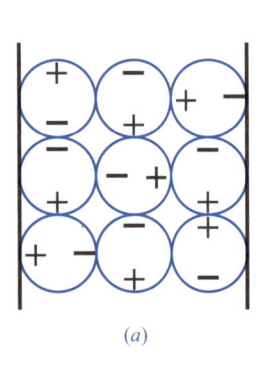

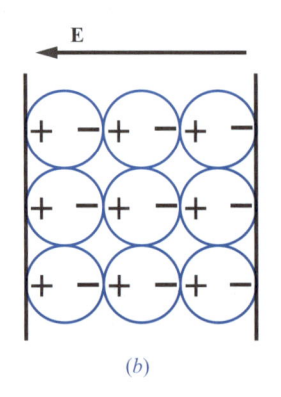

FIGURE 2-18

A material is placed between two electrodes that are separated by a distance d. An electric field is applied between the two electrodes. (*a*) Random orientation of the atoms before the application of the electric field. (*b*) Reorientation of the atoms after the application of the electric field.

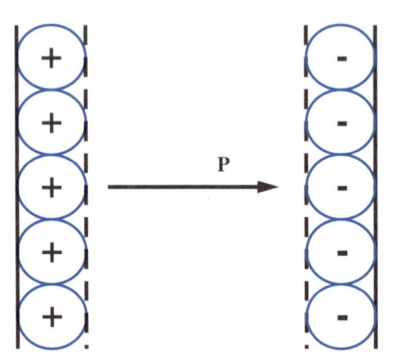

FIGURE 2-19

The reorientation of the atoms in a material due to the application of an electric field creates a polarization charge at the two edges whose density is ρ_P. This polarization charge creates a polarization field **P**.

negative charges. A ***nonpolar molecule*** does not have this dipole arrangement of charges unless an external electric field is applied. The positive and negative charges separate by a certain distance after the application of the electric field.

In some materials, the dipoles may reorient themselves such that a large number or even all of the atoms will realign themselves causing the electric field created by the dipoles to add to the applied electric field. In other materials, the reorientation may cause the dipole electric field to subtract from the applied field. This dipole field created by the atoms will be examined here.

After the application of the electric field between the two electrodes in Figure 2–18*b*, the atoms are reoriented. Since the distances depicted in this figure are of the atomic scale, it is possible to regroup the electric dipoles and suggest that the positive charge of one atom could unite with the negative charge of the adjacent atom in order to form a new distribution of electric dipoles as depicted in Figure 2–19. This regrouping of the electric dipoles will leave a thin layer of charge of the opposite sign at either edge of the material. This charge, which is due to the application of the electric field, is

2.7 Dielectric Materials

called the *polarization charge*. The polarization charge cannot be found in a vacuum. It is only due to the fact that the atoms had been reoriented due to the application of the electric field. We will define a polarization charge density using the symbol ρ_P as being the polarization charge per unit volume.

In the region between the two dashed lines, a positive nucleus of one atom "pairs" with an electron of the adjacent atom. The positive and negative charge centers overlap. However, in the region between the left electrode and the dashed line, there are more positively charged particles. In the region between the second dashed line and the right electrode, there are more negatively charged particles. This effectively states that there is a very narrow region of charge of one sign that has migrated to that edge of the dielectric while there is a narrow region of charge of the opposite sign that has migrated to the other edge of the dielectric. Between these two edges, a charge–neutral region exists. This displaced charge cannot be removed from the material; it is **bound** to the material. It is given the name of a *polarization charge*. Herein, we will just describe the polarization charge at the surfaces that is called the surface polarization charge. The density of this polarization charge has the symbol ρ_P and it is shown in Figure 2–19. This bound charge will set up a field that is called the *polarization field* **P**, which is defined as the dipole moment per unit volume. The polarization field is written via the relation

$$\mathbf{P} = \lim_{\Delta v \to 0} \left\{ \frac{1}{\Delta v} \sum_{j=1}^{N} \mathbf{p}_j \right\} \qquad (2.86)$$

where $\mathbf{p}_j = Q d \mathbf{u_d}$ is the dipole moment of an individual dipole. The units are C/m^2. Within the volume Δv, there are N atoms. With the notation given in (2.86), we see that the polarization field depends on position since we have let the differential volume Δv shrink to zero. In Figure 2–19, this would imply that the distance separating the two thin layers of polarization charge shrinks to zero. In analogy with Gauss's law, we can relate the polarization charge ρ_p to an electric field. This field is called the polarization **P** and we write

$$\rho_p = -\nabla \cdot \mathbf{P} \qquad (2.87)$$

Let us add the polarization charge density ρ_p to the real charge density ρ_v. The real charge density could come from a battery or from the ground. This will dramatically influence the resulting electric field that we calculated from (2.26).

$$\nabla \cdot \mathbf{E} = \frac{\rho_v + \rho_p}{\varepsilon_0} \qquad (2.88)$$

Replacing the polarization charge density in (2.88) with (2.87), we finally obtain

$$\nabla \cdot \mathbf{D} = \rho_v \qquad (2.89)$$

where

$$\mathbf{D} = \varepsilon_0 \mathbf{E} + \mathbf{P} \ (\text{C}/\text{m}^2) \qquad (2.90)$$

is called the *electric flux density* or the *displacement flux density*. The total *electric flux* ψ_e that passes through a surface equals the surface integral of the electric flux density integrated over the surface Δs.

$$\psi_e = \int_{\Delta s} \mathbf{D} \cdot \mathbf{ds} \ (\text{C}) \qquad (2.91)$$

Note that the displacement flux density has a significant meaning only when materials that can be polarized are discussed. In a vacuum, it is just equal to a constant ε_0 times the electric field.

Gauss's law, which was used to compute the electric field in a vacuum, can be employed to calculate the displacement flux density with the same restrictive limitations of symmetry requirements that were encountered previously. The procedure to develop this equation follows directly from an integration of (2.89) over the same volume. The volume integration of the divergence of the displacement flux density can be converted to a closed surface integral using the divergence theorem. The result of this is

$$\oint_{\Delta s} \mathbf{D} \cdot \mathbf{ds} = Q_{\text{enc}} \qquad (2.92)$$

Therefore, the total dielectric flux emanating from or terminating on a closed surface Δs is equal to the total charge that is enclosed within this surface.

A dielectric material is susceptible to being polarized. In many materials, this polarization is linearly proportional to the applied electric field if the electric field remains small. In these cases, we can write that $\mathbf{P} = \varepsilon_0 \chi_e \mathbf{E}$ where χ_e is the *electric susceptibility*. Finally, we obtain

$$\mathbf{D} = \varepsilon_0(1 + \chi_e)\mathbf{E} = \varepsilon_0 \varepsilon_r \mathbf{E} = \varepsilon \mathbf{E} \qquad (2.93)$$

The term ε_r is the *relative dielectric constant* for a material. Tabulated values of ε_r for various materials are given in Appendix B. In a vacuum, $\chi_e = 0$ and $\varepsilon_r = 1$ by definition.

2.7 Dielectric Materials

The expression (2.93) applies only for linear and isotropic materials. It is not difficult to create a material that does not satisfy this criterion. For example, the application of an external magnetic field to an ionized gas will make it anisotropic. Large amplitude signals that are applied to a material may cause the material to have a nonlinear response. This case could occur if the relative dielectric constant changed, say, due to the dielectric being modified where the modification was proportional to the square of the magnitude of the applied electric field $|E|^2$. Such nonlinear materials do exist and are currently under active investigation in the scientific and engineering communities. Dielectric materials can **breakdown**, or lose their insulating properties, if the electric field becomes too strong. In what follows, we will restrict our discussion to linear materials.

EXAMPLE 2.17

A dielectric slab is placed between two parallel plates. A battery is connected to one plate and the other plate is grounded. The area of each plate is equal to A and the charge on each plate is $\pm Q$. The separation of the plates is d. Sketch the following quantities between the plates:

(a) surface charge density ρ_s
(b) displacement flux density **D**
(c) electric field **E**
(d) polarization **P**
(e) the bound surface polarization charge density ρ_{ps}

Answer.

(a) The free charge Q can come from the battery or from the ground. It will be distributed on the surface of the metal plates creating a surface charge density $\rho_s = Q/A$.
(b) The displacement flux density D will be determined by the real charge from the battery or from the ground. It will not depend on whether a dielectric or a vacuum exists between the plates. It follows from Gauss's law that $D = \rho_s$.
(c) The electric field is $E = D/\varepsilon_0 \varepsilon_r$. Hence the electric field will be decreased within the dielectric below its value in the vacuum since $\varepsilon_r > 1$.
(d) The polarization field **P** will exist in the dielectric. Its value will be determined from (2.90).
(e) The bound surface polarization charge density ρ_{ps} can be evaluated from

$$\rho_{ps} = \frac{Q_p}{A} = -\frac{\mathbf{P} \cdot \mathbf{A}}{A} = \frac{\left(\frac{\varepsilon_r - 1}{\varepsilon_r}\right)Q}{A}$$

surface charge density ρ_s (a)

displacement flux density **D**   (b)

electric field intensity **E**   (c)

polarization **P**   (d)

surface polarization charge density ρ_{ps} (e)

2.8 Capacitance

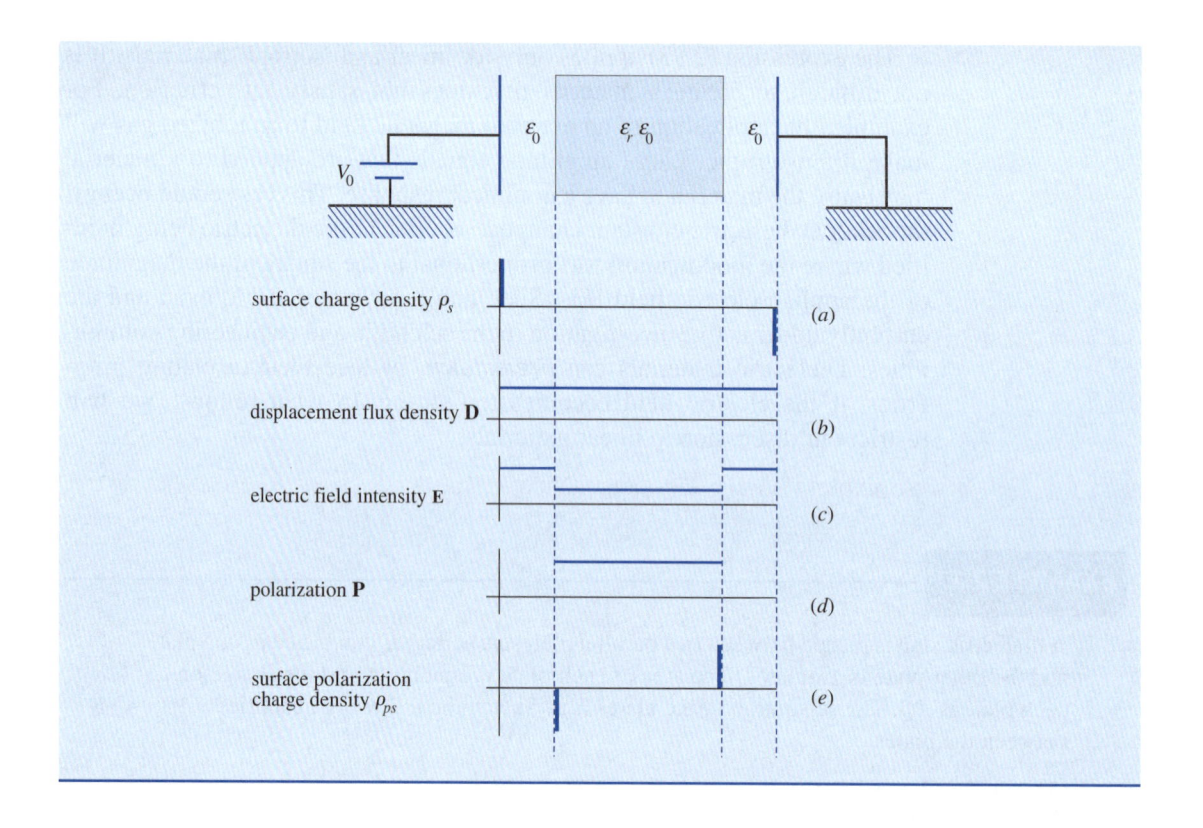

The electrical *capacitance* between two objects in space is defined as the ratio of the charge on one of the objects divided by the potential difference ΔV between the two objects. This is expressed as

$$C \equiv \frac{Q}{\Delta V} \text{ (F)} \qquad (2.94)$$

and it is measured in farads (F), where the unit F = C/V. This is the formal definition for the capacitance and, as we will see, it is possible to numerically calculate its value for objects that may have a very complicated shape. The term *self capacitance* implies that there is only one object and the term *mutual capacitance* is used to describe the capacitance between two separate objects. The procedure will be useful in practical situations such as finding the capacitance of various portions of an integrated circuit or of objects that have an odd shape. This will also be useful in developing various models for a transmission line that are described in Chapter 7. In this section, we'll just examine some very simple objects.

2.8 Capacitance

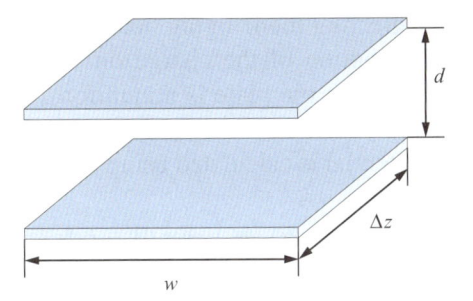

FIGURE 2–20

A parallel plate capacitor is depicted to the left. The plates, whose area is $A = w \times \Delta z$, are separated by a distance d. The region between the plates is a vacuum.

Your first encounter with a capacitor was probably in an introductory course dealing with electrical circuits in which you encountered a very simple expression for the capacitance of the parallel plate capacitor. In this book, we will obtain this expression using the terminology that we have already discussed. The area of each plate is equal to $A = w \times \Delta z$ and the two plates are separated by a distance d as shown in Figure 2–20. The choice for using these symbols for the dimensions is predicated on our future applications in this book. In addition, let us assume that a charge $+Q$ is uniformly distributed on the top plate and a charge $-Q$ is uniformly distributed on the bottom plate. This will result in a uniform charge density of $\rho_s = \pm Q / A$ being distributed on the two plates.

We assume that the transverse dimensions are much greater than the distance between the two plates. The electric field surrounding an infinite charged plate was obtained in Example 2.6, or we could obtain it using Gauss's law. We evaluate Gauss's law in equation (2.24) in order to obtain the electric field from one of the plates, say, the top plate that has a positive charge density

$$\oint_{\Delta s} \mathbf{E} \cdot \mathbf{ds} = \frac{Q_{\text{enc}}}{\varepsilon_0} \Rightarrow E(2A) = \frac{\rho_s A}{\varepsilon_0} \Rightarrow E = \frac{\rho_s}{2\varepsilon_0} \tag{2.95}$$

The electric field between the two plates and in the regions above and below the two plates is evaluated using the principle of superposition. The result is that the electric field is equal to zero in the regions above and below the two plates since the contributions from the two plates have the same magnitude but are in the opposite directions and therefore will cancel. The field in the external region but between the two plates is called a ***fringing field*** and it will be neglected since it is small in comparison with the field directly between the two plates. Using superposition, the electric field between the two plates becomes

$$E = \frac{\rho_s}{\varepsilon_0} \tag{2.96}$$

In addition to neglecting the fringing field, we are also assuming that there is a uniform distribution of the charge on the surface. The actual distribution of the charge on the plate is nonuniform and it will be numerically calculated later.

The electric potential is calculated using (2.54)

$$V_{ab} = \int_a^b \mathbf{E} \cdot \mathbf{dl} = Ed \tag{2.97}$$

Therefore, the capacitance of the parallel plate capacitor is calculated, using the definition (2.94), to be

$$C = \frac{Q}{V} = \frac{\rho_s A}{\left(\frac{\rho_s}{\varepsilon_0} d\right)} = \frac{\varepsilon_0 A}{d} \tag{2.98}$$

Equation (2.98) is the formula that you have encountered previously, but it has now been derived in terms of electromagnetic arguments. Most capacitors will have a dielectric placed between the two conducting plates. In these cases, you should replace ε_0 with ε in order to reflect this situation. A motivation for the insertion of a dielectric between the two plates is to ensure the separation distance remains the same even with the inclusion of the ever present gravitational force. One must be careful that the electric field between the two plates is always beneath the breakdown value of the dielectric material.

EXAMPLE 2.18

Calculate the self capacitance of a hollow metallic sphere whose radius is a.

Answer. Assume that there is a charge Q at the center of the sphere. The absolute potential at a radius a is found from (2.56) to be

$$V = \frac{Q}{4\pi\varepsilon_0 a}$$

The self capacitance is computed from (2.94) to be

$$C = \frac{Q}{V} = \frac{Q}{\left(\frac{Q}{4\pi\varepsilon_0 a}\right)} = 4\pi\varepsilon_0 a$$

It is interesting to calculate the self capacitance of the earth by assuming that it is a hollow sphere. Substituting the value of the radius of the Earth into this result, we compute

$$C = 4\pi\varepsilon_0 a = 4\pi\left(\frac{1}{36\pi} \times 10^{-9}\right)(6.37 \times 10^6) \approx 7 \times 10^{-4} \text{ F}$$

2.8 Capacitance

A unit of one farad is a very big number!

Hollow spheres are used as models to describe dust particles that can be found in integrated circuit manufacturing where they have a very deleterious effect on the final product. Upwards of 10,000 mobile electrons can attach themselves to these dust particles. In addition, charged dust particles are found in some of the rings that surround certain of the planets in our solar system such as Saturn. These negatively charged dust particles may have a mass that is greater than the surrounding positively charged ions; this leads to a current area of active research investigation.

EXAMPLE 2.19

Calculate the mutual capacitance of a coaxial cable whose length is $\mathcal{L}$ that consists of a cylindrical metallic rod whose radius is a, and that is surrounded concentrically with a metallic sleeve whose radius is b. There is a dielectric material separating the two conducting surfaces, and it has a relative dielectric constant ε_r.

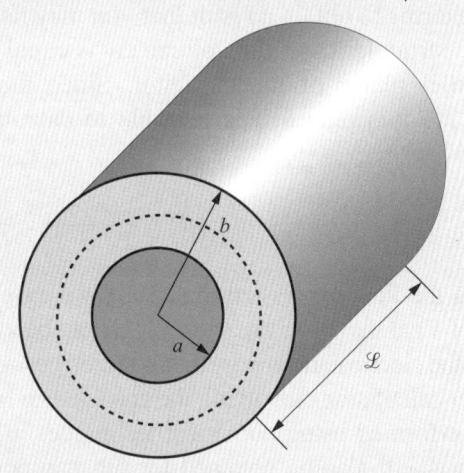

Answer. The displacement flux density between the two metallic surfaces can be calculated using Gauss's law (2.92) since there is significant symmetry in this example. The procedure is to assume initially that there is a linear charge density ρ_ℓ on the inner conductor. From (2.92), we calculate the displacement flux density as

$$\oint \mathbf{D} \cdot \mathbf{ds} = Q_{\text{enc}} \longrightarrow D(2\pi\rho\Delta\mathcal{L}) = \rho_\ell \mathcal{L}$$

The potential difference between the inner conductor and the outer conductor is computed from (2.54).

$$V_{ab} = \int_a^b \mathbf{E} \cdot \mathbf{dl} = \int_a^b \frac{\rho_\ell}{2\pi\varepsilon\rho} d\rho = \frac{\rho_\ell}{2\pi\varepsilon} \ln\left(\frac{b}{a}\right)$$

The total charge that is enclosed within the coaxial structure is $Q = \rho_\ell \mathcal{L}$. From (2.94), we write

$$C = \frac{Q}{V_{ab}} = \frac{\rho_\ell \mathcal{L}}{\frac{\rho_\ell}{2\pi\varepsilon}\ln\left(\frac{b}{a}\right)} = \frac{2\pi\varepsilon}{\ln\left(\frac{b}{a}\right)}\mathcal{L}$$

In addition to calculating the capacitance of a parallel plate capacitor, we can also find the electrostatic energy that is stored in this capacitor. In order to do this, we just have to evaluate the integral (2.67), which we rewrite here

$$W_e = \frac{\varepsilon_0}{2}\int_{\Delta v} E^2 \, dv \qquad (2.99)$$

Equation (2.99) is a very general expression for the electrostatic energy that is stored between the two objects. We will obtain an equation using electromagnetic terms to obtain a result that you may have already seen elsewhere.

In our particular case, the volume Δv is equal to the volume between the two parallel plates of the capacitor. Since the electric field is uniform between the two plates, it is possible to calculate the stored electrostatic energy. We write

$$W_e = \frac{\varepsilon_0}{2}\left(\frac{V}{d}\right)^2(Ad) = \frac{1}{2}\left(\frac{\varepsilon_0 A}{d}\right)V^2 = \frac{1}{2}CV^2 \qquad (2.100)$$

In writing the final expression in (2.100), we have recognized that the capacitance C of a parallel plate capacitor (2.98) can be identified. Therefore, we have obtained the electrostatic energy that is stored between the two parallel plates.

Further calculations involving the capacitance between conducting surfaces will be performed using numerical techniques in Chapter 4. In addition, the capacitance will also be obtained for other important structures that will be encountered in the later discussion of transmission lines. The insertion of a dielectric slab that does not completely fill the intervening space between the two parallel plates will require considerable care in the analysis. This calculation will be delayed until after the boundary conditions are examined.

2.9 Conclusion

Our study of electrostatic fields is based on the existence on a new entity that exists in nature. This entity is called an electric charge that has a distinct and discrete numerical value that may have either a positive or a negative sign associated with it. Charges that have the same sign are repelled from each

2.10 Problems

2.1.1. In terms of the units mass M, length L, time T, and charge Q, find the units of the permittivity of free space ε_0.

2.1.2. Find the force on a charge Q that is located at the point (6, 3). The charge has a value of $Q = -3$ C. The other charges are: $Q_1(2, 0) = +2$ C, $Q_2(2, 3) = -4$ C, and $Q_3(2, 6) = -3$ C. All dimensions are in meters.

2.1.3. For the charges given in Problem 2.1.2, find the force on the charge Q_2.

2.1.4. Four equal charges (Q) are placed at the corners of a square whose dimensions are ($a \times a$). Find the magnitude of the force on one of the charges.

2.1.5. Two small plastic balls, each with equal values of charge Q and mass M are constrained to slide on an insulating string. Find the separation of the two charges if the lower charge is constrained to one location on the string.

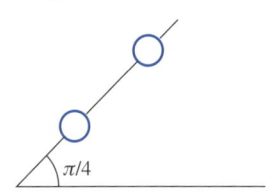

2.2.1. In terms of the units mass M, length L, time T, and charge Q, find the units of the electric field $\mathbf{E}$.

2.2.2. Find the electric field at the point P that is located at (5, 1) due to the charges $Q_1(1, 4) = +2$ C and $Q_2(1, 1) = +4$ C where the coordinates are measured in meters.

2.2.3. Find the electric field at the point P that is located at (5, 1) due to the charges $Q_1(1, 4) = +2$ C and $Q_2(1, 1) = -4$ C where the coordinates are measured in meters.

2.2.4. Four equal charges $Q = 1$ C are located at the corners of a square ($a \times a$) that is centered at the origin of a coordinate system in the x–y plane. The dimension $a = 1$m. Find the ratio of the magnitude of the electric fields at $z = 2a$ and $z = a$.

2.2.5. Two charges of equal magnitude $Q = 2(4\pi\varepsilon_0)$ C with the same sign are located at the points (1, 2) and at (5, 2) respectively. Find the electric field $\mathbf{E}$ on the line that would correspond to $x = 3$. Plot your results with MATLAB in the region ($-4 < y < 8$).

2.2.6. By measuring the potential difference between two small probes separated by a distance $\Delta \ell$, scientists are able to obtain data to plot electric field patterns caused by charges. By rotating the probes, the direction of the electric field can be obtained. From the following two measurements, find the location of the positive charge Q. At the origin, the probe measures a maximum electric field in the direction $\mathbf{u}_x - 3\mathbf{u}_y$. If the probes are moved to the point (0, 1), the maximum electric field is in the direction $\mathbf{u}_x - \mathbf{u}_y$.

2.3.1. Assume that $Q = (4\pi\varepsilon_0)$ C of charge is distributed uniformly along a line of length 2ℓ. Find an analytical solution for the electric field along a line that is perpendicular to the line of charge and is located at the center of the charged line.

2.3.2. Using MATLAB, perform the integration in Problem 2.3.1 and plot the results.

2.3.3. Assume that a total amount of charge equal to Q is distributed uniformly on the circular ring defined by $a \leq \rho \leq b$. Find an analytical expression for the electric field along the z axis. Use MATLAB to display the electric field distribution.

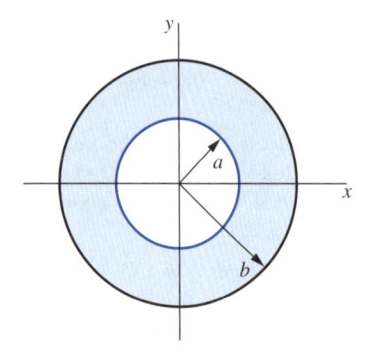

2.3.4. Find an analytical expression for the electric field along the z axis of a uniformly charged disc with radius a. Plot the field distribution.

2.3.5. Set up the integral using the ring configuration in Problem 2.3.3 to find the normal electric field from an infinite charged sheet. Evaluate this integral.

2.3.6. Find an analytical expression for the electric field along the z axis of a circular loop with a radius a, carrying a uniform charge of Q. Display the electric field distribution.

2.3.7. Find an analytical expression for the potential along the z axis of a circular loop in Problem 2.3.6 assuming it is zero in infinity. Display the potential profile.

2.4.1. Charge is distributed nonuniformly within a sphere of radius a as $\rho_v = \rho_0 r/a$. Using Gauss's law, calculate the electric field in the regions $r < a$ and $r > a$. Accurately sketch the electric field and label the axes.

2.4.2. A charge of $+Q$ C is distributed uniformly in the central region $0 \leq \rho \leq a$ and $-Q$ C is distributed in the external region $b \leq \rho \leq c$ of a coaxial cable. Find the electric field in all regions of the coaxial cable and sketch the results.

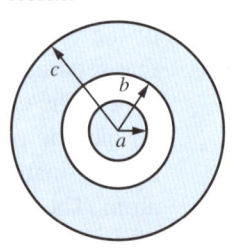

2.4.3. Repeat Problem 2.4.2 for a charge of $+2Q$ C uniformly distributed in the inner region and $-Q$ C distributed in the outer region of a coaxial cable. Find the electric field in all regions of the coaxial conductor and sketch the results.

2.4.4. Assume that a uniform charge density $+\rho_s$ C/m^2 exists on one infinite plane and $-\rho_s$ C/m^2 exists on the other infinite plane. The two planes are parallel and are separated by a distance d. Using Gauss's law, find the electric field in the regions between the two plates and external to the two plates.

2.4.5. Charges are placed on the concentric hollow spheres. The values are: $Q(r = a) = 2$ C; $Q(r = b) = -4$ C; and $Q(r = c) = 4$ C where $a < b < c$. Find and sketch the electric field in all regions $r \geq 0$.

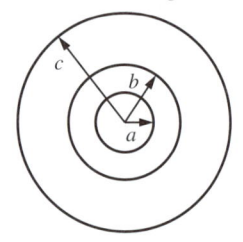

2.4.6. Find the voltage between two small electrodes A and B for the system described in Problem 2.4.5: A is fixed at the center ($r = 0$), while B is successively moved to a position on the spheres at $r = a$, then $r = b$ and finally at $r = c$.

2.5.1. Find the absolute potential at an arbitrary point $P(x, y, z = 0)$ due to charges $Q_1(2, 2) = +2(4\pi\varepsilon_0)$ C and $Q_2(4, 5) = -4(4\pi\varepsilon_0)$ C. Plot your results with MATLAB in the region ($0 < x < 6, 0 < y < 7$).

2.10 Problems

2.5.2. On a log–log graph, accurately plot the magnitude of the electric field E_r and the voltage V from a point charge as a function of the distance from the charge. Assume that E_r @ $r = 1$ m equals 1 V/m and V @ $r = 1$ m equals 1 V.

2.5.3. Calculate the work that is expended in moving a positive charge Q from point A to B along the indicated path. A charged infinite plane with a charge density ρ_s C/m² exists in the y–z plane at $x = 0$.

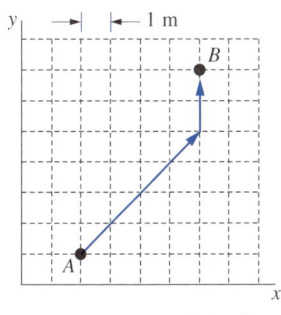

2.5.4. Given the electric field $\mathbf{E} = 4x\mathbf{u}_x - 2y\mathbf{u}_y$ V/m, find the voltage between the points $A(2, 0)$ and $B(0,2)$ integrating along: (a) straight line AB; (b) broken line AOB, where $O(0, 0)$.

2.5.5. Assume that there are two concentric cylinders with radii $a < \rho < b$. Show that the electric field between the two cylinders approaches a constant value as the separation distance $(b - a) \to 0$.

2.5.6. Assume that there are two concentric spheres with radii $a < r < b$. Show that the electric field between the two spheres approaches a constant value as the separation distance $(b - a) \to 0$.

2.5.7. A lightning rod provides a controlled path for a cloud to discharge itself to the ground in a safe way. Sketch the expected electric field distribution about the rod. Pay particular attention to the electric field that you would expect at the tip of the lightning rod.

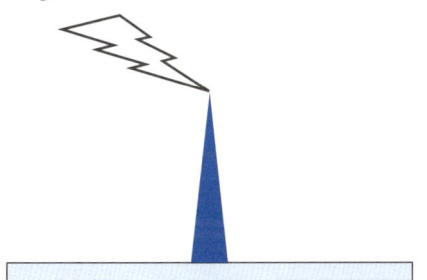

2.5.8. Compare the electric potential from a single charge, a dipole charge, and a quadrupole charge by accurately plotting the potential as a function of distance from the region of the charge. Use MATLAB and assume that the potential equals 1 V at a distance of 1 m from the charge.

2.6.1. Using MATLAB, evaluate the following integral numerically. Compare this result with an analytical calculation.

$$z = \int_0^1 (1 - x)\,dx$$

2.6.2. Using MATLAB, evaluate the following integral numerically. Compare your results with an analytical calculation.

$$z = \int_{y=0}^{y=1} \int_{x=0}^{x=1} xy\,dx\,dy$$

2.6.3. Using MATLAB, evaluate the following integral numerically. Compare your result with an analytical calculation.

$$w = \int_{z=0}^{z=2} \int_{y=0}^{y=3} \int_{x=0}^{x=4} xyz\,dx\,dy\,dz$$

2.6.4. Using MATLAB, evaluate the following integral numerically. Compare your result with an analytical calculation.

$$w = \int_0^{\frac{\pi}{2}} \sin(x)\,dx$$

2.6.5. Using MATLAB, calculate the total charge on a 1m² square plate. One corner of the plate is at the origin of a Cartesian coordinate system and there is a nonuniform charge density $\rho_s = (1 - x^2)(1 - y^2)$ on the plate. Compare the result with an analytical calculation.

2.7.1. Determine the permittivity of water if the relative dielectric constant of water is equal to 80.

2.7.2. A coaxial line is 20 cm long the radius of the inner conductor is 5 cm and the outer conductor is 10 cm. Calculate the electrostatic energy that is stored in the structure if the intermediate space is filled with a dielectric whose relative dielectric constant is 10 and a voltage is applied between the conductors creating an electric field

$$E = \frac{10^6}{\rho} V/m.$$

Compare this with the value that would be obtained if the intermediate space is a vacuum.

2.7.3. The electric field in a dielectric is 100 V/m. If the relative dielectric constant is equal to 5, calculate the displacement flux density in the dielectric.

2.7.4. The displacement flux density in a vacuum is equal to $1/36\pi$ C/m^2. Find the value of the electric field in this region.

2.8.1. Calculate the self capacitance of a circular disc whose radius is a. You are to assume that the charge is uniformly distributed on the disc.

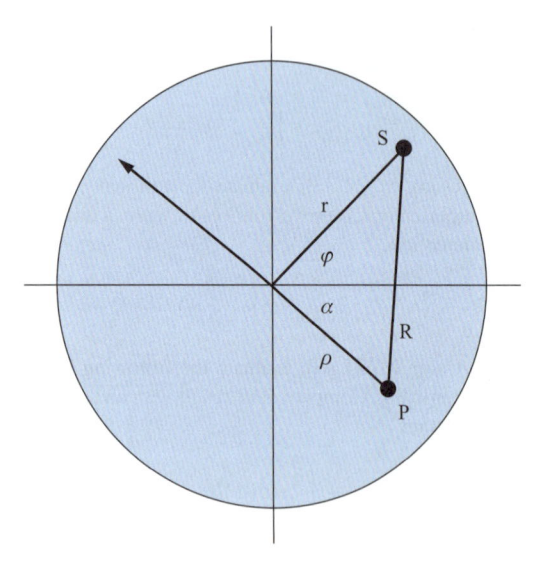

2.8.2. Calculate the self capacitance of a circular disc whose radius is a as shown in problem 2.8.1. You are to assume that the charge is nonuniformly distributed on the disc with the density distribution approximated with the distribution .

$$\rho_s(r) = \frac{\rho_s 1}{\sqrt{1 - (r/a)^2}}$$

2.8.3. Calculate the mutual capacitance between two parallel circular discs whose radii are a that are separated by a distance d. Neglect any fringing fields.

2.8.4. Calculate the total capacitance of the structure that consists of a dielectric inserted between two parallel plates that are separated by a distance $3d$. The dielectric whose thickness is d is inserted in the center of the structure. One can neglect any fringing fields and assume that the areas are all equal to A.

2.8.5. Calculate the capacitance of a coaxial cable that is 8 km long. The radius of the inner conductor is 10 mm in the radius of the outer conductor is 15 mm.

2.8.6. Calculate the numerical value of the mutual capacitance of a square parallel plate capacitor whose dimensions are 1 mm $\times$ 1 mm if the two plates are separated by a distance of 2 mm. The relative dielectric constant of the material between the two plates is equal to 2.

2.8.7. Determine the capacitance of the coaxial structure described in problem 2.7.2 assuming that a vacuum exists between the two conductors. In addition, determine the value of the applied voltage.

CHAPTER 3

Magnetostatic Fields

3.1 Electrical Currents 123
3.2 Fundamentals of Magnetic Fields 128
3.3 Magnetic Vector Potential and the Biot–Savart Law 138
3.4 Magnetic Forces 146
3.5 Magnetic Materials 157
3.6 Magnetic Circuits 162
3.7 Inductance 166
3.8 Conclusion 171
3.9 Problems 172

The important properties of time-independent static magnetic fields will be reviewed in this chapter. A magnetic field is significantly different from an electric field because it is impossible to isolate a North Pole from a South Pole of a magnet and treat them as independent magnetic charges as we were able to with electric fields. The magnetic field arises when electrical charge is in motion. Electrical charges in motion create an electrical current and this current creates a magnetic field. Several procedures to calculate the magnetic field from a current will be described. There will be a magnetic force exists between two current-carrying elements. Certain materials will have a profound impact on the characteristics of the magnetic fields. Finally, the electrical circuit element called the inductance and the topic of magnetic circuits will be introduced.

3.1 Electrical Currents

Imagine that a wire is connected to a battery and a resistor as shown in Figure 3–1a. The battery is a chemical source that provides particles with a positive charge and a negative charge. In the metallic wire, the ions are

FIGURE 3–1

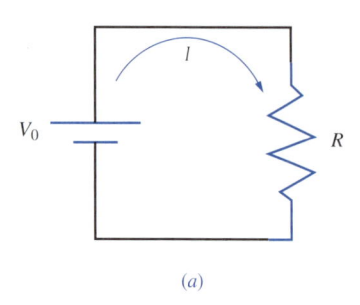

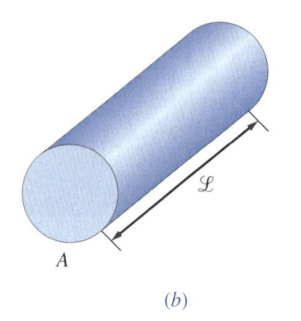

(*a*) A simple electrical circuit consisting of a battery and a resistor. (*b*) A current *I* flows through the wire whose area is equal to $A = \pi a^2$, where *a* is the radius of the wire.

stationary and a portion of the electrons are free to move. Benjamin Franklin gave us the convention that the direction of the flow of the current, however, should be in the direction of the motion of the positive particles. Between the two terminals of the battery there will be an electric field that will accelerate the electrons. Since the conductivity of a wire is significantly greater than the conductivity of the surrounding air, the motion of these accelerated electrons will follow the path of the wire. The resulting current in this circuit is called the **conduction current**. When we later describe electric fields that depend upon time, we will encounter another current that is called a **displacement current**.

The current that passes through the wire in Figure 3–1*b* can be computed from Ohm's law. Since we are more interested in local effects than in global effects, we would rather relate this current to a current density $J = I/A$ and a voltage difference ΔV across an incremental length $\mathscr{L}$, i.e., an electric field. The **resistance** R of the wire is given by $R = \mathscr{L}/(\sigma A)$ where σ is the conductivity of the wire. This definition of resistance follows from the intuition gained in circuits where the total resistance of a circuit is computed by adding the resistors in series (i.e., increased length of the wire $\mathscr{L}$) and the conductors in parallel (i.e., increased cross-sectional area A). The current density in a wire is therefore defined using **Ohm's law**.

$$J = \frac{I}{A} = \frac{\left(\frac{\Delta V}{R}\right)}{A} = \frac{\Delta V}{\mathscr{L}}\frac{\mathscr{L}}{A}\frac{1}{R} = \sigma E$$

$$\mathbf{J} = \sigma \mathbf{E} \ (\text{A/m}^2) \qquad (3.1)$$

This is a generalization of Ohm's law. One ampere of current at a point is defined as the passage of one coulomb of charge passing this point in one second. We are assuming that none of the parameters that appear in (3.1) depend upon the magnitude of any of the other parameters, a situation which could be found in nonlinear materials. One could think of a resistor whose resistance would change with increasing values of current caused by a heating of the resistor. Such nonlinear effects are important in practice but will not be considered here.

An alternative derivation for the conductivity σ follows from the definition of the current density $\mathbf{J} = \rho_v \mathbf{v}_{\text{drift}}$ where ρ_v is the electron volume charge density and $\mathbf{v}_{\text{drift}}$ is an average electron drift velocity. The drift velocity is proportional to the electric field $\mathbf{E}$, the proportionality constant being called the mobility μ_m of the material. Hence we write

$$\mathbf{J} = \rho_v \mathbf{v}_{\text{drift}} = \rho_v \mu_m \mathbf{E} = \sigma \mathbf{E} \ (\text{A/m}^2) \tag{3.2}$$

where the conductivity $\sigma = \rho_v \mu_m$.

The total current I that passes through the wire is computed from the integral of the current density $\mathbf{J}$ integrated over the cross sectional area ΔA of the wire. The current density $\mathbf{J}$ is a vector since it has both a magnitude and a direction.

$$\boxed{I = \int_{\Delta A} \mathbf{J} \cdot \mathbf{ds} \ (\text{A})} \tag{3.3}$$

If the current is distributed uniformly in a cylindrical wire whose radius is a, this integral can be performed easily, and we find that the total current I that passes through the wire is given by

$$I = J\pi a^2 \tag{3.4}$$

This is equal to the product of the current density times the cross-sectional area of the wire. However, if the current is distributed nonuniformly in the wire, this integration requires more care, as will be shown with an example. Later, we will encounter cases where the current is constrained to flow just on the surface of an object. These currents are called surface currents. Having now presented some fundamental definitions for the currents in terms of local current densities and cross sectional areas, we are prepared to explore various properties of magnetic fields that will be created by these currents.

EXAMPLE 3.1

Given an inhomogeneous current density

$$\mathbf{J} = (3y^2 z \mathbf{u_x} - 2x^3 z \mathbf{u_y} + z \mathbf{u_z})$$

find the total current passing through a square surface at $x = 1$ in the $\mathbf{u_x}$ direction. The dimensions of the surface are $1 \leq y \leq 2$, $1 \leq z \leq 2$.

Answer. The differential surface area is defined as $\mathbf{ds} = dydz\mathbf{u}_x$. Therefore, the scalar product of the current density and the differential surface area will yield a current only in the $\mathbf{u}_x$ direction. We write

$$I = \int_A \mathbf{J} \cdot \mathbf{ds} = \int_{y=1}^{y=2} \int_{z=1}^{z=2} 3y^2 z \, dz \, dy = \left. \frac{3y^3}{3} \right|_{y=1}^{y=2} \left. \frac{z^2}{2} \right|_{z=1}^{z=2}$$

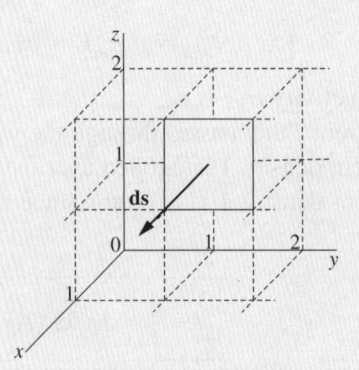

$$= 10.5 \text{ A}$$

EXAMPLE 3.2

Calculate the current that flows through a wire whose radius is a. The inhomogeneous current density in the wire is

$$\mathbf{J} = I_0 \left(\frac{\rho}{a} \right) \mathbf{u}_z$$

Nonuniform currents can be important in high frequency applications in which one encounters "skin effects."

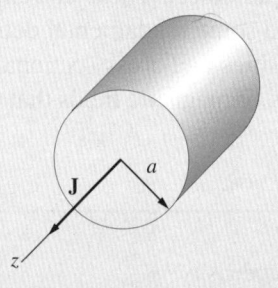

Answer. The current is calculated from the following integration:

$$I = \int_{\phi=0}^{\phi=2\pi} \int_{\rho=0}^{\rho=a} I_0 \left(\frac{\rho}{a} \right) \mathbf{u}_z \cdot \rho \, d\rho \, d\phi \, \mathbf{u}_z = I_0 (2\pi) \left(\frac{a^2}{3} \right)$$

3.1 Electrical Currents

The power that is dissipated within a conducting material can be calculated from the electric field and the current density that we have just encountered. The *power density* in a particular volume is defined as

$$p = \mathbf{J} \cdot \mathbf{E} \ (\text{W/m}^3) \tag{3.5}$$

The total power that is absorbed within the volume is calculated by integrating (3.5) over the entire volume Δv. This power is converted into another form and is given the name *Joule heating*. It is measured in SI units

$$1 \, \text{watt} \equiv \frac{1 \, \text{joule}}{1 \, \text{second}}$$

The reader has probably experienced the warming effects of Joule heating in cooking a meal on an electric stove or being warmed on a cold winter's night by an electric heater that is present in the room. Suffice it to say, this is a very important effect that has many practical applications.

EXAMPLE 3.3

Calculate the power that is dissipated within a resistor that has a uniform conductivity σ. The voltage between the two ends of the resistor is ΔV and a current I passes through the resistor.

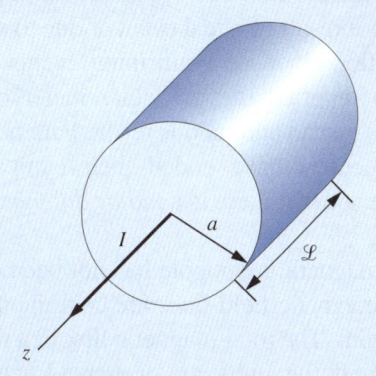

Answer. From (3.5), we obtain the power density. The total power loss is calculated from the integration of the power density over the volume.

$$P = \int_{\Delta v} \mathbf{J} \cdot \mathbf{E} \, dv = \int_{z=0}^{z=\mathscr{L}} \int_{\phi=0}^{\phi=2\pi} \int_{\rho=0}^{\rho=a} \left(\frac{I}{\pi a^2}\right)\left(\frac{\Delta V}{\mathscr{L}}\right) \rho \, d\rho \, d\phi \, dz = \Delta V I$$

3.2 Fundamentals of Magnetic Fields

The effects of magnetic fields were known for almost three millennia when it was discovered that certain stones would attract iron. A large deposit of these stones that are called "lodestones" was found in the district of Magnesia in Asia Minor. This mineral later became known as magnetite (Fe_3O_4) and it had some interesting properties. Early navigators used its north- and south-seeking characteristics in their early explorations. The first scientific study of magnetism was written in 1600 by William Gilbert. Little else was known about it until the early nineteenth century when Hans Christian Oersted discovered that an electric current in a wire affected a magnetic compass needle. This work together with the later works of Ampere, Gauss, Henry, Faraday, and others raised the magnetic field to equal partner status with the electric field. This elevation in stature was confirmed with the theoretical work of Maxwell.

In studying electric fields, we found that electric charges could be separated from each other such that a positive charge existed independently from a negative charge. Would the same separation of magnetic poles exist? Would it be possible to cut the earth at the equator and send one-half containing one of the pole faces (a "magnetic monopole") to a far off region and never see that pole again? As of the writing of this text, *a magnetic monopole has not yet been observed to exist in nature*. Several experimenters have searched for these elusive entities and in one five-year period, only one momentary deflection of a needle on a satellite had been recorded throughout the world. Whether this deflection was a result of a real event or due to some anomaly in the detector is open to question since no confirming events have ever been detected. The scientist who reported the particular meter deflection later stated that the deflection was just an anomalous result. Patience seems to be wearing thin and most experimenters have ended their campaign of "monopole sighting." Therefore, we see that there is a *major difference* between magnetic fields and electric fields.

Since the magnetic monopole has not been observed to exist in nature, we find that the magnetic field lines are continuous and do not originate nor terminate at a point. The total magnetic flux in a region is usually denoted with a symbol Ψ_m and the units are webers. Enclosing an arbitrary point with a closed surface, we can express this fact mathematically by stating that

$$\oint \mathbf{B} \cdot \mathbf{ds} = 0 \tag{3.6}$$

The term **B** that appears in this equation is the ***magnetic flux density***. The SI units of magnetic flux density are given in tesla (T). We may also see the

3.2 Fundamentals of Magnetic Fields

equivalent unit of (weber/meter^2) where $1\,\text{T} = 1\,\text{Wb/m}^2$. A magnetic flux density of one tesla is a very large value. For example, the equatorial magnetic field strength at sea level of the Earth is approximately 0.5×10^{-4} tesla. Another commonly employed unit for the magnetic flux density is the gauss (G) where 1 gauss = 10^{-4} tesla.

We can also write (3.6) in differential form by making use of the divergence theorem that relates a closed surface integral to a volume integral: $\oint \mathbf{B} \cdot d\mathbf{s} = \int_{\Delta v} \nabla \cdot \mathbf{B}\, dv$. In order for the closed surface integral to be equal to zero for any arbitrary volume Δv, the integrand must be identically equal to zero. Therefore, we write

$$\boxed{\nabla \cdot \mathbf{B} = 0} \qquad (3.7)$$

Equations (3.6) and (3.7) express the fact that the magnetic field closes upon itself and does not terminate on nor originate from an isolated magnetic monopole. If we cut a bar magnet in half with the hope of isolating one of the poles, we will just end up with two bar magnets of smaller physical size (see Figure 3–2). We can contrast the first postulate of magnetic fields with the first postulate of electrostatic fields.

$$\nabla \cdot \mathbf{E} = \frac{\rho_v}{\varepsilon_0}$$

Recall that it was physically possible to separate a positive charge from a negative charge.

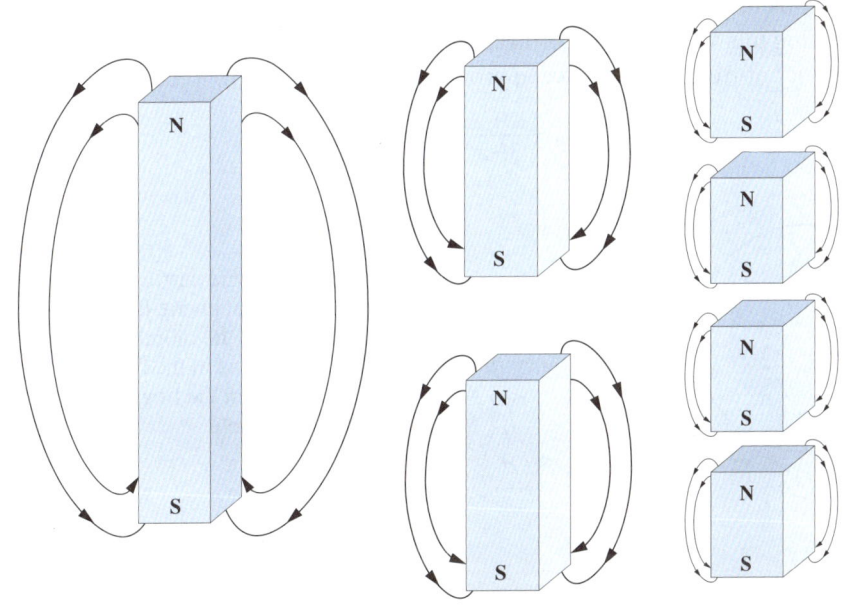

FIGURE 3–2

The physical cutting of a large permanent magnet creates a number of smaller permanent magnets. It is impossible to separate the north pole from the south pole of a magnet.

The second property of steady magnetic fields was discovered by Hans Christian Oersted in 1820. He observed that compass needles were deflected when an electrical current flowed through a nearby wire, and he concluded that the effect was due to the creation of a magnetic field by this current. Recall our earlier allusion to the model of an atom that consisted of an electron circulating about a positive nucleus. It was modeled as an electric dipole. The moving electron can also be interpreted as being a current. Therefore, the atom can be thought of as being a small electrical dipole and a small magnetic dipole.

The results of this experiment can be described with the following equation

$$\oint \mathbf{B} \cdot \mathbf{dl} = \mu_0 I_{enc} \tag{3.8}$$

where the total current that is enclosed within the closed loop is specified as I_{enc}. The constant μ_0 is the **permeability** of free space. In SI units μ_0 is defined to have the numerical value of

$$\mu_0 \equiv 4\pi \times 10^{-7} \quad (\text{H/m}) \tag{3.9}$$

We note that while the exact value of ε_0 is given by 8.854×10^{-12} F/m and the approximate values of ε_0 is given by $(1/36\pi) \times 10^{-9}$ F/m, the expression in (3.9) is the exact value for μ_0 and the approximate value is given by 12.566×10^{-7} H/m.

The magnetic flux density is perpendicular to the current density and follows the "right hand rule" convention in that the thumb of the right hand is in the direction of the current and the fingers are in the direction of the magnetic flux density as shown in Figure 3–3.

Using this value for the permeability of free space and the value for the permittivity of free space that we approximated earlier in (2.6) to be

$$\varepsilon_0 \approx \frac{1}{36\pi} \times 10^{-9} \text{ F/m} \tag{3.10}$$

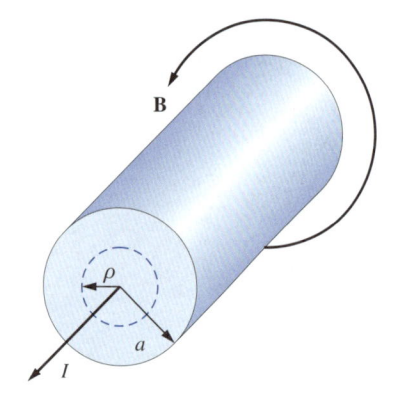

FIGURE 3–3

A cylindrical wire carries a current I that creates a magnetic field whose density is **B**. If the thumb of the right hand points in the direction of the current, then the fingers follow the magnetic field.

3.2 Fundamentals of Magnetic Fields

we may be intrigued by the numerical value that is computed from the expression $1/\sqrt{\mu_0 \varepsilon_0}$. This number has the same numerical value as the velocity of light. It also has the dimensions of a velocity, but the demonstration of this will be discussed later. We will also see later that this coincidence is more than fortuitous and it will lead to something very fundamental.

Equation (3.8) is called **Ampere's circuital law** or **Ampere's law,** and as we will see, it allows us to calculate magnetic flux density in many cases where there is considerable symmetry. Thus, we will see that Ampere's law is a magnetic analog of Gauss's law, simplifying problem solutions where there is symmetry. Several examples will be described in full detail in the following discussion since it is one of the fundamental methods of calculating the magnetic flux density caused by a current.

Equation (3.8) can also be written in differential form but this will require a vector operation. The left-hand side of (3.8) can be converted to a surface integral using Stokes's theorem.

$$\oint \mathbf{B} \cdot \mathbf{dl} = \int_{\Delta s} \nabla \times \mathbf{B} \cdot \mathbf{ds} \qquad (3.11)$$

Therefore, after equating the two surface integrals in (3.8) and (3.11), we obtain

$$\int_{\Delta s} \nabla \times \mathbf{B} \cdot \mathbf{ds} = \mu_0 \int_{\Delta s} \mathbf{J} \cdot \mathbf{ds} \qquad (3.12)$$

In order for these two surface integrals to be equal for any arbitrary surface, the integrands must be equal. This leads to

$$\boxed{\nabla \times \mathbf{B} = \mu_0 \mathbf{J}} \qquad (3.13)$$

This is the differential form of Ampere's law. The integral form and the differential form of Ampere's law are used in electromagnetic calculations.

Let us determine the magnetic field using (3.8) for the wire shown in Figure 3–3. It will be assumed that the wire is straight, it has no bends or kinks in it, and it is of infinite length. In this case, the magnetic field will be entirely in the $\mathbf{u}_\phi$ direction. We will follow the right-hand rule convention that is standard for determining the direction of the magnetic field in that if the current is pointing in the direction of the thumb of the *right hand*, then the magnetic flux density will be in the direction of the fingers.

The line integral in (3.8) can be evaluated easily if there is symmetry in the system. For the particular case of the wire shown in Figure 3–4, where significant cylindrical symmetry abounds, the path of the integration will follow a circle of radius ρ and the integral is written as

$$\oint \mathbf{B} \cdot \mathbf{dl} = \int_0^{2\pi} B_\phi \mathbf{u}_\phi \cdot \rho \, d\phi \, \mathbf{u}_\phi = 2\pi \rho B_\phi \qquad (3.14)$$

FIGURE 3–4

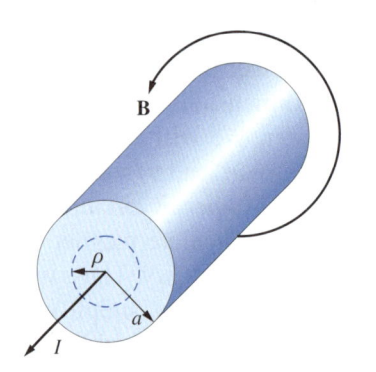

A current I passes through a cylindrical conductor. The magnetic flux density is to be computed for all values of the radius ρ.

The surface integral on the right side of (3.8) can also be evaluated. It is equal to the μ_0 times the current that is enclosed within the surface defined by the same radius ρ given in (3.14). For the moment, we will take the radius ρ to be greater than the radius a of the wire so the total current I that flows in the wire is enclosed within the surface. Hence the integral on the right side of (3.8) is just $\mu_0 I$. The magnetic flux density external to the wire is computed to be

$$B_\phi = \frac{\mu_0 I}{2\pi\rho} \text{ (T)} \qquad (3.15)$$

Using the same technique, we can also calculate the magnetic flux density within the wire. The procedure is the same as was used to calculate the field external to the wire; we first determine the current that is enclosed within a circle of radius ρ. If the current is distributed uniformly in the wire, the current density that flows through the wire is

$$\mathbf{J} = \frac{I}{\pi a^2}\mathbf{u}_z \text{ (A/m}^2\text{)}$$

The total current that is enclosed within the circle whose radius is ρ is given by

$$\int_{\Delta s} \mathbf{J} \cdot \mathbf{ds} = \int_{\phi=0}^{\phi=2\pi} \int_{\rho'=0}^{\rho'=\rho} \left(\frac{I}{\pi a^2}\right) \rho' d\rho' d\phi = \left(\frac{\rho}{a}\right)^2 I \qquad (3.16)$$

where ρ' is a dummy variable of integration. Equating this enclosed current multiplied by μ_0 with the expression given in (3.14), we finally obtain the magnetic flux density within the wire to be

$$B_\phi = \frac{1}{2\pi\rho}\left[\mu_0\left(\frac{\rho}{a}\right)^2 I\right] = \frac{\mu_0 I}{2\pi}\left(\frac{\rho}{a^2}\right) \qquad (3.17)$$

At the edge of the wire ($\rho = a$), the two solutions given by (3.15) and (3.17) agree as they must. The magnetic flux density is shown in Figure 3–5. At the center of the wire ($\rho = 0$), the magnetic flux density is equal to zero since no

3.2 Fundamentals of Magnetic Fields

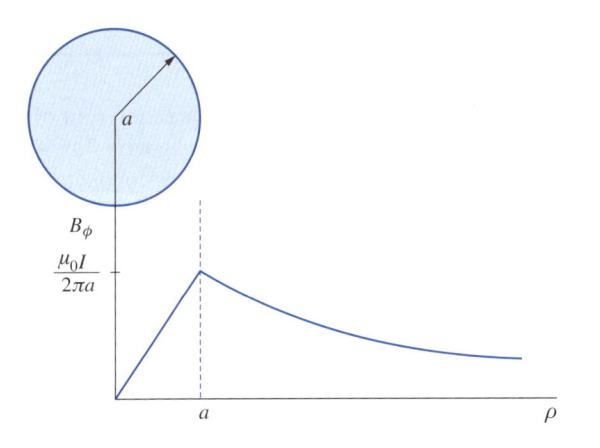

FIGURE 3-5

The radial dependence of the magnetic flux density as calculated in (3.15) and (3.17) is shown. There is a homogeneous current distribution within the wire.

current is enclosed within a circle whose radius is equal to zero. As more current is enclosed, the field increases until all of the current is enclosed at the edge ($\rho = a$). For radii greater than the radius of the wire ($\rho > a$), no additional current is enclosed, and the field decays geometrically as $1/\rho$.

EXAMPLE 3.4

Plot the magnetic flux density in the regions that are internal to the wire and external to the wire that carries a current out of the page.

Answer. The magnetic field is calculated using (3.15) and (3.17) and the result is illustrated in the following figure. The length of the arrows is proportional to the magnetic field strength.

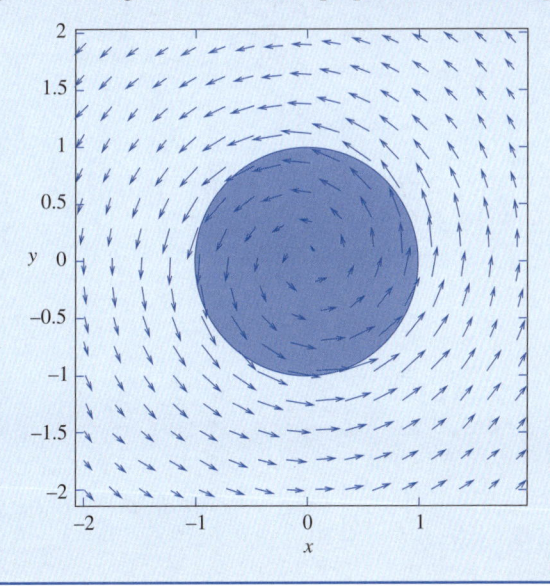

EXAMPLE 3.5

The center conductor of a coaxial cable carries a current I_0 in the $+\mathbf{u}_z$ direction (out of the page) and this current returns in the outer conductor. Calculate the magnetic flux density both within the coaxial cable and in the region external to the outer conductor.

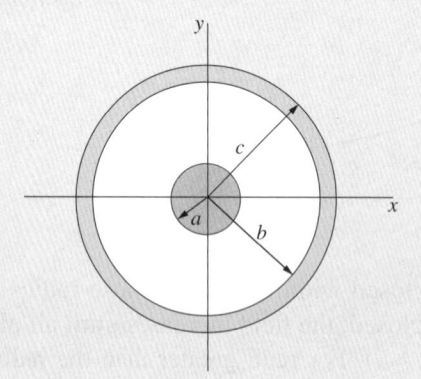

Answer. We will apply Ampere's law from (3.8) separately to each of the regions in the coaxial cable. Due to symmetry, the left-hand side of (3.8) will always have the value given in (3.14): $2\pi\rho B_\phi$. Hence, we write

For $0 \leq \rho \leq a$:

$$2\pi\rho B_\phi = \mu_0 \int_{\Delta s} \mathbf{J} \cdot \mathbf{ds} = \mu_0 \int_{\phi=0}^{\phi=2\pi} \int_{\rho'=0}^{\rho'=\rho} \left(\frac{I_0}{\pi a^2}\right) \rho' \, d\rho' \, d\phi' = \mu_0 I_0 \left(\frac{\rho}{a}\right)^2$$

$$B_\phi = \frac{\mu_0 I_0 \rho}{2\pi a^2} \quad \text{(T)}$$

For $a < \rho \leq b$:

$$2\pi\rho B_\phi = \mu_0 \int_{\Delta s} \mathbf{J} \cdot \mathbf{ds} = \mu_0 I_0$$

$$B_\phi = \frac{\mu_0 I_0}{2\pi\rho} \quad \text{(T)}$$

For $b < \rho \leq c$:

$$2\pi\rho B_\phi = \mu_0 \int_{\Delta s} \mathbf{J} \cdot \mathbf{ds}$$

$$= \mu_0 I_0 - \mu_0 \int_{\phi=0}^{\phi=2\pi} \int_{\rho'=b}^{\rho'=\rho} \left(\frac{I_0}{\pi(c^2-b^2)}\right) \rho' \, d\rho' \, d\phi = \mu_0 I_0 \left(1 - \frac{(\rho^2-b^2)}{(c^2-b^2)}\right)$$

3.2 Fundamentals of Magnetic Fields

$$B_\phi = \frac{\mu_0 I_0}{2\pi\rho}\left(1 - \frac{(\rho^2 - b^2)}{(c^2 - b^2)}\right) \text{ (T)}$$

For $\rho > c$:

$$2\pi\rho B_\phi = \mu_0 \int_{\Delta s} \mathbf{J} \cdot \mathbf{ds} = 0$$

$$B_\phi = 0$$

EXAMPLE 3.6

Use Ampere's law to find the magnetic field of a solenoid. A solenoid is constructed by winding wire uniformly around a cylindrical form such as a broomstick. There are N turns of wire in the length of the solenoid. Assume that the length d is much greater than its radius a.

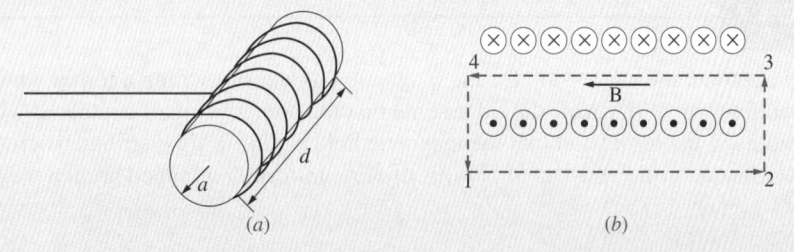

Answer. Apply Ampere's law from (3.8) to the loop that encloses the current that is coming out of the page. There will be four terms $[1 \to 2, 2 \to 3, 3 \to 4, 4 \to 1]$ that will contribute to the line integral defined in (3.8). However, we have assumed that dimensions of the solenoid satisfied the relation that $d \gg a$. This assumption will allow us to neglect any fringing fields at the two ends. Therefore, only two of the integrals $[1 \to 2$ and $3 \to 4]$ will contribute to our solution. This approximation implies that there is no component of magnetic field in the radial direction. The closed line integral approximately yields

$$\oint \mathbf{B} \cdot \mathbf{dl} \approx B_z(2d)$$

The surface integral gives us the current that is enclosed within the loop. Our use of the "approximately equal to" notation can be replaced with an equal sign if the current that is enclosed within the loop were an infinite current carrying slab out of the page. Then the integral from $1 \to 2$ would exactly equal the integral $3 \to 4$. Since there are N wires each carrying a current I into the paper, the surface integral yields

$$\mu_0 \int_{\Delta s} \mathbf{J} \cdot \mathbf{ds} = \mu_0 N I$$

Equating these two expressions, we find that the magnetic flux density at the center of the solenoid is given by

$$B_z = \frac{\mu_0 NI}{2d} \quad (\text{T})$$

Following the same procedure for the current that is going into the page in the top portion of the solenoid, we find that the magnetic flux density has the same magnitude and direction. Applying the principle of superposition, we find that the two fields add in the center of the solenoid and subtract in the external regions.

$$B_z = \frac{\mu_0 NI}{d} \quad (\text{T})$$

A more accurate calculation shows that the magnetic flux density given above is exact in the center of the solenoid and its value differs by 50% at the ends where symmetry disappears.

EXAMPLE 3.7

Find an approximate expression for the magnetic flux density within a toroid whose cross-sectional area is A. You should assume that the area A is small compared with a mean radius of the toroid and that the magnetic field is uniformly distributed across this cross-section. A toroid consists of N turns of wire uniformly wrapped around the torus.

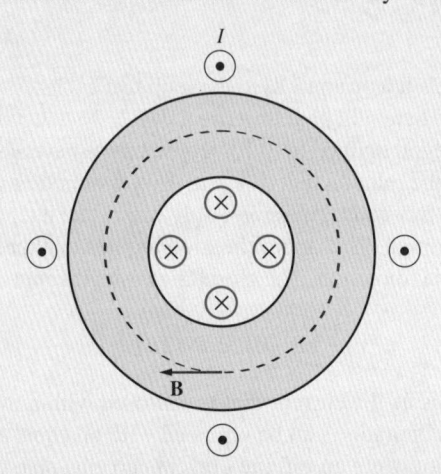

Answer. Within a mean circumference whose length is $\mathscr{L}$, there are NI amperes of current entering the paper. From Ampere's law in equation (3.8), we write

$$B_\phi \mathscr{L} \approx \mu_0 NI$$

which yields $B_\phi = \mu_0 NI / \mathscr{L}$

3.2 Fundamentals of Magnetic Fields

In order to emphasize this point that the magnetic flux density surrounding a current carrying region depends only on the current that is enclosed within the region, we will cast Ampere's law in Lagrangian mass variables as we had previously cast Gauss's law for electrostatics. The current that is enclosed within a radius ρ is given by

$$I_{enc} = \int_{\Delta s} \mathbf{J} \cdot \mathbf{ds} = \int_{\phi=0}^{\phi=2\pi} \int_{\rho'=0}^{\rho'=\rho} J_z(\rho')\rho' d\rho' d\phi = 2\pi \int_{\rho'=0}^{\rho'=\rho} J_z(\rho')\rho' d\rho' \quad (3.18)$$

In (3.18), ρ' is the variable of integration and we have assumed that the current depends only on the radial coordinate. The current is flowing in the z direction. The integration over the angular variable ϕ yielded the factor of 2π. The differential current dI_{enc} is given by

$$dI_{enc} = 2\pi\rho J_z(\rho) d\rho \quad (3.19)$$

We are treating the case where the magnetic flux density depends only on the coordinate ρ and the magnetic flux density is directed in the $\mathbf{u}_\phi$ direction. Hence (3.13) becomes

$$\frac{dB_\phi}{d\rho} = \mu_0 J_z(\rho) \quad (3.20)$$

Applying the chain rule to (3.20) and using (3.19), we write

$$\frac{dB_\phi}{d\rho} = \frac{dB_\phi}{dI_{enc}} \frac{dI_{enc}}{d\rho} = \frac{dB_\phi}{dI_{enc}} [2\pi\rho J_z(\rho)] = \mu_0 J_z(\rho)$$

or

$$\frac{dB_\phi}{dI_{enc}} = \frac{\mu_0}{2\pi\rho} \quad (3.21)$$

where the explicit dependence on the current density $J_z(\rho)$ has disappeared. The integral of (3.21) is explicitly stated as

$$B_\phi = \frac{\mu_0 I_{enc}}{2\pi\rho} \quad (3.22)$$

The application of (3.22) to the Figure 3–4 leads to the following results. The entire current I is enclosed within a circle whose radius $\rho > a$. Therefore, we obtain the same result as given in (3.15). The fraction of the current I that is enclosed within a circle whose radius $\rho < a$ is $I_{enc} = (\rho/a)^2 I$. Therefore, we obtain the same result given in (3.17). This calculation provides an additional interpretation of Ampere's law in that the current must be enclosed within the closed line integral.

EXAMPLE 3.8

There are two concentric hollow metallic cylinders. Calculate the magnetic flux density at all regions of space if there is a current I flowing out of the paper ($+z$) along the inner cylinder and the same current I returning into the paper ($-z$) along the outer cylinder.

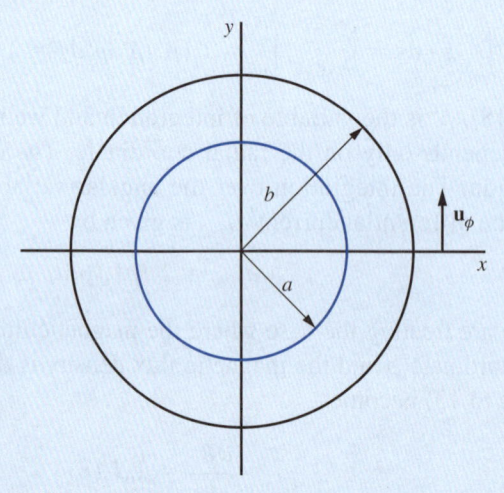

Answer. The current that is enclosed for the radius $\rho < a$ is equal to zero. Therefore, the magnetic flux density within the inner cylinder is equal to zero. In the region $a < \rho < b$, there is a current $I_{enc} = I$. From (3.22), the magnetic flux density in this region is equal to $B_\phi = \mu_0 I / (2\pi\rho)$. In the region $\rho > b$, the current that is enclosed is again equal to zero resulting in no magnetic flux density in this region.

It should be noted that all of the calculations using Ampere's law have required considerable symmetry. Unfortunately, there will be several problems in which this symmetry does not exist. In these cases, we will have to resort to more complicated analytical or numerical methods in order to obtain a solution for a particular problem. Some of these methods will be described in an upcoming section.

3.3 Magnetic Vector Potential and the Biot–Savart Law

There are several cases in practice where it is very difficult to find the magnetic flux density in terms of a current density. This is particularly true if there are difficulties invoking the symmetry arguments required for the application of Ampere's law that was discussed in the previous section. In the following, we will introduce techniques to find this magnetic flux density from a current distribution that has an arbitrary shape. This will be based on

3.3 Magnetic Vector Potential and the Biot–Savart Law

some mathematical relations and physical considerations. This will include the development of a new entity called the magnetic vector potential along with a derivation of the Biot–Savart law.

The nonexistence of magnetic monopoles allowed us to write that the magnetic flux density satisfied (3.7). For convenience, this is rewritten below

$$\nabla \cdot \mathbf{B} = 0 \tag{3.23}$$

The divergence of the magnetic flux density is now specified in (3.23). We still have freedom to examine other properties of it. In particular, we will define a vector **A** such that the magnetic flux density **B** can be expressed as the curl of this vector.

$$\mathbf{B} = \nabla \times \mathbf{A} \tag{3.24}$$

This vector will be given the symbol **A**, and it has the units of Tesla-meter or Webers/meter. This term is called the ***magnetic vector potential*** or just the ***vector potential***. The substitution of (3.24) into (3.23) leads to

$$\nabla \cdot \nabla \times \mathbf{A} = 0 \tag{3.25}$$

which is a repeated vector operation.

We will find that the magnetic flux density and the vector potential are somewhat similar to electric fields where it was found that the electric field could be obtained by taking the gradient of a scalar electric potential. This potential was found in terms of an electric charge distribution. Magnetic fields are related to a current density **J** via the differential form of Ampere's law (3.13).

$$\nabla \times \mathbf{B} = \mu_0 \mathbf{J} \tag{3.26}$$

Replacing the magnetic flux density using (3.24), we find that the vector potential can be obtained from the current density.

$$\nabla \times \nabla \times \mathbf{A} = \mu_0 \mathbf{J} \tag{3.27}$$

There is a vector relation for this repeated vector operation. In particular, we write (see Appendix A)

$$\nabla(\nabla \cdot \mathbf{A}) - \nabla^2 \mathbf{A} = \mu_0 \mathbf{J} \tag{3.28}$$

A vector is determined by two vector operations, namely its curl and its divergence. The curl of the vector **A** is specified in (3.24). We now choose the divergence of the vector **A** to be equal to zero. There are other choices that could be made but this will simplify our calculation. This is called a "Coulomb gauge" in the physics community and it has ramifications that are beyond the scope of this text. We shall let our colleagues in that community dwell on these finer points. With this assumption, (3.28) simplifies to

$$\nabla^2 \mathbf{A} = -\mu_0 \mathbf{J} \tag{3.29}$$

This equation is similar to Poisson's equation (2.61) that related an electric potential to a charge density. There is, however, a very significant difference in that (3.29) is a vector equation. The vector potential $\mathbf{A}$ is in the same direction as the current density $\mathbf{J}$. This means that there is a separate Poisson's type scalar equation for each component of the vector potential. Frequently the current is flowing in only one direction, which means that there will be only one component of the vector potential. We will find that there may be certain advantages in using this intermediate calculation. This is shown in Figure 3–6.

In Cartesian coordinates, (3.29) is written as

$$\nabla^2 A_x = -\mu_0 J_x, \qquad \nabla^2 A_y = -\mu_0 J_y, \qquad \nabla^2 A_z = -\mu_0 J_z \tag{3.30}$$

We can make use of the knowledge that we have gained from electrostatic fields and write down the solution for each of the components in (3.30).

In analogy with the electric potential (2.62), we write the solution for the vector potential as

$$\mathbf{A}(\mathbf{r}) = \frac{\mu_0}{4\pi} \int_{\Delta v} \frac{\mathbf{J}(\mathbf{r}')}{R} dv' \tag{3.31}$$

where $R = \sqrt{(x-x')^2 + (y-y')^2 + (z-z')^2}$ is the distance between the current element and the point where the vector potential is to be evaluated. The integration is to be performed over the volume Δv that contains the current density. Equation (3.31) is a vector equation that represents three scalar equations for the three components of the current density. This is illustrated

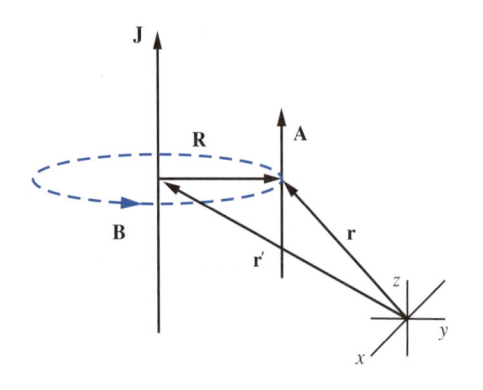

FIGURE 3–6

The orientation of the magnetic vector potential $\mathbf{A}(\mathbf{r})$ and magnetic flux density $\mathbf{B}(\mathbf{r})$ that surrounds a current element $\mathbf{J}(\mathbf{r}')$ is depicted.

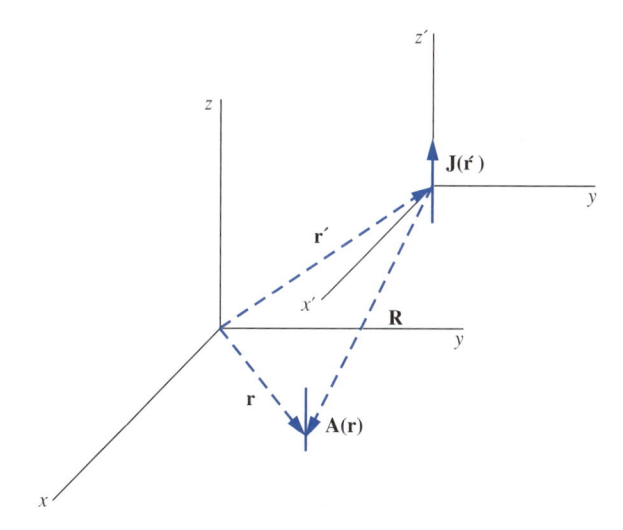

3.3 Magnetic Vector Potential and the Biot–Savart Law

FIGURE 3–7

Orientation of a current element in one location and the resulting vector potential in a different location is shown.

in Figure 3–7 where the vectors **r** and **r′** are directed to the vector potential and the current density respectively. The magnitude of the distance between these two vectors is given by $R = |\mathbf{r} - \mathbf{r}'|$.

EXAMPLE 3.9

Find the vector potential **A** and the magnetic flux density **B** caused by a length $2a$ of current $I\mathbf{dl}' = Idz'\mathbf{u}_z$ on a line that is perpendicular to the current carrying wire. Neglect any contributions caused by wires into the book.

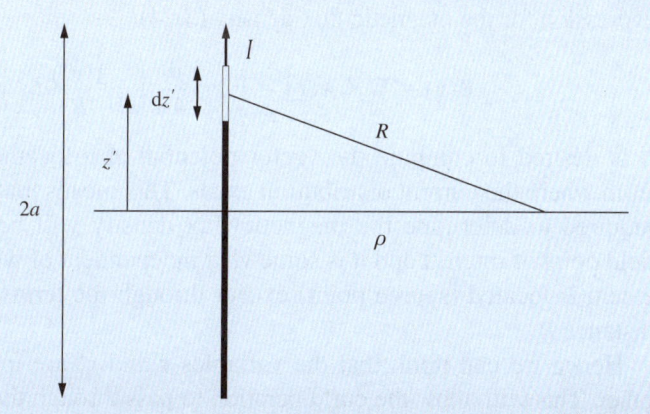

Answer. Since the magnetic flux density is to be determined at the midpoint of the line, we can invoke the argument of symmetry and do the calculation analytically. The vector potential at a distance R from the wire is found from (3.31). The volume integral becomes a line integral since we will assume that the current is distributed uniformly over the cross section of the wire. This states that

$$\mathbf{J}\,dv' = \mathbf{J}\,ds'dz' = I\,dz'\mathbf{u}_z$$

The integral (3.31) becomes

$$\mathbf{A} = \mathbf{u}_z \frac{\mu_0 I}{4\pi} \int_{-a}^{a} \frac{dz'}{\sqrt{z'^2 + \rho^2}} = \mathbf{u}_z \frac{\mu_0 I}{4\pi} \ln\left(\frac{\sqrt{a^2 + \rho^2} + a}{\sqrt{a^2 + \rho^2} - a}\right)$$

The magnetic flux density is computed from $\mathbf{B} = \nabla \times \mathbf{A}$. Note that the vector potential has only a $\mathbf{u}_z$ component that depends only on the variable ρ. From the definition of the curl operation in cylindrical coordinates (see Appendix A), the only nonzero contribution comes from the term $-\mathbf{u}_\phi(\partial A_z/\partial \rho)$. Therefore, the magnetic field due to a finite-length current-carrying wire is equal to

$$\mathbf{B} = \mathbf{u}_\phi \frac{\mu_0 I}{2\pi\rho}\left(\frac{a}{\sqrt{a^2 + \rho^2}}\right)$$

In the limit of an extremely long wire such that $a \gg \rho$, the term within the brackets approaches 1 and this results in

$$\mathbf{B} = \mathbf{u}_\phi \frac{\mu_0 I}{2\pi\rho}$$

This is the value that we previously obtained using Ampere's law (3.15).

Let us substitute the integral for the vector potential (3.31) into the expression for the magnetic flux density (3.24).

$$\mathbf{B}(\mathbf{r}) = \nabla \times \mathbf{A}(\mathbf{r}) = \nabla \times \left(\frac{\mu_0}{4\pi} \int_{\Delta v} \frac{\mathbf{J}(\mathbf{r}')}{R} dv'\right) \qquad (3.32)$$

It is desired to compute the vector potential at a location that is different from where the current distribution exists. This means that the curl operation required to determine the magnetic flux density will be performed at the field point of interest and it is somewhat independent of where the current element is located (source point) except through the terms that appear in the distance R.

Hence we can think that the variables $\mathbf{r}$ and $\mathbf{r}'$ are independent of each other. This will allow the curl operation to pass through the integral in (3.32), which is required when computing the magnetic field.

$$\mathbf{B}(\mathbf{r}) = \frac{\mu_0}{4\pi} \int_{\Delta v} \nabla \times \left(\frac{\mathbf{J}(\mathbf{r}')}{R}\right) dv' \qquad (3.33)$$

3.3 Magnetic Vector Potential and the Biot–Savart Law

This can be further simplified using the vector identity that relates the curl of a vector times a scalar quantity, both of which are spatially varying, to be (see Appendix A)

$$\nabla \times (a\mathbf{B}) = \nabla a \times \mathbf{B} + a \nabla \times \mathbf{B} \tag{3.34}$$

In (3.34), a is a scalar and $\mathbf{B}$ is a vector. Applying the vector identity (3.34) to (3.33), we identify these terms as

$$a = \frac{1}{R} \quad \text{and} \quad \mathbf{B} = \mathbf{J}(\mathbf{r}')$$

Since the current is defined with the variable $\mathbf{r}'$ and $\nabla \times \mathbf{J}(\mathbf{r}') = 0$ we obtain

$$\mathbf{B}(\mathbf{r}) = \frac{\mu_0}{4\pi} \int_{\Delta v} \nabla\left(\frac{1}{R}\right) \times \mathbf{J}(\mathbf{r}')dv' = -\frac{\mu_0}{4\pi} \int_{\Delta v} \frac{\mathbf{u_R}}{R^2} \times \mathbf{J}(\mathbf{r}')dv'$$

$$\mathbf{B}(\mathbf{r}) = \frac{\mu_0}{4\pi} \int_{\Delta v} \frac{\mathbf{J}(\mathbf{r}') \times \mathbf{u_R}}{R^2} dv' \tag{3.35}$$

The operation which is a derivative at the field point (unprimed variables) can be simplified with the relationship that

$$\nabla\left(\frac{1}{R}\right) = -\frac{\mathbf{u_R}}{R^2}$$

The vector product property that

$$\mathbf{A} \times \mathbf{B} = -\mathbf{B} \times \mathbf{A}$$

has also been employed in this derivation. The unit vector $\mathbf{u_R}$ is in the direction from the current element to the location where the magnetic field is to be computed.

If the current is localized to pass through a wire, it is possible to simplify the volume integral given in (3.35) to read

$$\boxed{\mathbf{B}(\mathbf{r}) = \frac{\mu_0}{4\pi} \oint \frac{I d\mathbf{l}' \times \mathbf{u_R}}{R^2}} \tag{3.36}$$

Note that a closed line integral has been used since the current in a wire has to pass through a closed loop, say, from one battery terminal through a wire and back into the battery through the other terminal. Equation (3.36) is called the **Biot–Savart law**.

EXAMPLE 3.10

Use the Biot–Savart law to find the magnetic flux density from a finite length of line with a current element $I\mathbf{dl'} = Idz'\mathbf{u_z}$ on a line that is perpendicular to the current carrying wire. Neglect any contributions caused by wires into the book.

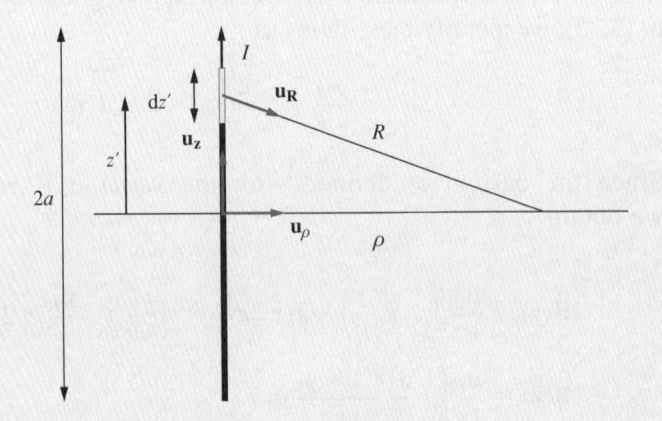

Answer. Before setting up the integral, let us first perform this vector product where each term can be clearly identified.

$$\mathbf{dl'} \times \mathbf{u_R} = dz'\mathbf{u_z} \times \left(\frac{\rho \mathbf{u_\rho} - z'\mathbf{u_z}}{\sqrt{\rho^2 + (z')^2}} \right) = \frac{\rho\, dz'}{\sqrt{\rho^2 + (z')^2}} \mathbf{u_\phi}$$

Note the "−" sign in one of the terms of the unit vector. We have to be careful that we follow the path from the current element to the point of observation. Hence the magnetic flux density can be calculated using the Biot–Savart law from (3.36) from which we write

$$\mathbf{B}(\rho) = \frac{\mu_0 I}{4\pi} \int_{-a}^{a} \frac{\rho\, dz'}{(\rho^2 + (z')^2)^{3/2}} \mathbf{u_\phi}$$

This integral can be performed analytically with the substitution $z' = \rho \tan\theta$ to finally yield

$$\mathbf{B}(\rho) = \mathbf{u_\phi} \frac{\mu_0 I}{4\pi} \frac{z'}{\rho\sqrt{\rho^2 + (z')^2}} \bigg|_{-a}^{a} = \mathbf{u_\phi} \frac{\mu_0 I}{2\pi\rho} \frac{a}{\sqrt{\rho^2 + a^2}}$$

As expected, this is the same result that was obtained in Example 3–9, although the integral here is more complicated.

EXAMPLE 3.11

Find the magnetic field on the axis that is perpendicular to the plane containing a circular loop of current. Use the Biot–Savart law.

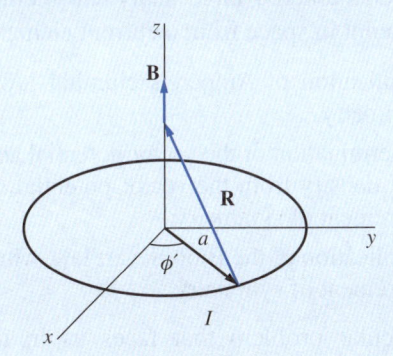

Answer. We must first identify the terms that appear in the Biot–Savart law in equation (3.36). We write $d\mathbf{l}' = a\,d\phi'\,\mathbf{u}_\phi$, $\mathbf{R} = -a\mathbf{u}_\rho + z\mathbf{u}_z$, and $R = \sqrt{a^2 + z^2}$. Therefore, we write

$$\mathbf{B}(z) = \frac{\mu_0 I}{4\pi} \oint \frac{(a\,d\phi'\,\mathbf{u}_\phi) \times (-a\mathbf{u}_\rho + z\mathbf{u}_z)}{(a^2 + z^2)^{3/2}} = \frac{\mu_0 I}{4\pi} \oint \frac{(a^2 d\phi'\,\mathbf{u}_z + az\,d\phi'\,\mathbf{u}_\rho)}{(a^2 + z^2)^{3/2}}$$

Due to symmetry, the term with the unit vector $\mathbf{u}_\rho$ will contribute zero to the magnetic field. The integration is performed along the path of the wire and yields a factor of 2

$$\mathbf{B}(z) = \frac{\mu_0}{2\pi} \frac{I\pi a^2}{(a^2 + z^2)^{3/2}} \mathbf{u}_z = \frac{\mu_0}{2\pi} \frac{\mathbf{m}}{R^3}$$

This result has been written to incorporate the area enclosed within the current loop. We will define the ***magnetic dipole moment*** as $\mathbf{m} = I\pi a^2 \mathbf{u}_z$. The magnitude of the magnetic dipole moment equals the current I, carried by the wire that forms the circumference of the loop, times the area enclosed within the loop πa^2. The unit vector is normal to the surface area of the loop using the right-hand convention.

This current loop in Example 3–11 is in agreement with the simple model of an atom that considers an atom to have a positive nucleus and an electron that revolves about the nucleus at a fixed radius a. This is usually called a ***magnetic dipole***. This simple model of an atom was previously used in Example 2–11 and led us to consider the atom as an electric dipole. This leads to a certain analogy between the electric field intensity $\mathbf{E}$ and the magnetic flux density $\mathbf{B}$ in that the electric dipole moment $\mathbf{p}$ is similar to the magnetic dipole moment $\mathbf{m}$.

Both involve a volume integration of either an electric charge density or an electric current density. In addition, both fields can be obtained from a vector differentiation of a potential, either a scalar potential for the electric field intensity or a vector potential for the magnetic flux density.

We have encountered three analytical methods to find the magnetic flux density at a point in space from a current element:

1. Application of Ampere's circuital law, which requires considerable symmetry.
2. Determination of the vector potential and the calculation of a magnetic flux density from the vector potential, which does not impose the requirement of symmetry.
3. Application of the Biot–Savart law, which also does not impose the requirement of symmetry.

The particular problem that faces us in the future will dictate which approach we should follow. Numerical methods are employed frequently to obtain the magnetic flux density in complicated geometries such as may be found in an electric motor or an electric generator. In fact, there are commercial products that have been developed to perform these calculations since they are so widely used.

3.4 Magnetic Forces

The first statement that we made regarding the behavior of stationary charged particles concerned the Coulomb force that existed between the particles. The force was created upon a stationary particle that had a charge q if the particle were in an electric field **E**. This force is given by

$$\mathbf{F}_{electric} = q\mathbf{E} \tag{3.37}$$

If the particle were in motion with a constant velocity **v** within a region that only contained an electric field, the particle would still experience the force that is given in (3.37).

However, if the particle is in motion with a velocity **v** in a region that contains only a magnetic field whose density is **B**, the force that acts upon the particle is given by

$$\mathbf{F}_{magnetic} = q(\mathbf{v} \times \mathbf{B}) \tag{3.38}$$

The resulting magnetic force **F** is perpendicular to both the magnetic flux density **B** and to the velocity **v** of the particle, and this is expressed with the vector product. In Figure 3–8, this magnetic force on a positively charged

3.4 Magnetic Forces

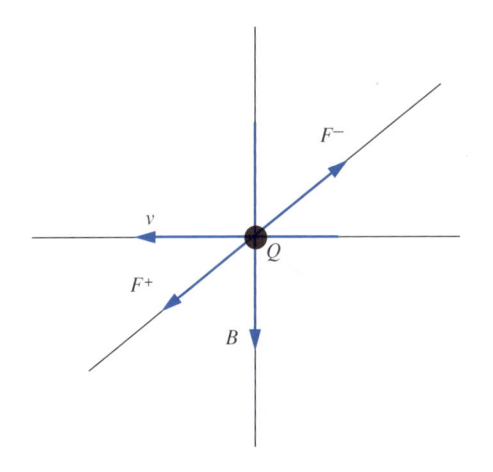

FIGURE 3–8

A charged particle entering a region containing a uniform magnetic field is deflected according to (3.38). The right-hand rule determines the direction of the force on the charge. This direction will be in the opposite direction depending upon the sign of the charge.

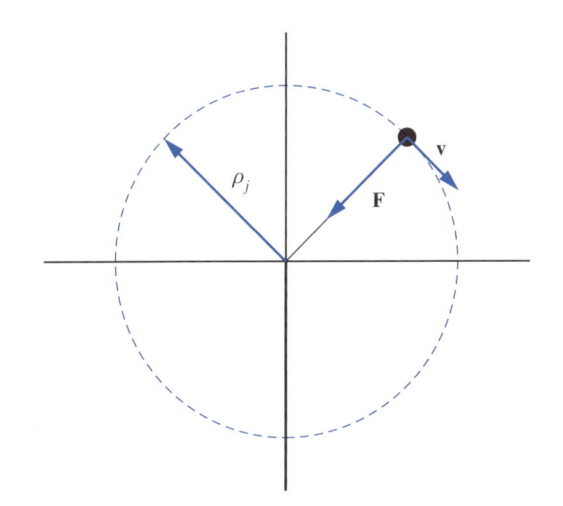

FIGURE 3–9

A charged particle moving with a constant velocity in a uniform magnetic field experiences a magnetic force that causes the particle to follow a circular trajectory. This figure would correspond to either a positively charged particle with the magnetic field coming out of the page or a negatively charged particle with the magnetic field going into the page.

particle and a negatively charged particle in a region of uniform magnetic field is illustrated. Since the sign of the charge of these two particles is different, the resulting forces will be in opposite directions. If the charged particle moves with a uniform velocity **v** through a uniform electric and magnetic field, the force is given by

$$\mathbf{F} = q(\mathbf{E} + \mathbf{v} \times \mathbf{B}) \text{ (N)} \tag{3.39}$$

This force, which is the sum of the electrostatic and the magnetostatic forces, is given the name **Lorentz force**.

In a region where the electric field is equal to zero, the charged particles will continue to experience the magnetic force given in (3.38). The resulting motion of the particles will be in a circular orbit as shown in Figure 3–9.

It is possible to find the radius of curvature of the motion for the charged particles as will be demonstrated in the following discussion. The particle that has a mass M_j will experience a centripetal force whose magnitude is given by

$$F_j = M_j a_j = M_j \frac{v_j^2}{\rho_j} \qquad (3.40)$$

where ρ_j is the radius of curvature and $a_j = v_j^2/\rho_j$ is the acceleration of the jth particle. The subscript j refers to the particular particle; $j =$ "$-$" for the negatively charged particles and $j =$ "$+$" for the positively charged particles. The positively charged particles could be singly charged or multiply charged positive ions. The negatively charged particles could either be electrons or ions to which one or more electrons have become attached, thus creating negative ions. In either case, the radius of curvature of the trajectory depends on the mass of the particle. The magnitude of the force caused by the magnetic field is given from (3.38) as

$$F_{\text{magnetic}} = q v_j B \qquad (3.41)$$

Equating the two forces given in (3.40) and (3.41) and solving for the radius of curvature ρ_j for the particle with the subscript j, we obtain

$$\rho_j = \frac{M_j v_j}{qB} \qquad (3.42)$$

This radius is called the **Larmor radius** or **gyroradius** of the charged particle. We note that the Larmor radius for the electrons moving with the same velocity through the same magnetic field as the ions will be significantly less than that for the ions due to the mass difference of

$$\frac{m_e}{M_j} = \left(\frac{1}{1836}\right) \times \left(\frac{1}{\text{atomic mass}}\right) \qquad (3.43)$$

The mass dependence of the Larmor radius suggests that it can be used as a diagnostic tool in order to determine the mass of an unknown material. After ionizing the unknown material and passing all of the ions through a uniform magnetic field with the same velocity **v**, the ions can be collected in a juxtaposed series of collectors. The location of each collector is determined by the Larmor radius of the different elements, and the presence or absence of ions in each collector can be monitored. We can include effects of an ion having more than a single charge also. A device that performs this monitoring is called a **mass spectrometer**. This has also been used to separate various isotopes from each other.

3.4 Magnetic Forces

EXAMPLE 3.12

Calculate the Larmor radius for an electron and an argon ion that pass through a magnetic field of 0.01 T. Both particles have been accelerated through a potential difference of one volt.

Answer. Before calculating the Larmor radius for either particle, the velocity of each particle must be computed. Since the particles have gained an energy of $q_e V = 1$ eV $= 1.602 \times 10^{-19}$ J, this energy will appear as kinetic energy, and we write

$$q_e \Delta V = \frac{m_e v_e^2}{2} = \frac{M_{Ar} v_{Ar}^2}{2}$$

$$v_e = \sqrt{\frac{2 q_e \Delta V}{m_e}} \quad \text{and} \quad v_{Ar} = \sqrt{\frac{m_e}{M_{Ar}}} \sqrt{\frac{2 q_e \Delta V}{m_e}}$$

The atomic mass of argon is 40, yielding a ratio of the masses to be

$$\frac{m_e}{M_{Ar}} = \left(\frac{1}{1836}\right) \times \left(\frac{1}{40}\right)$$

The Larmor radius for the electron is found from (3.42) to be

$$\rho_e = \frac{m_e v_e}{q_e B} = \frac{m_e \sqrt{\frac{2 q_e \Delta V}{m_e}}}{q_e B}$$

$$= \frac{(9.1 \times 10^{-31}) \sqrt{\frac{2(1.602 \times 10^{-19})(1)}{(9.1 \times 10^{-31})}}}{(1.602 \times 10^{-19})(10^{-2})} = 3.4 \times 10^{-4} \text{ m}$$

The velocity of the argon ion can be expressed in terms of the electron velocity.

$$v_{Ar} = \sqrt{\frac{m_e}{M_{Ar}}} \sqrt{\frac{2 q_e \Delta V}{m_e}} = \sqrt{\frac{m_e}{M_{Ar}}} v_e$$

Therefore, the Larmor radius for the argon ion is found from (3.42), which can also be expressed in terms of the electron Larmor radius.

$$\rho_{Ar} = \frac{M_{Ar} v_{Ar}}{q_e B} = \frac{M_{Ar} \sqrt{\frac{2 q_e \Delta V}{M_{Ar}}}}{q_e B} = \left(\frac{M_{Ar}}{m_e}\right) \left(\frac{m_e \sqrt{\frac{2 q_e \Delta V}{m_e}}}{q_e B}\right) \left(\sqrt{\frac{m_e}{M_{Ar}}}\right)$$

$$= \sqrt{\frac{M_{Ar}}{m_e}} \rho_e = \sqrt{40 \times 1836} \, \rho_e = 9.2 \times 10^{-2} \text{ m}$$

A comparison of the two Larmor radii indicates that the electrons are closely "tied" to the magnetic field lines and the ions are not. In several applications, the electrons are said to be "magnetized" and the ions are considered to be "unmagnetized."

FIGURE 3–10

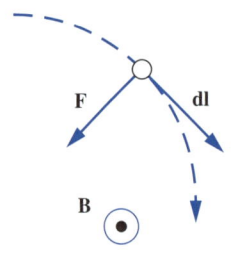

A positive charge moving a distance **dl** with a constant velocity **v** in a region containing a uniform magnetic flux density **B**.

It is interesting at this time to calculate the work that is performed by the charged particle as it passes through the region of magnetic field. Recall that this work ΔW is computed from the line integral

$$\Delta W = \int_a^b \mathbf{F} \cdot \mathbf{dl} \tag{3.44}$$

As shown in Figure 3–10 and from equation (3.38), we find that the force is perpendicular to the direction that it travels. This implies that the work that is computed from (3.44) will be zero.

EXAMPLE 3.13

Show that the incremental work ΔW performed in moving a positive charge Q with a velocity $\mathbf{v} = v_0 \mathbf{u_x}$ an incremental distance $\Delta x \mathbf{u_x}$ through a uniform magnetic field $\mathbf{B} = B_0 \mathbf{u_y}$ is equal to zero.

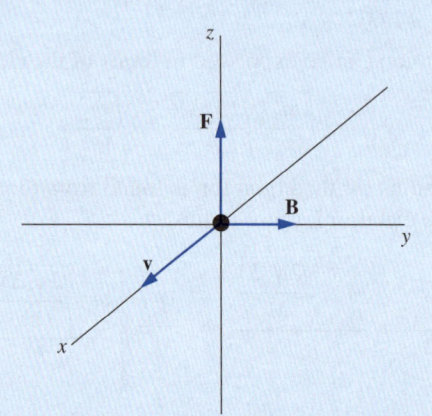

Answer. From (3.44) and (3.38), we write

$$\Delta W = \mathbf{F} \cdot \Delta x \mathbf{u_x} = Q(\mathbf{v} \times \mathbf{B}) \cdot \Delta x \mathbf{u_x}$$
$$= Q(v_0 \mathbf{u_x} \times B_0 \mathbf{u_y}) \cdot \Delta x \mathbf{u_x} = Q v_0 B_0 [(\mathbf{u_x} \times \mathbf{u_y}) \cdot \mathbf{u_x}] = 0$$

The term within the square brackets is identically equal to zero.

3.4 Magnetic Forces

FIGURE 3–11

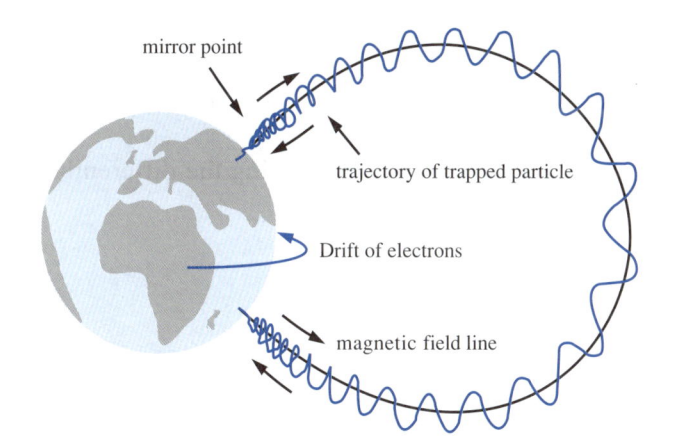

Schematic representation of the radiation belt. The first experimental detection of these charged particles was made on the satellite *Explorer 1* in 1958.

Earth and several of the other planets are examples that illustrate the effects of this force field upon charged-particle motion. See Figure 3–11. Particles are created by the collision of high-energy cosmic rays with low energy particles near the Earth as well as by complex acceleration processes due to the interaction of the solar wind (a stream of ionized particles flowing from the sun) and the Earth's magnetic field. These charged particles are trapped in the Earth's magnetic field. This entrapped region is called a radiation belt. As determined from the passage of the *Voyager* spacecraft on its more-than-twelve-year journey from Earth into the far reaches of the solar system, several planets[1] have magnetic fields that capture these charged particles coming from the sun. Since we might expect that there are almost an equal number of electrons and positively charged ions in this region, there is electrical neutrality, and this charged particle fluid is called a ***plasma***. The Earth's radiation belt is called the Van Allen belt in honor of Professor James Van Allen who originally discovered its existence using the satellite *Explorer 1* in 1958.

At the start of this section, we wrote the expression for the force on a charged particle that passed through a uniform magnetic flux density **B** in (3.38). A differential charge $dQ = \rho_v dv$ moving with a constant velocity constitutes a current. If this current flows in a closed path, (3.38) can be

---

[1] *Voyager* confirmed the presence of radiation belts at the planets Jupiter, Saturn, Uranus and Neptune. Within the sensitivity limits of the instruments, no radiation belts were detected at Venus or at Mars. This is indicative of the presence or absence of a magnetic field at these planets. Magnetic fields at Pluto, downgraded to dwarf planet status in 2006, are unknown. The satellite approached the edge of the solar system in 2003.

written as

$$d\mathbf{F}_{magnetic} = dQ(\mathbf{v} \times \mathbf{B}) = \rho_v(\mathbf{v} \times \mathbf{B})dv = \mathbf{J} \times \mathbf{B} ds dl = I \mathbf{dl} \times \mathbf{B} \quad (3.45)$$

The total force $\mathbf{F}_{magnetic}$ is computed by integrating the differential force over the path

$$\mathbf{F}_{magnetic} = -\int \mathbf{B} \times I\mathbf{dl} \quad (3.46)$$

where the "−" arises from the inversion of the vector product.

If we assume that the magnetic flux density is a constant, it can be taken outside of the integral sign. This leads to the closed line integral $\oint \mathbf{dl}$, which is equal to zero. This states that a closed loop will not move in a linear direction. If the magnetic field is not uniform in space, then the net force will not necessarily be equal to zero. Although the net translational force in a uniform magnetic field is equal to zero, there may be a torque that acts on the loop and causes it to rotate about an axis.

Before examining the torque that will exist on the loop, let us first examine the force that exists between two parallel wires, each of which carries a current as shown in Figure 3–12. We are going to calculate the force that exists between these two wires. Before presenting the formal derivation, let us postulate certain properties of the force that may exist on the two wires. Let the two wires lie in the x–z plane. In the first case, the currents are going in the same direction as shown in Figure 3–12a. The magnetic field created by wire 1 will be directed in the $+\mathbf{u}_y$ direction at the location of wire 2. The force on wire 2 as computed from (3.46) will be in the direction given by $\mathbf{F}_2 = -B_1\mathbf{u}_y \times I_2 dl\mathbf{u}_z$ or in the $-\mathbf{u}_x$ direction. This states that wire 2 will be *attracted* to wire 1. Similarly, the force on wire 1 caused by the magnetic field created by wire 2 will cause wire 1 to be attracted to wire 2. If the currents are going in opposite directions as depicted in Figure 3–12b, the magnetic force given in (3.46) will be in the direction that will cause the wires to *repel* each other.

These forces can also be argued from the following point of view. In the region between the two wires in Figure 3–12a, the magnetic fields caused by the two wires oppose each other and will therefore cancel. The magnetic fields will add in the regions external to this separation region. Hence we could think that there is a "pressure" on the wires to fill in this region since there is an old world axiom that "nature abhors a vacuum." The same argument could also be

3.4 Magnetic Forces

FIGURE 3–12

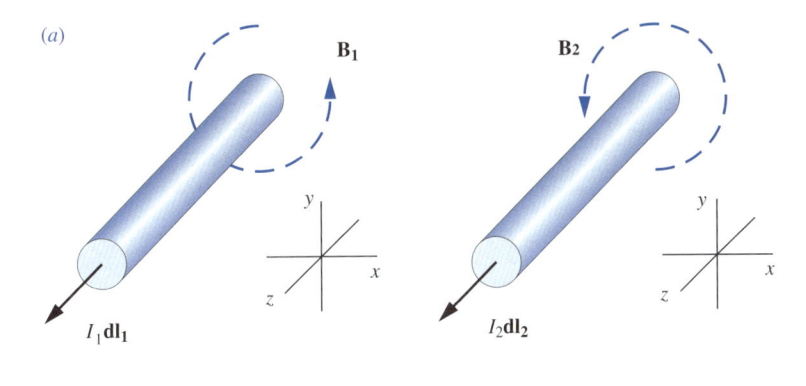

Currents are flowing through two parallel wires with lengths dl_1 and dl_2.
(*a*) The current in both wires is flowing in the same direction (attraction).
(*b*) The current in both wires is flowing in the opposite direction (repulsion).

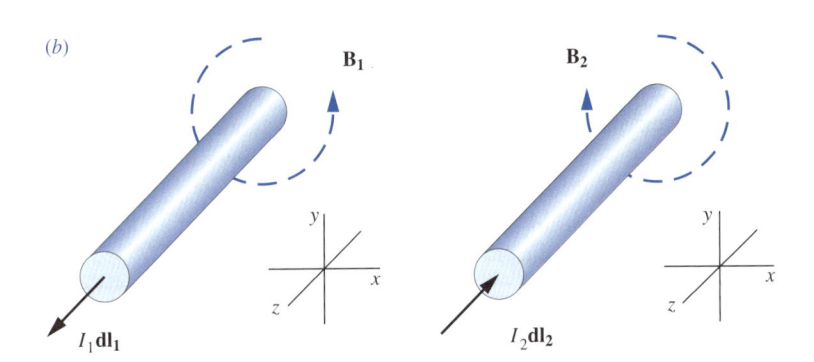

applied in describing the force for the situation depicted in Figure 3–12*b*, where the cancellation of the magnetic fields occurs in the region external to the two wires.

EXAMPLE 3.14

A lightning rod is a device that provides an attractive path for lightning to discharge through a wire so that a building is hopefully protected. In order to attract the lightning to the rod, the path must originally have a low resistance. However, after the lightning stroke starts to discharge through this rod, the resistance should suddenly increase in order to protect the rod so it can survive to another day. Discuss how this could be done with inexpensive parts.

(a) (b)

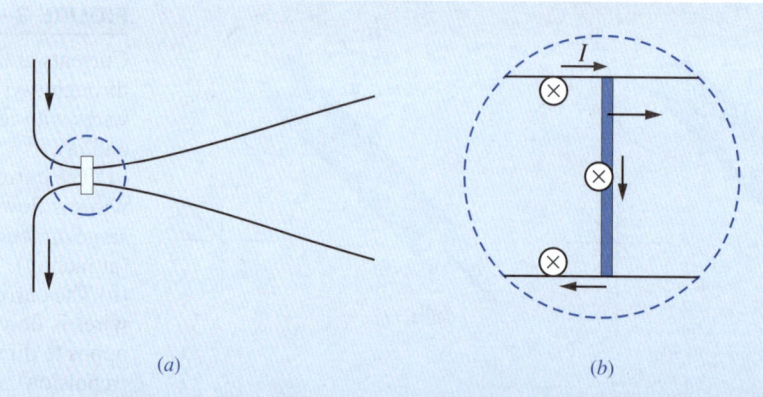

Answer. The lightning stroke will discharge through the shortest path between the two metal conductors as shown in (*a*). This discharge will act like a variable length conductor and will create a local magnetic field between the two metal conductors as depicted in (*b*). The direction of the resulting force (*I* **dl** × **B**) from (3.45) will cause this arc to move to the right where the path length becomes longer. Since the path length increases with motion of the arc to the right, the resistance will also increase. This device is called a *lightning arrestor*. The mechanism described here is also used in a device called a *rail gun* that may have applications of spewing forth "plasma bullets." Movie aficionados should also recognize the Jacob's ladders that appear in the laboratory of the old Dr. Frankenstein movies.

Now let us calculate the force between the two parallel current-carrying wires depicted in Figure 3–12 more formally. The magnetic field that should be incorporated into the force equation (3.46) is determined from the Biot–Savart law in equation (3.36). We write for the force on wire 1 caused by the magnetic field created by the current in wire 2 using the notation

$$\mathbf{F}_{12} = -I_1 \oint_{L_1} \mathbf{B}_{12} \times \mathbf{dl}_1 \tag{3.47}$$

From the Biot–Savart law, we find that the magnetic flux density at wire 1 caused by the current in wire 2 is given by

$$\mathbf{B}_{12} = -\frac{\mu_0 I_2}{4\pi} \oint_{L_2} \frac{\mathbf{u}_{R_{21}} \times \mathbf{dl}_2}{R_{21}^2} \tag{3.48}$$

Substitute (3.48) into (3.47) and derive this force, which is called *Ampere's force*, as

$$\boxed{\mathbf{F}_{12} = \frac{\mu_0 I_1 I_2}{4\pi} \oint_{L_1} \oint_{L_2} \frac{(\mathbf{u}_{R_{21}} \times \mathbf{dl}_2) \times \mathbf{dl}_1}{R_{21}^2}} \quad (\text{N}) \tag{3.49}$$

3.4 Magnetic Forces

FIGURE 3–13

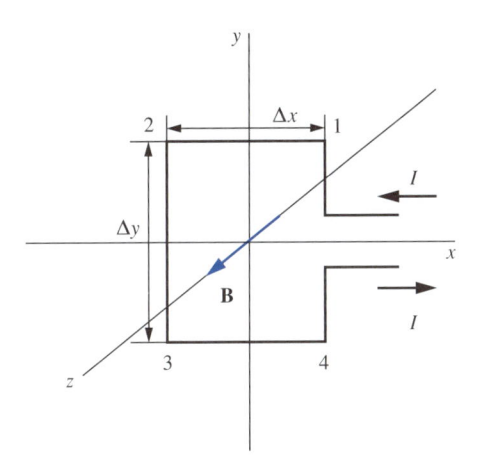

A rectangular current loop is inserted in a uniform magnetic field. The current in the loop flows in the counterclockwise direction.

The force on wire 2 can be computed by merely interchanging the subscripts 1 and 2.

In writing (3.49), we might be tempted to compare the force given by this equation with the Coulomb force equation given in Chapter 2. There are obvious similarities in that the force is proportional to the magnitudes $I_1 \mathbf{dl}_1$ and $I_2 \mathbf{dl}_2$ or Q_1 and Q_2 for Coulomb's law. Both equations are inversely proportional to the square of the separation distance. This is also similar to the gravitational force between two objects in that it is proportional to the masses on the two objects divided by the separation distance between them. Closed line integrals are used in (3.49) since one cannot construct isolated current elements experimentally. For the two infinite parallel wires depicted in Figure 3–12, they are closed at the place we call infinity.

Consider a current-carrying loop of wire as shown in Figure 3–13. We will assume, for simplicity, that $\mathbf{B} = B_0 \mathbf{u}_z$. The separation distance between the two wires that are closest to the x axis can be assumed to be infinitesimally small. We can consider that two parallel wires that each carry a current in the opposite direction. In addition, the other two wires also carry a current in the opposite directions. In Figure 3–12, it was shown that two parallel wires that carry currents in the opposite directions will have a repulsive force. Therefore, the net force on this closed loop will be equal to the sum of all these forces, which is equal to zero. This implies that there will be no net translation of this closed loop in any direction.

EXAMPLE 3.15

Formally demonstrate that the sum of the forces acting on the rectangular loop shown in Figure 3–12 is identically equal to zero, which implies that this loop will not translate in any direction.

Answer. The magnetic flux density is $\mathbf{B} = B_0 \mathbf{u_z}$. Using the definition of the magnetic force given in (3.46), we write the sum of the forces that act on the four sides as

$$\mathbf{F}_{magnetic} = -\int_1^2 \mathbf{B} \times I d\mathbf{l} - \int_2^3 \mathbf{B} \times I d\mathbf{l} - \int_3^4 \mathbf{B} \times I d\mathbf{l} - \int_4^1 \mathbf{B} \times I d\mathbf{l}$$

$$= -\int_{+\Delta x/2}^{-\Delta x/2} I B_0 \mathbf{u_z} \times dx \mathbf{u_x} - \int_{+\Delta y/2}^{-\Delta y/2} I B_0 \mathbf{u_z} \times dy \mathbf{u_y}$$

$$- \int_{-\Delta x/2}^{+\Delta x/2} I B_0 \mathbf{u_z} \times dx \mathbf{u_x} - \int_{-\Delta y/2}^{+\Delta y/2} I B_0 \mathbf{u_z} \times dy \mathbf{u_y} = 0$$

Recall that the sign of the integral is determined by the limits of the integration.

Although the summation of the forces is equal to zero, there will be a torque on the loop that will cause it to rotate. The forces on wires 1 and 3 are in opposite directions as shown in Figure 3–14. If the normal to the cross section of the loop is at a slight angle θ with respect to the applied magnetic field, this torque acting upon the loop can be calculated. In order to calculate the torque, we assume that loop is constrained to rotate about one axis only for simplicity. The torque on the loop is given by

$$\text{Torque} = F_1 \sin\theta \left(\frac{\Delta y}{2}\right) + F_3 \sin\theta \left(\frac{\Delta y}{2}\right) \quad (3.50)$$

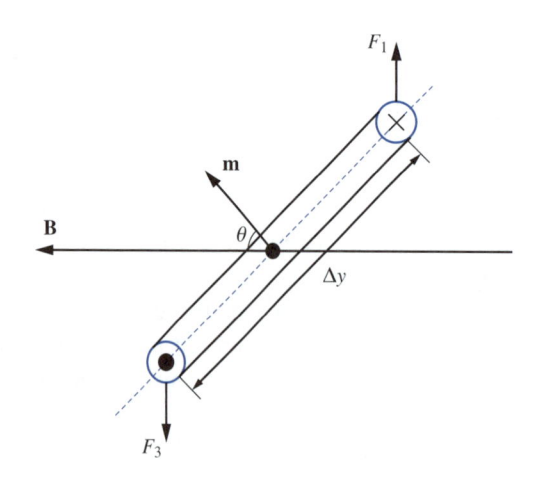

FIGURE 3–14

Two of the edges of the rectangular loop rotate across the magnetic field, while the other two are in the plane of the magnetic field and are unaffected by it.

3.5 Magnetic Materials

FIGURE 3–15

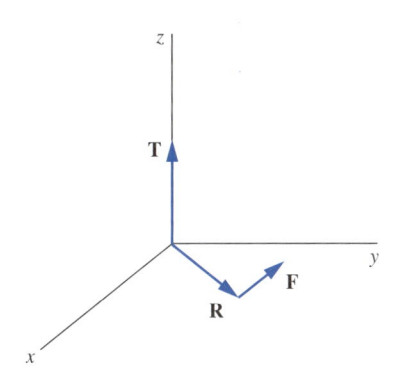

Definition of the torque **T** in terms of the force **F** and the lever arm **R**.

where

$$F_1 = IB_0 \Delta x \quad \text{and} \quad F_3 = IB_0 \Delta x$$

This leads to

$$\text{Torque} = IB_0 (\Delta x \Delta y) \sin \theta \qquad (3.51)$$

The area of the loop is equal to $\Delta \mathbf{s} = (\Delta x \Delta y)\mathbf{u_n}$ where the unit vector is normal to the surface area. If we multiply this area by the current I, we can recognize this is a representation for the magnetic dipole moment that was described in Example 3–11. Finally, we are able to write (3.51) in vector notation as

$$\mathbf{T} = \mathbf{m} \times \mathbf{B} \qquad (3.52)$$

where $\mathbf{T} = \mathbf{R} \times \mathbf{F}(Nm)$ is the torque causing the rectangular loop to rotate about its axis as shown in Figure 3–15. We have made liberal use of the definition of the magnetic dipole moment in that we have replaced a circular loop with a rectangular loop. This will have a dramatic effect when we consider the magnetic properties of materials.

3.5 Magnetic Materials

Having determined the magnetic field from a current carrying loop that could, in some sense, approximate an atom, we will now investigate the characteristics of a material made of a very large number of atoms and their corresponding magnetic dipoles. These dipoles will be assumed to be oriented randomly at the start of this discussion as shown in Figure 3–16. In addition to the magnetic dipole moments created by the electron orbiting about the positive nucleus, there is also a magnetic field created by the electron spinning about its own axis. This is a topic that we do not have to understand here.

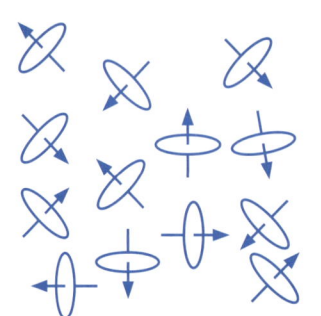

FIGURE 3–16

Random orientation of magnetic dipoles in a material.

The question that will now be answered is "What will happen to these magnetic fields from individual atoms if an external magnetic field is applied to the material?" The answer depends on the type of material that is being considered. There are three classes of materials that should be considered, and the classification is based on the reorientation of the magnetic dipoles under the influence of an external magnetic field.

In the first class of materials, the magnetic dipoles get reoriented such that their magnetic dipole moments **m** are in slight opposition to the applied magnetic field **B**. Under the influence of no external magnetic field, the atom's magnetic moment that is created by the electron rotating about the positive nucleus cancels the magnetic field created by the spin of the electron. The application of the external magnetic field perturbs the velocities of the orbiting electrons. Hence a small magnetic moment for the atom is created that, according to *Lenz's law*, will oppose the applied magnetic field. Lenz's law will be discussed in detail later. A dimensionless parameter that measures this reorientation is called the *magnetic susceptibility* and its symbol is χ_m. For this class of materials, it is usually very small and it is negative. A typical value for χ_m is of the order of $\chi_m \approx -10^{-5}$. This class of materials is called a *diamagnetic material*. Examples of diamagnetic materials include bismuth, copper, diamond, germanium, gold, lead, mercury, silicon, silver, and several inert gases. These materials exhibit no permanent magnetic field.

In the second class of materials, the magnetic moments created by the orbiting electrons and the spinning electrons do not cancel completely, leaving the atom with a small net magnetic moment. The application of an external magnetic field tends to align these magnetic moments in the direction of the applied magnetic field. This effect is also very small and nonpermanent. In this case, the magnetic susceptibility is small and positive (of the order of $\chi_m \approx +10^{-5}$). These materials are called *paramagnetic materials* and examples of these materials include aluminum, magnesium, oxygen, titanium, tungsten, and yttrium oxide.

3.5 Magnetic Materials

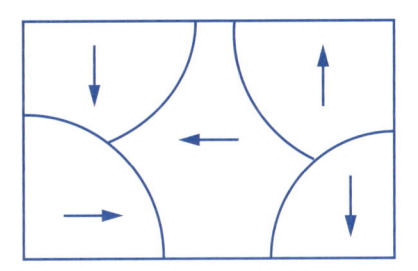

FIGURE 3–17

Domain structure of a ferromagnetic specimen. The magnetic moments of all of the atoms in each domain are pointed in the same direction.

The third class of magnetic materials is called a *ferromagnetic material* and it requires a different explanation. The materials that are in this class—iron, nickel, and cobalt—have $\chi_m \approx 250$ for nickel to $\chi_m \approx 4{,}000$ for pure iron. There are special alloys having χ_m values up to 100,000. The explanation that has experimental confirmation is that a ferromagnetic material consists of *domains* whose dimensions range from a few microns to 1 mm and that contain approximately 10^{15} to 10^{16} atoms. Each domain has all of the magnetic moments within it aligned in the same direction in the absence of an applied magnetic field, as shown in Figure 3–17. Separating one domain from the adjacent domain is a transition of approximately 100-atom thickness that is called a *domain wall*. The net magnetization of the randomly oriented domains is zero.

Under the influence of an external magnetic field, the domains that have their magnetic fields aligned with the applied field *grow* at the expense of the other domains. If the applied field is small, this process will reverse itself when the applied field is removed. However, if the applied field is strong enough, the domains will rotate in the direction of the applied field and the collective direction of the domains will remain fixed. This domain orientation will remain until a field of opposite orientation is applied. Thus, application of a high enough field results in a memory effect. This memory is useful in computer applications where the storage of information in various electronic components is important.

Before illustrating this process, we have to introduce the concept of magnetization of a material. The domains in the ferromagnetic material or the atoms in a diamagnetic or paramagnetic material each possess a magnetic moment that we will label as $\mathbf{m}_j$. It turns out that a ferromagnetic material will become a paramagnetic material if the temperature increases above some value that is called the Curie temperature. We will assume we are below this temperature in the following discussion. The total *magnetization* **M** that is the magnetic dipole moment per unit volume of a material in a volume Δv is defined as

$$\mathbf{M} \equiv \lim_{\Delta v \to 0} \frac{1}{\Delta v} \sum_{j=1}^{j=n} \mathbf{m}_j \qquad (3.53)$$

Its units are A/m. This magnetization **M** creates a current I_m that is bound to the domain of the atom. We can write an Ampere's circuital law for the domain or the atom as

$$I_m = \oint \mathbf{M} \cdot \mathbf{dl} = \int_{\Delta s} \mathbf{J}_m \cdot \mathbf{ds} \qquad (3.54)$$

In this integral, we have set the closed line integral of the magnetization **M** equal to the magnetization current I_m that is enclosed within this loop. This is not the real current I that you would draw from a battery. We, however, can analyze its effects. From Stokes's theorem, we write

$$\int_{\Delta s} \nabla \times \mathbf{M} \cdot \mathbf{ds} = \int_{\Delta s} \mathbf{J}_m \cdot \mathbf{ds} \qquad (3.55)$$

Therefore, the magnetization can be related to this magnetization current. Since the two integrands must be equal for (3.55) to be valid over any arbitrary surface Δs, we obtain

$$\nabla \times \mathbf{M} = \mathbf{J}_m \qquad (3.56)$$

Let us add the magnetization current density to the current density that was used in the differential form of Ampere's law (3.13) and write

$$\frac{1}{\mu_0} \nabla \times \mathbf{B} = \mathbf{J} + \mathbf{J}_m = \mathbf{J} + \nabla \times \mathbf{M} \qquad (3.57)$$

This can be rewritten as

$$\nabla \times \left(\frac{\mathbf{B}}{\mu_0} - \mathbf{M} \right) = \mathbf{J} \qquad (3.58)$$

We now define a new fundamental quantity, the *magnetic field intensity* **H**.

$$\mathbf{H} = \frac{\mathbf{B}}{\mu_0} - \mathbf{M} \text{ (A/m)} \qquad (3.59)$$

Ampere's circuital law can then be written as

$$\boxed{\oint \mathbf{H} \cdot \mathbf{dl} = I_{enc}} \qquad (3.60)$$

This states that a magnetic field intensity or *H field* can be created with a real applied current, and it is independent of whether a material is within the vicinity. Recall that this property is similar to the displacement flux density being independent of the dielectric. In a vacuum, the magnetization **M** is equal to zero, and the magnetic flux density is directly proportional to the magnetic field intensity. In magnetic materials, the magnetization is related

3.5 Magnetic Materials

to **H**, and we can let **M** = χ_m**H** where χ_m is the ***magnetic susceptibility***. Equation (3.59) can be written as

$$\mathbf{B} = \mu_0(1 + \chi_m)\mathbf{H} = \mu_0\mu_r\mathbf{H} = \mu\mathbf{H} \tag{3.61}$$

where μ_r is the ***relative permeability*** of the material. Except for the ferromagnetic materials iron, nickel, and cobalt, it is justified to assume that the relative permeability is equal to one. The values for several materials are included in Appendix B.

EXAMPLE 3.16

A magnetic flux density of $B = 0.05$ T appears in a magnetic material with $\mu_r = 50$. Find the magnetic susceptibility χ_m and the magnetic field intensity **H**.

Answer. The magnetic susceptibility χ_m is given by

$$\chi_m = \mu_r - 1 = 50 - 1 = 49$$

The magnetic field intensity **H** is computed from

$$H = \frac{B}{\mu_r\mu_0} = \frac{0.05}{50 \times 4\pi \times 10^{-7}} = 796 \text{ A/m}$$

In addition to having a large numerical value, the ferromagnetic materials also have another interesting behavior in that the dependence of the magnetic flux density on the magnetic field intensity is nonlinear. In addition, it remembers its previous value and the direction in which it reached that value. This can best be described from an examination of Figure 3–18a. In this figure, a sketch of the magnetic flux density **B** that could be obtained from an experimental measurement is plotted as a function of the magnetic field intensity **H**. The latter quantity starts at a value of **H** = 0 and increases as the magnetic flux density is measured. Initially, there is a linear increase in the magnetic flux density.

Let us start at the origin and slowly increase the magnetic field intensity **H**. This magnetic field intensity could be computed from (3.61) or measured in an experiment by uniformly wrapping a wire around a ferromagnetic material that is connected to a variable current supply. At small values of **H**, the magnetic flux density **B** will increase proportionally with **H**. As the value of **H** is further increased, almost complete domain rotation and domain wall motion will have occurred and a state of ***saturation*** will be achieved. If we now reduce the **H** field and eventually change its direction, the corresponding value of **B** does not follow the initial curve but follows a different path. This

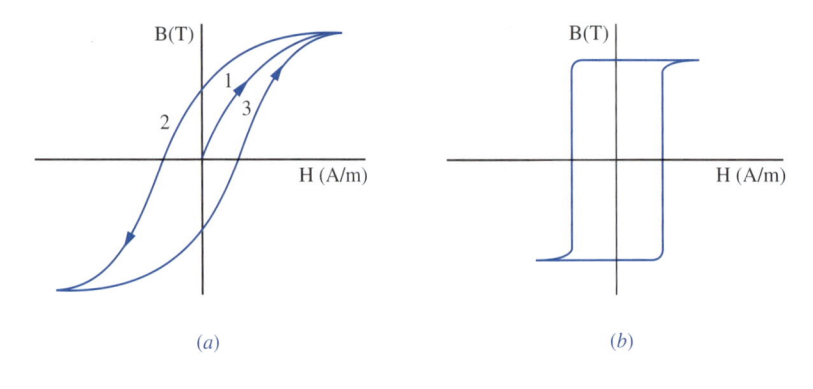

FIGURE 3–18

(*a*) Hysteresis curve in the **B**–**H** plane for a ferromagnetic material. The curve starts at the origin following curve 1 until it saturates. Curve 2 corresponds to decreasing values of **H** until it again saturates. Curve 3 corresponds to increasing values of **H**. (*b*) Idealized hysteresis curve useful for computer memories.

phenomenon is called *hysteresis*, which follows from the Greek word meaning "to lag." Even at a value of **H** = 0, there will be a residual magnetic flux density. Eventually a saturation region will be achieved in the opposite direction. Reducing **H** and reversing its direction eventually brings us back to the original saturation point. From this curve, we note that the relative permeability μ_r depends on the value of **H**. Frequently, average values are used in practice and several of these values are included in Appendix B. This total curve is called a *hysteresis curve*.

Since the magnetic material *remembers* the magnitude and the direction of the magnetic flux density (**B** > 0 or **B** < 0) at **H** = 0, it can be used as a memory element in a logic circuit. The critical factor that determines these values is the direction of the change of **H**; hence, whether some current is flowing in one direction or the other. Engineers and scientists who work with materials are able to optimize the hysteresis curve by making it almost rectangular in shape as shown in Figure 3–18*b*. This is of particular interest in computer applications such as a magnetic memory device. Other requirements may dictate the size of the hysteresis curve.

3.6 Magnetic Circuits

Because the relative permeability of ferromagnetic materials is much higher than the ambient regions in which they may be located, we can use this property to confine magnetic fields. This is similar to electric currents being confined by or actually flowing through conducting materials instead of the surrounding air since the conductivity of the conductor is so much higher

3.6 Magnetic Circuits

than the air. There are quantities in magnetic circuits that are similar to the voltages, currents and resistances that we find in the electric circuits. Much of the well-developed intuition and knowledge that we may already possess from previous courses in circuit theory can be brought to bear upon magnetic circuits. In particular, Kirchhoff's laws will apply as will be shown below. There is, however, one fly in the ointment. Although it is not a bad approximation to assume that the conductivity of the wire used in electric circuits is a constant that is independent of the amplitude or direction of the flow of the current, one has to be more careful in dealing with magnetic circuits. From the hysteresis curve shown in Figure 3–18a, we note that we have to be concerned not only with the amplitude but also with the direction of the flow of the magnetic flux when selecting an average value for a relative permeability that may be used in the calculation of a magnetic circuit.

In order to demonstrate this concept, we will examine the magnetic circuits shown in Figure 3–19. There are N turns of wire that are wrapped around an iron core that has a relative permeability $\mu_r \gg 1$. Hence, we may assume with a reasonable justification that the magnetic fields are confined within the magnetic material just as we considered the electric current to be confined within the high conductivity wire in an electric circuit. If we were to compare the ratios of the relative permeability of the magnetic material to the external region and the relative conductivity of the wire to the external region, the efficacy of using magnetic and electric circuit theory could be justified.

There is an additional assumption that is made concerning the very small gap that exists in the ferromagnetic material. The assumption is that the cross-sectional area of the gap is identical to the cross-sectional area of the magnetic material. This implies that there is no fringing magnetic field. In addition, the

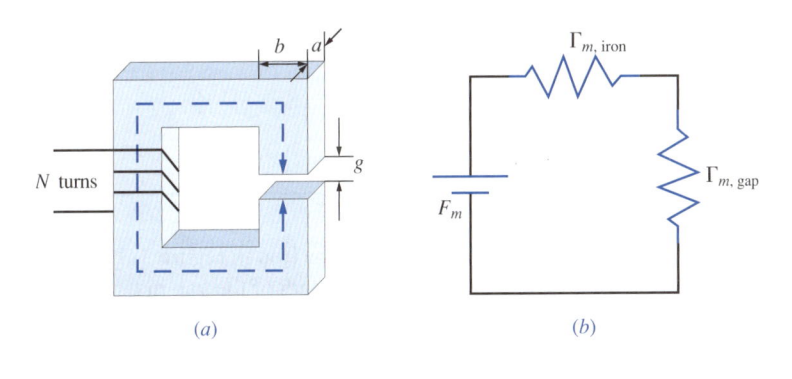

FIGURE 3–19

(*a*) A magnetic circuit. A mean length within the iron region is $\mathcal{L}$ and the gap length is g. The cross-sectional area of the iron is $A = a \times b$.
(*b*) Equivalent circuit for the magnetic circuit where F_m is the magnetomotive force, $\Gamma_{m, \text{iron}}$ is the reluctance of the iron, and $\Gamma_{m, \text{gap}}$ is the reluctance of the gap.

magnetic flux is distributed uniformly within the cross-sectional area of the iron, resulting in a constant magnetic flux density in the gap. In addition, we approximate the integration path of the closed line integral by assuming that it passes through the center of the iron-gap structure.

Let us apply Ampere's circuital law from (3.60) to the magnetic circuit shown in Figure 3–19. Since the path is understood, we can eliminate the vector notation and write

$$H(\mathcal{L} + g) = I_{enc} = NI \qquad (3.62)$$

Since there are N turns of wire, the current that is enclosed within this closed path $I_{enc} = NI$. In this magnetic circuit, the total magnetic flux Ψ_m passes through the magnetic circuit. This magnetic flux is defined as $\Psi_m = BA$ where $A = ab$ is the cross-sectional area of the iron. The magnetic flux density **B** is related to the magnetic field intensity **H** through the relation that $B = \mu_0 \mu_r H$. Therefore, (3.62) can be written as

$$\frac{B}{\mu_0 \mu_r}\mathcal{L} + \frac{B}{\mu_0}g = \Psi_m \left(\frac{\mathcal{L}}{A \mu_0 \mu_r} + \frac{g}{A \mu_0} \right) = NI \qquad (3.63)$$

The term NI plays the role of a ***magnetomotive force*** (***mmf***), and it will be given the symbol F_m. It has the units of ***ampere-turns***, and it creates the magnetic flux Ψ_m. The source of the ***mmf*** is external to the magnetic circuit. Therefore, we can consider (3.63) as being similar to Ohm's law that is applied to magnetic circuits. The term $\mathcal{L}/(A\mu_0\mu_r) + g/(A\mu_0)$ is called the ***reluctance*** $\Gamma_{m,iron} + \Gamma_{m,gap}$ of this circuit. In this case, the total magnetic flux passes through the iron and the gap, and we can consider that the two reluctances are in series. The units of $\Gamma_{m,r}$ are sometimes given in the units of ***rels***. This is very similar to an electrical circuit containing a battery and two resistors in series. Since the relative permeability of the iron core is of the order of 1000, the dominant contribution to the reluctance is seen to be caused by the narrow gap region.

EXAMPLE 3.17

Assume that the reluctance of the gap is equal to the reluctance of the iron in Figure 3–19. The iron has a relative permeability of $\mu_r = 1000$. Determine the length of the gap in terms of the length of the iron.

Answer. The reluctance of the two regions is defined as $\Gamma_{m,\,iron} = \mathcal{L}/A\mu_0\mu_r$ and $\Gamma_{m,\,gap} = g/A\mu_0$. The length of the gap g is found to be $g = \mathcal{L}/\mu_r = \mathcal{L}/1000$.

3.6 Magnetic Circuits

EXAMPLE 3.18

Write a set of coupled equations for the coupled circuit shown in (a). The iron has a cross-sectional area of A and a relative permeability μ_r. An equivalent circuit is given in (b).

Answer. In this circuit, the mean distances between the intersection points are given by $\mathscr{L}_1$, $\mathscr{L}_2$, and $\mathscr{L}_3$.

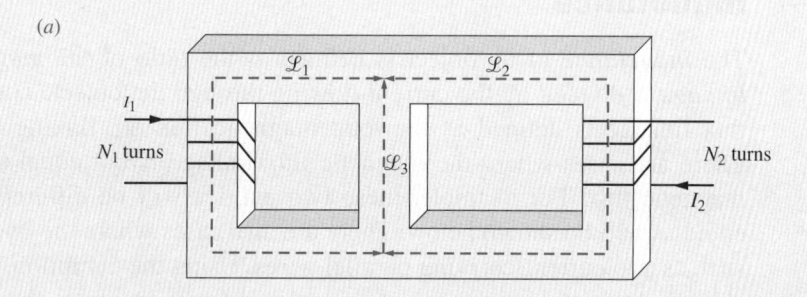

(a)

From (a), it is possible to draw an equivalent electrical circuit (b) that contains the following elements

$$\Gamma_{m1} = \frac{\mathscr{L}_1}{\mu_r \mu_0 A_1}, \quad \Gamma_{m2} = \frac{\mathscr{L}_2}{\mu_r \mu_0 A_2}, \quad \Gamma_{m3} = \frac{\mathscr{L}_3}{\mu_r \mu_0 A_3},$$

$$F_{m1} = N_1 I_1, \quad F_{m2} = N_2 I_2$$

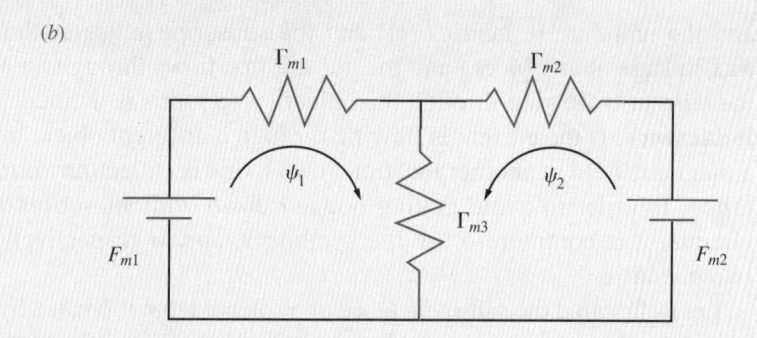

(b)

Due to the winding of the wire about the material and the direction of the current in these wires, the magnetic flux from both coils will be in a direction such that the magnetic fluxes will add in the center conductor. Hence, we write from our magnetic circuit using Kirchhoff's law that

$$F_{m1} = \Psi_1 (\Gamma_{m1} + \Gamma_{m3}) + \Psi_2 \Gamma_{m3}$$
$$F_{m2} = \Psi_1 \Gamma_{m3} + \Psi_2 (\Gamma_{m3} + \Gamma_{m2})$$

As we can see, the calculation that involves magnetic circuits is very similar to a calculation that we have performed already on electrical circuits. In the latter case, the electrical current was confined to flow in a high conductivity wire. In magnetic circuits, the magnetic flux is confined to the region of a high permeability iron.

3.7 Inductance

The *inductance* of an object is defined as the ratio of the ***magnetic flux linkage*** Λ divided by the current flowing through the object. The magnetic flux linkage is defined as the total magnetic flux Ψ_m linking the object. There are cases where the magnetic flux linkage is not equal to the total magnetic flux. For example, these two entities will be different in a solenoid as will be shown below. There are also cases where the two are equal such as two current-carrying parallel wires. Using the definition, we will be able to obtain the inductance of several important examples that are useful in electronic circuits and are found in the guided propagation of electromagnetic waves.

The inductance of an object is written as

$$L_{jk} \equiv \frac{\Lambda_j}{I_k} \; (\text{H}) \tag{3.64}$$

and the units are in henries (H) and the subscript indicates the magnetic flux linkage at an object j and the current that flows through an object k. If the current is flowing through the object ($j = k$), this is defined as the ***self-inductance***. If the current is flowing through a different object but creating a magnetic field at another location ($j \neq k$), this is called the ***mutual inductance***. In order to avoid getting bogged down with subscripts in simpler circuits, it is common to use the symbols L and M respectively for these two quantities.

Let us first find the self-inductance of a solenoid that is created by wrapping wire around a core as shown in Figure 3–20. We are assuming that the length of the solenoid d is much greater than its radius a, which allows us to approximate the magnetic flux density to be a constant over the cross-sectional area at its midpoint. There are N turns of wire that are distributed uniformly around the core. This core could either be air or a magnetic material. The magnetic flux density at the center of the solenoid was obtained in Example 3.6 and can be written as

$$B_z = \frac{\mu NI}{d} \tag{3.65}$$

3.7 Inductance

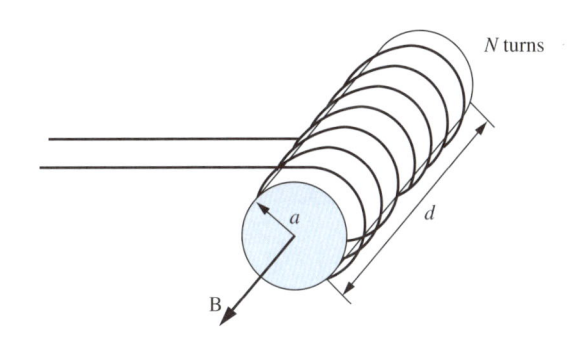

FIGURE 3–20

A solenoid consists of N turns of wire that are uniformly wrapped around a core whose radius is a and length d.

where we have incorporated the representation for the permeability $\mu = \mu_r \mu_0$. The total magnetic flux is just the product of the magnetic flux density times the cross-sectional area πa^2 of the solenoid. We obtain

$$\Psi_m = \frac{\mu N I}{d} \pi a^2 \tag{3.66}$$

The magnetic flux linkage Λ is equal to the magnetic flux linking all N turns of the solenoid or

$$\Lambda = N \Psi_m = \frac{\mu N^2 I}{d} \pi a^2 \tag{3.67}$$

Therefore, the self-inductance of the solenoid is calculated from (3.64) to be

$$L = \frac{\mu N^2}{d} \pi a^2 \quad (\text{H}) \tag{3.68}$$

We note that the value increases as the square of the number of turns increases. In addition, larger values of inductance can be obtained if an iron core is inserted within the coil.

The calculation of the self-inductance of a coaxial cable, which is required when we talk about transmission lines, can also be performed easily. In this case, the magnetic flux linkage Λ will be equal to the total magnetic flux Ψ_m. The length of the coaxial cable will be Δz. The radius of the inner conductor is a and the radius of the outer conductor is b as shown in Figure 3–21. Between the two conductors, there is a vacuum. We will assume that there is a uniform current that flows in the $+\mathbf{u}_z$ direction in the inner conductor and returns in the outer conductor of the coaxial cable. This assumed current will be equal to the current that is in the denominator of (3.64) and will therefore not appear in the final result for the inductance. The total magnetic flux that passes through the shaded region between the two conductors is computed by just integrating the magnetic flux density over

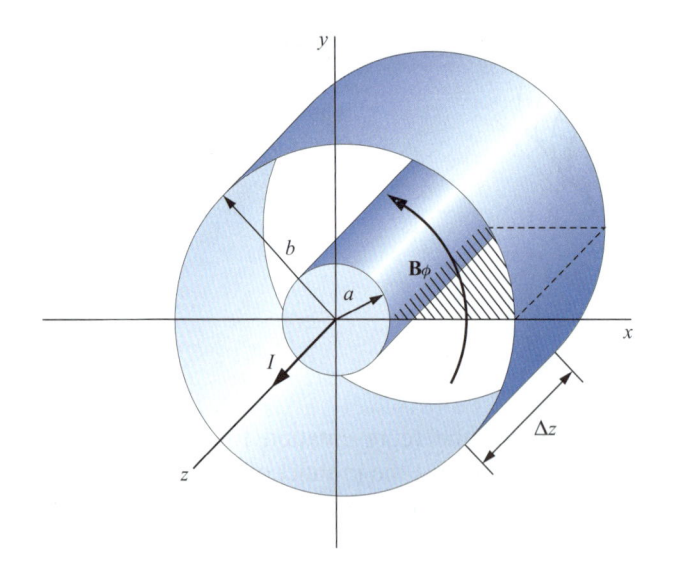

FIGURE 3–21

A section of a coaxial cable. The magnetic flux density passes through the shaded region between the two conductors.

the shaded area. Because of the cylindrical symmetry that is found in a coaxial cable, we can use Ampere's law in equation (3.8) in order to obtain the magnetic flux density between the two conductors.

Since we are only interested in the magnetic flux density external to the inner conductor, we can use a result that we have already obtained in (3.15), which we rewrite here

$$B_\phi = \frac{\mu_0 I}{2\pi\rho} \qquad (3.69)$$

The total magnetic flux that passes through the shaded region in Figure 3–21 is given by

$$\Lambda = \Psi_m = \int_{z=0}^{z=\Delta z} \int_{\rho=a}^{\rho=b} \frac{\mu_0 I}{2\pi\rho} d\rho dz = \frac{\mu_0 I}{2\pi} \ln\left(\frac{b}{a}\right)\Delta z \qquad (3.70)$$

The substitution of (3.70) into the definition for the inductance (3.64) leads to the self-inductance of a coaxial cable as

$$L = \frac{\mu_0}{2\pi} \ln\left(\frac{b}{a}\right)\Delta z \text{ (H)} \qquad (3.71)$$

The value of the inductance will increase as the length of the coaxial cable Δz increases.

3.7 Inductance

EXAMPLE 3.19

Calculate the self-inductance between two parallel planes. This object is frequently called a "microstrip" or "microstrip line", and it is important for the understanding of integrated circuits. The separation distance d can be assumed to be much less than the width w.

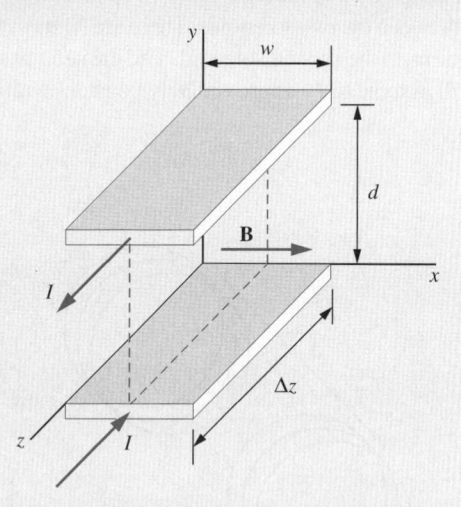

Answer. Since the current in each of the conductors is in the opposite direction, the magnetic field between the two conductors will be in the same direction. This magnetic field will pass through the rectangular area defined by the separation distance d and the length Δz. An approximate value for the magnetic flux density between the two conducting strips can be obtained from Ampere's law in equation (3.8). For one of the strips, we write $2wB \approx \mu_0 I$. Superposition applies, and if w is much larger than d, we can make the assumption that the magnetic flux density is uniform in the entire region between the two strips. Therefore, the total magnetic flux Ψ_m is given by

$$\Psi_m = \int_{z=0}^{z=\Delta z} \int_{y=0}^{y=d} \frac{\mu_0 I}{w} dy dz = \frac{\mu_0 I}{w} d \Delta z$$

In this case, the total magnetic flux will be equal to the magnetic flux linkage. The inductance is computed from (3.64) to be

$$L = \mu_0 \frac{d}{w} \Delta z \text{ (H)}$$

It is possible to calculate the inductance of a toroid using the same approximations that were used to calculate the inductance of a solenoid. This will be left for the problems. The calculation of the inductance between two parallel wires will be described in Appendix D since it requires certain mathematical approximations or numerical techniques. It is important for our consideration of transmission lines.

EXAMPLE 3.20

Calculate the mutual inductance M between two circular solenoids whose individual lengths are d and have surface areas s_1 and s_2 that are separated by a distance x where $x \ll d$. This latter assumption that is exaggerated in the figure allows us to assume that the magnetic flux density lines are parallel between the two solenoids. There are N_1 turns that have a current I_1 in the first solenoid and N_2 turns in the second solenoid. The dashed lines indicate the magnetic flux density which we will assume to be approximately constant within the solenoids.

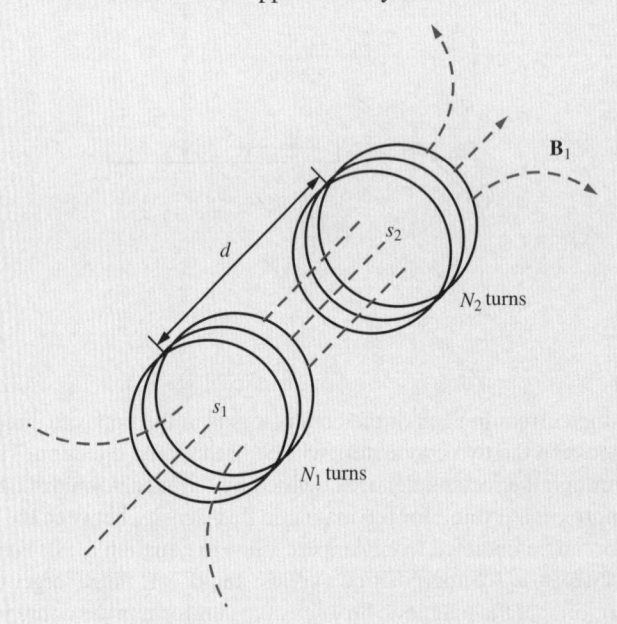

Answer. The magnetic flux density due to the first solenoid as determined in Example 3.6 is $B_1 = \dfrac{\mu_0 N_1 I_1}{d}$. The total magnetic flux from the first solenoid is equal to $B_1 s_1$. Assuming that this magnetic flux has the same value in the second solenoid, we obtain the total magnetic flux linked to the second solenoid as $\Lambda_2 = N_2 B_1 s_1$. Therefore, the mutual inductance M is given by

$$M = \frac{\Lambda_2}{I_1} = \frac{N_2 \dfrac{\mu_0 N_1 I_1 s_1}{d}}{I_1} = \frac{\mu_0 N_1 N_2 s_1}{d} \text{(H)}$$

The calculation of the mutual inductance is important in understanding transformers.

The magnetic energy that is stored in the elements that have just been described can be calculated using material that has been described in a course on circuit theory. In this case, we assume that a current source is connected to

3.8 Conclusion

the inductance and the value of the current I increases in time. There will be a voltage V that is generated across the terminals of the inductance. The magnetic energy W_m is just equal to the time integral of the electrical power

$$W_m = \int_{t'=0}^{t'=t} p\,dt' = \int_{t'=0}^{t'=t} IV\,dt' = \int_{t'=0}^{t'=t} I\left(L\frac{dI}{dt'}\right)dt' = L\int_{t'=0}^{t'=t} I\,dI = \frac{1}{2}LI^2 \quad (3.72)$$

For the solenoid, the magnetic energy is equal to

$$W_m = \frac{1}{2}\left(\frac{\mu N^2}{d}\pi a^2\right)\left(\frac{B_z d}{\mu N}\right)^2 \quad (3.73)$$

where we have replaced the current I with the magnetic flux density using (3.65). The volume Δv within the solenoid is equal to $d\pi a^2$. Therefore, (3.73) can be written as $W_m = 1/2\ B_z^2 \Delta v/\mu$. The total magnetic energy that is stored within a volume Δv is obtained from an integration of the magnetic energy density within that volume

$$\boxed{W_m = \int_{\Delta v}\left(\frac{B^2}{2\mu}\right)dv\ \text{(J)}} \quad (3.74)$$

3.8 Conclusion

Our study of time-independent magnetic fields has indicated that there are some similarities and some differences between electric and magnetic fields. The non-existence of a magnetic monopole implies that all north poles are paired with a south pole. Thus, enclosing a magnetic source with the equivalent of a Gaussian surface always yields a result of zero, as there will not be an imbalance of "magnetic charges." Both the Coulomb force for electric fields and Ampere's force for magnetic fields are proportional to the magnitudes of the charges or the current elements and inversely proportional to the square of the distances separating the entities. The source for the electric field is a stationary electrical charge and the source for a magnetic field is an electrical charge that is in motion with a constant velocity and creates a conduction current. The magnetic flux density can be directly obtained from the current using Ampere's law if there is the required symmetry in the problem; it can be obtained using the intermediate step of finding the vector potential first; or it can be obtained from the Biot–Savart integral. A model for an atom indicates that we can assume the atom can be thought of as being a small magnetic dipole consisting of a positive nucleus with an electron moving around the nucleus. The magnetic dipoles under the influence of an external magnetic field can have practical applications in developing magnetic circuits and the electrical circuit element of the inductance.

3.9 Problems

3.1.1. Calculate the total current that passes through the indicated surface. The current density $\mathbf{J} = 3x\mathbf{u}_x + 4x\mathbf{u}_y$ A/m².

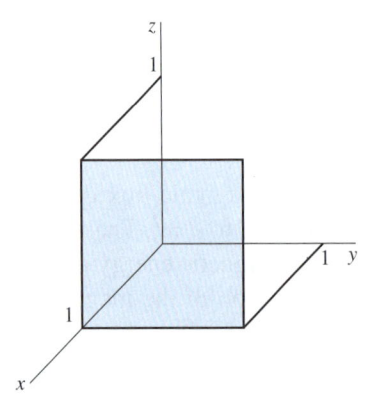

3.1.2. Calculate the total current that passes through the indicated surface. The current density $\mathbf{J} = 3x\mathbf{u}_x + 4x\mathbf{u}_z$ A/m².

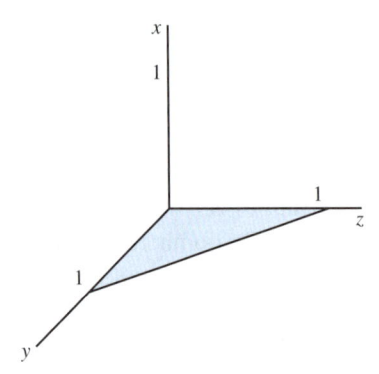

3.1.3. The current density in a cylindrical wire whose radius a is given by

$$\mathbf{J} = J_0\left(1 - \frac{\rho^2}{a^2}\right)\mathbf{u}_z$$

Compute the total current in the wire.

3.1.4. A lightning stroke with a current $I = 10$ kA strikes a post. Find the voltage V_{ab} between the legs of a man, situated on a distance $a = 5$ m from the post, if the step-size is $\mathscr{L} = 0.6$ m (see the figure). Assume that the ground is homogeneous with conductivity $\sigma = 0.02$ S/m.

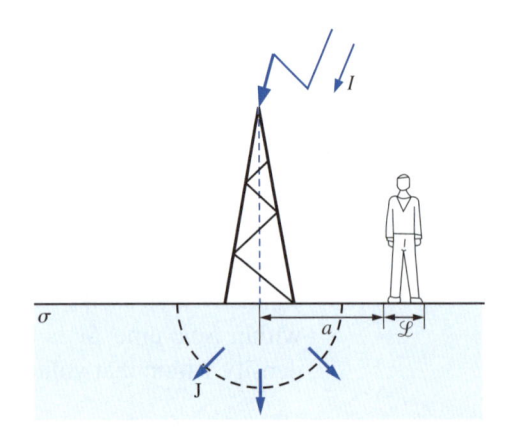

3.2.1. In terms of the units of mass M, length L, time T, and charge Q, find the units of the permeability μ_0 of vacuum.

3.2.2. In terms of the units of mass M, length L, time T, and charge Q, show that the units of $1/\sqrt{\mu_0 \varepsilon_0}$ have the same dimensions as a velocity.

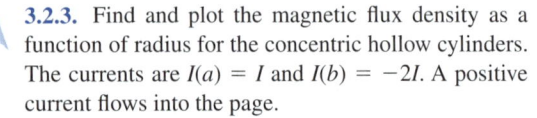

3.2.3. Find and plot the magnetic flux density as a function of radius for the concentric hollow cylinders. The currents are $I(a) = I$ and $I(b) = -2I$. A positive current flows into the page.

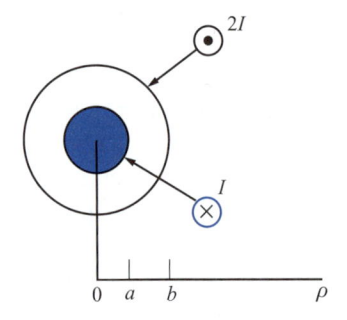

3.2.4. Find and plot the magnetic flux density as a function of radius for the concentric hollow cylinders. The currents are $I(a) = 4I$ and $I(b) = -2I$. A positive current flows into the page.

3.9 Problems

3.2.5. Due to a nonuniform conductivity of the material, the current density within a wire has a nonuniform distribution $\mathbf{J} = J_0(1-\rho^2/a^2)\mathbf{u}_z$ for $0 \leq \rho \leq a$ and $\mathbf{J} = 0$ for $\rho > a$. Find the magnetic flux density as a function of radius for $\rho > 0$.

3.2.6. If a long straight wire with a resistance of 1 Ω is connected to a 1 V battery, find the distance from the wire where the magnetic field equals 1 T.

3.2.7. An experimentalist desires to create a localized region where there is to be no magnetic field. Using a solenoid that contains one hundred uniformly-distributed turns that is connected to a power supply of 10 V, calculate the length of this region so the earth's magnetic field of 0.5 G = 5×10^{-5} T can be canceled. The resistance of the wire = 0.1 Ω.

3.3.1. Prove the vector identity $\Delta(1/R) = -\mathbf{u}_R/R^2$, where $\mathbf{u}_R$ is the unit vector in the direction between a current element and a point of observation.

3.3.2. In terms of the units of mass M, length L, time T, and charge Q, find the units of the vector potential $\mathbf{A}$ and the magnetic flux density $\mathbf{B}$.

3.3.3. Two infinitesimal current elements are displaced by a distance $2a$ on the x axis. Find the vector potential $\mathbf{A}$ at the point P that is at $y = b$. Find $\mathbf{B}$ at P.

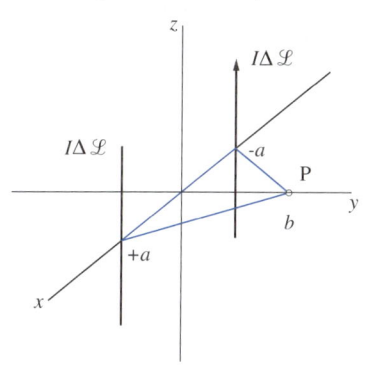

3.3.4. Use the Biot–Savart law to compute $\mathbf{B}$ at P for the configuration shown in Problem 3.3.3.

3.3.5. Show, using the Biot–Savart law that the magnetic flux density from a finite-length current element carrying a current I is given by

$$B = \frac{\mu_0 I}{4\pi\rho}(\cos\theta_1 - \cos\theta_2)$$

Find the magnetic flux density profile in case $\theta_2 = \pi - \theta_1$ (observation point on the axis of symmetry). Find the magnetic flux density in the limit of an infinitely long wire.

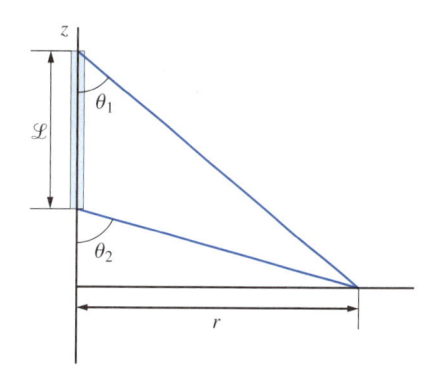

3.3.6. Show that the vector potential $\mathbf{A}$ of two parallel infinite straight wires carrying currents I in the opposite direction is given by

$$\mathbf{A} = \frac{\mu_0 I}{2\pi}\ln\left(\frac{\rho_2}{\rho_1}\right)\mathbf{u}_z$$

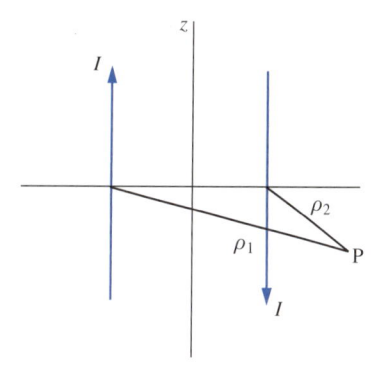

3.3.7. Use the Biot–Savart law to compute $\mathbf{B}$ at the point P in Problem 3.3.6.

3.3.8. Find the magnetic field along the z axis caused by the current in the small square loop.

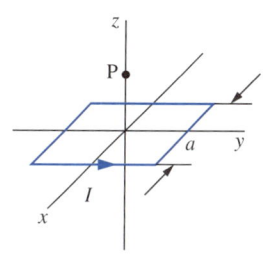

3.3.9. Find the approximate magnetic field at the center of a slender, rectangular loop that carries a current I. The dimensions of the loop are $\mathscr{L} \gg W$.

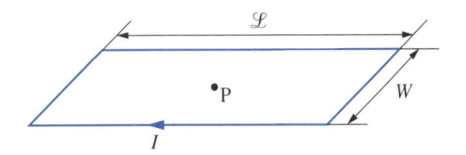

3.3.10. Two parallel current-carrying insulated wires are placed in close juxtaposition. Find the magnetic flux density at the center of the circle.

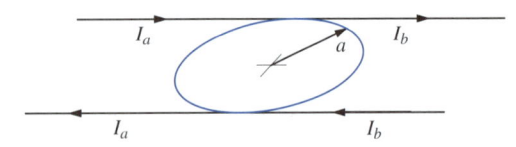

3.4.1. A wire lying on the x axis carries a current 10 A in the positive x direction. A uniform magnetic field of $\mathbf{B} = 5\mathbf{u}_x + 10\mathbf{u}_y$ T exists in this region. Find the force per unit length of this wire.

3.4.2. *Magnetohydrodynamics* (MHD) is based on passing charged particles (both electrons and positive ions) through a uniform magnetic field $\mathbf{B} = B_0\mathbf{u}_x$ with a velocity $\mathbf{v} = v_0\mathbf{u}_z$. The charged particles will separate due to their charge difference. Find the potential difference (called the *Hall voltage*) between two electrodes placed in the exhaust of a rocket engine.[2]

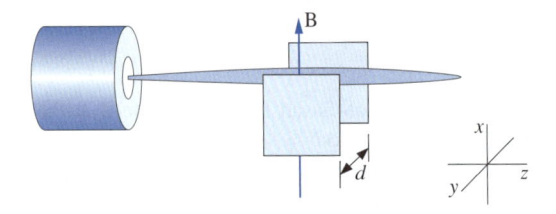

3.4.3. An electron is injected into a uniform magnetic field B_0. Find the value of the accelerating potential V_0 so that the electron will be captured in the detector that is at the end of the magnetic field at a distance d from the electron gun.

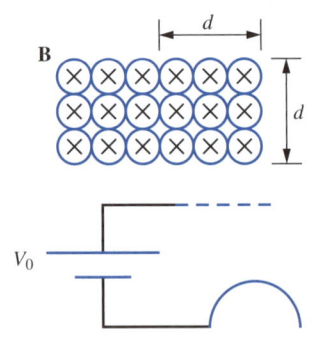

3.4.4. Describe the motion of a charged particle in a uniform electric and magnetic field such that the net force acting upon the particle is equal to zero.

3.4.5. Four infinitely long wires, each carrying a current I into the paper, are located at the indicated points. Find the direction of the net force on wire 1.

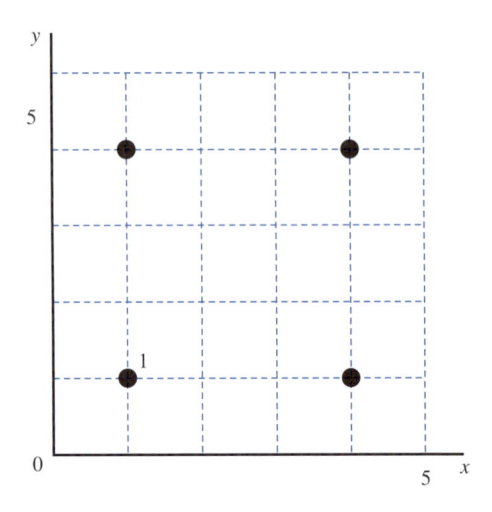

3.4.6. It is desired to move an axle of mass M up an inclined frictionless plane that makes an angle θ with respect to the horizontal plane. A uniform vertical magnetic field exists in the region of the inclined

---

[2] The Hall voltage can be measured if the exhaust of certain toy rocket motors is passed through a magnetic field of 1 T. Using a model C−60 and a surplus magnetron magnet, it is possible to generate 0.5 V for almost two seconds.

3.9 Problems

plane. What value must the battery V_0 be in order that the axle will remain at a fixed point?

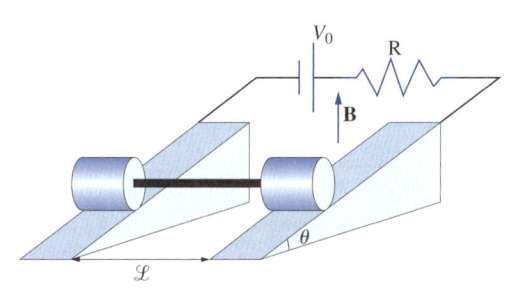

3.4.7. Find the force of repulsion between two conductors whose length is $\mathscr{L}$ of a planar transmission line with $b \gg d$. A current I_0 flows in the opposite directions in the two conductors. Ignore any fringing magnetic fields.

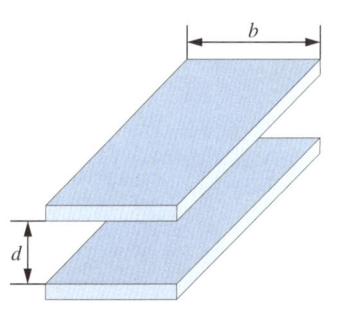

3.5.1. Let $\mathbf{B} = 0.1\mathbf{u}_x$ T everywhere and let the relative permeability $\mu_r = 2$ for $|x| \leq 1$ and $\mu_r = 1$ for $|x| > 1$. Plot $\mathbf{B}_x$, $\mathbf{M}_x$, and $\mathbf{H}_x$ everywhere.

3.6.1. For the magnetic circuit shown, calculate the magnetic flux in the gap. The cross-section and relative permeability of the iron is A and $\mu_r \gg 1$.

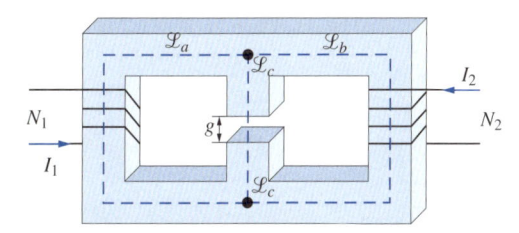

3.6.2. A magnetic core of square cross section is shown. Find the air-gap flux density when the total number of ampere-turns is 200. Assume the soft ferromagnetic material of the core has a constant relative permeability $\mu_r = \mu/\mu_0 = 4000$. The dimensions of the core are $\mathscr{L}_a = \mathscr{L}_b = 1$ m, $\mathscr{L}_c = 0.34$ m, $\mathscr{L}_g = 0.76$ mm, $A = 7.9 \times 10^{-3}$ m^2. Compensate for 5% air gap fringing.

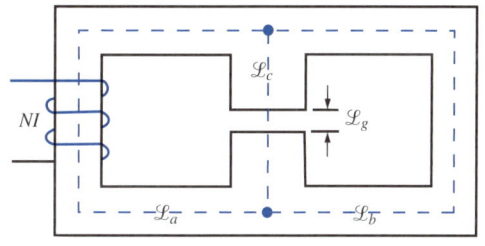

3.6.3. It is desired to produce an air-gap flux density of 0.2 T in the magnetic circuit of the previous problem. Find the mmf, that is, the total number of ampere-turns required.

3.6.4. Find the flux Ψ_b that flows through the outside path of the magnetic circuit of Problem 3.6.2.

3.7.1. Find the self-inductance L of the toroid shown in the figure below. Show that the self-inductance of a toroid is given by $L = \dfrac{\mu N^2 A}{\mathscr{L}_c}$. Evaluate L if $\mu = 1000\mu_0$, $N = 100$, $\mathscr{L}_c = 10$ cm, and cross section area $A = 1$ cm^2.

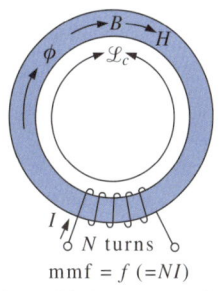

3.7.2. Find the self-inductance L of the toroid with air gap shown in the figure below. Assume $\mu = 1000\mu_0$, $N = 100$, $\mathscr{L}_i = 10$ cm, $\mathscr{L}_g = 0.1$ cm and cross section area $A = 1$ cm^2. Compensate for 5% air gap fringing (or: The effective air gap cross section area $A_g = 1.05A$.)

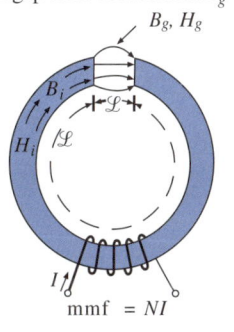

3.7.3. Find the self-inductance L_{11} and L_{22} and mutual inductance M for the transformer shown in the figure below. Evaluate L_{11} and L_{22} and M, if $\mu_r = 1000$, $N_1 = 100$, $N_2 = 1000$, $\mathscr{L}_c = 10$ cm and $A = 10$ cm^2. Find the coefficient of coupling, k, which is defined as the ratio of the mutual inductance, M, and the square root of the product of the two self-inductances involved in the coupling.

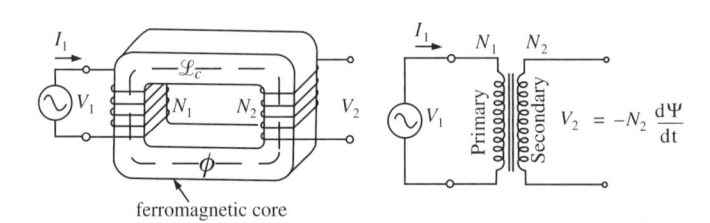

CHAPTER 4

Boundary Value Problems Using MATLAB

4.1 Boundary Conditions for Electric and Magnetic Fields 178
4.2 Poisson's and Laplace's Equations 186
4.3 Analytical Solution in One Dimension—
    Direct Integration Method 191
4.4 Numerical Solution of a One-Dimensional Equation—
    Finite Difference Method 201
4.5 Analytical Solution of a Two-Dimensional Equation—
    Separation of Variables 211
4.6 Finite Difference Method Using MATLAB 220
4.7 Finite Element Method Using MATLAB 226
4.8 Method of Moments Using MATLAB 241
4.9 Conclusion 251
4.10 Problems 252

Up to this point, we have examined electric fields and magnetic fields in an infinite space. The usual situation that one encounters in practice is a finite space where the fields transition from one region to another, each with different electrical material characteristics, or regions that are enclosed with specific boundaries. We first describe the transitioning of the fields from one region to another and then examine techniques to solve the boundary value problems. We will focus our attention on examining boundary value problems for electric fields although the techniques can also be used for magnetic fields.

In Chapter 2, we learned that a static electric field would be created from a charge distribution. In addition, it was possible to determine this static electric field from a scalar potential. We also showed there that the potential *V* could be obtained directly in terms of the charge distribution ρ_v via one partial differential equation: **Poisson's equation**. This equation reduces to **Laplace's equation** if the charge density in the region of interest is equal to zero. The general procedure for solving these equations for the cases where the potential *V* depends on only one or two spatial coordinates is given here. In this chapter, we will introduce analytical and numerical techniques that will allow us to examine such complicated problems. We also introduce modern numerical techniques for solving such problems (the *method of moments*, the *finite element method*, and the *finite difference method*) in this chapter and make extensive use of MATLAB in the process. Additionally, we will concentrate on problems formulated with Cartesian coordinates to emphasize the actual numerical solution technique and to simplify the specific mathematical calculations. Several MATLAB programs (M-files) that have been used in this book are available on the Web site *http://www.scitechpub.com/lonngren2e.htm* for the reader's benefit. These programs can be altered and easily customized by the user. They can also be translated into the reader's language of choice. They may not include *the* program for a particular task, but they do work.

4.1 Boundary Conditions for Electric and Magnetic Fields

The boundary conditions we will derive are valid for both time-independent and time-dependent electromagnetic fields. Also, to simplify the math, we will develop the boundary conditions in two dimensions and in the Cartesian coordinate system, although the development can be extended to three dimensions and any coordinate system. Electromagnetic fields can be separated into a component that is parallel to an interface and a component that is perpendicular to an interface. It is reasonable to construct a two-dimensional coordinate system where one of the axes is normal to the interface and the other axis is tangent to the interface between the two materials, as shown in Figure 4–1. The components for the displacement flux density and the magnetic flux density will be obtained separately using the physical intuition that we gained in the previous two chapters.

The boundary conditions involving the normal components of the displacement flux density **D** and the magnetic flux density **B** are shown in Figure 4–2. The interface is encapsulated with a "pillbox" that has a cross-sectional area Δs and a thickness that will satisfy the condition that the thickness $\Delta z \to 0$. We will first examine the displacement flux density.

4.1 Boundary Conditions for Electric and Magnetic Fields

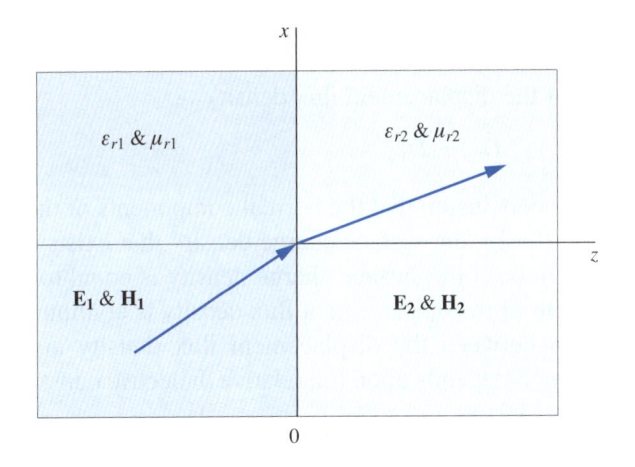

FIGURE 4–1

Two different materials are in juxtaposition. The relative dielectric and permeability constants are known for each material. The electromagnetic fields in one region are related to those in the other region. The interface is located at $z = 0$.

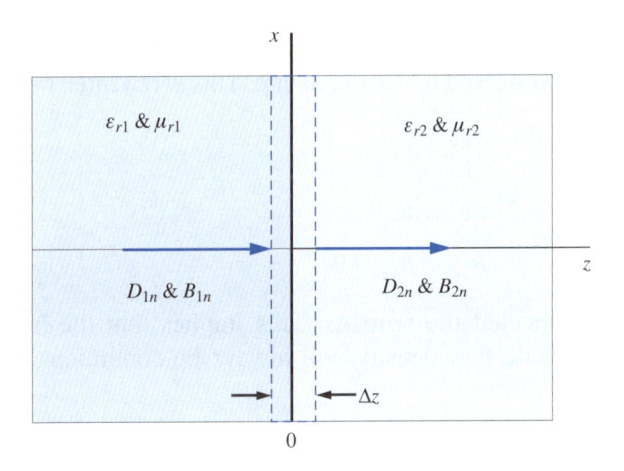

FIGURE 4–2

The interface between two materials is enclosed with a "pillbox" whose thickness $\Delta z \to 0$ and whose cross-sectional area is equal to Δs. There may be a surface charge density ρ_s (C/m^2) that is distributed at the interface.

We will assume a charge distributed along both sides of the interface that has a total surface charge density of ρ_s(C/m^2). The application of Gauss's law along the interface will allow us to derive the boundary condition for the normal component of the displacement flux density. We rewrite this law

$$\oint \mathbf{D} \cdot \mathbf{ds} = Q_{\text{enc}} \tag{4.1}$$

The charge that is enclosed within the pillbox is $Q_{\text{enc}} = \rho_s \Delta s$ where ρ_s is the surface charge density that will be assumed to be independent of position. Recall that the evaluation of a closed surface integral requires that the scalar product of the differential surface area Δs and the normal component of the displacement flux density **D** must be performed. This leads to $(D_{2n} - D_{1n})\Delta s$.

Therefore, the evaluation of (4.1) and the cancellation of the surface area Δs on each side of the equation leads to the following boundary condition for the normal components of the displacement flux density

$$D_{2n} - D_{1n} = \rho_s \tag{4.2}$$

This equation leads to the conclusion that the normal components of the displacement flux density differ by the surface charge density that exists at the interface of the two materials. If the surface charge density is equal to zero, then the normal component of the displacement flux density is continuous.

Using the relationship between the displacement flux density and the electric field intensity, which depends upon the relative dielectric constant of the material, we can write (4.2) as

$$\varepsilon_{r2}\varepsilon_0 E_{2n} - \varepsilon_{r1}\varepsilon_0 E_{1n} = \rho_s \tag{4.3}$$

A similar derivation can be used to find the normal component of the magnetic flux density. In this case, we invoke the experimental fact that magnetic monopoles have not been observed to exist in nature. This is rewritten below,

$$\oint \mathbf{B} \cdot \mathbf{ds} = 0 \tag{4.4}$$

Applying (4.4) to Figure 4–2, we write

$$B_{2n} - B_{1n} = 0 \tag{4.5}$$

where we have again canceled the term Δs. This implies that the normal components of the magnetic flux density will always be continuous at all points in the space.

The magnetic flux density is proportional to the permeability of the material times the magnetic field intensity. Therefore, (4.5) can also be written as

$$\mu_{r2}\mu_0 H_{2n} - \mu_{r1}\mu_0 H_{1n} = 0 \tag{4.6}$$

We will now derive the relation for the tangential components of the electromagnetic fields that are shown in Figure 4–3. For the tangential components of the electric field, we calculate the total work that is expended in moving around a closed path. This is written as

$$\oint \mathbf{E} \cdot \mathbf{dl} = 0 \tag{4.7}$$

The application of (4.7) to Figure 4–3 leads to the conclusion that the contributions at the top and at the bottom (the Δz components) can be neglected, if this portion of the path, which is perpendicular to the **E** field, is an equipotential surface. Along this surface, no work will be done in moving along the

4.1 Boundary Conditions for Electric and Magnetic Fields

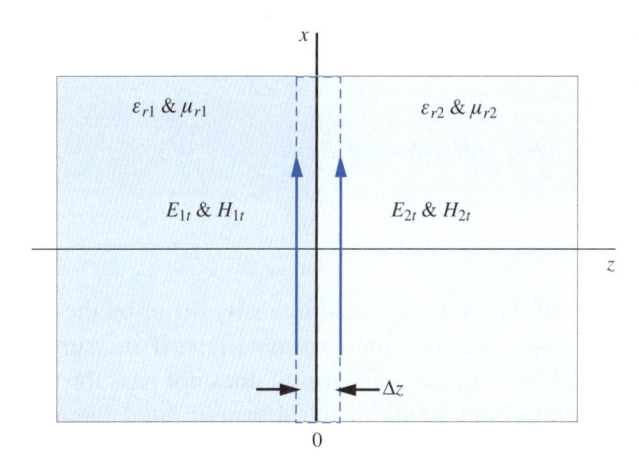

FIGURE 4-3

The tangential components of the electric field intensity and the magnetic field intensity in the two materials are indicated. The rectangular loop has the dimensions Δx and Δz. The coordinate y is directed out of the page.

path. However, the integration along the other two edges yields the result

$$(E_{2t} - E_{1t})\Delta x = 0$$

or

$$E_{2t} - E_{1t} = 0 \tag{4.8}$$

The tangential components of the electric field are continuous at an interface. Equation 4.8 can also be written in terms of the displacement flux density as

$$\frac{D_{2t}}{\varepsilon_{r2}} = \frac{D_{1t}}{\varepsilon_{r1}} \tag{4.9}$$

where the permittivity of free space ε_0 has been canceled from this equation.

The boundary condition for the tangential component of the magnetic field intensity **H** is also depicted in Figure 4–3. In this case, there could be a surface current $\mathbf{J}_s \cdot \mathbf{u}_y$ that is enclosed within the rectangular loop and directed out of the page. Therefore, we must evaluate Ampere's circuital law

$$\oint \mathbf{H} \cdot d\mathbf{l} = I_{enc} \tag{4.10}$$

The current that is enclosed within the loop is given by

$$I_{enc} = J_s \Delta x \tag{4.11}$$

and it resides entirely at the interface between the two materials. If this *surface current* is directed in the $\mathbf{u}_x$ direction, it will not pass through this rectangular loop and it will not affect this particular tangential component of the magnetic field intensity. Once again, we can neglect the integration over the top and bottom edges. The resulting integration yields the following result:

$$(H_{2t} - H_{1t})\Delta x = J_s \Delta x$$

or

$$H_{2t} - H_{1t} = J_s \tag{4.12}$$

The tangential components of the magnetic field intensity differ by the surface current density that passes between the two materials. If the surface current density is equal to zero or the surface current does not pass through this loop, then the tangential components of the magnetic field intensity will be continuous. This can also be written in terms of the magnetic flux density as

$$\frac{B_{2t}}{\mu_{r2}} - \frac{B_{1t}}{\mu_{r1}} = \mu_0 J_s \tag{4.13}$$

EXAMPLE 4.1

Calculate the change of electric field as it crosses an interface between the two dielectrics shown in the figure. There is no charge distributed on the surface between the two dielectrics.

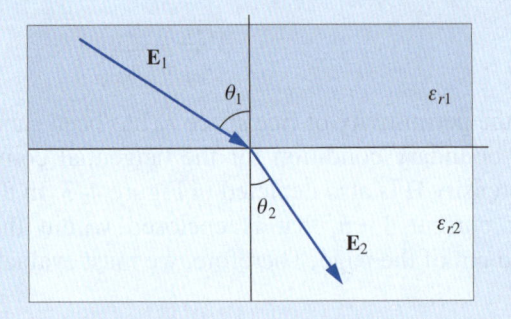

Answer. Since there is no surface charge between the two surfaces, we know that the normal components of the displacement flux density are continuous.

$$\varepsilon_{r1} E_1 \cos \theta_1 = \varepsilon_{r2} E_2 \cos \theta_2$$

The continuity of the tangential components of the electric field is given by

$$E_1 \sin \theta_1 = E_2 \sin \theta_2$$

4.1 Boundary Conditions for Electric and Magnetic Fields

The ratio of these two terms gives

$$\frac{\tan \theta_1}{\tan \theta_2} = \frac{\varepsilon_{r1}}{\varepsilon_{r2}}$$

This example illustrates that an electric field will be bent as it passes from one medium with a given dielectric constant to another medium with a different dielectric constant. This bending of electromagnetic waves, including light, is called refraction, and the relationship that we have found will give rise to an equivalent relationship which is called *Snell's law*.

EXAMPLE 4.2

Two homogeneous, linear, isotropic magnetic materials have an interface at $x = 0$. At the interface, there is a surface current with a density $\mathbf{J_s} = 20\mathbf{u}_y$ A/m. The relative permeability $\mu_{r1} = 2$ and the magnetic field intensity in the region $x < 0$ is $\mathbf{H}_1 = 15\mathbf{u}_x + 10\mathbf{u}_y + 25\mathbf{u}_z$ A/m. The relative permeability $\mu_{r2} = 5$ in the region $x > 0$. Find the magnetic field intensity in the region $x > 0$.

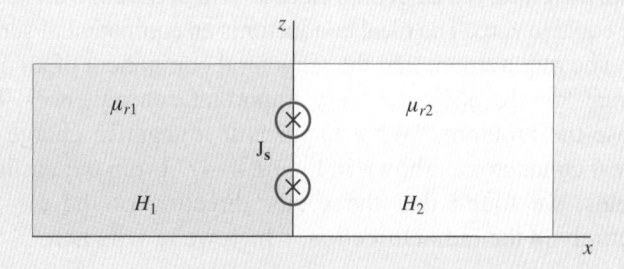

Answer. The magnetic field is separated into components and the boundary conditions are applied independently. For the normal component, we write

$$\mu_{r2}\mu_0 H_{2n} = \mu_{r1}\mu_0 H_{1n} \Rightarrow H_{2x} = \frac{\mu_{r1}}{\mu_{r2}}H_{1x} = \frac{2}{5}15 = 6 \text{ A/m}$$

For the tangential components, we write

$$H_{2t} - H_{1t} = J_s \Rightarrow H_{2z} = H_{1z} + J_y = 25 + 20 = 45 \text{ A/m}$$

There is no change in the y component of the magnetic field intensity. Therefore,

$$\mathbf{H}_2 = 6\mathbf{u}_x + 10\mathbf{u}_y + 45\mathbf{u}_z \text{ A/m}$$

If one of the materials is an ideal conductor, the tangential component of the electric field will be affected in the most dramatic fashion. Since we have determined that the tangential electric field must be continuous, there would also be a

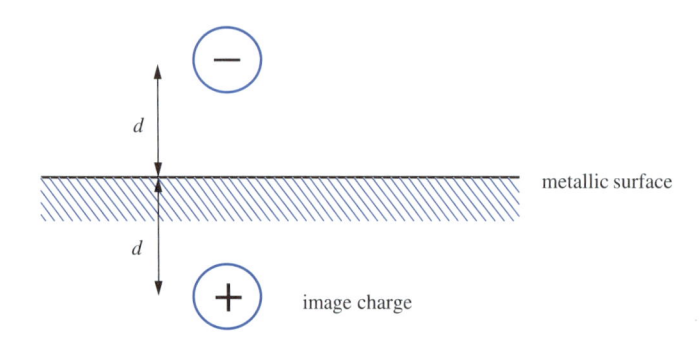

FIGURE 4-4

A charge a distance d above a grounded metallic surface induces a nonuniform charge distribution on the surface. The electric potential and electric field intensity can be calculated by assuming an image charge of the opposite sign exists beneath the surface and the metallic surface being removed.

tangential electric field at the edge of the ideal conductor. The application of Ohm's law implies that there would also be a very large tangential current that could approach infinity in a perfect conductor that has an infinite conductivity. There would then be an infinite amount of power dissipation within the ideal conductor that nature would not allow. The conclusion that must be drawn from this argument is that the tangential current *density* at the edge of the conductor; and, therefore, the tangential electric field at the edge of an ideal conductor must be equal to zero. The ideal conductor is an equipotential surface.

The requirement that the tangential component of an electric field must be equal to zero has some very important consequences. For example, let us pose the problem, "What happens if a negative charge is placed above an ideal conductor as shown in Figure 4–4?" From our earlier studies of electric fields, we found that the vector direction of the electric field would be entirely in the radial direction, which we rewrite here

$$\mathbf{E} = \frac{Q}{4\pi\varepsilon_0 r^2}\mathbf{u}_r \tag{4.14}$$

The tangential component of the electric field that follows from (4.14) will be equal to zero only at the point just beneath the charge as shown in Figure 4–4. What happens at the other points on the surface?

This last question can be answered by assuming that there is a charge of the opposite sign that is inhomogeneously distributed upon the surface of the ideal conductor. It is usually difficult to calculate this distribution, and there is a method to circumvent this calculation. This method assumes that there is an additional charge that is located an equidistance beneath the ideal conductor. This particular charge that is called an ***image charge*** will have a value that is equal to the negative value of the original charge that was placed above the ideal conductor. In order to calculate the resulting electric field, we remove this ideal conductor and use the principal of superposition since we're talking about a linear material. The tangential component of the electric field will then be equal to zero.

4.1 Boundary Conditions for Electric and Magnetic Fields

The use of image charges can be generalized to time-varying electric fields where one frequently encounters "image antennas" lying beneath the ground plane. The conductivity of this ground plane has been improved by burying copper wires in the ground that radiate in a radial direction away from the antenna.

EXAMPLE 4.3

Two grounded semi-infinite metallic plates are placed on the x axis and the y axis of a Cartesian coordinate system in order to form a 90° corner. A positive charge $4\pi\varepsilon_0$ is located at the point (a, a) where a is arbitrary. Using MATLAB, carefully plot equipotential contours surrounding this charge and the expected electric field.

Answer. There will be two negative-image charges located at the points $(a, -a)$ and $(-a, a)$. In addition, there will be a positive image charge at $(-a, -a)$ since the electric potential at all points of the metallic corner must be equal to zero. This system of four alternative charges is often called a *quadrupole*. The results of the calculation are shown below. We find that the electric field is always perpendicular to the metal surfaces. This indicates that the metallic surfaces are equipotential contours that have a value equal to zero.

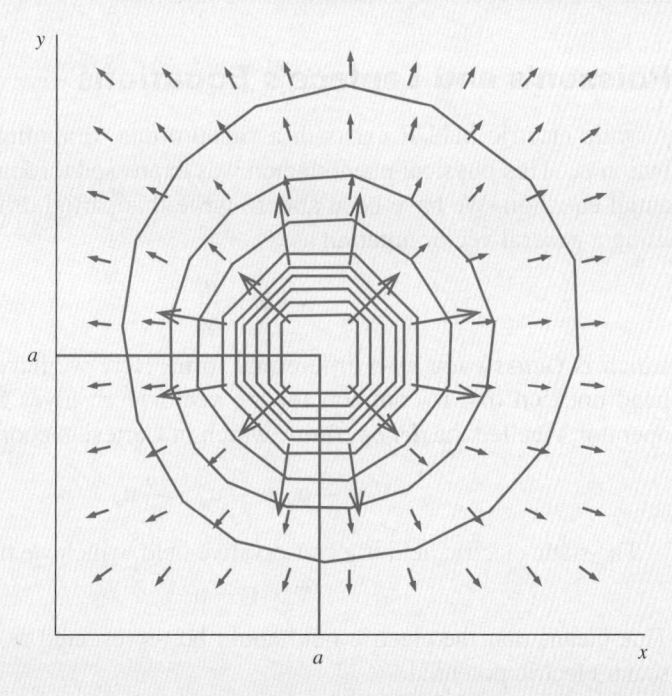

In high-energy particle experiments in which charged particles move through a metallic tube and are to collide together, the experimentalist must be aware of these image charges. The image charges can be found for cases only when the corner has an angle that is a submultiple of 2π.

TABLE 4–1 Summary of the boundary conditions for time independent and time dependent electromagnetic fields. The unit vector $\mathbf{u_n}$ is perpendicular to the interface between the two materials.

Tangential component of the electric field intensity $\mathbf{E}$	$(\mathbf{E}_2 - \mathbf{E}_1) \times \mathbf{u_n} = 0$
Normal component of the displacement flux density $\mathbf{D}$	$(\mathbf{D}_2 - \mathbf{D}_1) \cdot \mathbf{u_n} = \rho_s$
Tangential component of the magnetic field intensity $\mathbf{H}$	$(\mathbf{H}_2 - \mathbf{H}_1) \times \mathbf{u_n} = \mathbf{J_L}$
Normal component of the magnetic flux density $\mathbf{B}$	$(\mathbf{B}_2 - \mathbf{B}_1) \cdot \mathbf{u_n} = 0$
Tangential component of the electric field intensity $\mathbf{E}$ at a metallic surface. This directly follows from the first boundary condition and the fact that the current in an ideal conductor must be equal to zero.	$\mathbf{E} \times \mathbf{u_n} = 0$

The boundary conditions can also be written down in vector notation. A summary of these boundary conditions is presented in Table 4–1. In this table, the unit vector $\mathbf{u_n}$ is normal to the interface.

4.2 Poisson's and Laplace's Equations

A static electric field $\mathbf{E}$ exists in a vacuum due to a volume charge distribution ρ_v. This physical phenomenon was expressed through a partial differential equation. We have been able to write this partial differential equation using a general vector notation as

$$\nabla \cdot \mathbf{E} = \frac{\rho_v}{\varepsilon_0} \tag{4.15}$$

which is *Gauss's law* in a differential form. Here we have applied a shorthand notation that is common for the vector derivatives by using a vector operator ∇ called the *del operator*, which in Cartesian coordinates is

$$\nabla \equiv \frac{\partial}{\partial x}\mathbf{u_x} + \frac{\partial}{\partial y}\mathbf{u_y} + \frac{\partial}{\partial z}\mathbf{u_z} \tag{4.16}$$

The static electric field is a conservative field, which we have expressed as

$$\nabla \times \mathbf{E} = 0 \tag{4.17}$$

This means that the electric field could be represented as the gradient of a scalar electric potential V

$$\mathbf{E} = -\nabla V \tag{4.18}$$

Recall the vector identity

$$\nabla \times \nabla V = 0$$

4.2 Poisson's and Laplace's Equations

Combine (4.15) with (4.18) and obtain

$$\boxed{\nabla^2 V = -\frac{\rho_v}{\varepsilon_0}} \qquad (4.19)$$

where we have used the relation that $\nabla \cdot \nabla = \nabla^2$. Equation (4.19) is called **Poisson's equation**. If the charge density ρ_v in the region of interest were equal to zero, then Poisson's equation is written as

$$\boxed{\nabla^2 V = 0} \qquad (4.20)$$

Equation (4.20) is called **Laplace's equation**. Both of these equations have received considerable attention since equations of this type describe several physical phenomena, e.g., the temperature profile in a metal plate if one of the edges is heated locally, say with a blow torch.

In writing (4.19) or (4.20), we can think of employing the definition for ∇^2 that $\nabla^2 = \nabla \cdot \nabla$. This operation is based on interpreting the ∇ operator as a vector and a heuristic application of the scalar product of two vectors that leads to a scalar quantity. The resulting operator ∇^2 is called a **Laplacian operator**. Since each application of the ∇ operator yields a first order differentiation, we should expect that the Laplacian operator ∇^2 would lead to a second order differentiation.

The Laplacian operator depends on the coordinate system that is chosen for a calculation. The definitions for the operator ∇^2 operating on a function V in the three most commonly employed coordinate systems are written as follows:

1. Cartesian (x, y, z)

$$\nabla^2 V = \frac{\partial^2 V}{\partial x^2} + \frac{\partial^2 V}{\partial y^2} + \frac{\partial^2 V}{\partial z^2} \qquad (4.21)$$

   Using the definition for the del operator ∇ in Cartesian coordinates (4.16), we write

$$\nabla \cdot \nabla V = \left(\frac{\partial}{\partial x}\mathbf{u_x} + \frac{\partial}{\partial y}\mathbf{u_y} + \frac{\partial}{\partial z}\mathbf{u_z}\right) \cdot \left(\frac{\partial}{\partial x}\mathbf{u_x} + \frac{\partial}{\partial y}\mathbf{u_y} + \frac{\partial}{\partial z}\mathbf{u_z}\right) V$$

   From the definition of the scalar product for the unit vectors ($\mathbf{u_x} \cdot \mathbf{u_x} = 1$, $\mathbf{u_x} \cdot \mathbf{u_y} = 0$, etc.), the only terms that survive are the three terms given above.

2. Cylindrical (ρ, ϕ, z)

$$\nabla^2 V = \frac{1}{\rho}\frac{\partial}{\partial \rho}\left(\rho \frac{\partial V}{\partial \rho}\right) + \frac{1}{\rho^2}\frac{\partial^2 V}{\partial \phi^2} + \frac{\partial^2 V}{\partial z^2} \qquad (4.22)$$

3. Spherical (r, θ, ϕ)

$$\nabla^2 V = \frac{1}{r^2}\frac{\partial}{\partial r}\left(r^2\frac{\partial V}{\partial r}\right) + \frac{1}{r^2 \sin\theta}\frac{\partial}{\partial \theta}\left(\sin\theta\frac{\partial V}{\partial \theta}\right) + \frac{1}{r^2 \sin^2\theta}\frac{\partial^2 V}{\partial \phi^2} \qquad (4.23)$$

The choice of which form of this operation to actually employ in a calculation usually is dictated by any possible symmetry considerations inherent in the problem. For example, the calculation of the potential within a spherical ball would suggest the application of $\nabla^2 V$ in spherical coordinates rather than in the other representations. Definitions for the Laplacian operator exist for other coordinate systems than those mentioned above. A definition in a general orthogonal coordinate system can also be written. For complicated shapes and/or for very difficult problems, a numerical solution may have to be attempted. This is typically the procedure that has to be followed in practice. We will encounter these procedures in the next sections.

Similar Poisson's and Laplace's equations arise also for the magnetic field, where a magnetic vector potential **A** rather than an electric scalar potential V is involved. The nonexistence of magnetic monopoles allowed us to write that

$$\nabla \cdot \mathbf{B} = 0 \qquad (4.24)$$

This equation implies that the magnetic flux density **B** can be expressed as the curl of another vector. This vector is given the symbol **A** and it is called the ***magnetic vector potential*** or just ***vector potential***. This is a consequence of the vector identity

$$\nabla \cdot \nabla \times \mathbf{A} = 0$$

In electric fields, it was found that the electric field could be computed easily if the ***scalar potential*** were known. This was accomplished by taking the gradient of the scalar potential (along with a minus sign). Magnetic fields are related to the vector current density **J** via a curl operation, and this is significantly more complicated than the divergence operation through which the electric field is related to the charge density ρ_v. We suspect, therefore, that the potential that we seek should also be more complicated. This vector potential **A**, though, is in the *same direction* as the source current **J**. The vector magnetic potential is a very useful concept. We will explore some uses of **A** at this time, and other applications will be deferred until a later chapter.

The magnetic flux density **B** is determined from the vector potential **A** via the relation

$$\mathbf{B} = \nabla \times \mathbf{A} \qquad (4.25)$$

4.2 Poisson's and Laplace's Equations

where the units of **A** are tesla-meter. We combine this with **Ampere's law** in differential form

$$\nabla \times \mathbf{B} = \mu_0 \mathbf{J} \tag{4.26}$$

and obtain

$$\nabla \times \nabla \times \mathbf{A} = \mu_0 \mathbf{J} \tag{4.27}$$

The operation on the left-hand side of (4.27) can be reduced via a vector identity to yield

$$\nabla(\nabla \cdot \mathbf{A}) - \nabla^2 \mathbf{A} = \mu_0 \mathbf{J} \tag{4.28}$$

Two comments should be made concerning vectors at this point before proceeding. First, a vector is determined completely if its curl *and* its divergence are specified. We have specified $\nabla \times \mathbf{A}$ in (4.25). Since the divergence of the vector **A** can be specified to have *any* value, we will choose the value that will ease our calculation:

$$\nabla \cdot \mathbf{A} = 0 \tag{4.29}$$

This is the so-called **Coulomb gauge**. Second, the operation $\nabla^2 \mathbf{A}$ states that the Laplacian operator operates on all three vector components of the vector potential **A** *separately* and each component is set equal to the corresponding component of the current density **J**. Therefore, with (4.29), (4.28) simplifies to

$$\boxed{\nabla^2 \mathbf{A} = -\mu_0 \mathbf{J}} \tag{4.30}$$

This is similar to Poisson's equation (4.19) that was derived previously where the electric field was related to a charge density. In the case of magnetic fields, (4.30) represents three independent *scalar* equations for each of the components. In Cartesian coordinates, these equations are

$$\nabla^2 A_x = -\mu_0 J_x \quad \nabla^2 A_y = -\mu_0 J_y \quad \nabla^2 A_z = -\mu_0 J_z$$

At this stage we may wonder what has been gained by this little bit of vector manipulation since now we have a set of second-order uncoupled differential equations to solve instead of a coupled set of first-order differential equations. What has been gained is the good fortune of having the vector potential and the current density in the *same* direction. This implies that each component of the vector potential can be solved separately, and we have to solve at the most only three *scalar* equations. However, note that the Laplacian is separable only in Cartesian components. In the other coordinate systems we use in this text, this is not true. Be sure to use the proper form of the Laplacian (see Appendix A) when working in cylindrical and spherical coordinates.

EXAMPLE 4.4

A two-dimensional potential distribution can be approximated with the quadratic expression

$$V = -\frac{\rho_v}{4\varepsilon_0}(x^2 + y^2)$$

Show that this function satisfies Poisson's equation (4.19). Plot the graphs of the charge and the electric potential distributions.

Answer. Using the Laplacian operator in Cartesian coordinates, we find that $\partial^2 V / \partial x^2 = \partial^2 V / \partial y^2 = -\rho_v / 2\varepsilon_0$, which leads to Poisson's equation (4.19). The two-dimensional plot of the voltage distribution along with the calculated charge distribution is shown.

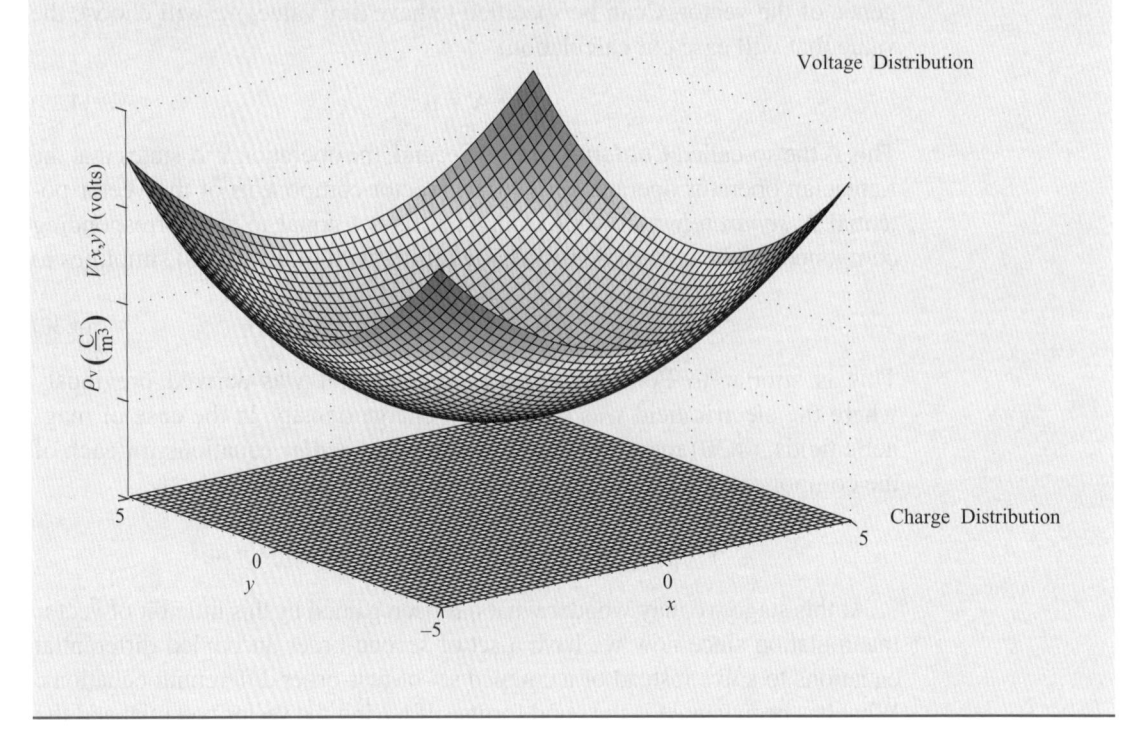

Having derived Poisson's and Laplace's equations for three-dimensional systems and having stated the definitions for ∇^2 in the three most widely used coordinate systems, we will now obtain analytical solutions for these equations. Rather than first attempting a general three-dimensional solution, we will simplify the discussion by assuming that the potential depends on only one coordinate rather than on all three coordinates. The procedure that we will describe in these simpler problems will be employed later in more difficult calculations. Several important results will be obtained, however, as we pass through this fairly difficult initial stage. Techniques that are germane to these more complicated problems will appear in the next sections.

4.3 Analytical Solution in One Dimension—Direct Integration Method

In order to conceptualize the method, we will calculate the potential variation between two infinite parallel metal plates located in a vacuum as shown in Figure 4–5a. This will require solving Laplace's equation in one dimension, and it will yield a result that approximates the potential distribution in a parallel plate capacitor where the separation between the plates is much less than any transverse dimension. Since the plates are assumed to be infinite, and the conductivity of these metal plates is very high, they can be assumed to be equipotential surfaces. We will assume that these metal plates have zero resistance, and thus we have no losses. Hence, in the y and the z

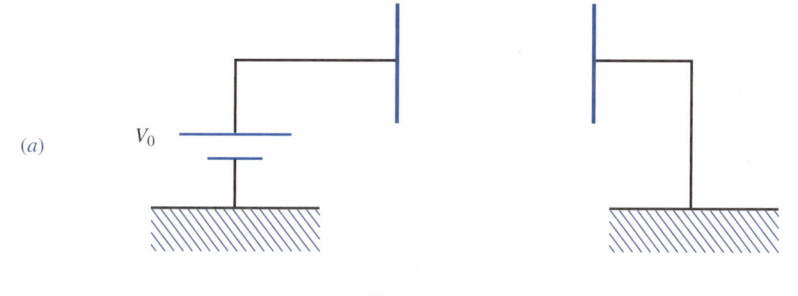

FIGURE 4–5

(*a*) Two infinite parallel plates located at $x = 0$ and at $x = x_0$. (*b*) Potential variation between the plates as determined from (4.33).

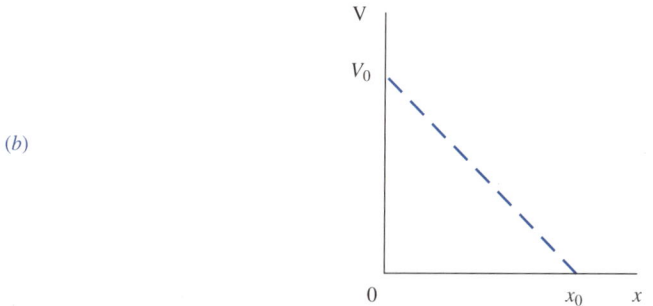

coordinates, we can postulate with a high degree of confidence that

$$\frac{\partial V}{\partial y} = \frac{\partial V}{\partial z} = 0$$

Since there is no variation of the potential in two of the three independent variables, we can let the remaining partial derivative that appears in Laplace's equation become an ordinary derivative. Hence, the one-dimensional Laplace's equation is

$$\frac{d^2 V}{dx^2} = 0 \qquad (4.31)$$

Let us assume the plate at $x = 0$ to be at a potential $V = V_0$ and that the plate at $x = x_0$ is connected to ground and therefore has the potential $V = 0$. These are the **boundary conditions** for this problem.

The solution of (4.31) is found by integrating this equation twice

$$\frac{dV}{dx} = C_1$$
$$V = C_1 x + C_2 \qquad (4.32)$$

where C_1 and C_2 are the constants of integration. These constants must be included in the solution at each step in the integration, and they will be determined from the boundary conditions that are imposed in the problem. Equation (4.32) is the most general solution of the ordinary differential equation (4.31) since it contains the two arbitrary constants of integration.

For the boundary conditions imposed by the battery ($V = V_0$ at $x = 0$) and the ground potential ($V = 0$ at $x = x_0$) in Figure 4–5a, we write

$$V_0 = C_1(0) + C_2$$
$$0 = C_1(x_0) + C_2$$

Solving these two algebraic equations for the unknown constants of integration leads to the values $C_1 = -V_0/x_0$ and $C_2 = V_0$. Therefore, we write the solution for the potential variation between the two parallel plates that satisfies the specified boundary conditions as

$$V = V_0 \left(1 - \frac{x}{x_0}\right) \qquad (4.33)$$

4.3 Analytical Solution in One Dimension—Direct Integration Method 193

The potential profile is shown in Figure 4–5b. Since this is an electrostatic potential, we can compute the electric field using (4.18). We find that

$$\mathbf{E} = -\frac{dV}{dx}\mathbf{u_x} = \frac{V_0}{x_0}\mathbf{u_x} \qquad (4.34)$$

a result that we could have also obtained using Gauss's law.

Note that this solution approach applies to any analogous physical situation that can be modeled in the same way. Thus, we could model a flow of heat from a fixed heat source on one end of a thermally conducting bar, a linear medium whose exterior edges were perfectly insulated, and whose other end was held at another lower temperature.

There are cases where we cannot specify the potential V at a boundary but can only specify an electric field that is normal to the boundary. Since the electric field is given by (4.18), this in one dimension gives a value for $\mathbf{E} = -(dV/dx)\mathbf{u_x}$ at that boundary.

EXAMPLE 4.5

The electric field normal to a surface at $x = 0$ in Figure 4–5a is arbitrarily specified to be $E_0 = -dV/dx$. Find the potential variation between the plates if the potential at $x = x_0$ is still given by $V = 0$.

Answer. From (4.32), the constant $C_1 = -E_0$ and $C_2 = -C_1 x_0 = E_0 x_0$. Therefore, the potential V is

$$V = E_0(x_0 - x)$$

which is similar to the result shown in Figure 4–5b.

In the above calculation and in the example, it was assumed that no additional charge was distributed between the plates. Let us now assume that a charge is distributed uniformly between the plates and has a *constant* density ρ_v. Other spatial distributions for this charge density could be encountered in practice, but let us not complicate the problem at this stage. For the problem that has been posed, we must solve a one-dimensional Poisson's equation that is written as

$$\frac{d^2V}{dx^2} = -\frac{\rho_v}{\varepsilon_0} \qquad (4.35)$$

where again we can neglect any variation in y or z since the plates have been assumed to be infinite in extent and are equipotential surfaces. This allows us to employ an ordinary derivative again. The two-step integration of (4.35) leads to

$$\frac{dV}{dx} = -\frac{\rho_v}{\varepsilon_0} x + C_3$$

$$V = -\frac{\rho_v}{\varepsilon_0} \frac{x^2}{2} + C_3 x + C_4 \qquad (4.36)$$

where C_3 and C_4 are the constants of integration. The constants of integration are found by specifying the boundary conditions and then solving the simultaneous algebraic equations

$$V_0 = -\frac{\rho_v}{\varepsilon_0} \frac{(0)^2}{2} + C_3(0) + C_4$$

$$0 = -\frac{\rho_v}{\varepsilon_0} \frac{(x_0)^2}{2} + C_3(x_0) + C_4$$

Solving these two equations, we find the constants to be

$$C_3 = -\frac{1}{x_0}\left(V_0 - \frac{\rho_v}{\varepsilon_0} \frac{x_0^2}{2}\right)$$

and $C_4 = V_0$. The substitution of these constants into the solution (4.36) yields the final potential to be

$$V = V_0 \left(1 + \frac{\rho_v}{\rho_c} \frac{x}{x_0}\right)\left(1 - \frac{x}{x_0}\right) \qquad (4.37)$$

where $\rho_c = 2\varepsilon_0 V_0 / x_0^2$. The electric field $\mathbf{E}$ is computed from (4.34) to be

$$\mathbf{E} = \frac{V_0}{x_0}\left(1 - \frac{\rho_v}{\rho_c}\left(1 - 2\frac{x}{x_0}\right)\right)\mathbf{u_x} \qquad (4.38)$$

The electric potential and electric field are depicted in Figure 4–6 for three values of the charge density.

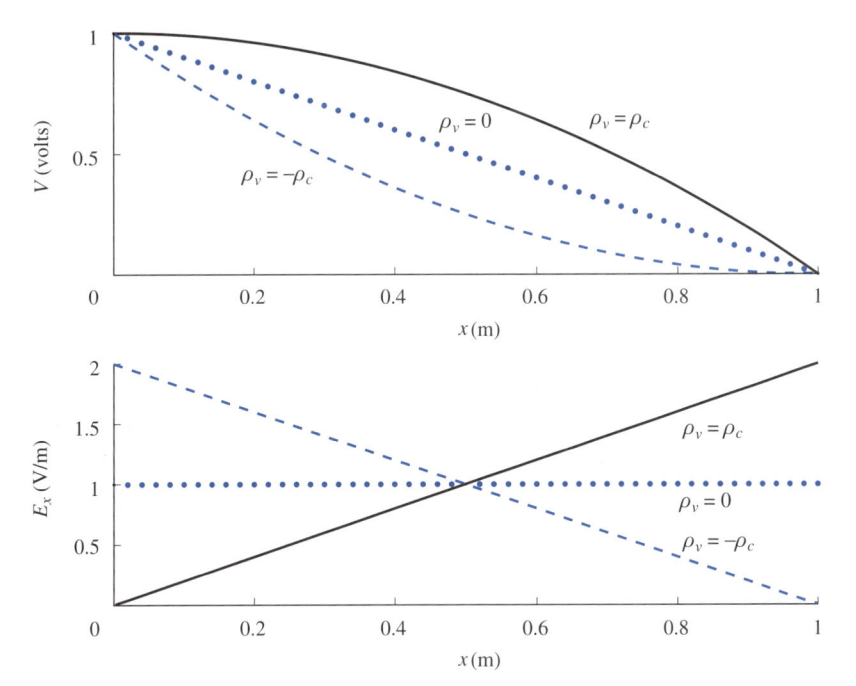

4.3 Analytical Solution in One Dimension—Direct Integration Method

FIGURE 4–6

The electric potential and the electric field for three different values of the charge density ρ_v as calculated in (4.37) and (4.38).

EXAMPLE 4.6

There is a sheath in a plasma that is very similar to the ***depletion layer*** in a pn junction. A plasma is an ionized gas that contains an equal density of positively charged ions n_i that is equal to the density of negatively charged electrons n_e. For instance, the conducting medium in a fluorescent bulb can be considered to be a plasma. A sheath connects the plasma to a foreign object such as a metal wall. Since the electron and ion densities are equal, the electric field in the plasma can be assumed to be equal to zero. If the plasma were grounded at some far-off place, we can assume that the absolute potential of the plasma is also equal to zero. This is a common state in nature since over 99% of the universe is in the plasma state.

Let us now consider a plasma that contains a metal plate to which a negative potential is applied suddenly at a time $t = 0$. At $t = 0^+$ after the switch is closed, electrons close to the plate are "blown" into and "lost" in the background plasma. Due to their heavier mass $M_i \geq 1836 m_e$, where m_e is the mass of an electron, the ions will not move at this early time. There is another narrow steady-state sheath that exists in such a plasma that is called the ***Debye sheath*** that "shields" the plasma from this metal plate and this narrow region of positive charge density from the plasma. We will neglect the Debye sheath in this example. This allows us to state the boundary conditions at the edge of the electron depleted region at $x = x_0$ to be $V = 0$ and $\mathbf{E} = 0$. The dimension x_0, however, is unknown and must be computed. Plot the resulting electric potential and electric field distribution.

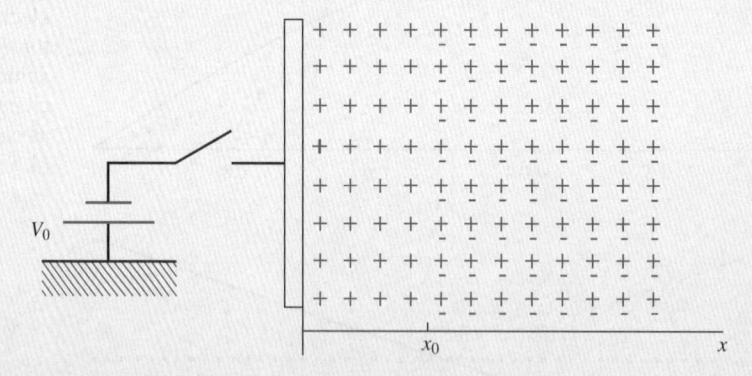

Answer. The system is modeled with a one-dimensional Poisson's equation in the region $0 \leq x \leq x_0$. Since the electrons have departed, only a bare cloud of ions with a uniform charge density $\rho_v = n_i q$ exists within this region. We write this equation as

$$\frac{d^2 V}{dx^2} = -\frac{n_i q}{\varepsilon_0}$$

This equation can be integrated twice to yield

$$V = -\frac{n_i q}{\varepsilon_0} \frac{x^2}{2} + Ax + B$$

where A and B are constants of integration. There are three unknown constants: A, B, and x_0. Remember that we *do not know* where x_0 is; we only know the boundary conditions that are applicable at that location. The three boundary conditions are

$$V(x = 0) = -V_0 \Rightarrow -V_0 = B$$

$$V(x = x_0) = 0 \Rightarrow 0 = -\frac{n_i q}{\varepsilon_0} \frac{x_0^2}{2} + Ax_0 + B$$

$$E_x(x = x_0) = 0 \Rightarrow 0 = \frac{n_i q}{\varepsilon_0} x_0 - A$$

The simultaneous solution of this set of three algebraic equations gives us the location x_0

$$x_0 = \sqrt{\frac{2\varepsilon_0 V_0}{n_i q}}$$

This dimension x_0 was originally called the ***transient sheath*** in plasma physics. As time increases and the ions start to move, a full set of nonlinear fluid equations has to be

4.3 Analytical Solution in One Dimension—Direct Integration Method

solved *numerically* in order to describe the physical behavior of the ion motion. This set consists of the equations of continuity and motion for the ions, an assumption for the distribution of the electrons and Poisson's equation to account for the charge non-neutrality. In plasma-processing applications where plasma ions are to be implanted into a metal surface, this dimension is called the ***ion matrix sheath***. These applications are found in integrated circuit manufacturing. This sheath evolution phenomena is an active area for research. Because of the practical importance resulting from this effect, the evolution of the implanting ions and the expanding sheath are described more fully in Appendix E.

EXAMPLE 4.7

Find and plot the potential distribution between two long concentric cylinders. The length of the cylinders is $\mathscr{L}$. The boundary conditions are: $V(\rho = a) = V_0$ and $V(\rho \geq b) = 0$.

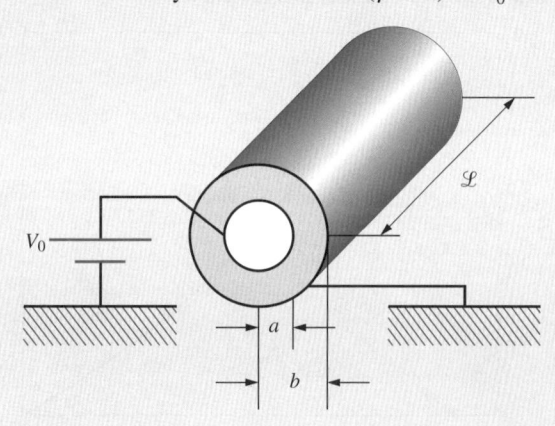

Answer. Because of the symmetry, we should employ Laplace's equation in cylindrical coordinates (4.22). We can assume that the cylinders are of infinite length and let $\partial V / \partial z = 0$. In addition, there is no variation in the ϕ direction, hence $\partial V / \partial \phi = 0$.

We have to solve the one-dimensional Laplace's equation in cylindrical coordinates, which becomes an ordinary differential equation in the independent variable ρ.

$$\frac{1}{\rho} \frac{d}{d\rho} \left(\rho \frac{dV}{d\rho} \right) = 0$$

After multiplying both sides of this equation by ρ, we find the first integration yields

$$\frac{dV}{d\rho} = \frac{C_1}{\rho}$$

which is the radial component of the electric field with a minus sign. The second integration yields

$$V = C_1 \ln \rho + C_2$$

Applying the boundary conditions, we write

$$V(\rho = a) \Rightarrow V_0 = C_1 \ln a + C_2$$
$$V(\rho = b) \Rightarrow 0 = C_1 \ln b + C_2$$

The constants of integration C_1 and C_2 are found from the simultaneous solution of these two equations.

$$C_1 = \frac{V_0}{\ln(a/b)} \quad \text{and} \quad C_2 = -C_1 \ln b$$

The potential variation between the two cylinders is finally written as

$$V = V_0 \left(\frac{\ln(\rho/b)}{\ln(a/b)} \right)$$

where $a < \rho < b$. The variation of the potential for the radius $b = 4a$ is shown below. Note that the potential within the inner cylinder is a constant.

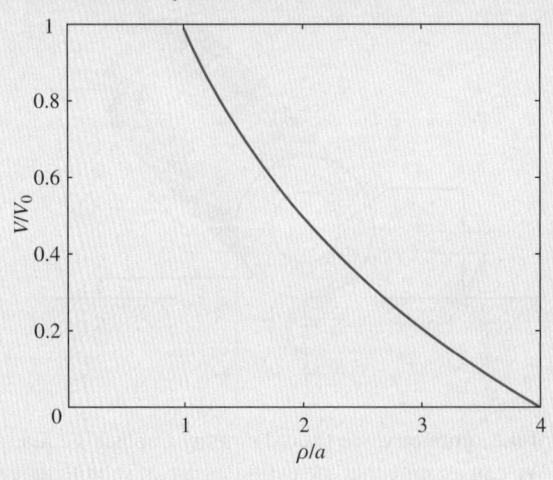

EXAMPLE 4.8

A problem that is found in the study of a plasma is to compute the effect of introducing an additional charge into a previously neutral plasma. The electron density n_e depends on the local potential V and can be described with a Maxwell–Boltzmann distribution

$$n_e(r) = n_0 e^{-\frac{qV(r)}{k_B T_e}}$$

where $k_B T_e$ defines the random thermal energy of the electrons (Boltzmann's constant $k_B = 1.38 \times 10^{-23}$ J/K and T_e is the electron temperature in K). The ion density $n_i = n_0$. Find the potential distribution caused by the introduction of one additional positively

4.3 Analytical Solution in One Dimension—Direct Integration Method

charged particle into a previously neutral plasma. Without loss of generality, we can assume that we initially have a spherical cloud of particles with no net charge, before we introduce an additional charge into the region.

Answer. Due to the spherical symmetry of the system, we can neglect any variation of potential in two of the coordinates (θ, ϕ) and write Poisson's equation in spherical coordinates with the assistance of (4.23), using only the dependence on the radial coordinate r, where the charge distribution is given by

$$\rho_v(r) = (n_e(r) - n_0)q$$

$$\frac{1}{r^2}\frac{d}{dr}\left(r^2 \frac{dV}{dr}\right) = -\frac{\rho_v}{\varepsilon_0} \equiv -\frac{n_0 q\left(e^{-\frac{qV(r)}{k_B T_e}} - 1\right)}{\varepsilon_0}$$

This differential equation is a nonlinear one, which in general has to be solved *numerically*—this is considered in the next section. To develop an analytical solution to this problem, we will convert the nonlinear differential equation to a linear representation. To do this we will linearize the equation around some point or region. There are many ways to approach this. For instance, we could use a Taylor series expansion of a function around some point. While this will not give us an exact answer for the most general case, it will provide accurate results close to the point of region of linearization. In this case, we will find a linear expression for the exponential term, then solve the resulting equation analytically.

$$-\frac{\rho_v}{\varepsilon_0} = -\frac{n_0 q}{\varepsilon_0}\left\{\left[1 - \frac{qV}{k_B T_e} + \cdots\right] - 1\right\} = \frac{n_0 q^2}{\varepsilon_0 k_B T_e}V$$

or

$$\frac{1}{r^2}\frac{d}{dr}\left(r^2 \frac{dV}{dr}\right) = \frac{V}{\lambda_D^2}$$

where $\lambda_D = \sqrt{\varepsilon_0 k_B T_e / n_0 q^2}$ is the **Debye length**. We have used a small potential expansion for the exponential term: $\exp(\pm x) \approx 1 \pm x$. The solution of this equation is facilitated if we define a new variable $W = rV$. Substitute this variable into this equation to obtain

$$\frac{d}{dr}\left(r\frac{dW}{dr} - W\right) = \frac{rW}{\lambda_D^2}$$

or

$$\frac{d^2 W}{dr^2} = \frac{W}{\lambda_D^2}$$

The solution of this equation, which is finite when $\rho \to \infty$, is

$$W = e^{-\frac{r}{\lambda_D}} \quad \text{or} \quad V = \frac{1}{r} e^{-\frac{r}{\lambda_D}}$$

This states that the effect of the additional charge will be "screened" away in a few Debye lengths. In other words, this says we can add a small amount of excess charge in a larger neutral region, and not upset the overall balance of the larger region. Understanding this process will turn out to be very useful when we also try to understand the doping process in semiconductor materials.

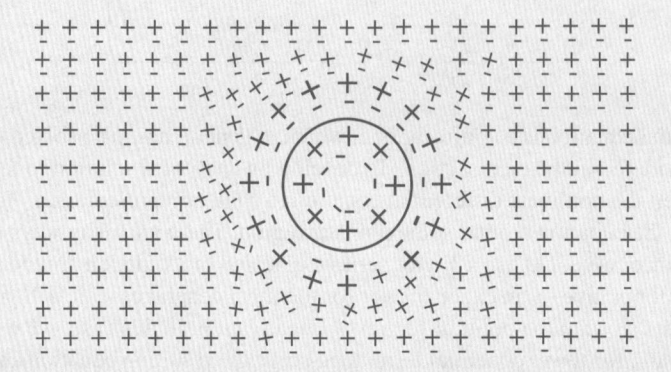

We can think of the Debye length as being analogous to a "time constant" in an electric circuit that has a potential that decays to zero in a few time constants.

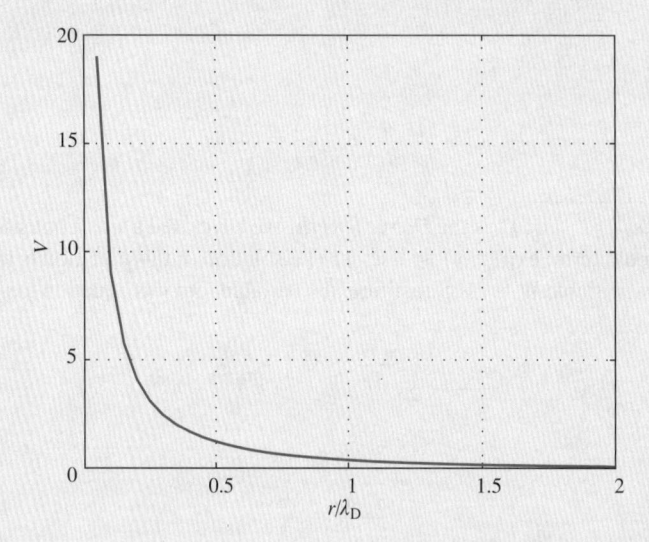

These examples illustrate the methodical procedure that should be followed in order to solve analytically either Poisson's or Laplace's equations in one dimension.

1. Choose the most appropriate representation for the Laplacian ∇^2 based on any symmetry that may be found in the problem. Certain derivatives may also be equal to zero due to the symmetry.
2. Perform the integrations of the differential equation in order to obtain the most general solution for the potential, being very careful to include all of the arbitrary constants of integration.
3. Let this general representation for the potential satisfy the boundary conditions of the problem. This will specify the values of the arbitrary constants of integration found in step 2.

The reader should be made aware that almost all of the electrostatics problems that can be solved analytically have already been solved, and the method of solution and the answers usually appear as examples or problems in some textbook. Further calculations of electrostatic potential problems could be performed for two or three dimensional configurations. These typically will involve mathematical techniques, such as separation of variables that will be encountered later.

4.4 Numerical Solution of a One-Dimensional Equation—Finite Difference Method

To solve the one-dimensional Laplace and Poisson equations numerically, we will use the technique known as the ***Finite Difference Method*** (FDM). We will develop an implementation of this technique using MATLAB, although the technique can be implemented using other software packages. To use the FDM, we will replace the derivatives in differential equations with difference expressions, turning differential equations into finite difference equations. We will use Figure 4-7 to help us understand this approach to defining a derivative.

We know that one definition of the derivative of a function $f(x)$ is the difference between two values of the function divided by the difference between the two values of x, as that difference approaches zero. Mathematically, we write this as:

$$\lim_{h \to 0} \frac{f(x+h) - f(x)}{h}$$

It turns out that this is equivalent to finding the slope of the curve at $f(x)$. Thus, we can find either the derivative at a point on a curve or the slope of

FIGURE 4–7

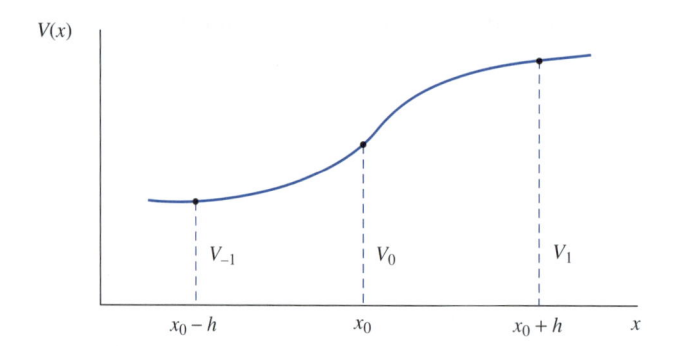

Voltage as a function of position. The finite difference equations will be derived with reference to this figure.

the tangent line at that point by the calculation of a difference divided by the distance between the two points on the curve. Therefore, in theory, we can make our difference approximate the derivative as accurately as we desire.

$$\left.\frac{dV}{dx}\right|_{x_0} = \frac{V_1 - V_0}{h} \quad \text{(Forward difference method)} \tag{4.39}$$

$$\left.\frac{dV}{dx}\right|_{x_0} = \frac{V_0 - V_{-1}}{h} \quad \text{(Backward difference method)} \tag{4.40}$$

$$\left.\frac{dV}{dx}\right|_{x_0} = \frac{V_1 - V_{-1}}{2h} \quad \text{(Central difference method)} \tag{4.41}$$

We can interpret the central difference method as being an average of the other two methods. In order to ascertain which method is *better*, we estimate the errors that might be expected to be found in each method. The errors can be estimated by expanding the voltage in a **Taylor series expansion** about the point x_0

$$V(x_0 + h) = V(x_0) + \frac{h}{1!}\left.\frac{dV}{dx}\right|_{x_0} + \frac{h^2}{2!}\left.\frac{d^2V}{dx^2}\right|_{x_0} + \frac{h^3}{3!}\left.\frac{d^3V}{dx^3}\right|_{x_0} + \cdots \tag{4.42}$$

If we neglect the third derivative and higher order terms, we write

$$\left.\frac{dV}{dx}\right|_{x_0} = \frac{V(x_0 + h) - V(x_0)}{h} - \frac{h}{2}\left.\frac{d^2V}{dx^2}\right|_{x_0} \tag{4.43}$$

4.4 Numerical Solution of a One-Dimensional Equation—Finite Difference Method

A comparison of (4.43) with (4.39) shows that this is equivalent to the forward difference method with the additional term

$$\frac{h}{2} \frac{d^2 V}{dx^2}\bigg|_{x_0} \quad (4.44)$$

This term is an error term. There are, of course, additional higher order terms that could be included. These additional terms are multiplied by the parameter h to a higher order power. If the parameter h can be made sufficiently small, (4.43) could be useful.

In a similar manner, we write the Taylor series expansion $V(x_0 - h)$ about the point x_0.

$$V(x_0 - h) = V(x_0) - \frac{h}{1!} \frac{dV}{dx}\bigg|_{x_0} + \frac{h^2}{2!} \frac{d^2 V}{dx^2}\bigg|_{x_0} - \frac{h^3}{3!} \frac{d^3 V}{dx^3}\bigg|_{x_0} + \cdots \quad (4.45)$$

From (4.45), we compute

$$\frac{dV}{dx}\bigg|_{x_0} = \frac{V(x_0) - V(x_0 - h)}{h} + \frac{h}{2} \frac{d^2 V}{dx^2}\bigg|_{x_0} \quad (4.46)$$

for the backward difference method. The error term is also given by (4.46) plus higher order terms.

Subtracting (4.45) from (4.42) yields

$$V(x_0 + h) - V(x_0 - h) = 2h \frac{dV}{dx}\bigg|_{x_0} + 2 \frac{h^3}{6} \frac{d^3 V}{dx^3}\bigg|_{x_0} \quad (4.47)$$

or

$$\frac{dV}{dx}\bigg|_{x_0} = \frac{V(x_0 + h) - V(x_0 - h)}{2h} - \frac{h^2}{6} \frac{d^3 V}{dx^3}\bigg|_{x_0} \quad (4.48)$$

In (4.48) the error is of the order of h^2. This is in the **central difference method**. The error in using this method will be *smaller* than in either of the other two methods, and it will be the one employed throughout the rest of the chapter.

Using the central difference method, we find the representation for the second derivative to be

$$\left.\frac{d^2 V}{dx^2}\right|_{x_0} = \frac{\left.\frac{dV}{dx}\right|_{x_0+\frac{h}{2}} - \left.\frac{dV}{dx}\right|_{x_0-\frac{h}{2}}}{h} = \frac{\frac{V_1 - V_0}{h} - \frac{V_0 - V_{-1}}{h}}{h}$$

or

$$\left.\frac{d^2 V}{dx^2}\right|_{x_0} = \frac{V_1 - 2V_0 + V_{-1}}{h^2} \tag{4.49}$$

EXAMPLE 4.9

Find the potential distribution between two surfaces if $V(x = 0) = 0$ and $V(x = 1) = 3$. There is no charge distribution in the space $0 \leq x \leq 1$.

Answer. Let us use only three points for the first iteration: $x_0 = 0$, $x_1 = 0.5$, and $x_2 = 1$. Using the central difference method with a step size $h = 0.5$, we write Laplace's equation as

$$\left.\frac{d^2 V}{dx^2}\right|_{x_1} = \frac{V_2 - 2V_1 + V_0}{0.5^2} = 0$$

The boundary conditions imply $V_0 = V(x_0 = 0) = 0$ and $V_2 = V(x_2 = 1) = 3$. Hence

$$V_1 = 0.5(V_0 + V_2) = 1.5$$

which demonstrates the ***principle of the average value*** for the middle point x_1. The second iteration with the smaller steps size $h = 0.25$ is applied to five points in the same interval. The boundary conditions, which are now $V_0 = 0$ and $V_4 = 3$, lead to three simultaneous equations

$$V_1 = 0.5(V_0 + V_2); \quad V_2 = 0.5(V_1 + V_3); \quad V_3 = 0.5(V_2 + V_4)$$

In this case, the boundary conditions specify V_0 and V_4, and the voltage V_2 was calculated in the previous iteration. The solutions for the three intermediate points and the two end points are

$$V_0 = 0; \quad V_1 = 0.75; \quad V_2 = 1.50; \quad V_3 = 2.25; \quad V_4 = 3$$

A comparison of these computed values is in agreement with the analytical solution $V(x) = 3x$ obtained in Example 4–6. The MATLAB calculation produces the following results:

4.4 Numerical Solution of a One-Dimensional Equation—Finite Difference Method 205

$V =$	0	NaN	1.5000	NaN	3.0000
$V =$	0	0.7500	1.5000	NaN	3.0000
$V =$	0	0.7500	1.5000	2.2500	3.0000

The voltage distribution is shown below.

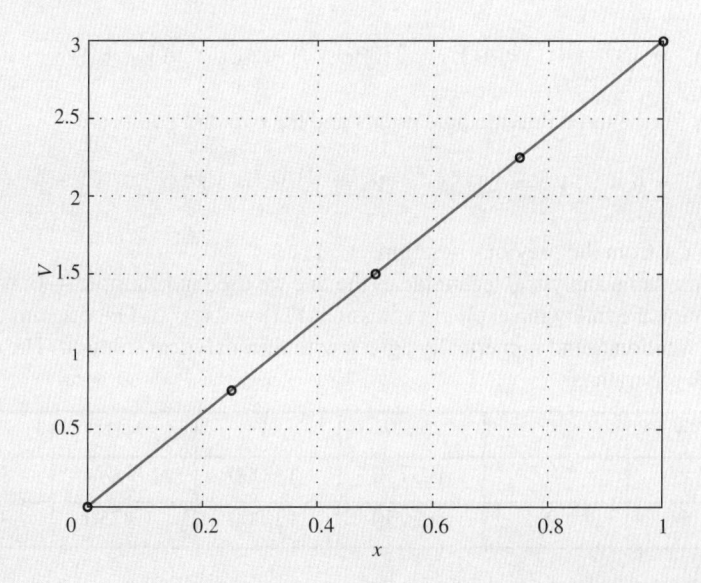

EXAMPLE 4.10

Repeat Example 4–9 with a uniform charge distribution $\rho_v = -4\varepsilon_0$ in the space $0 \leq x \leq 1$. Find the potential distribution between two surfaces if $V(x = 0) = 0$ and $V(x = 1) = 3$.

Answer. Using the central difference method, we write Poisson's equation as

$$\left.\frac{d^2V}{dx^2}\right|_{x_1} = \frac{V_2 - 2V_1 + V_0}{0.5^2} = 4$$

for the first iteration. The boundary conditions imply $V_0 = V(x = 0) = 0$ and

$V_2 = V(x = 1) = 3$. Hence

$$\frac{V_2 - 2V_1 + V_0}{0.5^2} = 4(3 - 2V_1 + 0) = 4 \Rightarrow V_1 = 1$$

The second iteration with the boundary conditions $V_0 = V(x = 0) = 0$ and $V_4 = V(x = 1) = 3$ leads to

$$4(V_2 - 2V_1 + V_0) = 4; \quad 4(V_3 - 2V_2 + V_1) = 4; \quad 4(V_4 - 2V_3 + V_2) = 4$$

The solutions for the three intermediate points and the two end points are

$$V_0 = 0; \quad V_1 = 0.375; \quad V_2 = 1; \quad V_3 = 1.875; \quad V_4 = 3$$

The term $V_2 = 1$ from the previous iteration.

Applying the same analytical technique as the one we used in Example 4–6, we find for the exact solution the following explicit expression: $V(x) = 2x^2 + x$. Once again, the approximate solution found numerically agrees with the analytical solution. The output of the MATLAB program is

V =	0	NaN	1	NaN	3
V =	0	0.3750	1.0000	NaN	3.0000
V =	0	0.3750	1.0000	1.8750	3.0000

The plot of the results is shown below.

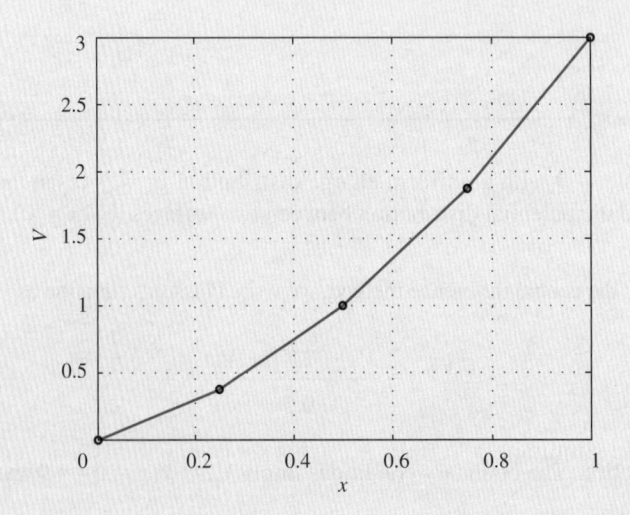

4.4 Numerical Solution of a One-Dimensional Equation—Finite Difference Method

The critical restriction on the mesh size is that the first point must be at the center and its numerical value is determined by the value at the two boundaries. This will restrict the number of internal points N to have only certain values that are prescribed by the following prescription:

$$1;\quad 3;\quad 7;\quad 15;\quad 31;\quad 63;\quad \ldots [2^N - 1]$$

Let us call this the array size.

We may have noted that the length h that appears in our application of (4.49) has changed from 0.5 to 0.25. In the next step, it will be reduced to 0.125 and then 0.0625 and so on. We can also evaluate (4.49) and keep h as a prescribed value, but as we will see, the calculation will have to be repeated several times. The numbers will converge, hopefully in a reasonable time, to the correct answer.

In order to introduce the procedure, we will redo the calculation of the one-dimensional Laplace's equation that we have just performed but now make the *a priori* assumption that $h = 0.25$. We set the three values internal to the fixed boundaries as initially being equal to zero. Hence, we write

$$V_1(0) = 0;\quad V_2(0) = 0;\quad V_3(0) = 0$$

The boundary values will remain fixed at all iterations, namely $V_0 = 0$ and $V_4 = 3$.

In our first iteration denoted with the (1), using (4.49), we write

$$\frac{V_0 - 2V_1(1) + V_2(0)}{0.25^2} = 0$$

$$\frac{V_1(1) - 2V_2(1) + V_3(0)}{0.25^2} = 0$$

$$\frac{V_2(1) - 2V_3(1) + V_4}{0.25^2} = 0$$

In the second equation, we include the value for $V_1(1)$ that has just been obtained from the previous equation since it is now known. A similar argument holds for $V_3(1)$ in the third equation. In fact, this will be a general pattern. The simultaneous solution of this set of equations leads to

$$V_1(1) = 0;\quad V_2(1) = 0;\quad V_3(1) = 1.5$$

In order to compute the values at the second iteration, we use the values from the first iteration and sequentially write

$$\frac{V_0 - 2V_1(2) + V_2(1)}{0.25^2} = 0$$

$$\frac{V_1(2) - 2V_2(2) + V_3(1)}{0.25^2} = 0$$

$$\frac{V_2(2) - 2V_3(2) + V_4}{0.25^2} = 0$$

From this set, we compute

$$V_1(2) = 0; \quad V_2(2) = 0.75; \quad V_3(2) = 1.875$$

The third iteration is

$$\frac{V_0 - 2V_1(3) + V_2(2)}{0.25^2} = 0$$

$$\frac{V_1(3) - 2V_2(3) + V_3(2)}{0.25^2} = 0$$

$$\frac{V_2(3) - 2V_3(3) + V_4}{0.25^2} = 0$$

We obtain

$$V_1(3) = 0.375; \quad V_2(3) = 1.125; \quad V_3(3) = 2.0625$$

We could keep going using our calculator, but let us stop here. A more interesting question arises at this point. Is the answer *correct*? We can check this easily by dividing the parameter h by two and redoing the calculation again and again. If the numbers are the same or seem to asymptotically approach the same value, we are finished.

EXAMPLE 4.11

Write a MATLAB program to evaluate and plot the first six iterations of the solution of the one-dimensional Laplace's equation by FDM. The boundary conditions are: $V(1) = 0$ and $V(5) = 3$.

Answer. The iterations are indicated with the integer k. The analytical solution $V(x) = 0.75(x - 1)$ is shown with a dashed line for a comparison

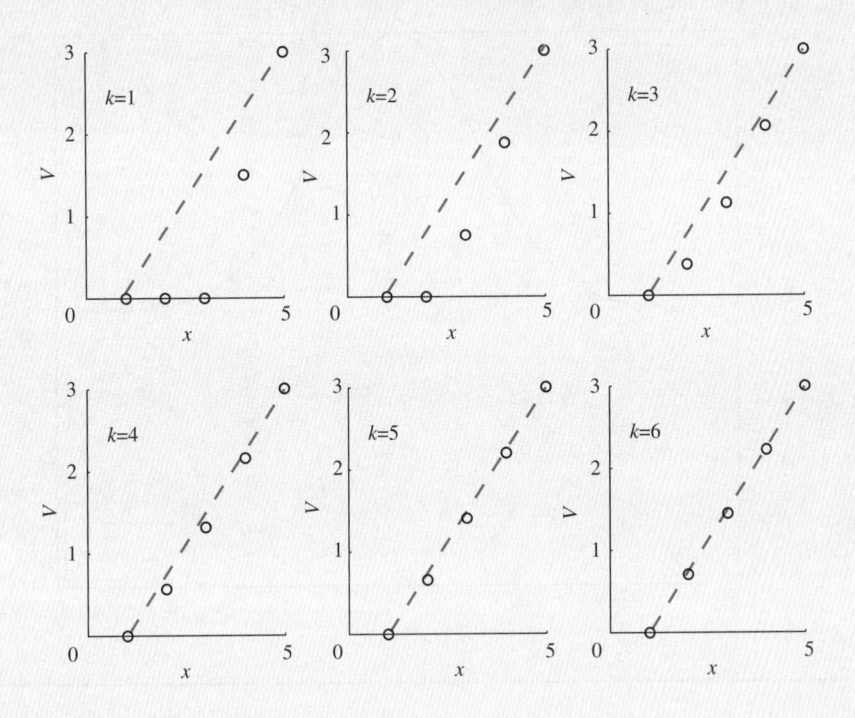

4.4 Numerical Solution of a One-Dimensional Equation—Finite Difference Method

in the figures below. For this case six iterations are sufficient.

EXAMPLE 4.12

Using MATLAB, plot the voltage $V(x) = e^{-x^2}$ and the electric field $E_x(x) = -dV/dx$ as a function of x in the range $-3 \leq x \leq 3$.

Answer. The difference operation *diff(V)./diff(x)* sequentially performs and stores the values

$$\left\{ \frac{V(2) - V(1)}{h}, \frac{V(3) - V(2)}{h}, \ldots, \frac{V(n) - V(n-1)}{h} \right\}$$

There are only $(n-1)$ values of the derivatives. Therefore, if we wish to plot the results, we plot V in the range x_a to x_b at increments of h and dV/dx in the range $x_a + h/2$ to $x_b - h/2$ with the same increment h. The plots of the potential and the electric fields are shown in the figure below.

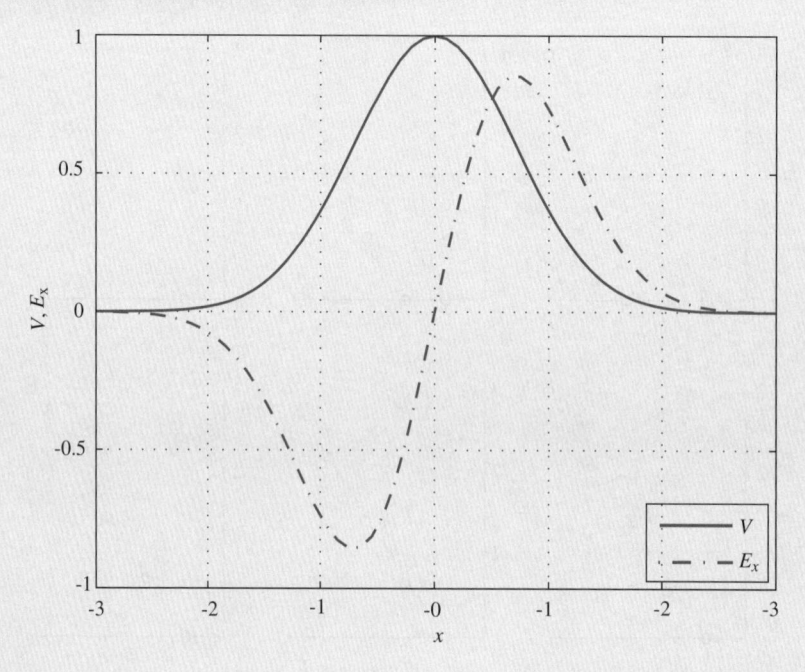

EXAMPLE 4.13

Consider again Example 4–8, which has been solved analytically after a linearization. Write a MATLAB program to calculate and plot the potential distribution using the function *ode45* for obtaining numerical solutions of differential equations.

Answer. We first transform Poisson's equation into a standard form with the substitution $dV/dr = U$. This results in two coupled first-order differential equations.

$$\frac{dV}{dr} = U$$

$$\frac{dU}{dr} = -\frac{2}{r}U + D(e^{-CV} - 1)$$

where $C = q/k_B T_e$ and $D = n_0 q/\varepsilon_0$. Choosing numerical values of $C = 0.25$ and $D = 4$, we normalize the Debye length $\lambda_D = 1/\sqrt{CD} = 1$. Assuming that the value of the potential V at $r = 0$ is 10 and that the electric field $E = -dV/dr$ at the same location has a value of 1, we are able to evaluate numerically the potential distribution in space.

The numerical results are shown below.

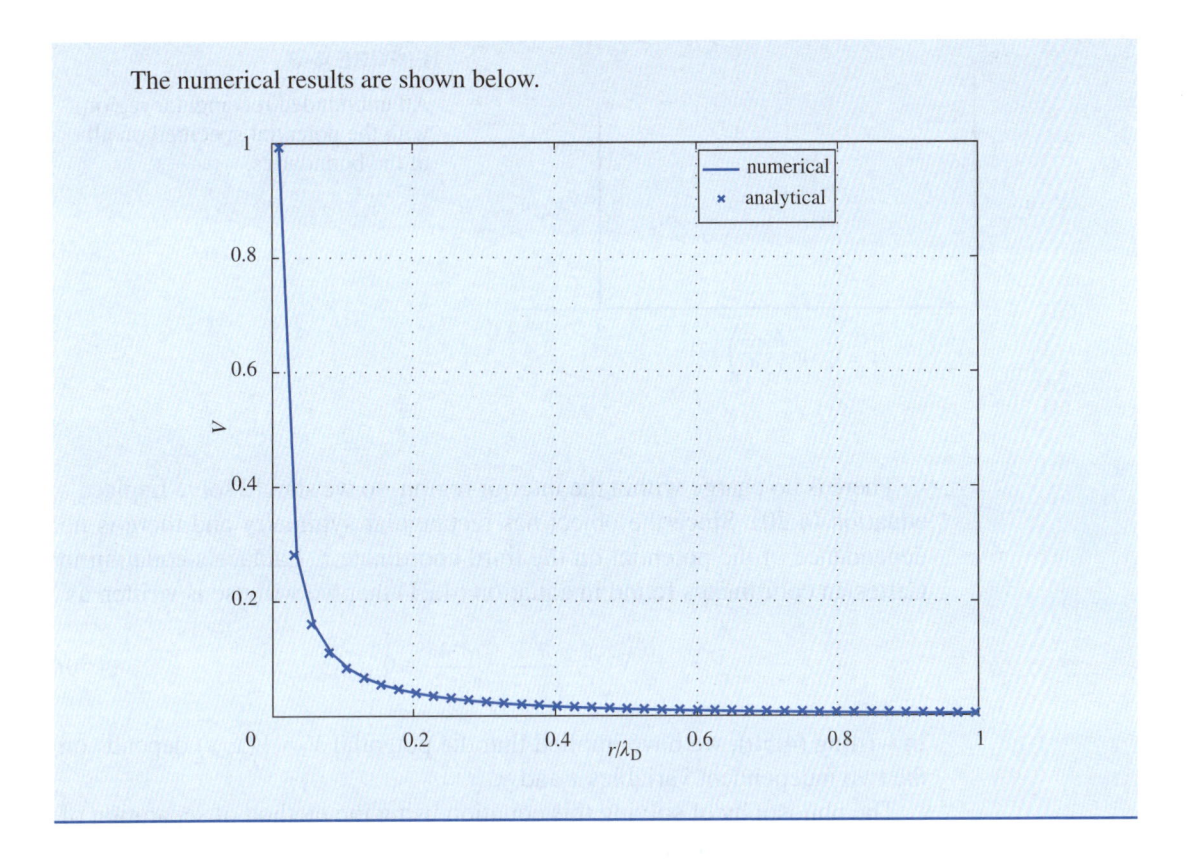

(Plot: V vs r/λ_D, showing numerical curve and analytical points ×, decaying from 1 at 0 to ~0 by 0.4)

4.5 Analytical Solution of a Two-Dimensional Equation—Separation of Variables

In this section, we will introduce a methodical procedure to effect the solution of the type of problem that is governed by Poisson's or Laplace's equations in higher dimensions. The technique that will be employed to obtain this solution is the powerful *method of separation of variables* using the *Fourier series expansion method*.

It is pedagogically convenient to introduce the technique using a simple example and then carefully work through the details. The problem that we will examine initially is to calculate the potential distribution within a charge-free *unbounded* region illustrated in Figure 4–8 where the potential is prescribed on all four edges. In our example, we specify that the potential on two of the edges is equal to zero, the potential approaches zero on the third edge, which is taken to be at $y \to \infty$, and the potential has a particular distribution on the fourth edge.

FIGURE 4–8

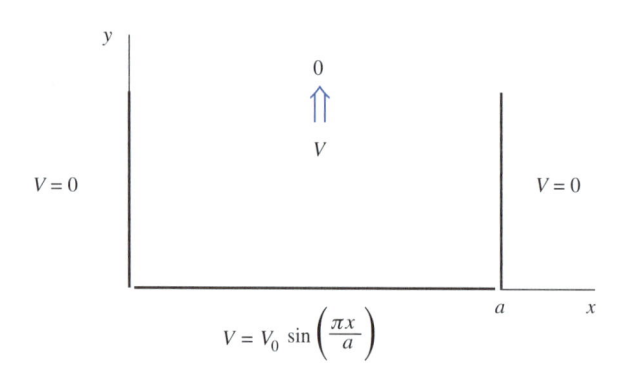

An unbounded rectangular region with the potential specified on all of the boundaries.

There is no charge within the interior region, so we should solve Laplace's equation (4.20). Since the object has rectangular symmetry and there is no dependence of the potential on the third coordinate z, Laplace's equation in Cartesian coordinates found in equation (4.21) that we will use is written as

$$\nabla^2 V = \frac{\partial^2 V}{\partial x^2} + \frac{\partial^2 V}{\partial y^2} = 0 \qquad (4.50)$$

In writing (4.50), we have implied that the potential $V = V(x, y)$ depends on the two independent variables x and y.

The philosophy of solving this equation using the method of separation of variables is to assert that the potential $V(x, y)$ is equal to the product of two terms, $X(x)$ and $Y(y)$, which separately are functions of only one of the independent variables x and y. The potential is then given by

$$\boxed{V(x, y) = X(x)Y(y)} \qquad (4.51)$$

This is a critical assertion and our solution depends on it being a correct assumption. We may wonder if other functional forms would work at this stage. They might or they might not. The resulting solutions that would be obtained using different combinations might physically make no sense or they might not satisfy the boundary conditions. Therefore, we will follow in the footsteps of those pioneering giants who have led us through the dark forest containing problems of this genre and just use (4.51) and not concern ourselves with these questions. "If it ain't broke, why fix it?" will be our motto.

Substitute (4.51) into (4.50) and write

$$Y(y)\frac{d^2 X(x)}{dx^2} + X(x)\frac{d^2 Y(y)}{dy^2} = 0 \qquad (4.52)$$

4.5 Analytical Solution of a Two-Dimensional Equation—Separation of Variables

Note that the terms to be differentiated only involve one independent variable. Hence, the partial derivatives can be replaced with ordinary derivatives and this will be done in the subsequent development.

The next step in this methodical procedure is to divide both sides of this equation by $V(x, y) = X(x)Y(y)$. Our friends in mathematics may stand up in horror at this suggestion! As we will later see, one of these terms could be *zero* at one or more points in space. Recall what a calculator or computer tells us when we do this "evil" deed of dividing by zero! With this warning in hand and with a justified amount of trepidation, let us see what does result from this action. In our case, the end will justify the means. We find that

$$\frac{1}{X(x)} \frac{d^2 X(x)}{dx^2} + \frac{1}{Y(y)} \frac{d^2 Y(y)}{dy^2} = 0 \tag{4.53}$$

The first term on the left side of (4.53) is **independent** of the variable y. As for the variable y, it can be considered to be a constant that we will take to be $-k_y^2$. Using a similar argument, the second term on the left side of (4.53) is **independent** of the variable x, and it also can be replaced with another constant that will be written as $+k_x^2$. Therefore, (4.53) can be written as two **ordinary differential equations** and one **algebraic equation**

$$\frac{d^2 X(x)}{dx^2} + k_x^2 X(x) = 0 \tag{4.54}$$

$$\frac{d^2 Y(y)}{dy^2} - k_y^2 Y(y) = 0 \tag{4.55}$$

$$k_x^2 - k_y^2 = 0 \tag{4.56}$$

A pure mathematician would have just written these three equations down by inspection in order to avoid any problems with dividing by zero, which we have glossed over so cavalierly.

The two second-order ordinary differential equations can be solved easily. We write that

$$X(x) = C_1 \sin(k_x x) + C_2 \cos(k_x x) \tag{4.57}$$

$$Y(y) = C_3 e^{k_y y} + C_4 e^{-k_y y} \tag{4.58}$$

where C_1 to C_4 are constants to be determined. Let us now determine these constants from the boundary conditions imposed in Figure 4–8. From (4.51), we note that the potential $V(x, y)$ is determined by multiplying the solution $X(x)$ with $Y(y)$. Therefore, we can specify the constants by examining each term *separately*. For any value of y at $x = 0$, the potential $V(0, y)$ is equal to zero. The only way that we can satisfy this requirement is to let the constant $C_2 = 0$ since $\cos 0 = 1$. Nothing can be stated about the constant C_1 from this particular boundary condition since $\sin 0 = 0$. For any value of x and in the limit of $y \to \infty$, the potential $V(x, y \to \infty) = 0$. This specifies that the constant $C_3 = 0$ since the term $\exp(k_y y) \to \infty$ as $y \to \infty$. The constant C_4 remains undetermined from the application of this boundary condition. The potential on the third surface $V(a, y)$ is also specified to be zero at $x = a$, from which we conclude that $k_x = n\pi/a$ since $\sin(n\pi) = 0$. From (4.56), we also write that the constants $k_y = k_x$. With these values for the constants, our solution $V(x, y) = X(x)Y(y)$ becomes

$$V(x, y) = C_1 C_4 e^{-\frac{n\pi y}{a}} \sin\left(\frac{n\pi x}{a}\right) \qquad (4.59)$$

For this example, the integer n will take the value of $n = 1$ in order to fit the fourth boundary condition at $y = 0$. Finally, the product of the two constants $[C_1 C_4]$, which is just another constant, is set equal to V_0. The potential in this channel finally is given by

$$V(x, y) = V_0 e^{-\frac{\pi y}{a}} \sin\left(\frac{\pi x}{a}\right) \qquad (4.60)$$

The variation of this potential in space is shown in Figure 4–9 for the values of $a = 1$, $V_0 = 10$.

An examination of Figure 4–9 will yield some important physical insight into the variation of the potential. First, the potential V only approaches zero as the coordinate $y \to \infty$. Second, the boundary conditions at $x = 0$ and at

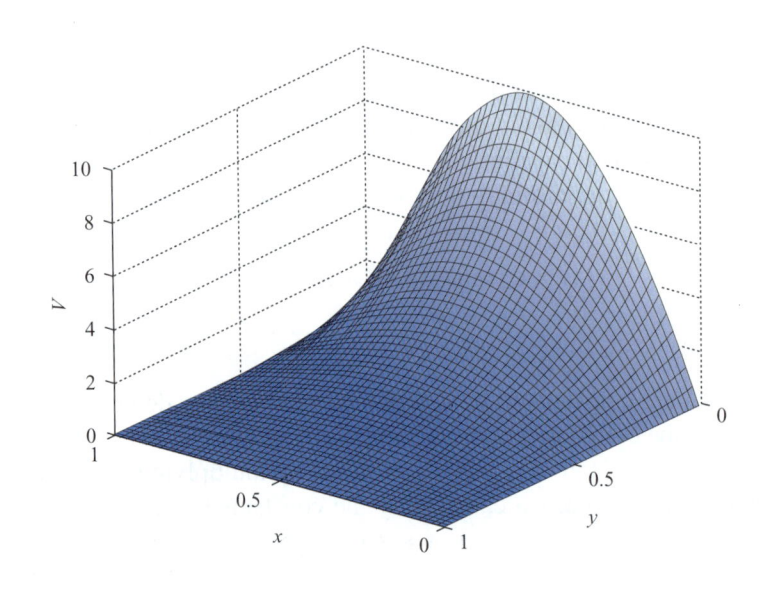

FIGURE 4–9

Variation of the potential within the region for the prescribed boundary conditions depicted in Figure 4–8.

4.5 Analytical Solution of a Two-Dimensional Equation—Separation of Variables

$x = a$ were that the potential V equaled a constant that, in this case, was equal to zero. Recall from Chapter 2 that $E_y = -\partial V / \partial y$. This implies that the component of electric field E_y must also be equal to zero along these two surfaces. We can conclude that the *tangential* component of the electric field adjacent to an equipotential surface will be equal to zero. This conclusion will be of importance in several later calculations.

The procedure that we have conducted is the determination of the solution of a partial differential equation. Let us recapitulate the procedure before attacking a slightly more difficult problem.

1. The proper form of the Laplacian operator $\nabla^2 V$ for the coordinate system of interest was chosen. This choice was predicated on the symmetry and the boundary conditions of the problem.
2. The potential $V(x, y)$ that depended on two independent variables was *separated* into two dependent variables that individually depended on only one of the independent variables. This allowed us to write the partial differential equation as a set of ordinary differential equations and an algebraic equation by assuming that the solution could be considered as a product of the individual functions of the individual independent variables.
3. Each of the ordinary differential equations that was solved led to several constants of integration. The solutions of the ordinary differential equation were multiplied together to obtain the general solution of the partial differential equation.
4. The arbitrary constants of integration that appeared when the ordinary differential equations were solved were determined such that the boundary conditions would be satisfied. The solution for a particular problem has now been obtained. Note that this step is similar to the methodical procedure that we employed in the one-dimensional case.

Let us examine the potential distribution in a *bounded* space as depicted in Figure 4–10. The procedure will be the same as for the unbounded case

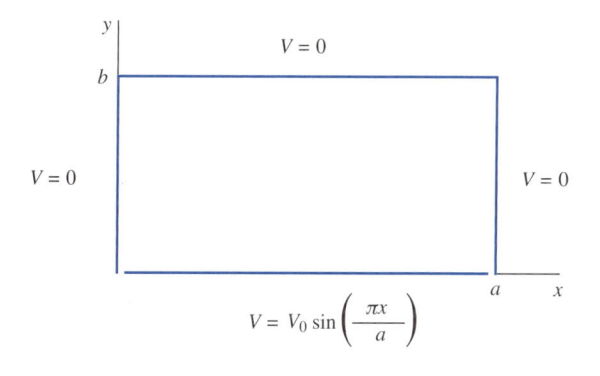

FIGURE 4–10

A boundary value problem for a bounded surface.

treated above. In this case, the potential is required to be equal to zero on three of the boundaries, and it has a sinusoidal variation on the remaining boundary. As before, the solution of Laplace's equation for $X(x)$ and $Y(y)$ is given again by (4.57) and (4.58).

We are able to predict the functional characteristics of the basic *eigenfunction*. This is a German word that means "characteristic function." The values for k_x and k_y are called *eigenvalues* or "characteristic values." In this case again, the eigenvalue $k_y = k_x$ is determined from (4.56). We may also find this function referred to as a "proper function."

The constants are once again determined by the boundary conditions. From (4.57), the constant $C_2 = 0$ again since the potential $V = 0$ at $x = 0$. The constant $k_x = n\pi/a$ ($n = 0, 1, 2, \ldots$) since the potential $V = 0$ at $x = a$ — these are the eigenvalues of the problem. From (4.58), we write

$$C_3 e^{k_y b} + C_4 e^{-k_y b} = 0$$

since the potential $V = 0$ at $y = b$. This yields a relationship between C_3 and C_4. Therefore, the potential within the enclosed region specified in Figure 4–10 can be written as

$$V = \left[C_1 C_3 e^{\frac{n\pi b}{a}} \right] \left[e^{\frac{n\pi(y-b)}{a}} - e^{-\frac{n\pi(y-b)}{a}} \right] \sin\left(\frac{n\pi x}{a}\right)$$

or

$$V(x, y) = \left[2 C_1 C_3 e^{\frac{n\pi b}{a}} \right] \sinh\left(\frac{n\pi(y-b)}{a}\right) \sin\left(\frac{n\pi x}{a}\right) \quad (4.61)$$

The constants within the square brackets will be determined from the remaining boundary condition at $y = 0$.

The boundary condition at $y = 0$ states that $V = V_0 \sin(\pi x/a)$ for $0 \leq x \leq a$. Hence the integer $n = 1$ and

$$C_1 C_3 = \frac{V_0}{2 e^{\frac{\pi b}{a}} \sinh\left(-\frac{\pi b}{a}\right)} \quad (4.62)$$

The potential finally is written as

$$V = \frac{V_0 \sinh\left(\frac{\pi(b-y)}{a}\right) \sin\left(\frac{\pi x}{a}\right)}{\sinh\left(\frac{\pi b}{a}\right)} \quad (4.63)$$

4.5 Analytical Solution of a Two-Dimensional Equation—Separation of Variables 217

FIGURE 4–11

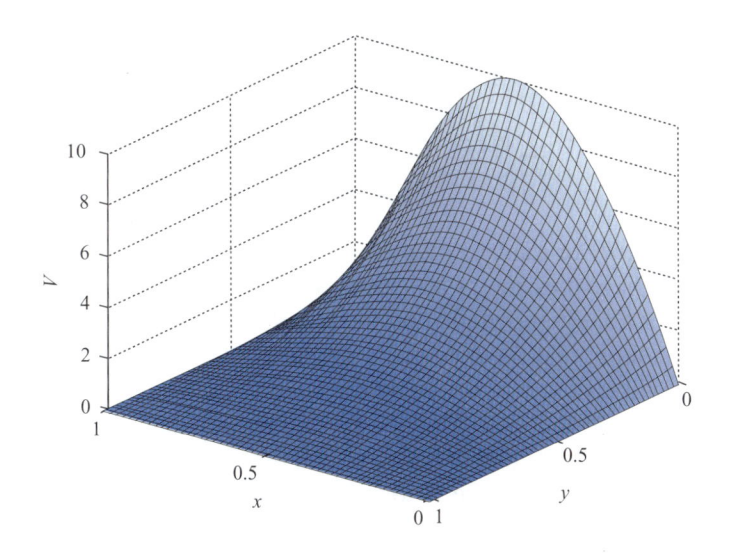

Normalized potential profile within the region described in Figure 4–10. Note that the potential is equal to zero on three edges.

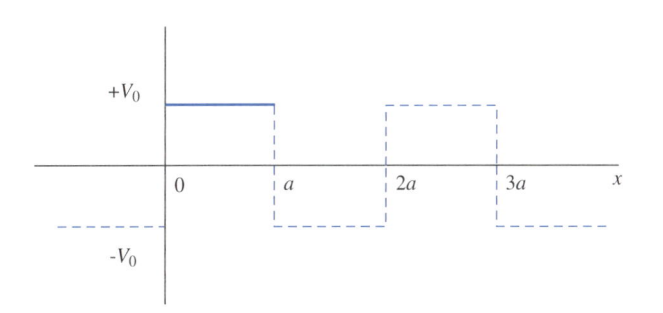

FIGURE 4–12

Periodic potential represents the constant potential $V = V_0$ within the region $0 \leq x \leq a$.

This is shown in Figure 4–11. Note that we do satisfy the imposed boundary condition that the potential equals zero on three edges.

In the two examples that were treated above, we assumed that the boundary condition at $y = 0$ had a nonuniform distribution. This was an academic-type distribution rather than a realistic one, but we were able to "carry out the details" to the very end without having to introduce more complicated mathematics. However, we should look at the real world where we might expect that a *more realistic* distribution for the potential at $y = 0$ in Figure 4–12 would be to assume that the potential would be a *constant*, say $V = V_0$. The boundary conditions on the other three edges could remain the same in realistic situations. Let us carry through the details for this particular boundary condition.

Since the other boundary conditions have not been altered, the general solution of Laplace's equation can be written as a **superposition** of particular solutions given by (4.53)

$$V(x, y) = \sum_{n=1}^{\infty} d_n \frac{\sinh\left(\frac{n\pi(b-y)}{a}\right)\sin\left(\frac{n\pi x}{a}\right)}{\sinh\left(\frac{n\pi b}{a}\right)} \qquad (4.64)$$

because of the **linearity** of Laplace's equation, where n is an integer ($n = 1, 2, 3, \ldots$). In writing this expression as a summation of an infinite series of sinusoidal functions, we are being guided by the fact that *each* term does satisfy the boundary condition that the potential $V = 0$ at $x = 0$ and at $x = a$. Hence the infinite sum will also satisfy the boundary conditions. The coefficients c_n will be chosen to yield the *best fit* of the remaining boundary condition at $y = 0$ that has now been specified to be a constant potential $V = V_0$.

We may recognize (4.64) as the **Fourier sine series** and the constants d_n as the **Fourier coefficients**. The coefficients c_n and d_n are defined for a general periodic function $F(x) \equiv V(x, 0)$ with a period L:

$$F(x) = \frac{c_0}{2} + \sum_{n=1}^{\infty}\left(c_n \cos\left(\frac{2n\pi x}{L}\right) + d_n \sin\left(\frac{2n\pi x}{L}\right)\right) \qquad (4.65)$$

from the relations

$$c_n = \frac{2}{L}\int_{-L/2}^{L/2} F(x)\cos\left(\frac{2n\pi x}{L}\right)dx, \ d_n = \frac{2}{L}\int_{-L/2}^{L/2} F(x)\sin\left(\frac{2n\pi x}{L}\right)dx \qquad (4.66)$$

The potential V is known to have a constant value only in the region $0 \leq x \leq a$. Outside of this region, it is not specified and could have any value that we choose in order to ease our mathematical difficulties. In this case, the period of the wave is $L = 2a$. Our choice for the potential at the boundary is to assume that it is an **odd** function in the variable x. This means that there will only be sine functions in the expansion. Therefore, $c_n = 0$. The coefficients d_n with reference to Figure 4–12 are calculated from

$$d_n = 2\frac{V_0}{a}\int_0^a \sin\left(\frac{n\pi x}{a}\right)dx$$

This integral leads to

$$d_n = \begin{cases} \dfrac{4V_0}{n\pi} & n = \text{odd} \\ 0 & n = \text{even} \end{cases} \qquad (4.67)$$

4.5 Analytical Solution of a Two-Dimensional Equation—Separation of Variables

The potential is given by (4.65) with the coefficients defined in (4.67).

$$V(x, y) = \frac{4V_0}{\pi} \sum_{n=1, 3, 5, \ldots}^{\infty} \frac{\sinh\left(\frac{n\pi(b - y)}{a}\right)\sin\left(\frac{n\pi x}{a}\right)}{n\sinh\left(\frac{n\pi b}{a}\right)} \quad (4.68)$$

EXAMPLE 4.14

Write a MATLAB program that calculates and plots the two-dimensional potential variation for the potential source shown in Figure 4–12. Use 5 terms for the potential.

Answer. The graph below shows the potential variation resulting from summing the first five terms of the Fourier series representing the potential of Figure 4–12.

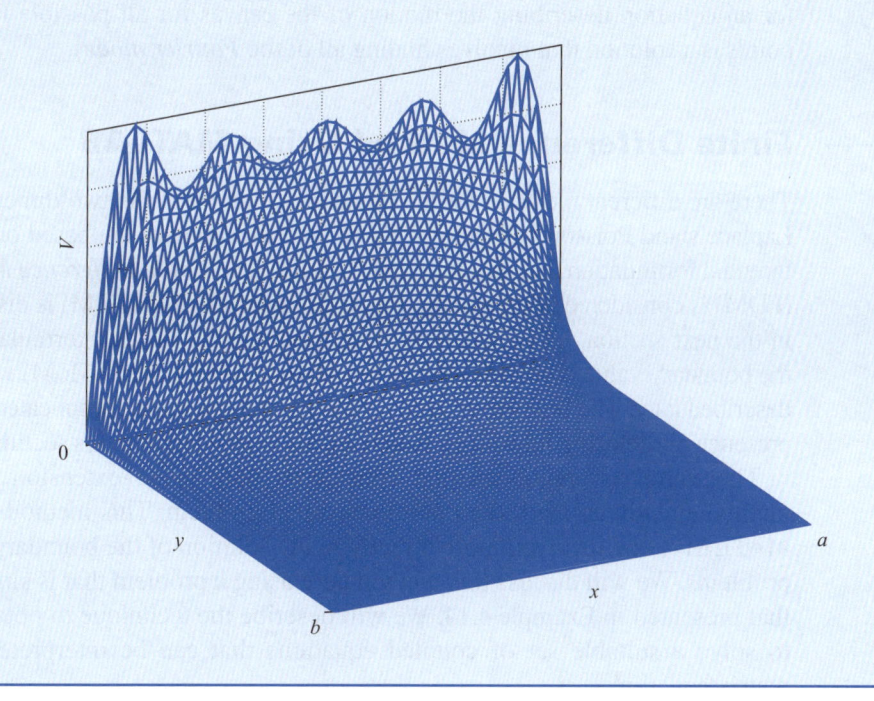

Certain general comments can be made about the potential variation that is shown in the figure in the previous example, especially when it is compared with the potential profile in Figure 4–12, which is just the first Fourier term of the Fourier series. The fit to a constant value at $y = 0$ is better if more modes are included in the expansion. The fit at $x = 0$ and at $x = a$ will not be possible since the function is double-valued there. However, for $y = 0$,

$V = V_0$. If we had included more terms in the expansion, we would have observed a very rapid oscillation at $x = 0$ and $x = a$ along the $y = 0$ line. This effect is given the name *Gibb's phenomenon*, and is a topic for further consideration in an advanced calculus course.

In this section, we found that it was desirable to sum *all* of the terms in the Fourier series in order to get a valid representation for the potential profile. As a general rule, we can say that the more terms that are included in the summation, the better the representation for the potential. The question then arises, "Is there something unique about each of the terms in the series?" We can answer this question by watching a gymnast jumping on a trampoline. If the gymnast lands in the middle of the trampoline, the perturbation in the canvas will be different from what it would be if the landing were at a point that is away from the center or if two gymnasts were jumping in tandem. There are different *modes* for the oscillation. The mathematical structure of the solution for an equation describing the motion of the canvas for all possible landing points is a solution that involves finding all of the *Fourier modes*.

4.6 Finite Difference Method Using MATLAB

There are different methods for the numerical solution of the two-dimensional Laplace's and Poisson's equations. Some of the techniques are based on a differential formulation that was introduced earlier. The *Finite Difference Method* (FDM) is considered here, and the *Finite Element Method* (FEM) is discussed in the next section. Other techniques are based on the integral formulation of the boundary value problems such as the *Method of Moments* (MoM) which is described later. The boundary value problems become more complicated in the presence of dielectric interfaces, which are also considered in this section.

The finite difference method considered here is an extension of the method already applied to a one-dimensional problem. This method allows MATLAB to be involved more directly in the solution of the boundary value problems. We will discuss this method here using a problem that is similar to that presented in Example 4.14. We will describe the technique to obtain and to solve a suitable set of coupled equations that can be interpreted as a matrix equation.

The algorithm that we use is based on the approximation (4.49) for the second derivatives in Cartesian coordinates. In this case, we assume a square grid with a step size h in both directions for a two-dimensional calculation.

$$\nabla^2 V(x, y) = \frac{1}{h^2}[V(x + h, y) + V(x - h, y) + V(x, y + h) + V(x, y - h) - 4V(x, y)]$$

4.6 Finite Difference Method Using MATLAB

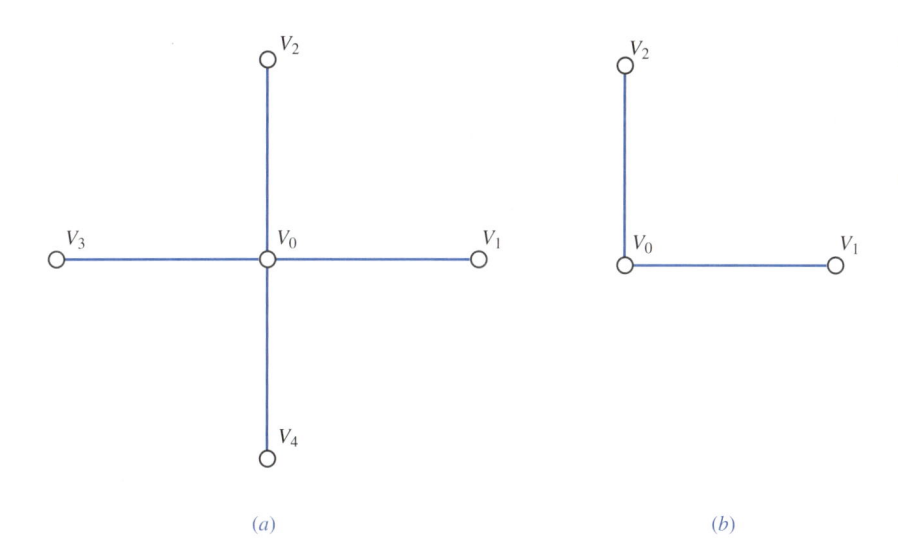

FIGURE 4–13

(a) The general five-point scheme.
(b) The three-point scheme at the corner.

This leads to the following "star shape" representation for a two-dimensional Laplace's equation (4.51) as shown in Figure 4–13a

$$V_0 = \frac{V_1 + V_2 + V_3 + V_4}{4} \qquad (4.69)$$

The voltage at the center is approximated as being the *average* of the voltages at the four tips of the star. For the three-dimensional case, the square is replaced with a cube and a seven-point scheme is applied. In this case, the coefficient 1/4 in (4.69) simply is replaced with 1/6.

For the special case of a corner point, this five-point scheme has to be modified to a three-point one as shown in Figure 4–13b. In this case the principle of the average value simply gives

$$V_0 = \frac{V_1 + V_2}{2} \qquad (4.70)$$

This principle can be applied iteratively for the computation of the potential in the points of the square grid as shown in Figure 4–14. This method is also called a **relaxation method**. After computing the first iteration, we determine the potential at the other points within the nine-point mesh. This will involve two more iterations as shown in Figure 4–15. In the second iteration, all of the potentials at the locations indicated by a solid circle ● in Figure 4–15a are now known. The values indicated by a square ■ are to be computed in this iteration using (4.69). In the third iteration, the values of the potential indicated by the solid circles ● and squares ■ are known from the previous

FIGURE 4–14

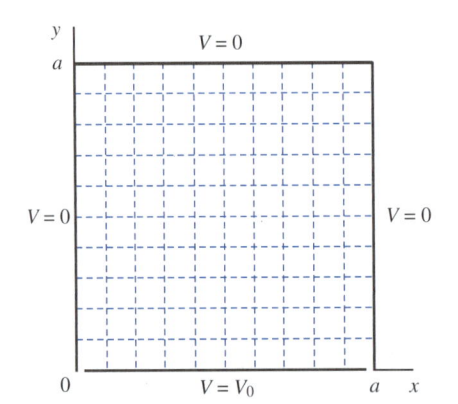

The square grid in two dimensions in Cartesian coordinates.

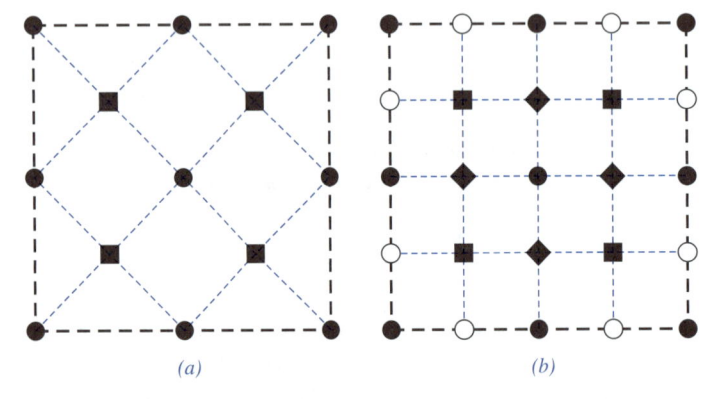

FIGURE 4–15

The second and third iterations. (*a*) The values of the potential indicated by the solid circles ● are known from either the boundary conditions or as a result of the first iteration. The values at the locations of the solid squares ■ are computed in the second iteration. (*b*) The potentials at the boundaries indicated by the hollow circles ○ are assumed to be known. The potentials at the locations indicated by the diamonds ◆ are computed in the third iteration.

two iterations or as initial values in the calculation. Again employing (4.69), the values of the potential at the locations indicated by the diamonds ◆ can be computed. In this mesh, it is assumed that the potentials at the boundaries are already given in the statement of the problem. Hence, the potentials at the locations indicated by the hollow circles ○ are also known as shown in Figure 4–15*b*.

This iterative procedure can continue until the computed values at all of the points in the decreasing meshes become closer to each other. The ***accuracy*** of the calculation can be ensured by repeating the calculation with a different initial mesh size. A mesh with a shape and orientation that is different than the one used here could also be employed in a numerical calculation. This is particularly useful in calculations involving unusual shapes. It is also possible to scale the various dimensions in order to use this particular mesh.

4.6 Finite Difference Method Using MATLAB

A critical *restriction* is also found on the square mesh size in that the first point must be in the center of the square. This point will be evaluated from the four boundaries of the square. This will restrict the number of internal points N of the square to contain the following number of points

$$1^2;\quad 3^2;\quad 7^2;\quad 15^2;\quad 31^2;\quad 63^2;\quad \ldots [2^N - 1]^2$$

This is called the *array size*.

EXAMPLE 4.15

Given that the potential at the four sides of the square region have the values: $V(0, y) = V(a, y) = V(x, a) = 0$, $V(x, 0) = V_0 = 10$ V, plot the potential internal to the boundaries. Use an array size of 31×31. From (4.70), the potentials at the four corners are (0, 0, 5, 5).

Answer. The results of the numerical calculation are shown below. Note that the solution satisfies the boundary conditions and also the values at the corners.

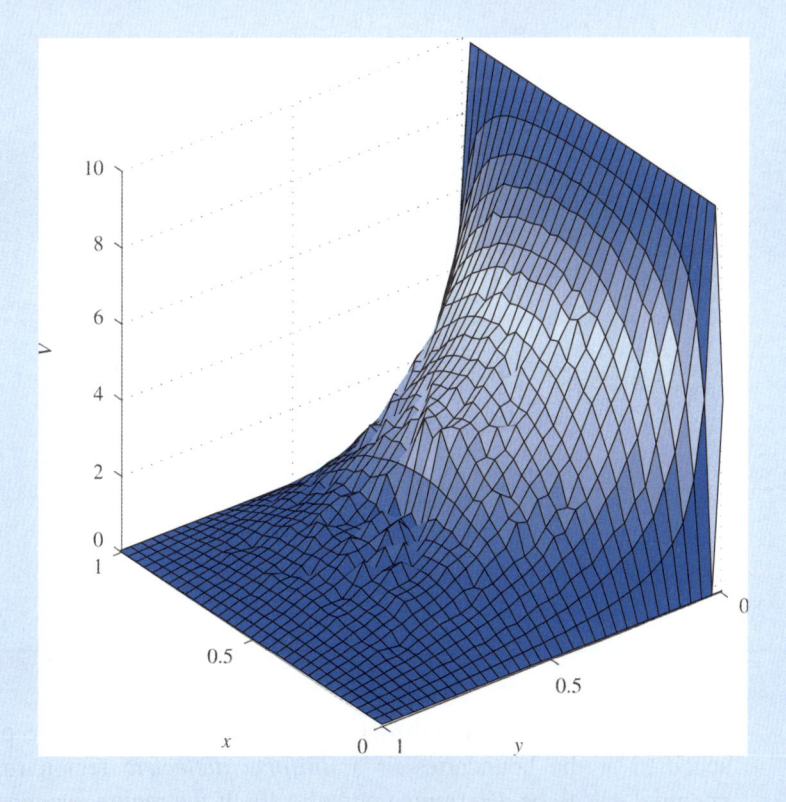

In the above discussion we assumed that the region was a square and used the star given in (4.69) and depicted in Figure 4–15. This technique seemed to work well. However, this technique can be extended to an area with a more complex shape. There are techniques that can be employed to enhance the rate of convergence to the final solution. For example, if one of the boundaries did not have a constant value, it might be advantageous to use a different mesh configuration. It is not a large step to get into examples that are beyond the scope of this text. We will let others tread in those waters.

EXAMPLE 4.16

The potential in a region from $0 \leq x \leq 2$ and $0 \leq y \leq 2$ is described with the expression $V(x, y) = V_0 \exp[-(x^2 + y^2)]$. Calculate and plot the volume charge density $\rho_v(x, y)$ that would be calculated from Poisson's equation assuming that $\varepsilon_0 = 1$.

Answer. Here the *del2* function is applied. The results are shown in the figure below.

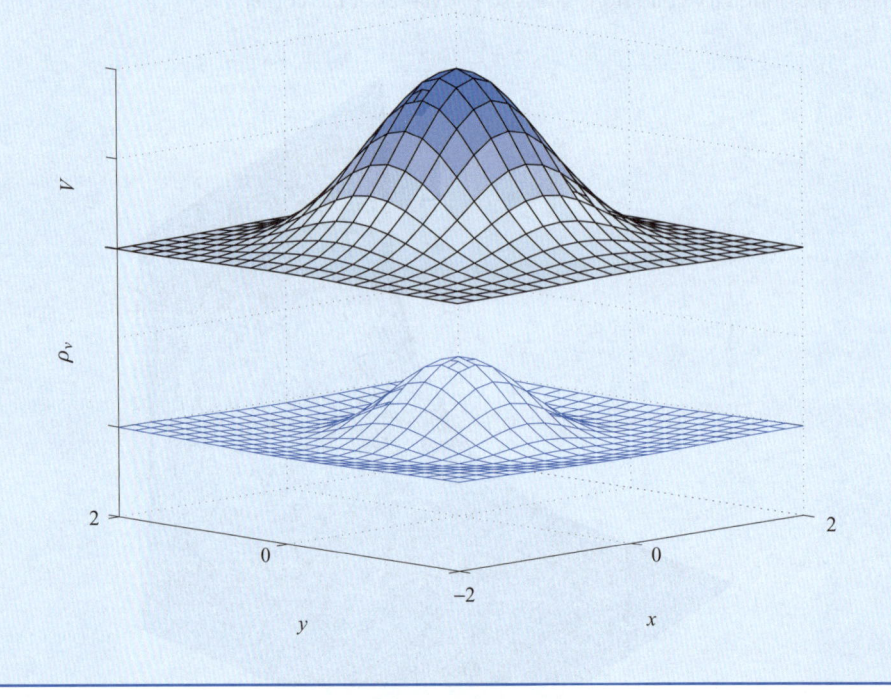

In the material described so far, we have assumed that the potential was specified at the boundaries of a ***uniform dielectric*** region for which the potential was to be determined numerically. If the region contains two dielec-

4.6 Finite Difference Method Using MATLAB

FIGURE 4–16

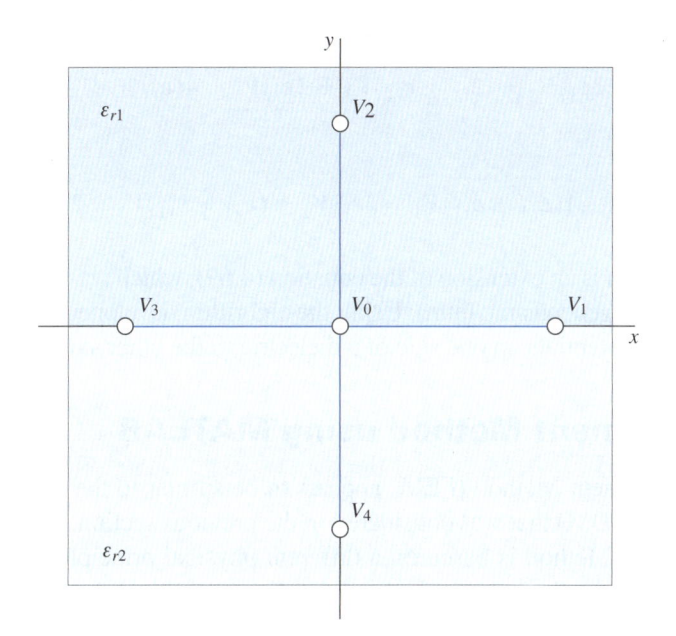

Interface of two dielectrics. Points of the five-point star are in the two dielectrics.

trics as shown in Figure 4–16, we have to obtain an algorithm that will allow us to evaluate the potential on both sides of the *dielectric interface*. This is related to the material that was described in the first section of this chapter.

In order to calculate the boundary condition for the interface of the two dielectrics, we make use of Gauss's law. This is written as

$$\oint \varepsilon \mathbf{E} \cdot \mathbf{ds} = Q_{\text{enc}} = 0 \qquad (4.71)$$

where we have assumed that there is no surface charge density at the interface. With reference to Figure 4–16, (4.71) can be written as

$$\oint \varepsilon \mathbf{E} \cdot \mathbf{ds} = \Delta z \oint \varepsilon \mathbf{E} \cdot \mathbf{dl} = -\Delta z \oint \varepsilon \frac{dV}{dn} dl = 0 \qquad (4.72)$$

where we have replaced the electric field with the derivative of the potential that is normal to the surface. The term Δz is the distance in the third coordinate. The surface integral has become a contour integral times this distance Δz that is directed out of the page. In terms of Figure 4–16, we write

$$\oint \varepsilon_r \frac{dV}{dn} dl = \frac{V_1 - V_0}{h}\left(\varepsilon_{r2}\frac{h}{2} + \varepsilon_{r1}\frac{h}{2}\right) + \frac{V_2 - V_0}{h}(\varepsilon_{r1}h)$$
$$+ \frac{V_3 - V_0}{h}\left(\varepsilon_{r2}\frac{h}{2} + \varepsilon_{r1}\frac{h}{2}\right) + \frac{V_4 - V_0}{h}(\varepsilon_{r2}h) \qquad (4.73)$$

Rearranging terms, we rewrite (4.73) as

$$(\varepsilon_{r1} + \varepsilon_{r2})V_1 + 2\varepsilon_{r1}V_2 + (\varepsilon_{r1} + \varepsilon_{r2})V_3 + 2\varepsilon_{r2}V_4 - 4(\varepsilon_{r1} + \varepsilon_{r2})V_0 = 0 \quad (4.74)$$

or

$$V_0 = \frac{1}{4(\varepsilon_{r1} + \varepsilon_{r2})}[(\varepsilon_{r1} + \varepsilon_{r2})V_1 + 2\varepsilon_{r1}V_2 + (\varepsilon_{r1} + \varepsilon_{r2})V_3 + 2\varepsilon_{r2}V_4] \quad (4.75)$$

This equation is an extension of the equation (4.69), which was written when $\varepsilon_{r1} = \varepsilon_{r2}$ (homogeneous medium). Using the algorithm developed in (4.75), we can relate the potentials on one side of a dielectric to the other side.

4.7 Finite Element Method using MATLAB

The Finite Element Method (FEM) appears to be similar to the Finite Difference Method (FDM) that was considered in the previous section. However, the Finite Element Method is based on a different physical principle than the Finite Difference Method. Instead of subdividing the area into small squares with a side h, this technique subdivides the area into small triangles. We will find that the potential within the triangle can be specified in terms of the potentials at the nodes of the triangle. It should be noted that shapes other than triangles have also been employed in this technique but we shall just focus on the triangular shape in this introduction. As we will observe, this method is more flexible in its application. For example, the calculation of the potential profile in the region between two concentric rectangular metallic surfaces can be handled with this technique. The technique has a strong foundation using matrix manipulations.

In Figure 4–17a, one quarter of the cross section of a rectangular coaxial line is shown. Figure 4–17b depicts the appropriate modeling with the finite triangular elements. The mesh may be irregular in that the grid is denser in the vicinity of the corners where a more rapid variation of the potential is to be expected. The scalar potential V between the two conductors satisfies Laplace's equation (4.14)

$$\boxed{\nabla^2 V = 0} \quad (4.76)$$

and we will define this area as S.
There are two different conditions at the boundaries.

1. On the boundary $L = L_1 + L_2$, the voltage is specified. This is called a **Dirichlet boundary condition**;
2. Because of the symmetry inherent in this problem, we require that the normal derivative of the voltage be equal to 0 on the plane of symmetry. This is called a **Neumann boundary condition**.

4.7 Finite Element Method using MATLAB

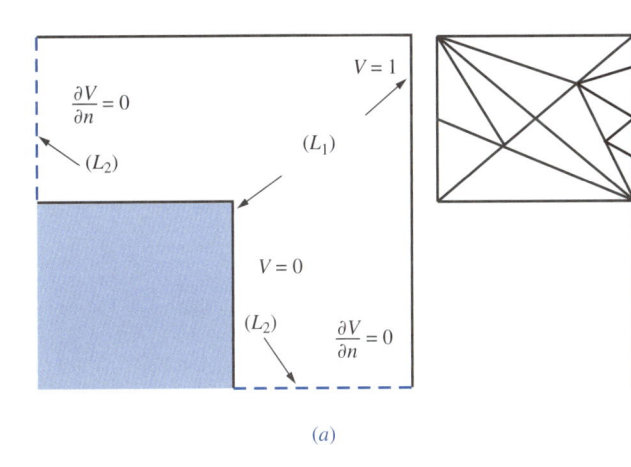

FIGURE 4–17
(*a*) Cross section of 1/4 of a square coaxial line is shown.
(*b*) A finite element mesh is used to subdivide the square coaxial cable.

(*a*)   (*b*)

A solution of (4.76) is

$$V^{(e)}(x, y) = a + bx + cy \tag{4.77}$$

where there are three unknown coefficients (a, b, c) and we have introduced a superscript notation to indicate that this potential is within the triangle. This particular solution assumes that the potential within the triangle has a planar profile. If we had selected a shape other than a triangle, we could have chosen a more complicated expression with more unknown coefficients as a possible solution. We would require that the choice still satisfy Laplace's equation. These sophistications will be found in more advanced treatments of the subject.

In order to determine these coefficients, we employ the known voltages at the three nodes of the triangle (1, 2, 3) in Figure 4–18. This approximation replaces the actual solution with a *piecewise* smooth function that is based on a linear interpolation. The coefficients (a, b, c) can be determined from the given **node potentials** (V_1, V_2, V_3) provided that the coordinates of the nodes are known quantities: (x_j, y_j) (j = 1, 2, 3).

Let us assume that the voltages at the nodes are already known

$$V^{(e)}(x_j, y_j) = V_j \quad (j = 1, 2, 3) \tag{4.78}$$

This can be written in matrix notation where the only unknowns are the coefficients (a, b, c)

$$\begin{bmatrix} 1 & x_1 & y_1 \\ 1 & x_2 & y_2 \\ 1 & x_3 & y_3 \end{bmatrix} \begin{bmatrix} a \\ b \\ c \end{bmatrix} = \begin{bmatrix} V_1 \\ V_2 \\ V_3 \end{bmatrix} \tag{4.79}$$

FIGURE 4–18

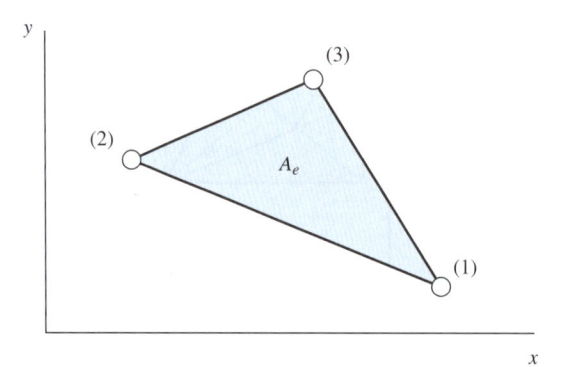

Triangular finite element.

The magnitude of the determinant of the square matrix in (4.79) is equal to twice the area of the triangle $2A_e$.

The substitution of the solution of the simultaneous equations (4.79) into (4.77) yields

$$V^{(e)}(x, y) = [1 \ x \ y] \begin{bmatrix} 1 & x_1 & y_1 \\ 1 & x_2 & y_2 \\ 1 & x_3 & y_3 \end{bmatrix}^{-1} \begin{bmatrix} V_1 \\ V_2 \\ V_3 \end{bmatrix} \quad (4.80)$$

The potential within the triangular region is specified by the coefficients that are determined by the potentials at the known locations of the three nodes of the triangle.

After performing the matrix multiplication of the first two matrices, we arrive at the following expression for the potential at an arbitrary point within the triangle as being a linear combination of the potentials at the nodes of the triangle

$$V^{(e)}(x, y) = \sum_{i=1}^{3} V_i \alpha_i(x, y) \quad (4.81)$$

where the coefficients that depend on the node locations and the coordinates of the observation point are found to be

$$\left.\begin{aligned} \alpha_1(x, y) &= \frac{1}{2A_e}[(x_2 y_3 - x_3 y_2) + (y_2 - y_3)x + (x_3 - x_2)y] \\ \alpha_2(x, y) &= \frac{1}{2A_e}[(x_3 y_1 - x_1 y_3) + (y_3 - y_1)x + (x_1 - x_3)y] \\ \alpha_3(x, y) &= \frac{1}{2A_e}[(x_1 y_2 - x_2 y_1) + (y_1 - y_2)x + (x_2 - x_1)y] \end{aligned}\right\} \quad (4.82)$$

4.7 Finite Element Method using MATLAB

Using the explicit expressions, we find that these three linear functions satisfy the following interpolation criteria

$$\alpha_i(x_j, y_j) = \begin{cases} 0, & i \neq j \\ 1, & i = j \end{cases} \tag{4.83}$$

EXAMPLE 4.17

(a) Find the voltage distribution within the triangular element if the voltages at the three nodes have the following values $V_1 = 8$ @ $(4, 0)$; $V_2 = 0$ @ $(0, 0)$; and $V_3 = 0$ @ $(4, 3)$.
(b) Determine the area of the triangle that is defined by the three nodes.

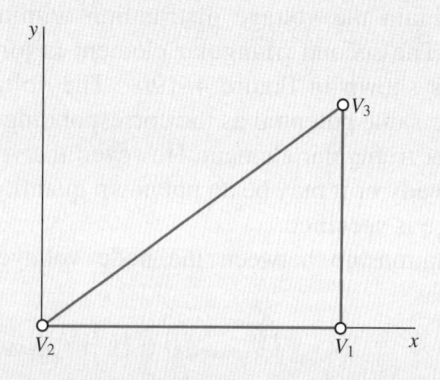

Answer. (a) The voltage distribution within the element is computed from (4.80)

$$V^{(e)}_{1,2,3}(x, y) = [1 \ x \ y] \begin{bmatrix} 1 & 4 & 0 \\ 1 & 0 & 0 \\ 1 & 4 & 3 \end{bmatrix}^{-1} \begin{bmatrix} 8 \\ 0 \\ 0 \end{bmatrix} = [1 \ x \ y] \begin{bmatrix} 0 & 1 & 0 \\ 1/4 & -1/4 & 0 \\ -1/3 & 0 & 1/3 \end{bmatrix} \begin{bmatrix} 8 \\ 0 \\ 0 \end{bmatrix}$$

$$= [1 \ x \ y] \begin{bmatrix} 0 \\ 2 \\ -8/3 \end{bmatrix} = 2x - \frac{8}{3}y$$

The solution satisfies Laplace's equation (4.76) and the boundary conditions (4.78).
(b) The area of the triangle is determined from magnitude of the determinant of the square matrix given in (4.79).

$$A_e = \frac{1}{2} \begin{vmatrix} 1 & 4 & 0 \\ 1 & 0 & 0 \\ 1 & 4 & 3 \end{vmatrix} = 6$$

FIGURE 4–19

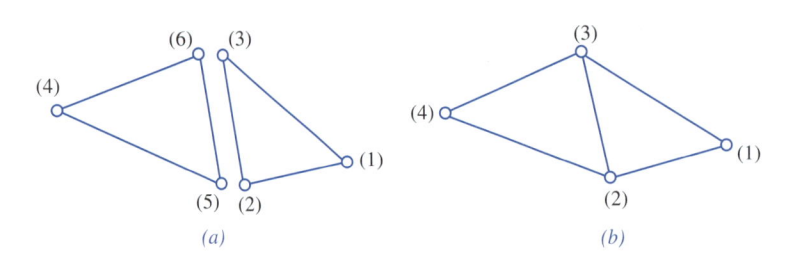

Two triangular elements are coupled together.
(a) Before coupling.
(b) After coupling.

In order to employ this method, we must first ascertain what would happen if there are two triangular elements as shown in Figure 4–19a. The voltages at the three nodes 1, 2, and 3 are already known and have been used to obtain the voltage distribution within this particular triangular element. The second triangular element is joined together with the first element as shown in Figure 4–19b. The voltages at the nodes 5 and 6 assume the same potential as the corresponding nodes 2 and 3 respectively of the other triangular element. However, the voltage at the node 4 may be known already or it may be an unknown quantity. We will first assume that this voltage is specified.

The relationship between the node voltages in Figure 4–19 can be expressed as

$$[V_{uncoupled}] = [C][V_{coupled}] \tag{4.84}$$

that corresponds to the following equation

$$\begin{bmatrix} V_{uncoupled\ 1} \\ V_{uncoupled\ 2} \\ V_{uncoupled\ 3} \\ V_{uncoupled\ 4} \\ V_{uncoupled\ 5} \\ V_{uncoupled\ 6} \end{bmatrix} = \begin{bmatrix} 1 & 0 & 0 & 0 \\ 0 & 1 & 0 & 0 \\ 0 & 0 & 1 & 0 \\ 0 & 0 & 0 & 1 \\ 0 & 1 & 0 & 0 \\ 0 & 0 & 1 & 0 \end{bmatrix} \begin{bmatrix} V_{coupled\ 1} \\ V_{coupled\ 2} \\ V_{coupled\ 3} \\ V_{coupled\ 4} \end{bmatrix} \tag{4.85}$$

EXAMPLE 4.18

Determine the potential distribution within the two adjacent triangular regions if the potential is specified at all nodes

$V_1 = 8$ @ (4, 0); $V_2 = 0$ @ (0, 0); $V_3 = 0$ @ (4, 3);   and $V_4 = 8$ @ (0, 3).

4.7 Finite Element Method using MATLAB

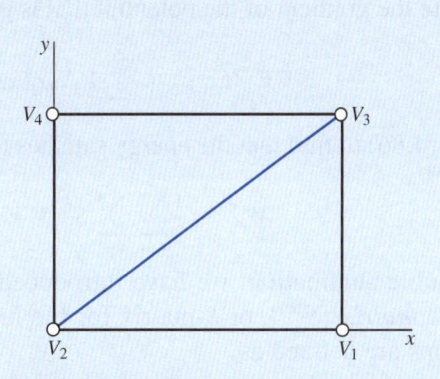

Answer. The voltage distribution in the triangle (1, 2, 3) was determined in Example 4–17 to be $V_{1,2,3}^{(e)}(x, y) = 2x - (8/3)y$. The potential distribution within the adjacent triangle (2, 3, 4) is

$$V_{2,3,4}^{(e)}(x, y) = [1 \; x \; y] \begin{bmatrix} 1 & 0 & 0 \\ 1 & 4 & 3 \\ 1 & 0 & 3 \end{bmatrix}^{-1} \begin{bmatrix} 0 \\ 0 \\ 8 \end{bmatrix} = [1 \; x \; y] \begin{bmatrix} 1 & 0 & 0 \\ 0 & 1/4 & -1/4 \\ -1/3 & 0 & 1/3 \end{bmatrix} \begin{bmatrix} 0 \\ 0 \\ 8 \end{bmatrix}$$

$$= [1 \; x \; y] \begin{bmatrix} 0 \\ -2 \\ 8/3 \end{bmatrix} = -2x + \frac{8}{3}y$$

The results that have been obtained in Examples 4–17 and 4–18 should not be too surprising. In these two examples, we have specified the voltage at each of the nodes before actually doing the computation in order to obtain the voltage distribution within the triangles. However, there will be cases where it is impossible to *a priori* specify these values. For example, the voltage at an intermediate location between two known potentials is an unknown quantity. In this case, we must follow a slightly different procedure in order to obtain the voltage distribution between these extremities.

The procedure that we must follow in this case is to examine the electrostatic energy within the triangular elements. It was shown in Chapter 2 that the electrostatic energy that can be stored in a volume with a cross section A_e is

$$\boxed{W = \int_{A_e} \frac{\varepsilon}{2} |\nabla V|^2 ds} \qquad (4.86)$$

In Appendix C, it is shown that this integral or "functional" has a *minimum value* for the actual solution of this boundary value problem. The *necessary condition* for this minimum leads to Laplace's equation for the potential V in the area A_e. We use this energy relationship (4.86) to further develop the finite element method.

We write the gradient of the potential that is given in (4.81)

$$\nabla V^{(e)}(x, y) = \sum_{i=1}^{3} V_i \nabla \alpha_i(x, y) \qquad (4.87)$$

and using (4.86) to find that the energy satisfies the following quadratic form

$$W^{(e)} = \frac{1}{2} \sum_{i=1}^{3} \sum_{j=1}^{3} S_{i,j}^{(e)} V_i V_j \qquad (4.88)$$

In this double summation, we have introduced a matrix that is known as **Dirichlet's matrix** $[S^{(e)}]$, or S-matrix for the finite elements, and the individual terms are defined as

$$S_{i,j}^{(e)} = \varepsilon \int_{A_e} \nabla \alpha_i \cdot \nabla \alpha_j \, \mathbf{ds} \qquad (4.89)$$

This Dirichlet matrix identified by the symbol $S_{i,j}^{(e)}$ is different from the scattering matrix that may be talked about in conjunction with Chapter 8, which is also identified by $S_{i,j}$.

We assume that the permittivity $\varepsilon = \varepsilon_r \varepsilon_0$ is constant within the element that has an area A_e. Explicit expressions for these terms are presented in Appendix C. In terms of these elements, the electrostatic energy in equation (4.88) can be written as

$$W^{(e)} = \frac{1}{2} [V]^T [S^{(e)}][V] \qquad (4.90)$$

where the superscript "T" indicates that you should take the transpose of the V-matrix.

EXAMPLE 4.19

Find the electrostatic energy that is within the triangular element shown in Example 4–17. The voltages at the three nodes have the following values: $V_1 = 8$ @ (4, 0); $V_2 = 0$ @ (0, 0); and $V_3 = 0$ @ (4, 3).

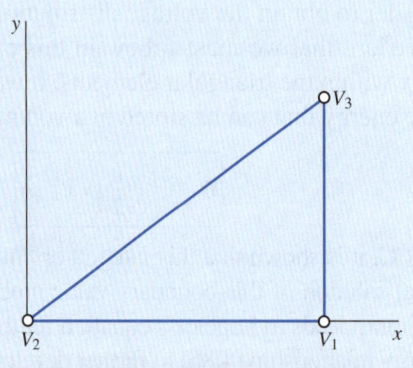

4.7 Finite Element Method using MATLAB

Answer. In order to calculate the energy, we must first calculate the numerical values for the explicit expressions for the $[S^{(e)}]$ matrix elements. The area enclosed within the triangle is $A_e = 6$.

$$S_{1,1}^{(e)} = \frac{\varepsilon}{4A_e}[(y_2 - y_3)^2 + (x_2 - x_3)^2] = \frac{\varepsilon}{4 \times 6}[(0-3)^2 + (0-4)^2] = \frac{25\varepsilon}{24}$$

$$S_{1,2}^{(e)} = S_{2,1}^{(e)} = \frac{\varepsilon}{4A_e}[(y_2 - y_3)(y_3 - y_1) + (x_2 - x_3)(x_3 - x_1)]$$

$$= \frac{\varepsilon}{4 \times 6}[(0-3)(3-0) + (0-4)(4-4)] = -\frac{9\varepsilon}{24}$$

$$S_{1,3}^{(e)} = S_{3,1}^{(e)} = \frac{\varepsilon}{4A_e}[(y_3 - y_2)(y_2 - y_1) + (x_3 - x_2)(x_2 - x_1)]$$

$$= \frac{\varepsilon}{4 \times 6}[(3-0)(0-0) + (4-0)(0-4)] = -\frac{16\varepsilon}{24}$$

$$S_{2,2}^{(e)} = \frac{\varepsilon}{4A_e}[(y_3 - y_1)^2 + (x_3 - x_1)^2] = \frac{\varepsilon}{4 \times 6}[(3-0)^2 + (4-4)^2] = \frac{9\varepsilon}{24}$$

$$S_{2,3}^{(e)} = S_{3,2}^{(e)} = \frac{\varepsilon}{4A_e}[(y_3 - y_1)(y_1 - y_2) + (x_3 - x_1)(x_1 - x_2)]$$

$$= \frac{\varepsilon}{4 \times 6}[(3-0)(0-0) + (4-4)(4-0)] = 0$$

$$S_{3,3}^{(e)} = \frac{\varepsilon}{4A_e}[(y_1 - y_2)^2 + (x_1 - x_2)^2] = \frac{\varepsilon}{4 \times 6}[(0-0)^2 + (4-0)^2] = \frac{16\varepsilon}{24}$$

This is a written as the symmetric matrix

$$[S^{(e)}] = \frac{\varepsilon}{24}\begin{bmatrix} 25 & -9 & -16 \\ -9 & 9 & 0 \\ -16 & 0 & 16 \end{bmatrix}$$

The energy is computed from (4.80)

$$\frac{W_e}{\varepsilon} = \frac{1}{2\varepsilon}[V]^T[S^{(e)}][V] = \frac{1}{2}[8\ 0\ 0]\begin{bmatrix} 25/24 & -9/24 & -16/24 \\ -9/24 & 9/24 & 0 \\ -16/24 & 0 & 16/24 \end{bmatrix}\begin{bmatrix} 8 \\ 0 \\ 0 \end{bmatrix}$$

$$= \frac{1}{2}[8\ 0\ 0]\begin{bmatrix} 25/3 \\ -3 \\ -16/3 \end{bmatrix} = \frac{100}{3} = 33.333$$

The total energy of the ensemble of all triangular elements in the mesh can be calculated as just being the *sum* of the energy of each of the individual elements

$$W = \sum_e W^{(e)} \qquad (4.91)$$

The inclusion of additional triangular elements implies that the electrostatic energy of the entire system will change. The procedure uses the coupling that was previously described. We will find the global S-matrix of the coupled system. Recall that the potentials of the common nodes will be identical.

We will develop the procedure using the assumption that we have separately obtained the S-matrices of the two triangular meshes $[S^{(1)}]$ and $[S^{(2)}]$ as shown in Figure 4–19. The decoupled potentials can be written as a column matrix. The transpose of this matrix (Figure 4–19a) is

$$[V]_d = [V_1, V_2, V_3, V_4, V_5, V_6]_d^T,$$

where the subscript "d" means "decoupled." If the S-matrices of the two elements are $[S^{(1)}] = [S_{i,j}]$ ($i, j = 1, 2, 3$) and $[S^{(2)}] = [S_{i,j}]$ ($i, j = 4, 5, 6$), then the global S-matrix of the decoupled system is the following block-diagonal square matrix

$$[S^{(e)}]_d = \begin{bmatrix} [S^{(1)}] & [0] \\ [0] & [S^{(2)}] \end{bmatrix} \qquad (4.92)$$

The column matrix of the coupled potentials (Figure 4–19b) is

$$[V] = [V_1, V_2, V_3, V_4]^T$$

The boundary conditions from Figure 4–19 can be written using the rectangular coupling matrix $[C]$ as in (4.85). In this particular case, we obtain

$$[S] = \begin{bmatrix} S_{1,1}^{(1)} & S_{1,2}^{(1)} & S_{1,3}^{(1)} & 0 \\ S_{2,1}^{(1)} & S_{2,2}^{(1)} + S_{5,5}^{(2)} & S_{2,3}^{(1)} + S_{5,6}^{(2)} & S_{5,4}^{(2)} \\ S_{3,1}^{(1)} & S_{3,2}^{(1)} + S_{6,5}^{(2)} & S_{3,3}^{(1)} + S_{6,6}^{(2)} & S_{6,4}^{(2)} \\ 0 & S_{4,5}^{(2)} & S_{4,6}^{(2)} & S_{4,4}^{(2)} \end{bmatrix} \qquad (4.93)$$

Note that two types of superscript numeration are used: (1) local for decoupled elements and (2) global for coupled elements.

4.7 Finite Element Method using MATLAB

The coupled potentials $[V]$ can be split into two parts: (a) the *unknown* potentials $[V]_u$ and (b) the *known* potentials $[V]_k$ or

$$[V] = [[V]_u \ [V]_k]^T \tag{4.94}$$

The same separation holds for the matrix $[S^{(e)}]$, which is split into four parts

$$[S^{(e)}] = \begin{bmatrix} [S]_{u,u} & [S]_{u,k} \\ [S]_{k,u} & [S]_{k,k} \end{bmatrix} \tag{4.95}$$

We use the minimization of the energy of requirement that was derived in Appendix C in order to find in relation between the known and the unknown potentials

$$\frac{\partial W^{(e)}}{\partial [V_j]_u} = \frac{\partial \left(\frac{1}{2}[V]^T[S^{(e)}][V]\right)}{\partial [V_j]_u} = 0 \tag{4.96}$$

We find that the unknown potentials can be written as

$$[V]_u = -[S]_{u,u}^{-1}[S]_{u,k}[V]_k \tag{4.97}$$

provided that the matrix $[S]_{u,u}$ is non-singular. The last matrix equation shows that the unknown potentials can be represented as a linear combination of the known potentials although the coefficients may implicitly depend upon the geometry of the mesh.

EXAMPLE 4.20

Using the results of Example 4–19, determine the unknown potential V_4. The values of the other three potentials are known.

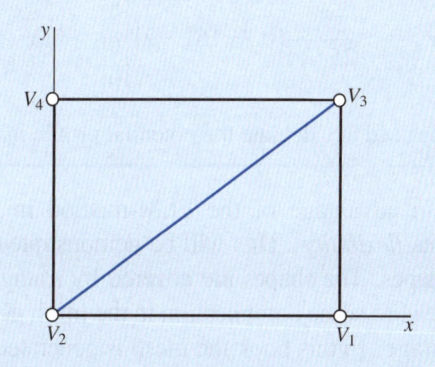

Answer. From Example 4–19 we obtained the S-matrix of the first triangle

$$[S^{(1)}] = \frac{\varepsilon}{24}\begin{bmatrix} 25 & -9 & -16 \\ -9 & 9 & 0 \\ -16 & 0 & 16 \end{bmatrix}$$

Following the same procedure, we obtain the S-matrix of the second triangle

$$[S^{(2)}] = \frac{\varepsilon}{24}\begin{bmatrix} 25 & -16 & -9 \\ -16 & 16 & 0 \\ 9 & 0 & 9 \end{bmatrix}$$

To directly apply (4.93) for this particular case, we have to change the indices as follows: $(1, 2, 3, 4, 5, 6) \rightarrow (3, 2, 1, 3, 4, 2)$. This produces the following coupled S-matrix of the rectangle—a system of two triangles $(1, 2, 3)$ and $(2, 3, 4)$:

$$[S^{(e)}] = \begin{bmatrix} [S]_{kk} & [S]_{ku} \\ [S]_{uk} & [S]_{uu} \end{bmatrix} = \frac{\varepsilon}{24}\begin{bmatrix} S_{33}^{(1)}+S_{33}^{(2)} & S_{32}^{(1)}+S_{32}^{(2)} & S_{31}^{(1)} & S_{34}^{(2)} \\ S_{23}^{(1)}+S_{23}^{(2)} & S_{22}^{(1)}+S_{22}^{(2)} & S_{21}^{(1)} & S_{24}^{(2)} \\ S_{13}^{(1)} & S_{12}^{(1)} & S_{11}^{(1)} & 0 \\ S_{43}^{(2)} & S_{42}^{(2)} & 0 & S_{44}^{(2)} \end{bmatrix}$$

$$= \frac{\varepsilon}{24}\begin{bmatrix} 25 & -9 & -16 & 0 \\ -9 & 25 & 0 & -16 \\ -16 & 0 & 25 & -9 \\ 0 & -16 & -9 & 25 \end{bmatrix}$$

Applying equation (4.97), we obtain the unknown potential V_4

$$V_4 = -\frac{24}{25\varepsilon}\frac{\varepsilon}{24}[0 \ -16 \ -9]\begin{bmatrix} 8 \\ 0 \\ 0 \end{bmatrix} = 0 \text{ V}$$

This computed potential is used to calculate the potential profile in this triangular element.

The main advantage of the FEM-method in comparison with the FDM-method is its *flexibility*. This will be demonstrated by applying it to areas with different shapes. The shapes are covered by triangles in the domain of interest. This may lead to some complication in the mesh generation, which is, of course, a disadvantage. In this book the mesh is generated manually for the purpose of illustrating the problem solving approach. There are many procedures for automatic mesh generation that are available. These can be found both in commercial

4.7 Finite Element Method using MATLAB

software packages, such as COMSOL Multiphysics, and in software routines on the internet. They mainly differ in the ease of their implementation, and the ability to handle such factors as the introduction of inhomogeneous material properties or the speed and difficulty of meshing up a complicated object. Mechanical and aerospace engineers have used this technique for decades and there is an extensive literature base for those who wish to investigate this approach further.

In summary, the procedure of using this technique has five stages:

1. Generation of the mesh.
2. Inclusion of the surface and volume sources.
3. Construction of the matrices for every element.
4. Collection of all the elements of the $[S^{(e)}]$-matrix.
5. Solution of the resulting matrix equation.

In order to illustrate the FEM-method, we consider a simple example.

EXAMPLE 4.21

Find the potential at the point 1 using the FEM.

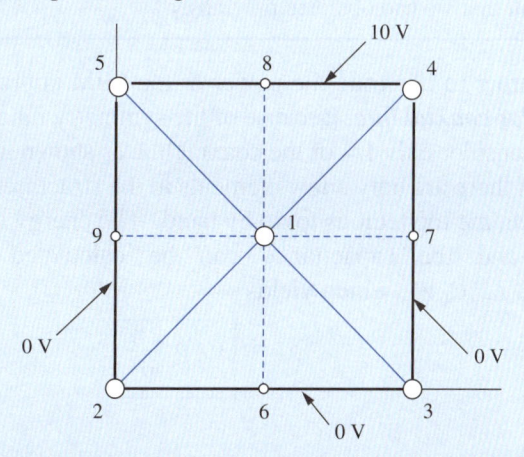

Answer. To solve this problem using the FEM approach, consider the symmetry of the problem and define the region with $N_N = 5$ nodes denoted as # = 1, 2, 3, 4, 5 and $N_E = 4$ equal triangular elements #1-4-5, 1-2-5, 1-2-3, and 1-3-4. The only unknown potential is V_1, while the other four potentials have the numerical values $V_2 = V_3 = 0$, and

$$V_4 = V_5 = \frac{10 + 0}{2} = 5 \text{ V}$$

Using the FEM approach, we first calculate the matrices using the known potentials at the points # = 2, 3, 4, 5. We find that the global sub-matrices required for the calculation are

$$[S]_{u,u} = [4S_{1,1}]; \quad [S]_{u,k} = [2S_{1,2}, 2S_{1,3}, 2S_{1,4}, 2S_{1,5}]$$

The elements of the local $[S^{(e)}]$-matrices are calculated. The area of each of the triangular elements is equal to 1 and the coordinates of the nodal points can be obtained from the figure. We finally obtain

$$S_{1,1} = \varepsilon; \qquad S_{1,2} = S_{1,3} = S_{1,4} = S_{1,5} = -\frac{\varepsilon}{2}.$$

This leads to

$$[A] = -\frac{1}{4\varepsilon}[-\varepsilon, -\varepsilon, -\varepsilon, -\varepsilon] = \frac{1}{4}[1, 1, 1, 1]$$

and since $[V]_u = [V_1]$ and $[V]_k = [V_2, V_3, V_4, V_5]^T$, we obtain

$$V_1 = \tfrac{1}{4}(V_2 + V_3 + V_4 + V_5) = \tfrac{1}{4}(0 + 0 + 5 + 5) = 2.5 \text{ V}$$

This problem can be also evaluated using the FDM approach. We can write the potential at the point 1 as an *average value* of the potentials at the other four points (# = 6, 7, 8, 9)

$$V_1 = \tfrac{1}{4}(V_6 + V_7 + V_8 + V_9) = \tfrac{1}{4}(0 + 0 + 10 + 0) = 2.5 \text{ V}$$

This is the same result that we had obtained previously.

In order to illustrate the power of the FEM approach, we examine a *rectangular coaxial line*. Because of the symmetry inherent in this problem, we shall consider only 1/4 of the coaxial line as shown in the Figure 4–17. Even though there are only a few elements in the structure, the actual mathematics will become too tedious to do by hand. The energy of the capacitor is calculated and the capacitance can be calculated from the expression $W = (1/2)C_0V^2$, which yields

$$\boxed{C_0 = \frac{2W}{V^2}} \qquad (4.98)$$

EXAMPLE 4.22

Write a MATLAB program to:
1. Calculate the potentials in the three free nodes with # = 4, 5, 9.
2. Calculate the total electrostatic energy W stored between the plates and find the capacitance C_0 of 1 m of the line. Use a simple regular mesh shown in the figure and the following numerical values for the parameters: $a = c = 1$ cm, $b = d = 2$ cm, $V = 10$ V on the outer conductor, and $V = 0$ V on the inner conductor. In this case, there are $N_N = 11$ nodes and $N_E = 12$ elements.

4.7 Finite Element Method using MATLAB

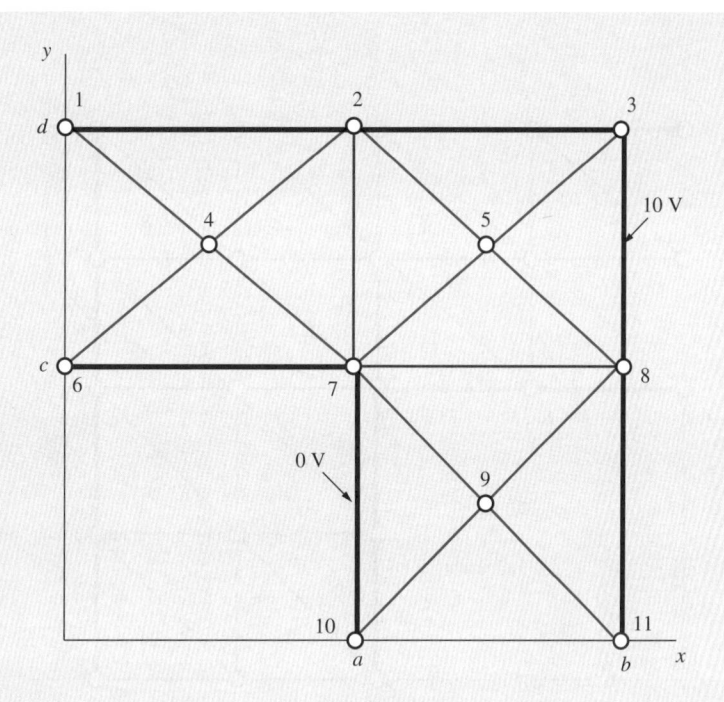

Answer.

1. The potential is found by solving the matrix equation (4.97) using a mesh that is manually generated. The unknown potentials calculated with the MATLAB program are:

$$V_4 = 5\text{ V}; \quad V_5 = 7.5\text{ V}; \quad V_9 = 5\text{ V}$$

2. The calculated normalized capacitance $C = C_0 / \varepsilon_0$ per unit length of the line calculated from (4.98) is $C = 11.0$. The value, found in the literature for this square coaxial line with $b = 2a$, is $C = 10.2341$. The accuracy can be improved by increasing the number of the nodes and elements which will be shown in the next example.

EXAMPLE 4.23

Solve the same boundary value problem as in the previous example but apply the finer mesh shown in the figure below. There are $N_N = 21$ nodes and $N_E = 24$ triangular elements. In addition, plot the equipotential contours and the electric field between the two surfaces.

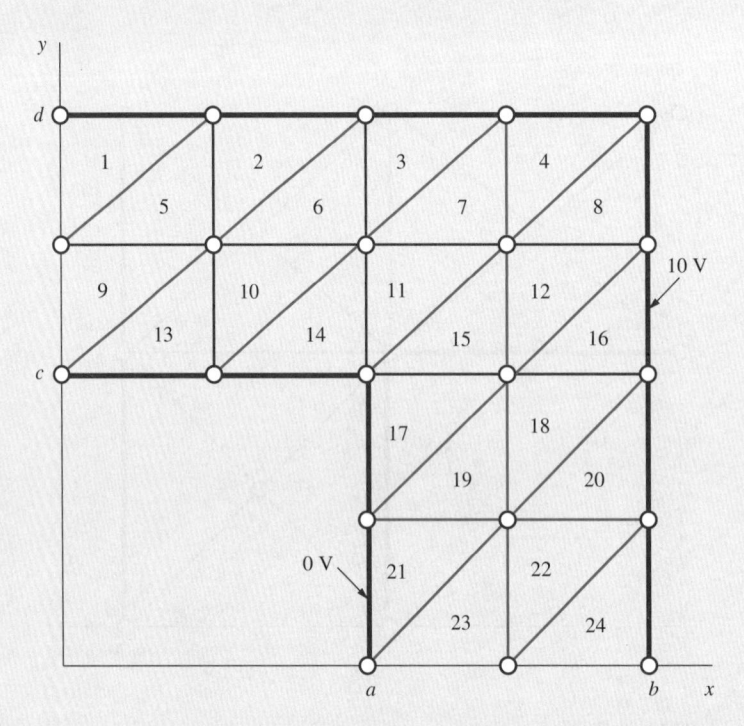

Answer.

1. The potential is found by solving the matrix equation (4.97) using a mesh that is manually generated. The unknown potentials obtained from the MATLAB program are:

   $V_6 = 5.111$ V; $\quad V_7 = 5.2222$ V; $\quad V_8 = 5.7778$ V; $\quad V_9 = 7.8889$ V;

   $V_{14} = 5.7778$ V; $\quad V_{17} = 5.2222$ V; $\quad V_{20} = 5.1111$ V;

2. The calculated normalized capacitance for a unit length of the line is $C = 10.8444$ F. The relative error achieved here is smaller than in the previous example.

3. The equipotential contours are plotted below using the *contour* function. The electric field is determined using the *gradient* function and displayed using the *quiver* function.

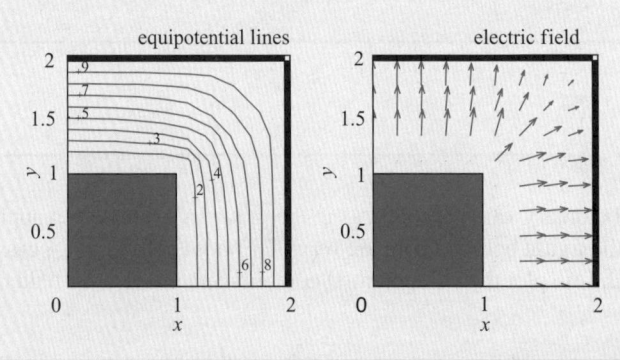

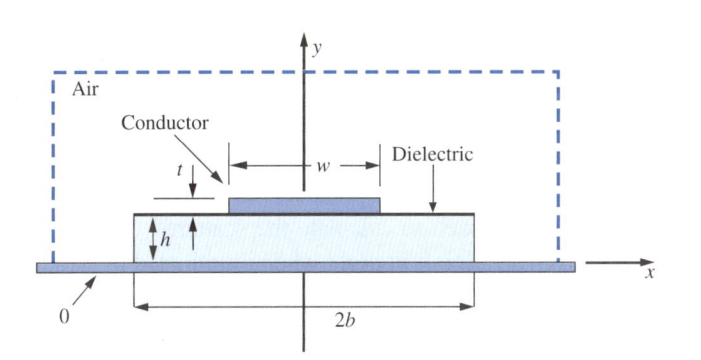

FIGURE 4–20

A microstrip transmission line.

The previous two examples were solutions of a *closed* electrostatics problem. Closed problems can easily be handled by the FEM method. A general conclusion can be drawn from these examples in that the accuracy can be improved by using a finer mesh structure. However, increasing the accuracy also increases the computational time.

An *open* electrostatic problem is shown in Figure 4–20. This would correspond to an open **microstrip line**. A dielectric with a dielectric constant $\varepsilon_r > 1$ separates the conductors and also increases the capacitance of the line. It is not appropriate to use either the FEM approach or the FDM approach to calculate the capacitance or the stored electrostatic energy for structures of this type. This is because boundary conditions at infinity will have to be invoked at the open surfaces. It is possible to introduce an **absorbing boundary condition** at these surfaces. However, it is usually better to follow a different path of using the **method of moments** (MoM), which will be introduced in the next section.

4.8 Method of Moments Using MATLAB

In the previous chapter, we found that the electric potential V could be computed from a *known* charge distribution. This was accomplished using the integral

$$V(x, y, z) = \frac{1}{4\pi\varepsilon_0} \iiint_{\Delta v} \frac{\rho_v(x', y', z')}{R} dx' dy' dz' \qquad (4.99)$$

where R is the distance between the charge located at the source point (x', y', z') and the observation point (x, y, z). If the charge distribution is known, then the potential can be computed easily. We note that (4.99) can be approximated by a summation, hence the integral can be evaluated numerically.

There are cases, however, where the potential may actually be known and the charge distribution may be *unknown*. Static fields abound with such problems.

An example would be the determination of an unknown surface charge distribution on a conductor if the potential of the conductor were specified.

The technique that will be introduced is called the **Method of Moments** (MoM). This technique will be very powerful in calculating the capacitance of various metallic objects. It is also useful in calculating the capacitance of a transmission line, especially the more complicated transmission structures. Finally, we can use this solution approach to calculate the scattered electromagnetic waves that are reflected from an aircraft in flight when illuminated by a radar beam.

Consider the configuration shown in Figure 4–21. Four charges are located in space. A Cartesian coordinate system is also introduced, and the location of the centers of the four charges are specified with reference to this coordinate system. The potential at two of the charges (Q_1 and Q_2) is specified to be $V_1 = V_2 = -1$ and the potential at the other two (Q_3 and Q_4) is specified to be $V_3 = V_4 = +1$. The value of the individual charges is unknown. In order to obtain a unique solution for the values of these four charges, we must be able to write down four equations that will describe the potential at the four defined locations. We assume that the region is a vacuum and we can use superposition. We write four linear equations for the potentials at the four points

$$-1 = \frac{1}{4\pi\varepsilon_0}\left(\frac{Q_1}{|\mathbf{r}_1 - \mathbf{r}_1|} + \frac{Q_2}{|\mathbf{r}_1 - \mathbf{r}_2|} + \frac{Q_3}{|\mathbf{r}_1 - \mathbf{r}_3|} + \frac{Q_4}{|\mathbf{r}_1 - \mathbf{r}_4|}\right)$$

$$-1 = \frac{1}{4\pi\varepsilon_0}\left(\frac{Q_1}{|\mathbf{r}_2 - \mathbf{r}_1|} + \frac{Q_2}{|\mathbf{r}_2 - \mathbf{r}_2|} + \frac{Q_3}{|\mathbf{r}_2 - \mathbf{r}_3|} + \frac{Q_4}{|\mathbf{r}_2 - \mathbf{r}_4|}\right)$$

$$+1 = \frac{1}{4\pi\varepsilon_0}\left(\frac{Q_1}{|\mathbf{r}_3 - \mathbf{r}_1|} + \frac{Q_2}{|\mathbf{r}_3 - \mathbf{r}_2|} + \frac{Q_3}{|\mathbf{r}_3 - \mathbf{r}_3|} + \frac{Q_4}{|\mathbf{r}_3 - \mathbf{r}_4|}\right)$$

$$+1 = \frac{1}{4\pi\varepsilon_0}\left(\frac{Q_1}{|\mathbf{r}_4 - \mathbf{r}_1|} + \frac{Q_2}{|\mathbf{r}_4 - \mathbf{r}_2|} + \frac{Q_3}{|\mathbf{r}_4 - \mathbf{r}_3|} + \frac{Q_4}{|\mathbf{r}_4 - \mathbf{r}_4|}\right)$$

(4.100)

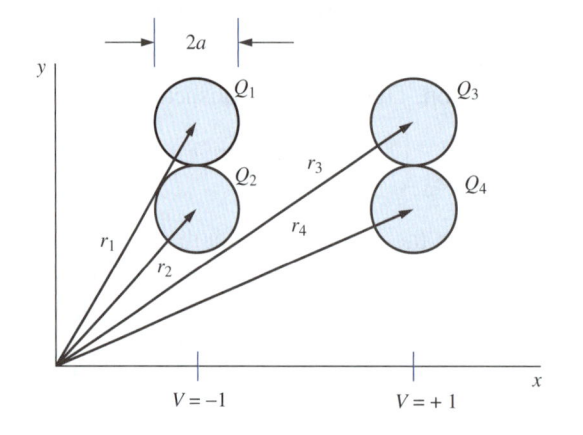

FIGURE 4–21

Four charges distributed in space. The potential at the indicated points are assumed to be $V = -1$ and $V = +1$.

4.8 Method of Moments Using MATLAB

This can be written using the summation sign as

$$V_i = \frac{1}{4\pi\varepsilon_0} \sum_{j=1}^{4} \frac{1}{|\mathbf{r}_i - \mathbf{r}_j|} Q_j \quad (j = 1, 2, 3, 4)$$

Remember that MATLAB originally was created in order to solve problems of the type

$$[P][Q] = [V] \quad (4.101)$$

where $[V]$ is the column vector of the known potentials, $[Q]$ is the column vector of the unknown charges, and $[P]$ is the square matrix of coefficients.

The four equations in (4.100) can also be written in matrix notation

$$[P] = \frac{1}{4\pi\varepsilon_0} \begin{bmatrix} \frac{1}{|\mathbf{r}_1 - \mathbf{r}_1|} & \frac{1}{|\mathbf{r}_1 - \mathbf{r}_2|} & \frac{1}{|\mathbf{r}_1 - \mathbf{r}_3|} & \frac{1}{|\mathbf{r}_1 - \mathbf{r}_4|} \\ \frac{1}{|\mathbf{r}_2 - \mathbf{r}_1|} & \frac{1}{|\mathbf{r}_2 - \mathbf{r}_2|} & \frac{1}{|\mathbf{r}_2 - \mathbf{r}_3|} & \frac{1}{|\mathbf{r}_2 - \mathbf{r}_4|} \\ \frac{1}{|\mathbf{r}_3 - \mathbf{r}_1|} & \frac{1}{|\mathbf{r}_3 - \mathbf{r}_2|} & \frac{1}{|\mathbf{r}_3 - \mathbf{r}_3|} & \frac{1}{|\mathbf{r}_3 - \mathbf{r}_4|} \\ \frac{1}{|\mathbf{r}_4 - \mathbf{r}_1|} & \frac{1}{|\mathbf{r}_4 - \mathbf{r}_2|} & \frac{1}{|\mathbf{r}_4 - \mathbf{r}_3|} & \frac{1}{|\mathbf{r}_4 - \mathbf{r}_4|} \end{bmatrix} \quad (4.102)$$

This matrix is symmetric because the potential between the charge and the point of observation depends upon the magnitude of the distance R between the two points.

The diagonal terms of this matrix ($i = j$) appear to give us problems since they become very large. These terms are called ***singularities***, and we remove these singularities with an approximation. The approximation makes the assumption that the potential at these singular points is evaluated at the edge of the spherical charge that has a radius a and not at the center. It maintains that potential throughout the interior of the spherical charge. The diagonal elements of the matrix $[P]$ will then be given by

$$P_{i,i} = \frac{1}{4\pi\varepsilon_0 a} \quad (4.103)$$

EXAMPLE 4.24

Find the values of the charges that will cause the potentials as shown in the Figure 4–21 if the coordinates of the points in the plane $z = 0$ are $Q_1(2, 3)$; $Q_2(2, 2)$; $Q_3(5, 3)$; $Q_4(5, 2)$. Assume that the diameter of the charges is $2a = 1$ m.

Answer. The matrix $[P]$ in (4.102) and (4.103) has the elements

$$[P] = \frac{1}{4\pi\varepsilon_0} \begin{bmatrix} \frac{1}{1/2} & \frac{1}{1} & \frac{1}{3} & \frac{1}{\sqrt{1^2+3^2}} \\ \frac{1}{1} & \frac{1}{1/2} & \frac{1}{\sqrt{1^2+3^2}} & \frac{1}{3} \\ \frac{1}{3} & \frac{1}{\sqrt{1^2+3^2}} & \frac{1}{1/2} & \frac{1}{1} \\ \frac{1}{\sqrt{1^2+3^2}} & \frac{1}{3} & \frac{1}{1} & \frac{1}{1/2} \end{bmatrix}$$

$$= \frac{1}{4\pi\varepsilon_0} \begin{bmatrix} 2 & 1 & 0.3333 & 0.3162 \\ 1 & 2 & 0.3162 & 0.3333 \\ 0.3333 & 0.3162 & 2 & 1 \\ 0.3162 & 0.3333 & 1 & 2 \end{bmatrix}$$

The column vector for the potential is $[V] = [-1, -1, 1, 1]^T$ where the T indicates the transpose. Solving the matrix equation (4.101) leads to

$$Q_1 = Q_2 = -Q_3 = -Q_4 = -0.4254(4\pi\varepsilon_0)$$

We could continue on with the individual charges that have been presented up to this point; however, it is more meaningful to examine cases where the charge is *distributed* upon various surfaces. If the charge were distributed on a line as shown in Figure 4–22, it would be prudent to describe the

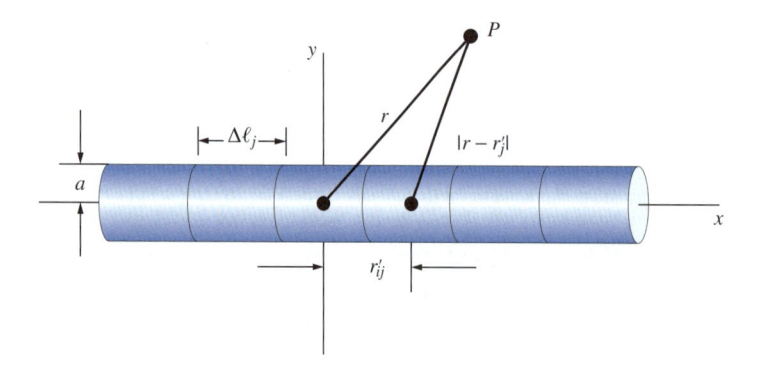

FIGURE 4–22

The potential at point P results from charges $\rho_\ell \Delta \ell_j$ located at the centers of the jth section.

charge distribution with a linear charge density ρ_ℓ C/m. The charge on a particular element j would be $\Delta Q_j = 2\pi a \Delta \ell_j \rho_{\ell,j}$ ($j = 1, 2, ..., N$) and it would be located at the center of the jth section. We proceed using the same method that has just been described. The identical problem with the singularity that was discussed above with individual charges will also be encountered in cylindrical coordinates.

In this case, the column vector of the unknown charges is chosen to be $[Q/\Delta L] = [\rho_{\ell,1}, \rho_{\ell,2}, ..., \rho_{\ell,N}]^T$ while the column vector of the known potentials is written as $[V] = [V_1, V_2, ..., V_N]^T$. The off-diagonal terms of the square matrix are written as

$$P_{i,j} = \frac{2\pi a}{4\pi\varepsilon_0} \frac{\Delta \ell_j}{|x_i - x_j|} = \frac{a \Delta \ell_j}{2\varepsilon_0 |x_i - x_j|} \quad (4.104)$$

The singularities in the diagonal terms of the matrix will also be encountered here but they also can be removed. We evaluate the potential at the surface of the cylinder and assert that it is also equal to the potential at the center, which is the "singular point." The evaluation of this potential V_j at the surface of this cylindrical section shown in Figure 4–23 is calculated using the integral

$$V_j = \frac{1}{4\pi\varepsilon_0} \int_{x' = -\frac{\Delta \ell_j}{2}}^{x' = +\frac{\Delta \ell_j}{2}} \int_{\phi' = 0}^{\phi' = 2\pi} \frac{\rho_\ell a \, d\phi' \, dx'}{\sqrt{a^2 + (x')^2}}$$

The integral can be performed and we find for the diagonal elements

$$P_{j,j} = \frac{2\pi a}{4\pi\varepsilon_0} \ln\left[x' + \sqrt{a^2 + (x')^2}\right]\Bigg|_{x' = -\frac{\Delta \ell_j}{2}}^{x' = +\frac{\Delta \ell_j}{2}} = \frac{a}{2\varepsilon_0} \ln\left[\frac{\frac{\Delta \ell_j}{2} + \sqrt{a^2 + \left(\frac{\Delta \ell_j}{2}\right)^2}}{-\frac{\Delta \ell_j}{2} + \sqrt{a^2 + \left(\frac{\Delta \ell_j}{2}\right)^2}}\right]$$

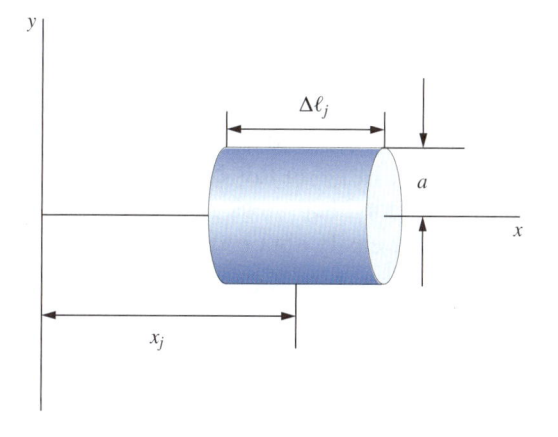

FIGURE 4–23

The jth section of a linearly charged line.

If we make the approximation that the radius a is much less than the length of the section $\Delta\ell_j$ ($a \ll \Delta\ell_j$), this simplifies to

$$P_{j,j} \cong \frac{a}{\varepsilon_0} \ln\left[\frac{\Delta\ell_j}{a}\right] \tag{4.105}$$

EXAMPLE 4.25

Find the charge distribution on the cylindrical conductor whose radius is $a = 0.01$ m and whose length is $\mathscr{L} = 1$ m. The potential on the surface is $V = 1$ V. You may assume that the charge is distributed uniformly in each section. Assume that the number of the sections is $N = 5$ and the step size is $\Delta\ell = 0.20$ m.

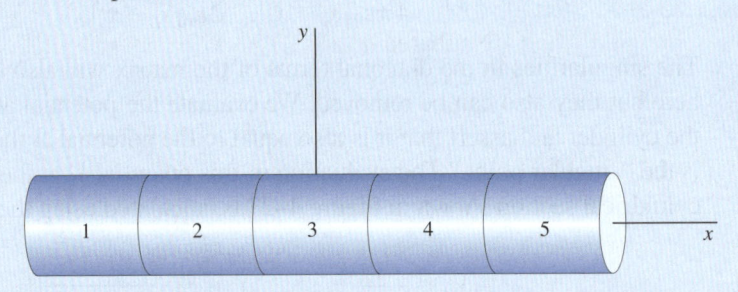

Answer. The matrix equation relating the potentials to the charges is (4.101), where the off-diagonal and the diagonal elements are given by (4.104) and (4.105) respectively. The solution for the unknown charge distribution is

$$\rho_{\ell,1} = \rho_{\ell,5} = 0.2556\varepsilon_0, \qquad \rho_{\ell,2} = \rho_{\ell,4} = 0.2222\varepsilon_0, \qquad \rho_{\ell,3} = 0.2170\varepsilon_0$$

Note that the charge density in the center of the line is smaller than at either end. We should expect this nonuniform distribution since there is a loss of symmetry at either end.

The charge could have been distributed upon a surface resulting in an inhomogeneous surface charge density ρ_s. In this case, we subdivide the surface into small rectangular areas Δs_i ($i = 1, 2, \ldots, N$) This is shown in Figure 4–24. In this case, the column vector of the unknown charges is taken $[Q/\Delta s] = [\rho_{s,1}, \rho_{s,2}, \ldots, \rho_{s,N}]^T$ and the column vector of the known potentials is $[V] = [V_1, V_2, \ldots, V_N]^T$. In this case, $N = M \times M$ is the number of the square sections with individual areas $\Delta s_j = a^2$. The mutual-coupling terms of the square matrix are easily obtained again to be

$$P_{i,j} = \frac{1}{4\pi\varepsilon_0} \frac{\Delta s_j}{|\mathbf{r}_i - \mathbf{r}_j|} \tag{4.106}$$

You should expect at this stage that we will encounter again a singularity in the diagonal terms. In order to remove the singularity, we replace the

4.8 Method of Moments Using MATLAB

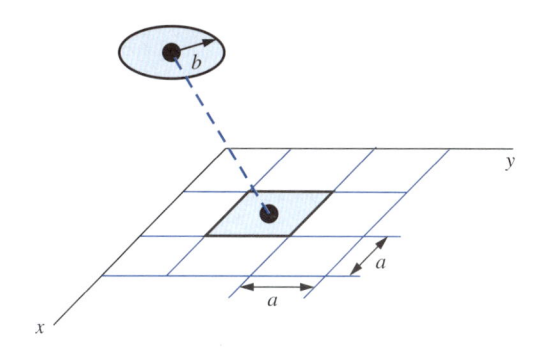

FIGURE 4–24

Charge is distributed on a surface and has a density ρ_s. Singular elements are replaced with discs with the same area.

small rectangular subarea whose area is Δs with a circular region that contains the same incremental charge. This implies that the area of the circle πb^2 and the square subarea a^2 are equal. With this assumption, the radius of the circle b can be calculated to be $b = a/\sqrt{\pi}$. We assume that this charge that is distributed within the circle is localized at the center of the circle. We then compute the potential at the perimeter of the circle from this localized charge and assume that it has this value throughout the circle. The evaluation of the potential leads to the following approximation for the diagonal terms:

$$P_{i,i} = \frac{1}{4\pi\varepsilon_0} \int_0^{2\pi} \int_0^b \frac{\rho' d\rho' d\phi'}{\rho'} = \frac{1}{4\pi\varepsilon_0}(2\pi b) = \frac{b}{2\varepsilon_0} = \frac{1}{2\varepsilon_0}\sqrt{\frac{\Delta s_i}{\pi}} \quad (4.107)$$

where the radius of the circle is written in terms of the area of the grid element.

A typical problem that is encountered in this case would be the calculation of the capacitance of a parallel plate capacitor. In such a calculation, the two plates are each subdivided into N subareas as shown in Figure 4–24. Here the capacitance C is calculated from the equation

$$C = \frac{Q}{V} \quad (4.108)$$

where Q is the total charge stored on the top plate and V is the voltage difference between the plates.

EXAMPLE 4.26

Two charged parallel square plates with dimensions $\mathscr{L} \times \mathscr{L} = 1 \text{ m}^2$ are separated by a distance of $d = 0.1$ m. Each side is subdivided into $N = 64$ equal subareas. The potential of the top plate is $+5$ V, and the potential of the bottom plate is -5 V.

(a) Find and plot the charge density distribution using the MoM.
(b) Find the capacitance C of this charged conductor system. Compare the limiting case with the known simple classical solution $C/\varepsilon_0 = \mathscr{L}^2/d = 10$.

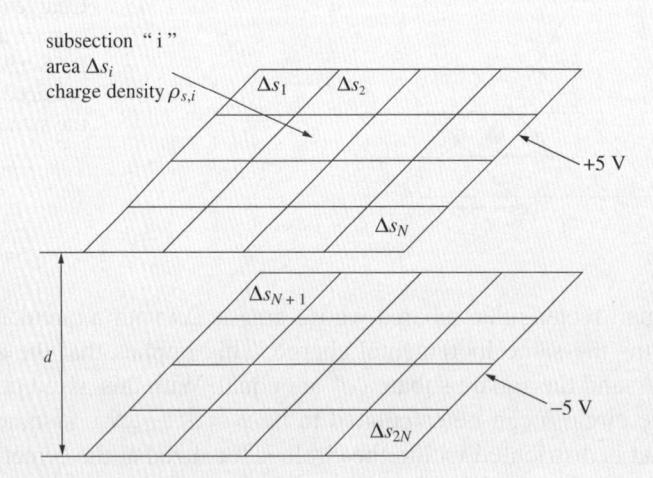

Answer.

(a) The column vector of the known potential is

$$[V] = [+5, +5, \ldots, +5; -5, -5, \ldots, -5]^T$$

The matrix equation that must be solved is again (4.101) with a matrix $[P]$ described by (4.106) and (4.107). The solution of this matrix equation for the surface charge distribution is plotted in the following figure.

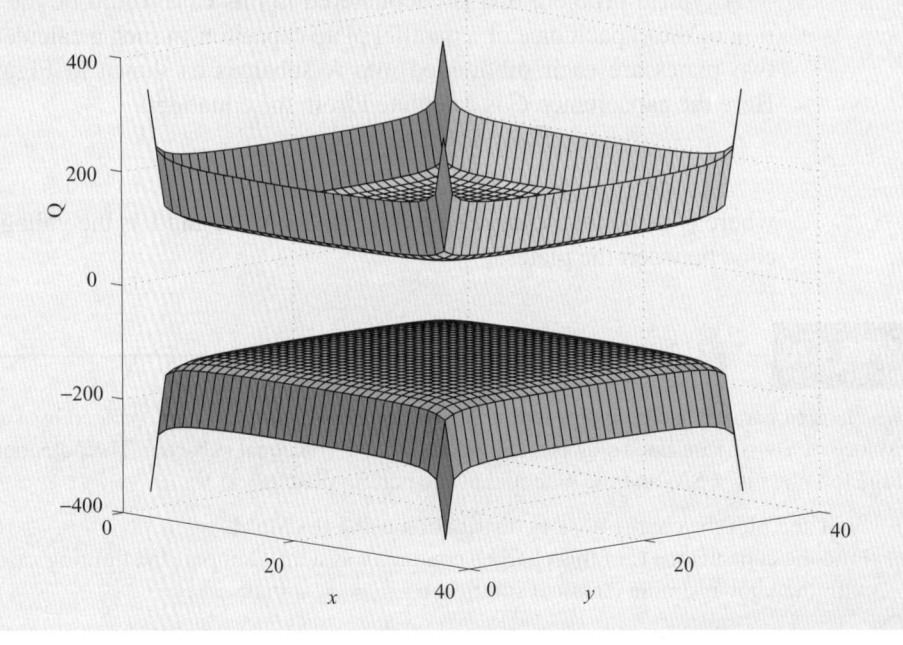

4.8 Method of Moments Using MATLAB

(b) The normalized capacitance $C_0 = C/\varepsilon_0$ is calculated to yield a value of $C_0 = 13.3811$ for the value of the number of sections $N = 64$. This value is larger than that predicted from the elementary formula $C/\varepsilon_0 = \mathscr{L}^2/d = 10$. This formula assumed that the separation distance was significantly less than the area of the plate and all fringing fields at the edges could be neglected. The accuracy can be improved by subdividing the area into smaller subareas. It is also possible to determine the inhomogeneous charge distribution caused by perturbations in the plates or not having the plates exactly parallel as was assumed here.[1]

Let us also apply the MoM to a slightly different topic, that of ascertaining the expected one-dimensional charge distribution from a known potential profile. The potential profile and the resulting charge distribution could be very nonuniform as in, for example, the depletion layer of a pn junction. In this case, we assume that there is a sequence of *sheets* of charge as shown in Figure 4–25. The incremental charge density on each sheet j is uniform across the plane of that particular sheet and it has a value $\rho_{s,j}$. The separation between each sheet also will be assumed to be uniform with a separation distance of d.

The electric field surrounding an infinite plane of constant surface charge density $\rho_{s,j}$ (C/m^2) is given by

$$E_j = \frac{\rho_{s,j}}{2\varepsilon}$$

The electric potential at a point z_i at a distance $z = |z_i - z_j|$ from the charged sheet j is found from the integral of the electric field

$$V_j = -\frac{\rho_{s,j}}{2\varepsilon} z \quad (4.109)$$

where the constant of integration in (4.109) is set equal to zero. The potentials at the two extremities $z_1 = -2d$ and $z_5 = +2d$ in Figure 4–25 have a value that is equal to one half of the value that is given in (4.109). This additional factor of half arises since we are only interested in the electric field that is confined within the region of interest. The singularities that were previously encountered in the diagonal terms are removed automatically since the factor of zero is introduced in (4.109). In this case the column vector of

---

[1] Bai, E. W., and Lonngren, K. E., "Capacitors and the method of moments," *Computers and Electrical Engineering*, vol. 30, 2004, pp. 223–229.

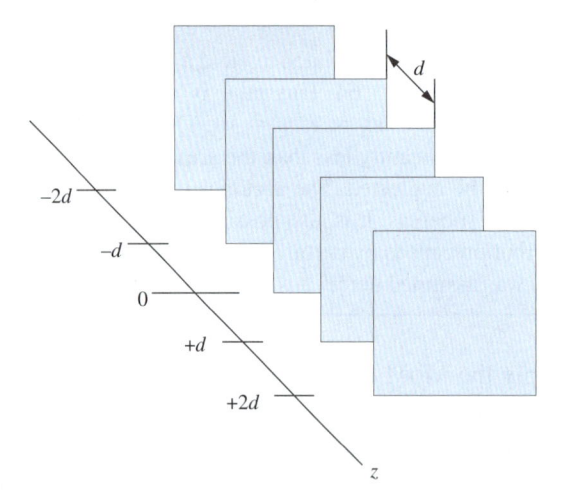

FIGURE 4–25

Charged sheet model to represent the pn junction.

the source is $[Q/\Delta s] = [\rho_{s,1}, \ldots, \rho_{s,5}]^T$. The elements in the square matrix $[P]$ are obtained from (4.109) to be

$$P_{i,j} = \begin{cases} -\dfrac{d}{2\varepsilon}|z_i - z_j|, & j \neq 1 \text{ or } 5 \\ -\dfrac{d}{4\varepsilon}|z_i - z_j|, & j = 1 \text{ or } 5 \end{cases} \quad (4.110)$$

or as a matrix

$$[P] = -\frac{d}{2\varepsilon}\begin{bmatrix} 0/2 & 1 & 2 & 3 & 4/2 \\ 1/2 & 0 & 1 & 2 & 3/2 \\ 2/2 & 1 & 0 & 1 & 2/2 \\ 3/2 & 2 & 1 & 0 & 1/2 \\ 4/2 & 3 & 2 & 1 & 0/2 \end{bmatrix}$$

Let us now specify the values of the potential as indicative of a linear variation in space to be $[V] = [-2, -1, 0, 1, 2]^T$. Solving the matrix equation (4.101), we find that

$$[\rho_{s,1}, \rho_{s,2}, \rho_{s,3}, \rho_{s,4}, \rho_{s,5}]^T = \frac{2\varepsilon}{d}[-1, 0, 0, 0, +1]^T. \quad (4.111)$$

This result expresses the fact that the electric field that is proportional to the negative gradient of the linearly varying potential is a constant that is determined by the charge densities that are localized at either edge. Other potential distributions could be employed. This would lead to a different charge distribution.[2]

Other potential distributions can be assumed. In studying a pn junction, you may encounter the terms "linear graded junction" or "quadratic graded junction." Of course, a different choice yields a different charge distribution.

This is a brief introduction to the method of moments. As we have seen, the technique is very useful in determining a source distribution such as a charge in terms of a response function such as voltage between two plates containing a homogeneous dielectric. Further discussion of this technique is usually reserved for more advanced courses.

4.9 Conclusion

The calculation of the electric and magnetic fields in a region that is subdivided into two parts requires an understanding of the boundary conditions for the fields. The actual solution of the resulting boundary value problem has caused us to examine several techniques. These techniques are either analytical or numerical. The numerical techniques have been highlighted using some features inherent in MATLAB.

Five techniques that are encountered frequently when we attempt to solve more complicated problems in electromagnetic theory were introduced in this chapter. The first two are analytical methods. The latter three techniques involve the application of numerical methods

1. Direct integration of a one-dimensional equation.
2. Fourier series expansion of a two-dimensional equation.
3. Finite difference method.
4. Finite element method.
5. Method of moments.

The methods are described in both one and two dimensions. The first two of these numerical methods are based on differential equations for the electric potential (Poisson's and Laplace's equations), while the last one is based on an integral equation for the unknown charge distribution. The numerical techniques that were introduced here are also applicable to three-dimensional problems that may be encountered later. The numerical programs that have

---

[2] K. E. Lonngren, P. V. Schwartz, E. W. Bai, W. C. Theisen, R. L. Merlino, and R. T. Carpenter, "Extracting Double Layer Charge Density Distributions using the Method of Moments," *IEEE Transactions on Plasma Science*, vol. 24, pp. 278–280, (1994).

been written using MATLAB are available at the Web site *http://www.scitechpub.com/lonngren2e.htm*.

As presented in this chapter, all of the methods assume linearity, which leads to **linear superposition principles**. However, you will frequently encounter **nonlinearity** in nature, which will lead to significant alterations in your method of obtaining a solution. Numerical questions concerning other specific programming languages, convergence requirements, numerical errors, aliasing, etc., are better left to later courses. As problems arise, hopefully solutions can be found.

4.10 Problems

4.1.1. Find the electric field in the region $x > 0$ if $\mathbf{E} = 2\mathbf{u}_x + 2\mathbf{u}_y$ V/m in the region $x < 0$. There is no surface charge density.

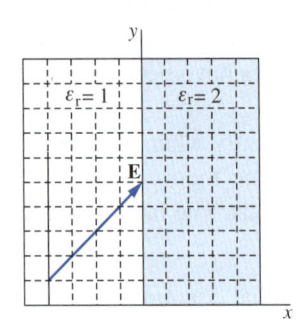

4.1.2. Repeat Problem 4.1.1 with a surface charge density $\rho_s = 1$ C/m².

4.1.3. Find the electric field in the region $x > 0$. There is no surface charge density. The magnitude of the electric field is 5 V/m.

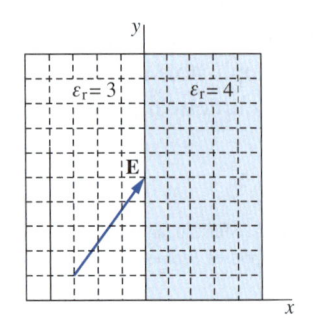

4.1.4. Repeat Problem 4.1.3 with a surface charge density $\rho_s = 0.5$ C/m².

4.1.5. Find the magnetic flux density in the region $x > 0$ if $\mathbf{B} = 4\mathbf{u}_x + 4\mathbf{u}_y$ T in the region $x < 0$. The surface current equals zero.

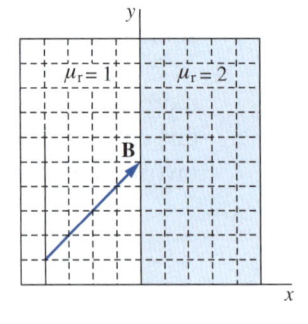

4.1.6. Repeat Problem 4.1.5 if the surface current density $\mathbf{J}_s = (9\mathbf{u}_z + 9\mathbf{u}_y)$ A/m.

4.1.7. Can the indicated electric field exist? If not, suggest how an additional electric field will permit the existence of this field.

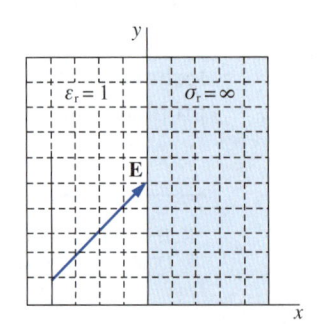

4.10 Problems

4.3.1. Find the potential distribution $V(r)$ by solving Laplace's equation analytically for the region between two concentric hollow spheres (spherical capacitor). Apply a spherical coordinate system with the following boundary conditions: $V = 0$ at $r = a$ and $V = V_0$ at $r = b$. Simplify the calculation with symmetry arguments.

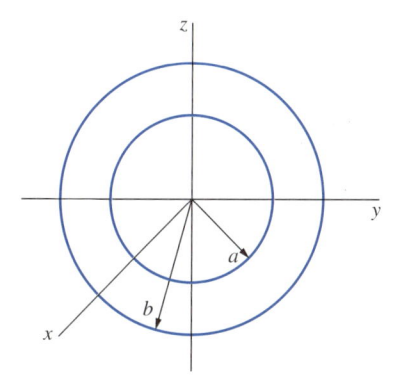

4.3.2. Find the capacitance C of the spherical capacitor in Problem 4.3.1.

4.3.3. Find the potential distribution $V(\theta)$ by solving Laplace's equation analytically for the region between two hollow coaxial cones. A potential $V = V_1$ is assumed at $\theta = \theta_1$ and $V = 0$ at $\theta = \theta_2 = \pi - \theta_1$. The vertices of the cones are insulated at $\rho = 0$. Simplify the calculation with symmetry arguments.

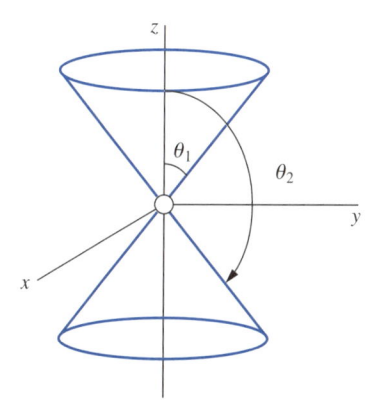

4.3.4. Find the potential $V(x)$ in the region $0 < x < 1$ satisfying the boundary conditions $V(0) = 3$ and $V(1) = 0$.

4.3.5. Find the potential $V(x)$ in the region $0 < x < 2$ if the electric field is normal with a constant value $\mathbf{E}_x = 4$.

4.3.6. Find the potential $V(x)$ in the region $0 < x < 1$. Assume that a charge distributed uniformly there has a density $\rho_v = -4\varepsilon$. The potential satisfies the boundary conditions: $V(0) = 3$ and $V(1) = 0$.

4.3.7. Find the normal electric field $\mathbf{E}_x(x)$ in Problem 4.3.6.

4.3.8. Find the capacitance C_0 of unit length of the cylindrical capacitor—two long concentric cylinders with radii a and b ($b > a$). The boundary conditions for the potential are $V(\rho = a) = V_0$ and $V(\rho = b) = 0$.

4.3.9. Compute using an analytical integration, the potential $V(r)$ at the point P that is a distance $r = 1$ m from the midpoint of a narrow finite strip that has a length $2\mathscr{L} = 2$ m. A charge of 1C is distributed uniformly on the strip.

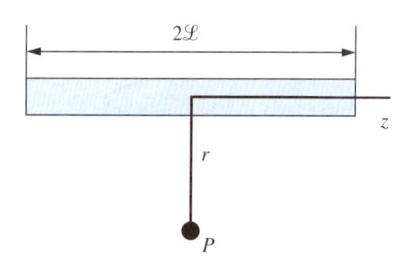

4.3.10. Repeat Problem 4.3.9 with a nonuniform charge distribution $\rho_\ell(z) = \rho_{\ell 0}[1 - |z|/\mathscr{L}]$ where $z = 0$ is at the midpoint of the strip.

4.3.11. Repeat Problem 4.3.9 but compute the electric field $\mathbf{E}_r(\mathbf{r})$ instead of the potential.

4.3.12. Compare the numerical and analytical integrations that lead to the potential $V(z)$ along the z axis from the charged circular loop with a diameter $2a$. Charge Q is distributed uniformly upon the loop.

4.3.13. Compare the results obtained from a numerical and an analytical integration for the electric field $\mathbf{E}_z(z)$ in the previous problem. Find the solution in the limiting case $a \to 0$.

4.3.14. Compare the results of a numerical and analytical analysis for the potential $V(z)$ along the z axis from the charged circular plate with a diameter $2a$. Charge Q is distributed uniformly upon the plate.

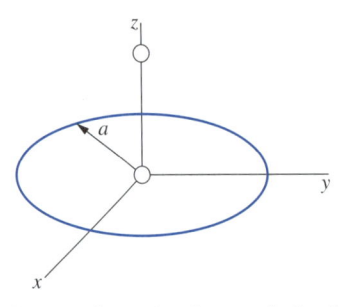

4.3.15. Compare the results of a numerical and analytical analysis for the potential $\mathbf{E}_z(z)$ in the previous problem. Find the solution in the limiting case $a \to \infty$.

4.3.16. Compare the results of a numerical and analytical analysis for the potential $V(z)$ between two parallel discs ($0 < z < c$) that have a large enough radius a such that the electric field is constrained to be entirely between them. This implies that the fringing fields are neglected. The boundary conditions are $V(0) = 0$ and $V(c) = V_0$. Find the normal electric field $\mathbf{E}_z$, the flux density $\mathbf{D}_z$, and the surface charge density ρ_s on both plates.

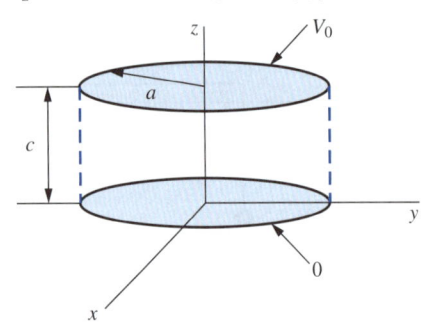

4.5.1. For the indicated boundary conditions that are specified in the figure, find the potential distribution $V(x, y)$ within the enclosed region by solving Laplace's equation and expanding one boundary condition in a Fourier series expansion.

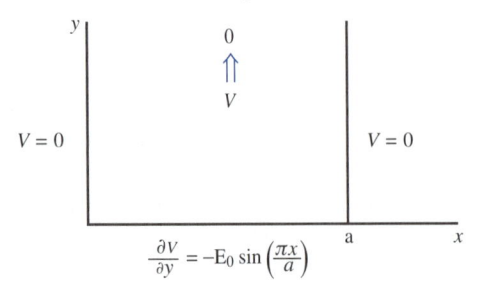

4.5.2. Find the potential within the channel given in Problem 4.5.1 using the boundary conditions at $x = 0$, $x = a$, and $y = \infty$ as stated there. The boundary condition at $y = 0$ is given as $\mathbf{E} = E_0 \mathbf{u}_y$ where E_0 is a constant.

4.5.3. Find the potential within the channel given in Problem 4.5.1 using the boundary conditions at $x = 0$, $x = a$, and $y = \infty$ as stated there. The boundary condition at $y = 0$ is: $V_0 = +V_0$ for $0 < x < (a/2)$ and $V = -V_0$ for $(a/2) < x < a$.

4.5.4. Find the components of the electric field (E_x, E_y) in Problem 4.5.3.

4.5.5. Find the expression for a potential $V = V_0 (x/a)$ that describes the potential variation in the region $0 > x > a$.

4.5.6. Find the potential distribution $V(x, y)$ in the region $y > 0$. Because of the periodicity of the boundary condition, expand boundary in a Fourier series. This problem models a VLSI circuit where conductors are implanted on an insulating material. The thickness of the metal strip can be neglected.

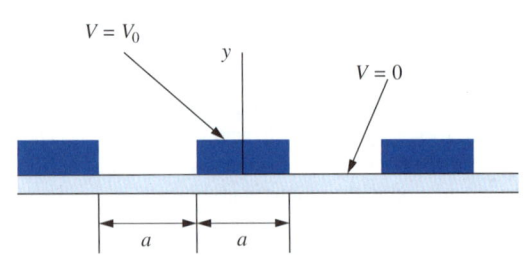

4.5.7. For the indicated boundary conditions that are specified in the figure, find the potential distribution $V(x, y)$ within the enclosed region by solving Laplace's equation. Plot the potential distribution ($a = 1$ m, $V_0 = 10$ V).

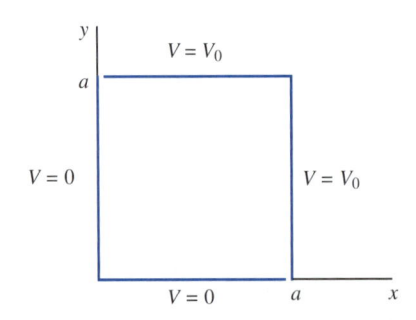

4.10 Problems

4.5.8. Using a product solution in Laplace's equation in cylindrical coordinates $V(\rho, \phi, z) = R(\rho)\Phi(\phi)Z(z)$, show that the term $R(\rho)$ satisfies the ordinary differential equation

$$\rho^2 \frac{d^2R}{d\rho^2} + \rho\frac{dR}{d\rho} + ((\lambda\rho)^2 - n^2)R = 0$$

The separation constant is assumed to be $\alpha = n^2$ where n is a positive integer and the constant λ is real. The solutions for two of the dependent variables are $\Phi_n(\phi) = e^{\pm jn\phi}$ and $Z_n(z) = e^{\pm \lambda z}$. The solution for this equation is $R_n(\rho) = J_n(\lambda\rho)$, and it can be obtained in terms of an infinite power series. Find the series expansion of the last function that is known as a **Bessel function** of a first kind, nth order. Plot the first three functions ($n = 0, 1, 2; 1 = 1$) using MATLAB.

4.5.9. Using a product solution in Laplace's equation in spherical coordinates $V(r, \theta, \phi) = R(r)\Theta(\theta)\Phi(\phi)$, show that the term $R(r)$ satisfies the ordinary differential equation

$$r^2 \frac{d^2R}{dr^2} + 2r\frac{dR}{dr} - n(n+1)R = 0$$

The separation constant is assumed to be $\alpha = n(n+1)$ where $n \geq 0$ is an integer and the term $\Theta(\theta)$ is a constant (rotational symmetry). Find two particular solutions $R_n(r)$ of this equation.

4.5.10. Show that the corresponding ordinary differential equation for $\Theta(\theta)$ in Problem 4.5.9 is

$$\sin\theta \frac{d^2\Theta}{d\theta^2} + \cos\theta \frac{d\Theta}{d\theta} + [n(n+1)\sin\theta]\Theta = 0$$

Let $\cos\theta = \xi$ and find an expression of the solution $\Theta_n(\theta) = P_n(\xi)$ of this equation, which is a polynomial of nth order that is known as a **Legendre polynomial**. Plot the first three functions ($n = 0, 1, 2$) using MATLAB.

4.5.11. Find particular solutions of Laplace's equation $V_n(r, \theta)$ that increase with the distance r for the following three cases: a) $n = 0$ (charge), b) $n = 1$ (dipole), and c) $n = 2$ (quadrupole) using the results from Problem 4.5.9 and Problem 4.5.10.

4.5.12. Compare the solution of Laplace's equation in cylindrical coordinates $f_1(\rho) = J_0(\lambda\rho)$ found in Problem 4.5.8 for $n = 0$ with the solution of Laplace's equation in Cartesian coordinates $f_2(\rho) = \cos(\lambda x)\cos(\lambda y)$ using a second order approximation in the power series for small values of the argument ($\lambda\rho < 1$).

4.7.1. Find the conditions under which a voltage distribution

$$V(x, y) = a + bx + cy + dx^2 + exy + fy^2$$

where a, b, c, d, e, and f are constants is a solution of Laplace's equation.

4.7.2. (a) Determine the area of a triangle that is defined by the three points: (0, 0), (0, 2), (2, 2) using the square matrix in (4.79). Verify this result using MATLAB.

(b) Determine the voltage distribution within the triangular region if the voltage at the nodes has the values $V(0, 0) = 0$; $V(0, 2) = 0$; and $V(2, 2) = 2$.

4.7.3. Verify that the coefficients that are presented in (4.82) are correct.

4.7.4. Obtain the matrix coefficients that are used in Example 4.19.

4.7.5. Determine the electrostatic energy that is stored within the triangular region defined in Problem 4.7.2.

4.7.6. Using the energy minimization technique, calculate the voltage at the node that is located at (0, 2) if the other voltages are specified as in 4.7.2. Find the voltage distribution within this triangular region.

4.7.7. Find the potential distribution in the triangular region between the two charged surfaces using the FEM method with $a = 1$ m and $V_0 = 16$ V. Display the potential profile.

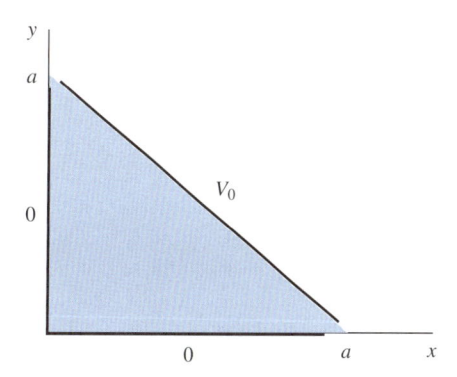

4.8.1. Find the surface charge density ρ_s in a square region $a \times a$ ($a = 2$ m) between two charged surfaces with potentials 0 and V_0 ($V_0 = 10$ V) using the MoM method applied to the boundary integral equation. Then by simple integration find the potential in every internal point of the mesh. Display the potential profile.

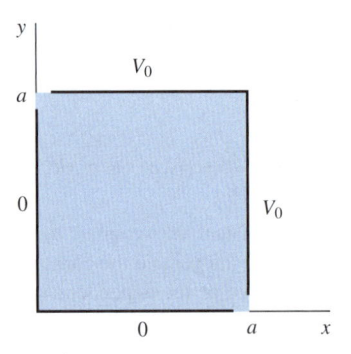

CHAPTER 5

Time-Varying Electromagnetic Fields

5.1	Faraday's Law of Induction	257
5.2	Equation of Continuity	270
5.3	Displacement Current	274
5.4	Maxwell's Equations	280
5.5	Poynting's Theorem	285
5.6	Time-Harmonic Electromagnetic Fields	290
5.7	Conclusion	293
5.8	Problems	294

The subject of time-varying electromagnetic fields will be the central theme throughout the remainder of this text. Here and in the following chapters we will generalize to the time-varying case the static electric and magnetic fields that were reviewed in Chapter 2 and Chapter 3. In doing this, we must first appreciate the insight of the great nineteenth century theoretical physicist James Clerk Maxwell who was able to write down a set of equations that described electromagnetic fields. These equations have survived unblemished for almost two centuries of experimental and theoretical questioning. The equations are now considered to be on an equal footing with the equations of Isaac Newton and many of Albert Einstein's thoughts on relativity.[1] We will concern ourselves here and now with what was uncovered and explained at the time of Maxwell's life.

5.1 Faraday's Law of Induction

The first time-varying electromagnetic phenomenon that usually is encountered in an introductory course dedicated to the study of time-varying electrical circuits is the determination of the electric potential across an inductor

---

[1] We can only speculate about what these three giants would say and write if they met today at a café and had only one "back of an envelope" between them.

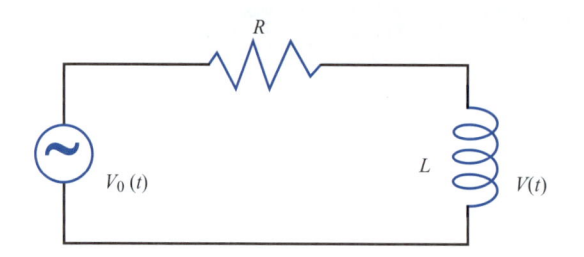

FIGURE 5–1

A simple electrical circuit consisting of an AC voltage source $V_0(t)$, an inductor, and a resistor. The voltage across the inductor is $V(t)$.

that is inserted in an electrical circuit. A simple circuit that exhibits this effect is shown in Figure 5–1.

The voltage across the inductor is expressed with the equation

$$V(t) = L\frac{dI(t)}{dt} \tag{5.1}$$

where L is the inductance, the units of which are henries, $V(t)$ is the time-varying voltage across the inductor, and $I(t)$ is the time-varying current that passes through the inductor. The actual dependence that these quantities have on time will be determined by the voltage source. For example, a sinusoidal voltage source in a circuit will cause the current in that circuit to have a sinusoidal time variation or a temporal pulse of current will be excited by a pulsed voltage source. In Chapter 3, we recognized that a time-independent current could create a time-independent magnetic field and that a time-independent voltage was related to an electric field. These quantities are also related for time-varying cases as will be shown here.

The actual relation between the electric and the magnetic field components is computed from an experimentally verified effect that we now call **Faraday's law**. This is written as

$$V(t) = -\frac{d\Psi_m(t)}{dt} \tag{5.2}$$

where $\Psi_m(t)$ is the total time-varying magnetic flux that passes through a surface. This law states that a voltage $V(t)$ will be induced in a closed loop that completely surrounds the surface through which the magnetic field passes. The voltage $V(t)$ that is induced in the loop is actually a voltage or potential difference $V(t)$ that exists between two points in the loop that are separated by an infinitesimal distance. The distance is so small that we can actually think of the loop as being closed. The polarity of the induced voltage will be such that it opposes the change of the magnetic flux, hence a minus sign appears in (5.2). The voltage can be computed from the line integral

5.1 Faraday's Law of Induction

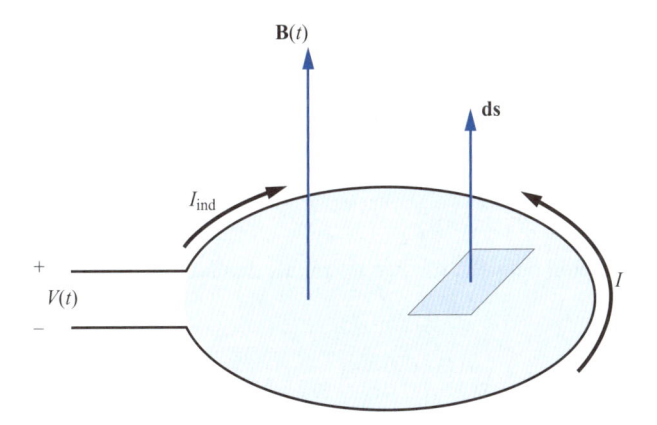

FIGURE 5-2

A loop through which a time-varying magnetic field passes.

of the electric field between the two points. This effect is also known as **Lenz's law**.

A schematic representation of this effect is shown in Figure 5–2. Small loops as indicated in this figure, and which have a cross-sectional area Δs, are used to detect and plot the magnitude of time-varying magnetic fields in practice. We will assume that the loop is sufficiently small or that we can let it shrink in size so that it is possible to approximate Δs with the differential surface area |**ds**|. The vector direction associated with **ds** is normal to the plane containing the differential surface area. If the stationary orientation of the loop **ds** is perpendicular to the magnetic flux density **B**(*t*), zero magnetic flux will be captured by the loop and *V*(*t*) will be zero. By rotating this loop about a known axis, it is also possible to ascertain the vector direction of **B**(*t*) by correlating the maximum detected voltage *V*(*t*) with respect to the orientation of the loop. Recall from our discussion of magnetic circuits that we used the total magnetic flux Ψ_m. This is formally written in terms of an integral. In particular, the magnetic flux that passes through the loop is given by

$$\Psi_m(t) = \int_{\Delta s} \mathbf{B} \cdot \mathbf{ds} \qquad (5.3)$$

As we will see later, either the magnetic flux density or the surface area or both could be changing in time. The scalar product reflects the effects arising from an arbitrary orientation of the loop with respect to the orientation of the magnetic flux density. It is important to realize that the loops that we are considering may not be wire loops. The loops could just be closed paths.

EXAMPLE 5.1

Let a stationary square loop of wire lie in the x–y plane that contains a spatially homogeneous time-varying magnetic field.

$$\mathbf{B}(\mathbf{r}, t) = B_0 \sin \omega t \, \mathbf{u}_z$$

Find the voltage $V(t)$ that could be detected between the two terminals that are separated by an infinitesimal distance.

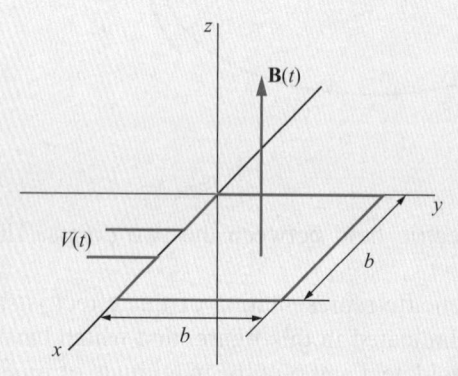

Answer. The magnetic flux that is enclosed within the loop is given by

$$\Psi_m = \int_{\Delta s} \mathbf{B} \cdot \mathbf{ds} = \int_{y=0}^{b} \int_{x=0}^{b} (B_0 \sin \omega t \, \mathbf{u}_z) \cdot (dx dy \, \mathbf{u}_z) = B_0 b^2 \sin \omega t$$

The induced voltage in the loop is found from Faraday's law (5.2)

$$V(t) = -\frac{d\Psi_m(t)}{dt} = -\omega B_0 b^2 \cos \omega t$$

EXAMPLE 5.2

Let a stationary loop of wire lie in the x–y plane that contains a spatially inhomogeneous time-varying magnetic field given by

$$\mathbf{B}(r, t) = B_0 \cos\left(\frac{\pi \rho}{2b}\right) \cos \omega t \, \mathbf{u}_z$$

where the amplitude of the magnetic flux density is $B_0 = 2$ T, the radius of the loop is $b = 0.05$ m, and the angular frequency of oscillation of the time-varying magnetic field is $\omega = 2\pi f = 314$ s^{-1}. The center of the loop is at the point $\rho = 0$. Find the voltage $V(t)$ that could be detected between the two terminals that are separated by an infinitesimal distance.

5.1 Faraday's Law of Induction

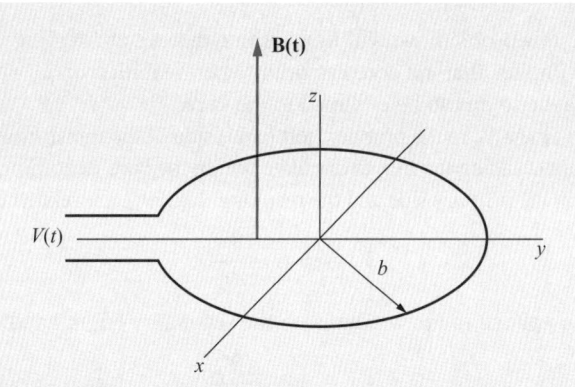

Answer. The magnetic flux Ψ_m that is enclosed within the "closed" loop is given by

$$\Psi_m = \int_{\Delta s} \mathbf{B} \cdot \mathbf{ds} = \int_{\phi=0}^{2\pi} \int_{\rho=0}^{b} \left[B_0 \cos\left(\frac{\pi \rho}{2b}\right) \cos \omega t \, \mathbf{u_z}\right] \cdot [\rho \, d\rho \, d\phi \, \mathbf{u_z}]$$

The integral over ϕ yields a factor of 2π while the integral over ρ is solved via integration by parts. The result is

$$\Psi_m = (B_0 \cos \omega t)(2\pi)\left[\frac{4b^2}{\pi^2}\left(\frac{\pi}{2} - 1\right)\right] = \frac{8b^2}{\pi}\left(\frac{\pi}{2} - 1\right) B_0 \cos \omega t$$

The induced voltage in the loop is obtained from Faraday's law (5.2)

$$V(t) = -\frac{d\Psi_m(t)}{dt} = \frac{8b^2}{\pi}\left(\frac{\pi}{2} - 1\right) B_0 \omega \sin \omega t = 228.2 \sin(314t) \text{ V}$$

EXAMPLE 5.3

Another application of Faraday's law is the explanation of how an ideal ***transformer*** operates. Find the voltage that is induced in loop two if a time-varying voltage V_1 is connected to loop one.

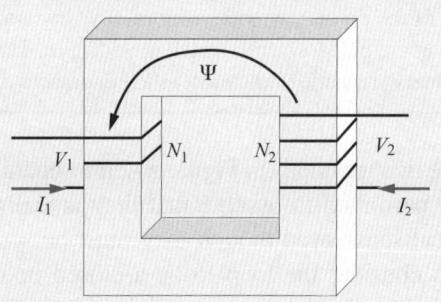

Answer. To solve this problem we will first assume that we are dealing with an ideal transformer. This implies that the core has infinite permeability, or $\mu_r = \infty$. This will cause all of the magnetic flux to be confined to the core.

Application of voltage V_1 to the primary (left hand) side of the transformer will cause a current I_1 to flow, also causing the flux to circulate within the core, according to the right hand rule. The voltage on the primary side and the resulting flux, Ψ_m, are related by the equation

$$V_1 = -N_1 \frac{d\Psi_m}{dt}$$

This flux induces a voltage in the windings on the secondary (right hand) side equal to

$$V_2 = -N_2 \frac{d\Psi_m}{dt}$$

Dividing the second equation by the first equation yields

$$\frac{V_2}{V_1} = \frac{N_2}{N_1}$$

Because this is an ideal lossless transformer, all of the instantaneous power delivered to the primary will be available at the secondary, or $P_1 = P_2$, which means $(I_1)(V_1) = (I_2)(V_2)$.

Given this relationship, as well as the relationship of the third equation, we can also show that

$$\frac{I_1}{I_2} = \frac{N_2}{N_1}$$

Finally, we can use the relationships for primary and secondary resistance, that is

$$R_{pri} = \frac{V_1}{I_1} \quad \text{and} \quad R_{sec} = \frac{V_2}{I_2}$$

to show that the ratio of the primary resistance to the secondary resistance equals the square of the turns ratio or

$$\frac{R_{pri}}{R_{sec}} = \left(\frac{N_1}{N_2}\right)^2$$

Note that even though a real transformer has some loss, it is a highly efficient device, typically having an efficiency of 95–98%, as a properly designed transformer has very low core losses. Also, the resistance of the primary winding is normally very small, as is the secondary winding, unless we are using the transformer to match impedances. Thus, compared to other electrical devices, transformers are some of the most efficient devices available.

Since the two terminals in Figure 5–2 are separated by a very small distance, we will be permitted to assume that they actually are touching, at least in a mathematical sense even though they must be separated physically. This will allow us to consider the loop to be a closed in various integrals that follow but still permit us to detect a potential difference between the two terminals. The magnetic flux $\Psi_m(t)$ can be written in terms of the magnetic flux density

$\mathbf{B}(\mathbf{r}, t) = \mathbf{B}$ and the voltage $V(t)$ can be written in terms of the electric field $\mathbf{E}(\mathbf{r}, t) = \mathbf{E}$. This yields the result

$$\oint \mathbf{E} \cdot \mathbf{dl} = -\frac{d}{dt} \int_{\Delta s} \mathbf{B} \cdot \mathbf{ds} \tag{5.4}$$

It is worth emphasizing the point that this electric field is the component of the electric field that is tangential to the loop since this is critical in our argument. In addition, (5.4) includes several possible mechanisms in which the magnetic flux could change in time. Either the magnetic flux density changes in time, the cross-sectional area changes in time, or there is a combination of the two mechanisms. These will be described below.

Although both the electric and magnetic fields depend on space and time, we will not explicitly state this fact in every equation that follows. This will conserve time, space, and energy if we now define and later understand that $\mathbf{E}(\mathbf{r}, t) \equiv \mathbf{E}$ and $\mathbf{B}(\mathbf{r}, t) \equiv \mathbf{B}$ in the equations. This short-hand notation will also allow us to remember more easily the important results in the following material. In this notation, the independent spatial variable $\mathbf{r}$ refers to a three-dimensional position vector where

$$\mathbf{r} = x\mathbf{u_x} + y\mathbf{u_y} + z\mathbf{u_z} \tag{5.5}$$

in Cartesian coordinates. The independent variable t refers to time. We must keep this notation in our minds in the material that follows.

The closed-line integral appearing in (5.4) can be converted into a surface integral using Stokes's theorem. We obtain for the left side of (5.4)

$$\oint \mathbf{E} \cdot \mathbf{dl} = \int_{\Delta s} \nabla \times \mathbf{E} \cdot \mathbf{ds} \tag{5.6}$$

Let us assume initially that the surface area of the loop does not change in time. This implies that Δs is a *constant*. In this case, the time derivative can then be brought inside the integral

$$\int_{\Delta s} (\nabla \times \mathbf{E}) \cdot \mathbf{ds} = -\int_{\Delta s} \frac{\partial \mathbf{B}}{\partial t} \cdot \mathbf{ds} \tag{5.7}$$

The two integrals will be equal over any arbitrary surface area if and only if the two integrands are equal. This means that

$$\nabla \times \mathbf{E} = -\frac{\partial \mathbf{B}}{\partial t} \tag{5.8}$$

The integral representation (5.4) and the differential representation (5.8) are equally valid in describing the physical effects that are included in Faraday's law. These equations are also called *Faraday's law of induction* in honor of their discoverer. Faraday stated the induction's law in 1831 after making the assumption that a new phenomenon called an *electromagnetic field* would surround every electric charge.

EXAMPLE 5.4

A small rectangular loop of wire is placed next to an infinitely long wire carrying a time-varying current. Calculate the current $i(t)$ that flows in the loop if the conductivity of the wire is σ. To simplify the calculation, we will neglect the magnetic field created by the current $i(t)$ that passes through the wire of the loop.

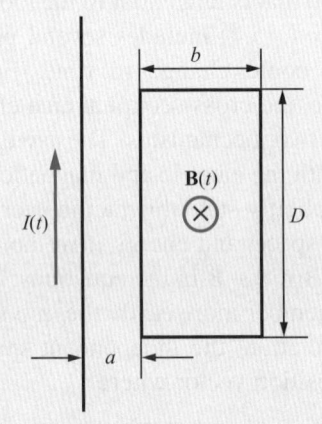

Answer. From the left-hand side of (5.4), we write the induced current $i(t)$ as

$$\oint \mathbf{E} \cdot \mathbf{dl} = \oint \frac{\mathbf{J} \cdot \mathbf{dl}}{\sigma} = \frac{i(2D + 2b)}{\sigma A} = i(t)R$$

where A is the cross-sectional area of the wire and R is the resistance of the wire. The current density is assumed to be constant over the cross-section of the wire. The magnetic flux density of the infinite wire is found from Ampere's law

$$B(t) = \frac{\mu_0 I(t)}{2\pi \rho}$$

The right-hand side of (5.4) can be written as

$$-\frac{d}{dt} \int_{\Delta s} \mathbf{B} \cdot \mathbf{ds} = -\frac{d}{dt} \left[\int_{z=0}^{D} \int_{\rho 5a}^{a+b} \frac{\mu_0 I(t)}{2\pi \rho} d\rho dz \right]$$

$$= -\frac{d}{dt} \left[D \frac{\mu_0 I(t)}{2\pi} \ln\left(\frac{b+a}{a}\right) \right]$$

$$= -\left[\frac{\mu_0 D}{2\pi} \ln\left(\frac{b+a}{a}\right) \right] \frac{dI(t)}{dt}$$

Hence, the current $i(t)$ that is induced in the loop is given by

$$i(t) = -\frac{1}{R} \left[\frac{\mu_0 D}{2\pi} \ln\left(\frac{b+a}{a}\right) \right] \frac{dI(t)}{dt}$$

5.1 Faraday's Law of Induction

The term in the brackets corresponds to a term that is called the ***mutual inductance M*** between the wire and the loop. If we had included the effects of the magnetic field created by the current $i(t)$ in the loop, an additional term proportional to $di(t)/dt$ would appear in the equation. In this case, a term called the ***self-inductance L*** of the loop would be found. Hence a differential equation for $i(t)$ would have to be solved in order to incorporate the effects of the self-inductance of the loop.

It is also important to calculate the mutual inductance between two coils. This can be accomplished by assuming that a time-varying current in one of the coils would produce a time-varying magnetic field in the region of the second coil. This will be demonstrated in the following example.

EXAMPLE 5.5

Find the normalized mutual inductance M/μ_0 of the system consisting of two similar parallel wire loops with a radius $R = 50$ cm if their centers are separated by a distance $h = 2R$ (Helmholtz's coils).

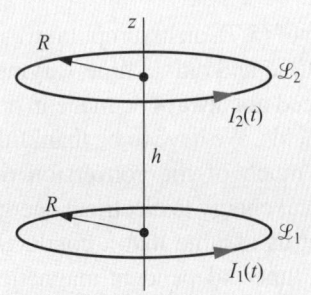

Answer. The magnetic flux density can be replaced with the magnetic vector potential. Using $\mathbf{B} = \nabla \times \mathbf{A}$ and Stokes's theorem, we can reduce the surface integral for the magnetic flux to the following line integral:

$$\Psi_{m12} = \int_{\Delta s_2} (\nabla \times \mathbf{A}) \cdot \mathbf{ds}_2 = \oint_{\mathcal{L}_2} \mathbf{A} \cdot \mathbf{dl}_2$$

The magnetic vector potential is a solution of the vector Poisson's equation. With the assumption that the coil has a small cross-section, the volume integral can be converted to the following line integral:

$$\mathbf{A} = \frac{\mu_0 I_1}{4\pi} \oint_{\mathcal{L}_1} \frac{\mathbf{dl}_1}{R_{12}}$$

where R_{12} is a distance between the source (1) and the observer (2). The definition of the mutual inductance

$$M = \frac{\Psi_{m12}}{I_1}$$

yields the following double integral for the normalized mutual inductance

$$\frac{M}{\mu_0} = \frac{1}{4\pi} \oint_{\mathscr{L}_2} \oint_{\mathscr{L}_1} \frac{d\mathbf{l}_1 \cdot d\mathbf{l}_2}{R_{12}}$$

For the case of two identical parallel loops, we obtain

$$\frac{M}{\mu_0} = \frac{R^2}{2} \int_0^{2\pi} \frac{\cos\phi \, d\phi}{\sqrt{h^2 + 2R^2 - 2R^2\cos\phi}}$$

For the special case ($h = 2R$), this single integral can be further simplified.

$$\frac{M}{\mu_0} = R \times \left(\frac{1}{2\sqrt{2}} \int_0^{2\pi} \frac{\cos\phi \, d\phi}{\sqrt{3 - \cos\phi}} \right) = R \times \text{constant}$$

This shows that the mutual inductance increases linearly with the increasing of the wire radius R. The integral for the constant C can be solved *numerically* by applying the MATLAB function *quad*. Assuming a radius $R = 0.5$ m and a constant $C = 0.1129$, the normalized mutual inductance is computed to be $M/\mu_0 = 0.0564$.

In the derivation of (5.7), an assumption was made that the area Δs of the loop did not change in time and only a time-varying magnetic field existed in space. This assumption need not always be made in order for electric fields to be generated by magnetic fields. We have to be thankful for the fact that the effect can be *generalized* since much of the conversion of electric energy to mechanical energy or mechanical energy to electrical energy is based on this phenomenon.

In particular, let us assume that a conducting bar moves with a velocity **v** through a uniform time-independent magnetic field **B** as shown in Figure 5–3. The wires that are connected to this bar are parallel to the magnetic field and are connected to a voltmeter that lies far beneath the plane of the moving bar. From the Lorentz force equation, we can calculate the force **F** on the freely mobile charged particles in the conductor. Hence, one end of the bar will become positively charged, and the other end will have an excess of negative charge.

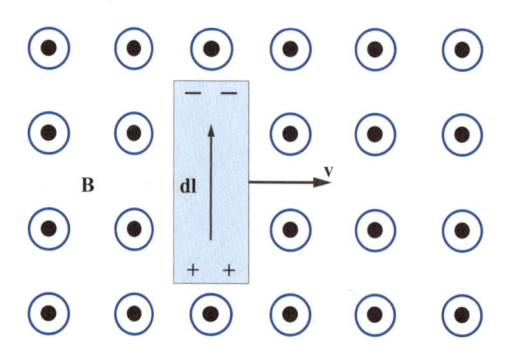

FIGURE 5–3

A conducting bar moving in a uniform time-independent magnetic field that is directed out of the paper. Charge distributions of the opposite sign appear at the two ends of the bar.

5.1 Faraday's Law of Induction

Since there is a charge separation in the bar, there will be an electric field that is created in the bar. Since the net force on the bar is equal to zero, the electric and magnetic contributions to the force cancel each other. This results in an electric field that is

$$\mathbf{E} \equiv \frac{\mathbf{F}}{q} = \mathbf{v} \times \mathbf{B} \tag{5.9}$$

This electric field can be interpreted to be an induced field acting in the direction along the conductor that produces a voltage V, and it is given by

$$V = \int_a^b (\mathbf{v} \times \mathbf{B}) \cdot \mathbf{dl} \tag{5.10}$$

EXAMPLE 5.6

A *Faraday disc generator* consists of a circular metal disc rotating with a constant angular velocity $\omega = 600 \text{ s}^{-1}$ in a uniform time-independent magnetic field. A magnetic flux density $\mathbf{B} = B_0 \mathbf{u}_z$ where $B_0 = 4$ T is parallel to the axis of rotation of the disc. Determine the induced open-circuit voltage that is generated between the brush contacts that are located at the axis and the edge of the disc whose radius is $a = 0.5$ m.

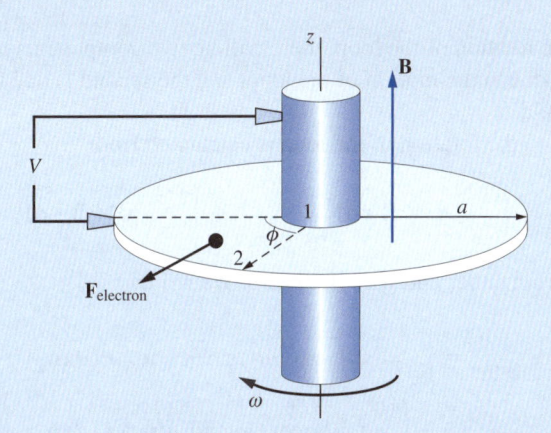

Answer. An electron at a radius ρ from the center has a velocity $\omega \rho$ and therefore experiences an outward directed radial force $-q\omega \rho B_0$. The Lorentz force acting on the electron is

$$-q[\mathbf{E} + (\mathbf{v} \times \mathbf{B})] = 0$$

At *equilibrium*, we find that the electric field can be determined from the Lorentz force equation to be directed radially inward and it has a magnitude $\omega \rho B_0$. Hence we write

$$V = -\int_1^2 (\mathbf{v} \times \mathbf{B}) \cdot \mathbf{dl} = -\int_0^a [(\omega \rho \mathbf{u}_\phi) \times B_0 \mathbf{u}_z] \cdot d\rho \mathbf{u}_\rho$$

$$= \omega B_0 \int_a^0 \rho d\rho = -\frac{\omega B_0 a^2}{2} = -300 \text{ V}$$

which is the potential generated by the Faraday disc generator.

If the bar depicted in Figure 5–3 were moving through a time-dependent magnetic field instead of a constant magnetic field, then we would have to add together the potential caused by the motion of the bar and the potential caused by the time-varying magnetic field. This implies that the principle of superposition applies for this case. This is a good assumption in a vacuum or in any linear medium.

EXAMPLE 5.7

A rectangular loop rotates through a time-varying magnetic flux density $\mathbf{B} = B_0 \cos(\omega t) \mathbf{u}_y$. The loop rotates with the same angular frequency ω. Calculate the induced voltage at the terminals.

Answer. Due to the rotation of the loop, there will be two components to the induced voltage. The first is due to the motion of the loop, and the second is due to the time-varying magnetic field.

The voltage due to the rotation of the loop is calculated from

$$V_{\text{rotation}} = \int_{b/2}^{-b/2} \mathbf{v} \times \mathbf{B} \cdot \mathbf{dl} \bigg|_{\text{bottom edge}} + \int_{-b/2}^{b/2} \mathbf{v} \times \mathbf{B} \cdot \mathbf{dl} \bigg|_{\text{top edge}}$$

The contributions from either end will yield zero. We write

$$V_{\text{rotation}} = \int_{b/2}^{-b/2} v B_0 \cos(\omega t) \sin\theta (-\mathbf{u}_x) \cdot (dx \mathbf{u}_x)$$

$$+ \int_{-b/2}^{b/2} v B_0 \cos(\omega t) \sin\theta (\mathbf{u}_x) \cdot (dx \mathbf{u}_x)$$

The angle $\theta = \omega t$, and the velocity $v = \omega a$. Hence the term of the induced voltage due to the rotation of the loop yields

$$V_{\text{rotation}} = \omega B_0 ba(2\omega t)$$

We recognize that the area of the loop is equal to $\Delta s = 2ab$.

5.1 Faraday's Law of Induction

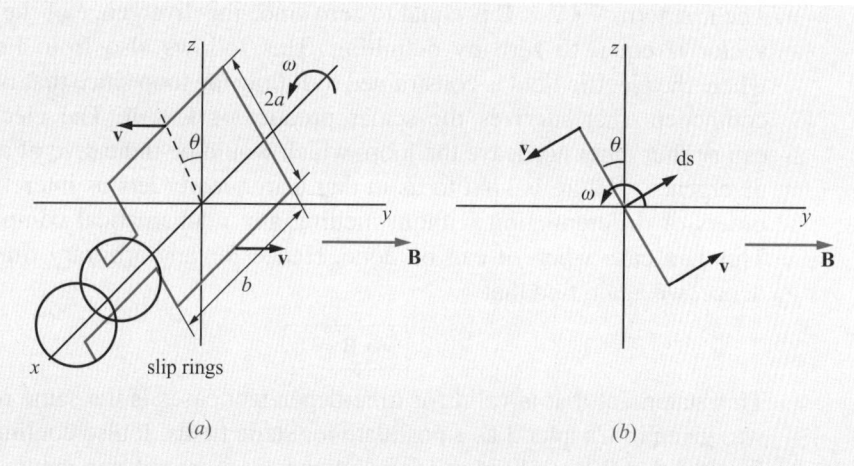

(a)   *(b)*

From (5.4), we can compute the voltage due to the time-varying magnetic field. In this case, we note that

$$\mathbf{s} = [\cos(\omega t)\mathbf{u_y} + \sin(\omega t)\mathbf{u_z}]dx\,a$$

Therefore, the voltage due to time variation of the magnetic field is

$$V_{\text{time varying}} = -\int_{\Delta s} \frac{\partial \mathbf{B}}{\partial t} \cdot \mathbf{ds}$$

$$= \omega B_0 \int_{x5-b/2}^{b/2} \int_{z5-a}^{a} \sin(\omega t)\mathbf{u_y} \cdot [\cos(\omega t)\mathbf{u_y} + \sin(\omega t)\mathbf{u_z}]\,dz\,dx$$

The integration leads to

$$V_{\text{time varying}} = \omega B_0 ba \sin(2\omega t)$$

The total voltage is given by the sum of the voltage due to rotation and the voltage due to the time variation of the magnetic field.

$$V = V_{\text{rotation}} + V_{\text{time varying}} = 2\omega B_0 ba \sin(2\omega t)$$

This results in the generation of the second harmonic.

We can apply a repeated vector operation to Faraday's law of induction that is given in (5.8) to obtain an equation that describes another feature of time-varying magnetic fields. We take the divergence of both sides of (5.8) and interchange the order of differentiation to obtain

$$\nabla \cdot \nabla \times \mathbf{E} = -\nabla \cdot \frac{\partial \mathbf{B}}{\partial t} = -\frac{\partial}{\partial t}(\nabla \cdot \mathbf{B}) \quad (5.11)$$

The first term $\nabla \cdot \nabla \times \mathbf{E}$ is equal to zero since the divergence of the curl of a vector is equal to zero by definition. This follows also from Figure 5–2 where the electric field is constrained to follow the loop since that is the only component that survives the scalar product of $\mathbf{E} \cdot \mathbf{dl}$. The electric field can neither enter nor leave the loop, which would be indicative of a nonzero divergence. Nature is kind to us in that it frequently lets us interchange the orders of differentiation without inciting any mathematical complications. This is a case where it can be done. Hence for any arbitrary time dependence, we again find that

$$\boxed{\nabla \cdot \mathbf{B} = 0} \tag{5.12}$$

This statement that is valid for time-dependent cases is the same result that was given in Chapter 3 as a postulate for static fields. It also continues to reflect the fact that we have not found magnetic monopoles in nature and time varying magnetic field lines are continuous.

Let us integrate (5.12) over an arbitrary volume Δv. This volume integral can be converted to a closed-surface integral using the divergence theorem

$$\int_{\Delta v} \nabla \cdot \mathbf{B} \, dv = \oint \mathbf{B} \cdot \mathbf{ds} \tag{5.13}$$

from which we write

$$\boxed{\oint \mathbf{B} \cdot \mathbf{ds} = 0} \tag{5.14}$$

Equation (5.14) is also valid for time-independent electromagnetic fields.

5.2 Equation of Continuity

Before obtaining the next equation of electromagnetics, it is useful to step back and derive the equation of continuity. In addition, we must understand the ramifications of this equation. The equation of continuity is fundamental at this point in developing the basic ideas of electromagnetic theory. It can also be applied in several other areas of engineering and science, so the process of understanding this theory will be time well spent.

In order to derive the equation of continuity, let us consider a model that assumes a stationary number of positive charges are located initially at the center of a transparent box whose volume is $\Delta v = \Delta x \Delta y \Delta z$. These charges are at the prescribed positions within the box for times $t \leq 0$. A one-dimensional view of this box consists of two parallel planes, and it is shown in Figure 5–4a. We will neglect the Coulomb forces between the individual

5.2 Equation of Continuity

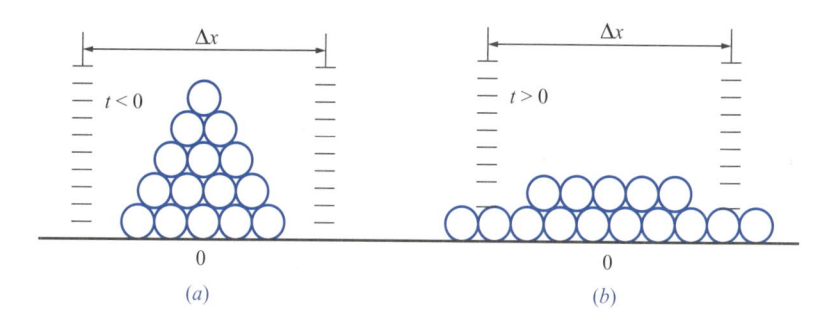

FIGURE 5–4
(a) Charges centered at $x = 0$ at time $t < 0$ are allowed to expand at $t = 0$.
(b) As time increases some of these charges may pass through the screens at $x = \pm \Delta x / 2$.

charges. If we are uncomfortable with this assumption of noninteracting charged particles, we could have assumed alternatively that the charges were just noninteracting gas molecules or billiard balls and derived a similar equation for these objects. The resulting equation could then be multiplied by a charge q that would be impressed on an individual entity. For times $t \geq 0$, the particles or the charges can start to move and actually leave through the two screens as time increases. This is depicted in Figure 5–4b. The magnitude of the cross-sectional area of a screen is equal to $\Delta s = \Delta y \Delta z$. The charges that leave the "screened-in region" will be in motion. Hence, those charges that leave the box from either side will constitute a current that emanates from the box. Due to our choice of charges within the box having a positive charge, the direction of this current I will be the same direction as the motion of the charge.

Rather than just examine the small number of charges depicted in Figure 5–4, let us assume that there is now a large number of them. We will still neglect the Coulomb force between charges. The number of charges will be large enough so it is prudent to describe the charge within the box with a charge density ρ_v where $\rho_v = \Delta Q / \Delta v$ and ΔQ is the total charge within a volume Δv. Hence, the decrease of the charge density ρ_v acts as a source for the total current I that leaves the box as shown in Figure 5–5.

A temporal *decrease* of charge density within the box implies that charge leaves the box since the charge is neither destroyed nor does it recombine with charge of the opposite sign. The total current that leaves the box through any portion of the surface is due to a decrease of the charge within the box. The magnitude of this current is expressed by

$$I = -\frac{dQ}{dt} \tag{5.15}$$

These are real charges, and the current that we are describing is not the displacement current that we will encounter later.

FIGURE 5-5

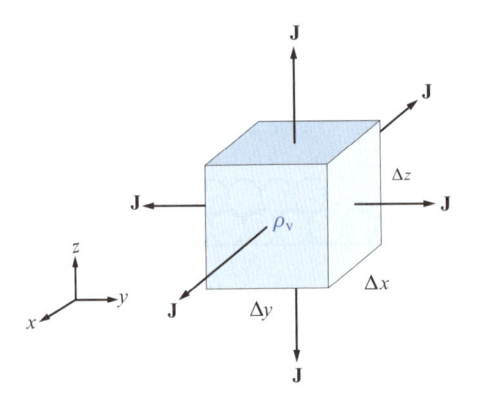

Charge within the box leaves through the walls.

Equation (5.15) can be rewritten as

$$\oint_{\Delta s} \mathbf{J} \cdot \mathbf{ds} = -\int_{\Delta v} \frac{\partial \rho_v}{\partial t} dv \tag{5.16}$$

where we have taken the liberty of summing up the six currents that leave the six sides of the box that surrounds the charges; this summation is expressed as a closed-surface integral. The closed-surface integral given in (5.16) can be converted to a volume integral using the divergence theorem. Hence (5.16) can be written as

$$\int_{\Delta v} \nabla \cdot \mathbf{J} dv = -\int_{\Delta v} \frac{\partial \rho_v}{\partial t} dv \tag{5.17}$$

Since this equation must be valid for any arbitrary volume, we are left with the conclusion that the two integrands must be equal, from which we write

$$\boxed{\nabla \cdot \mathbf{J} + \frac{\partial \rho_v}{\partial t} = 0} \tag{5.18}$$

Equation (5.18) is the *equation of continuity* that we are seeking. Note that this equation has been derived using very simple common sense arguments. However, we can show the same result by a more rigorous argument proving that this expression holds under all known circumstances. Although we have derived it using finite-sized volumes, the equation is valid at a point. Its importance will be noted in the next section where we will follow in the footsteps of James Clerk Maxwell.

We recall from our first course that dealt with circuits that the Kirchhoff's current law stated that the net current entering or leaving a node was equal to zero. Charge is neither created nor destroyed in this case. This is shown in Figure 5–6. The dashed lines represent a closed surface that surrounds the node. The picture shown in Figure 5–5 generalizes this node to three dimensions.

5.2 Equation of Continuity

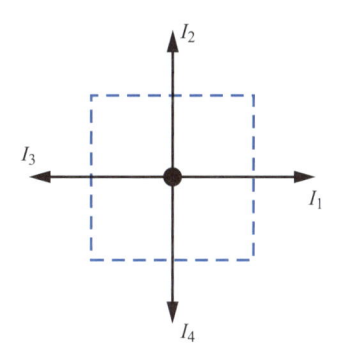

FIGURE 5-6

A closed surface (represented by dashed lines) surrounding a node.

EXAMPLE 5.8

Charges are introduced into the interior of a conductor during the time $t < 0$. Calculate how long it will take for these charges to move to the surface of the conductor so the interior charge density $\rho_v = 0$ and interior electric field $\mathbf{E} = 0$.

Answer. Introduce Ohm's law $\mathbf{J} = \sigma \mathbf{E}$ into the equation of continuity

$$\sigma(\nabla \cdot \mathbf{E}) = -\frac{\partial \rho_v}{\partial t}$$

The electric field is related to the charge density through Poisson's equation

$$\nabla \cdot \mathbf{E} = \frac{\rho_v}{\varepsilon}$$

Hence we obtain the differential equation

$$\frac{d\rho_v}{dt} + \frac{\sigma}{\varepsilon}\rho_v = 0$$

whose solution is

$$\rho_v = \rho_{v0} e^{-\left(\frac{\sigma}{\varepsilon}\right)t}$$

The initial charge density ρ_{v0} will decay to $[1/e \approx 37\%]$ of its initial value in a time $\tau = \varepsilon/\sigma$, which is called the **relaxation time**. For copper, this time is

$$\tau = \frac{\varepsilon_0}{\sigma} = \frac{\frac{1}{36\pi} \times 10^{-9}}{5.8 \times 10^7} \approx 1.5 \times 10^{-19} \text{ s}$$

Other effects that are not described here may cause this time to be different. Relaxation times for insulators may be hours or days.

EXAMPLE 5.9

The current density is $\mathbf{J} = e^{-x^2} \mathbf{u_x}$. Find the time rate of increase of the charge density at $x = 1$.

Answer. From the equation of continuity (5.18), we write

$$\frac{\partial \rho_v}{\partial t} = -\nabla \cdot \mathbf{J} \Rightarrow \frac{\partial \rho_v}{\partial t} = -\frac{\partial J_x}{\partial x} = 2x e^{-x^2}\bigg|_{x=1} = 0.736 \text{ A/m}^3$$

EXAMPLE 5.10

The current density in a certain region may be approximated with the function

$$\mathbf{J} = J_0 \frac{e^{-t/\tau}}{r} \mathbf{u}_r$$

in spherical coordinates. Find the total current that leaves a spherical surface whose radius is a at the time $t = \tau$. Using the equation of continuity, find an expression for the charge density $\rho_v(r, t)$.

Answer. The total current that leaves the spherical surface is given by

$$I = \oint \mathbf{J} \cdot d\mathbf{s}\bigg|_{r=a,\, t=\tau} = 4\pi a^2 \left(\frac{J_0 e^{-t/\tau}}{a}\right)\bigg|_{t=\tau} = 4\pi a J_0 e^{-1} \text{ (A)}.$$

In spherical coordinates, the equation of continuity that depends only upon the radius r is written as

$$\frac{\partial \rho_v}{\partial t} = -\frac{1}{r^2}\frac{\partial}{\partial r}(r^2 J_r) = -\frac{1}{r^2}\frac{\partial}{\partial r}\left(r^2 J_0 \frac{e^{-t/\tau}}{r}\right) = -J_0 \frac{e^{-t/\tau}}{r^2}$$

Hence, after integration the charge density is given by

$$\rho_v = \int \frac{\partial \rho_v}{\partial t} dt \Rightarrow \rho_v = J_0 \frac{\tau e^{-t/\tau}}{r^2} \text{ (C/m}^3\text{)}$$

where the arbitrary constant of integration is set equal to zero.

5.3 Displacement Current

Our first encounter with time-varying electromagnetic fields yielded Faraday's law of induction in equation (5.2). The next encounter will illustrate the genius of James Clerk Maxwell. Through his efforts in the nineteenth

5.3 Displacement Current

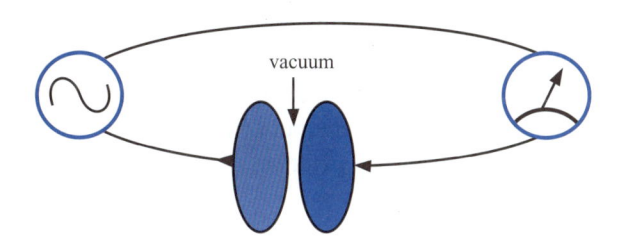

FIGURE 5-7

An elementary circuit consisting of an ideal parallel plate capacitor connected to an AC voltage source and an AC ammeter.

century, we are now able to answer a fundamental question that would arise when analyzing a circuit in the following *gedanken experiment*. Let us connect two wires to the two plates of an ideal capacitor consisting of two parallel plates separated by a vacuum and an AC voltage source as shown in Figure 5–7. An AC ammeter is also connected in series with the wires in this circuit, and it measures a constant value of AC current I. Two questions might enter our minds at this point.

1. How can the ammeter read any value of current since the capacitor is an open circuit and the current that passes through the wire would be impeded by the vacuum that exists between the plates?
2. What happens to the time-varying magnetic field that is created by the current and surrounds the wire as we pass through the region between the capacitor plates?

The answer to the first question will require that we first reexamine the equations that we have obtained up to this point and then interpret them, guided by the light that has been turned on by Maxwell. In particular, let us write the second postulate of steady magnetic fields—Ampere's law. This postulate states that a magnetic field **B** is created by a current **J**. It is written here

$$\nabla \times \mathbf{B} = \mu_0 \mathbf{J} \tag{5.19}$$

Let us take the divergence of both sides of this equation. The term on the left-hand side

$$\nabla \cdot \nabla \times \mathbf{B} = 0 \tag{5.20}$$

by definition. Applying the divergence operation to the term on the right-hand of (5.19), we find that

$$\mu_0 \nabla \cdot \mathbf{J} = 0 \tag{5.21}$$

This, however, is not compatible with the equation of continuity (5.18) that we have just shown to be true under all circumstances.

To get out of this dilemma, Maxwell postulated the existence of another type of current in nature. This current would be in addition to the conduction current discussed in Chapter 3 and a convection current that would be created by charge passing through space with a constant drift velocity. The new current with a density $\mathbf{J_d}$ is called a ***displacement current***, and it is found by incorporating the equation for the displacement flux density $\mathbf{D}$ into the equation of continuity by use of Gauss's law. Hence

$$\nabla \cdot \mathbf{J} + \frac{\partial \rho_v}{\partial t} = 0$$

$$\nabla \cdot \mathbf{J} + \frac{\partial (\nabla \cdot \mathbf{D})}{\partial t} = \nabla \cdot \left(\mathbf{J} + \frac{\partial \mathbf{D}}{\partial t} \right) = 0 \tag{5.22}$$

where we have freely interchanged the order of differentiation. The displacement current density is identified as

$$\boxed{\mathbf{J_d} = \frac{\partial \mathbf{D}}{\partial t}} \tag{5.23}$$

This is the current that passes between the two plates of the capacitor in our gedanken experiment that was performed at the beginning of this section.

The time-varying conduction current that passes through the wire causes a build-up of charges of opposite signs on the two plates of the capacitor. The time variation of these charges creates a time-varying electric field between the plates.[2] The time-varying displacement current will pass from one plate to the other, and an answer to the first question has been obtained. The conduction current in the wire becomes a displacement current between the plates. This displacement current does not exist in a time-independent system.

The postulate for magnetostatics will have to be modified to incorporate this new current and any possible time-varying magnetic fields. It becomes

$$\boxed{\nabla \times \frac{\mathbf{B}}{\mu_0} = \mathbf{J} + \frac{\partial \mathbf{D}}{\partial t}} \tag{5.24}$$

We can also answer the second question. With the inclusion of the displacement current that passes between the capacitor plates, we can assert that the time-varying magnetic field that surrounds the conduction current-carrying wire will be equal in magnitude and direction to the time-varying magnetic field that surrounds the capacitor.

Let us integrate both sides of (5.24) over the cross-sectional area specified by the radius ρ at two locations in Figure 5–8. The first integral will be at a

---

[2] Recall that in a vacuum $\mathbf{D} = \varepsilon_0 \mathbf{E}$. If a dielectric is inserted between the plates, we must use $\mathbf{D} = \varepsilon \mathbf{E}$.

5.3 Displacement Current

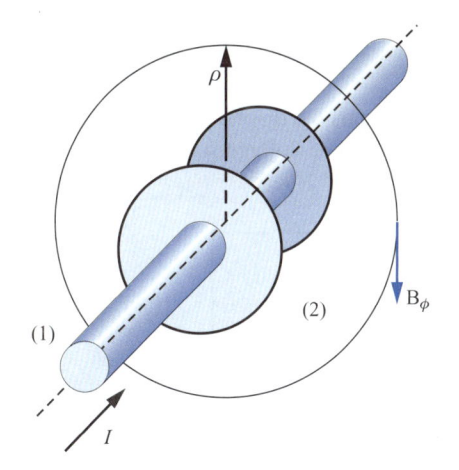

FIGURE 5–8

Two parallel plates in a capacitor separate two wires. The circle whose radius is ρ could be surrounding the wire (1) at either edge or between the plates (2). The radius of the wire is a and that of the plate is b.

location surrounding the wire and the second will be between the two capacitor plates

$$\int_{\Delta s} \left(\nabla \times \frac{\mathbf{B}}{\mu_0} \right) \cdot \mathbf{ds} = \int_{\Delta s} \left(\mathbf{J} + \frac{\partial \mathbf{D}}{\partial t} \right) \cdot \mathbf{ds} \tag{5.25}$$

or using Stokes's theorem, we write

$$\oint \frac{\mathbf{B}}{\mu_0} \cdot \mathbf{dl} = \int_{\Delta s} \left(\mathbf{J} + \frac{\partial \mathbf{D}}{\partial t} \right) \cdot \mathbf{ds} \tag{5.26}$$

The left-hand side of the integral in (5.26) yields

$$2\pi\rho \frac{B_\phi}{\mu_0}$$

At location (1) in Figure 5–8, the displacement current equals zero, and we are left with the integral

$$2\pi\rho \frac{B_\phi}{\mu_0} = \int_{\Delta s} \mathbf{J} \cdot \mathbf{ds} \tag{5.27}$$

At location (2) in Figure 5–8, the conduction current equals zero, and we are left with the integral

$$2\pi\rho \frac{B_\phi}{\mu_0} = \int_{\Delta s} \frac{\partial \mathbf{D}}{\partial t} \cdot \mathbf{ds} \tag{5.28}$$

The next example will demonstrate that these currents are *identical*, hence the magnetic flux densities will be the same at the same radius ρ.

EXAMPLE 5.11

Verify that the conduction current in the wire equals the displacement current between the plates of the parallel plate capacitor in the circuit. The voltage source has $V_c = V_0 \sin \omega t$.

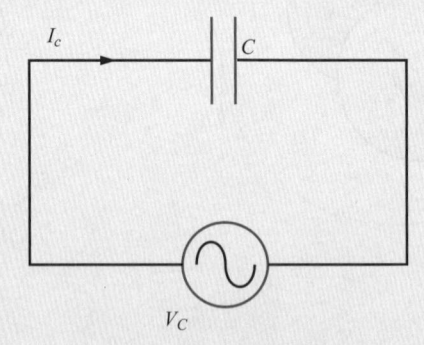

Answer. The conduction current in the wire is given by

$$I_c = C \frac{dV_c}{dt} = CV_0 \omega \cos \omega t$$

The capacitance of the parallel plate capacitor is given by

$$C = \frac{\varepsilon A}{d}$$

where A is the area of the plates that are separated by a distance d. The electric field between the plates is given by $E = V_c/d$. The displacement flux density equals

$$D = \varepsilon E = \varepsilon \frac{V_0}{d} \sin \omega t$$

The displacement current is computed from

$$I_d = \int_A \frac{\partial \mathbf{D}}{\partial t} \cdot \mathbf{ds} = \left(\frac{\varepsilon A}{d}\right) V_0 \omega \cos \omega t = CV_0 \omega \cos \omega t = I_c$$

EXAMPLE 5.12

The magnetic flux density in a vacuum is given by

$$\mathbf{B} = B_0 \cos(2x) \cos(\omega t - \beta y) \mathbf{u_x} = B_x \mathbf{u_x}$$

Find the displacement current, the displacement flux density, and the volume charge density associated with this magnetic flux density.

Answer. We write

$$\mathbf{J_d} = \frac{\partial \mathbf{D}}{\partial t} = \frac{1}{\mu_0} \nabla \times \mathbf{B}$$

$$= \frac{1}{\mu_0} \begin{vmatrix} \mathbf{u_x} & \mathbf{u_y} & \mathbf{u_z} \\ \frac{\partial}{\partial x} & \frac{\partial}{\partial y} & \frac{\partial}{\partial z} \\ B_x & 0 & 0 \end{vmatrix} = -\frac{\beta B_0}{\mu_0} \cos(2x)\sin(\omega t - \beta y)\mathbf{u_z}$$

The displacement flux density **D** is found from the displacement current as

$$\mathbf{D} = \int \mathbf{J_d}\, dt = \int \left[-\frac{\beta B_0}{\mu_0} \cos(2x)\sin(\omega t - \beta y)\mathbf{u_z} \right] dt$$

$$= \frac{\beta B_0}{\omega \mu_0} \cos(2x)\cos(\omega t - \beta y)\mathbf{u_z}$$

Noting that the magnitude of the displacement flux density is not a function of z, we find the charge density given by

$$\rho_v = \nabla \cdot \mathbf{D} \Rightarrow \frac{\partial D_z}{\partial z} = 0$$

EXAMPLE 5.13

In a lossy dielectric medium with a conductivity σ and a relative permittivity ε_r, there is a time-harmonic electric field $E = E_0 \sin \omega t$. Compare the magnitudes of the following terms: (a) the conduction current density J_c and (b) the displacement current density J_d.

Answer. The conduction current density can be found from Ohm's law $J_c = \sigma E = \sigma E_0 \sin \omega t$, while the displacement current density can be calculated from (5.23) $J_d = \partial D/\partial t = \varepsilon E_0 \omega \cos \omega t$. The ratio of their magnitudes is

$$\frac{|J_c|}{|J_d|} = \frac{\sigma}{\omega \varepsilon_0 \varepsilon_r}$$

For materials that have a relative dielectric constant that is close to one, this fraction will depend mainly on the conductivity of the material and the frequency of the electromagnetic signal. The conduction current is dominant at low frequencies in a conductor, and the displacement current will be dominant in a dielectric at high frequencies. This latter effect will be discussed further in the next chapter.

5.4 Maxwell's Equations

Everything that we have learned up to this point can be summarized in Maxwell's[3] four differential equations, which are rewritten below as

$$\nabla \times \mathbf{E} = -\frac{\partial \mathbf{B}}{\partial t} \tag{5.29}$$

$$\nabla \times \mathbf{H} = \mathbf{J} + \frac{\partial \mathbf{D}}{\partial t} \tag{5.30}$$

$$\nabla \cdot \mathbf{D} = \rho_v \tag{5.31}$$

$$\nabla \cdot \mathbf{B} = 0 \tag{5.32}$$

These four equations, along with a set of relations called the **constitutive relations**

$$\left. \begin{array}{l} \mathbf{D} = \varepsilon \mathbf{E} \\ \mathbf{B} = \mu \mathbf{H} \\ \mathbf{J} = \sigma \mathbf{E} \end{array} \right\} \tag{5.33}$$

describe electromagnetic phenomena. The constitutive relations relate the electromagnetic fields to the material properties in which the fields exist. We will see that the propagation of electromagnetic waves such as light is described with Maxwell's equations.

Nonlinear phenomena can also be described with this set of equations through any nonlinearity that may exist in the constitutive relations. For example, certain optical fibers used in communication have a dielectric constant that depends nonlinearly on the amplitude of the wave that propagates in the fiber. It is possible to approximate the relative dielectric constant in the fiber with the expression $\varepsilon_r \approx [1 + \alpha |E|^2]$, where α is a constant that has the dimensions of $[V/m]^{-2}$. In writing (5.33), we have also assumed that the materials are isotropic and hysteresis can be neglected. The study of the myriad effects arising from these phenomena is of interest to a growing number of engineers and scientists throughout the world. However, we will not concern ourselves with these problems here other than to be aware of their existence.

---

[3] It is common in engineering talks to place a standard six-foot stick man next to a drawing of a machine in order to indicate its size. For example, a man standing adjacent to a truck in a modern coal mine would still be beneath the center of the axle, which would mean that this truck is huge. To emphasize the importance and size of these four equations, the reader could think of the symbol ∇ as being an inverted pyramid arising out of the grains of sand of the remaining words and equations in this text and several others.

5.4 Maxwell's Equations

EXAMPLE 5.14

Show that the two "divergence" equations are implied by the two "curl" equations and the equation of continuity.

Answer. To show this, we must remember the vector identity $\nabla \cdot \nabla \times \zeta \equiv 0$ where ζ is any vector. Hence

$$\nabla \cdot \nabla \times \mathbf{E} \equiv 0 = -\frac{\partial}{\partial t}(\nabla \cdot \mathbf{B})$$

implies $\nabla \cdot \mathbf{B}$ = constant. This constant equals zero since there are no isolated sources nor sinks at which the magnetic flux density can originate nor terminate. This implies that magnetic monopoles do not exist in time varying systems either.

We write similarly

$$\nabla \cdot \nabla \times \mathbf{H} \equiv 0 = \nabla \cdot \mathbf{J} + \frac{\partial}{\partial t}(\nabla \cdot \mathbf{D}) = -\frac{\partial \rho_v}{\partial t} + \frac{\partial}{\partial t}(\nabla \cdot \mathbf{D})$$

This implies that $\nabla \cdot \mathbf{D} = \rho_v$.

EXAMPLE 5.15

In a conducting material, we may assume that the conduction current density is much larger than the displacement current density. Show that Maxwell's equations can be cast in the form of a *diffusion equation* in this material.

Answer. In this case, (5.29) and (5.30) are written as

$$\nabla \times \mathbf{E} = -\frac{\partial \mathbf{B}}{\partial t} \quad \text{and} \quad \nabla \times \mathbf{H} = \mathbf{J} = \sigma \mathbf{E}$$

where the displacement current has been neglected. Take the curl of the second equation

$$\nabla \times \nabla \times \mathbf{H} = \sigma \nabla \times \mathbf{E}$$

Expand the left-hand side with a vector identity and substitute the first equation into the right-hand side.

$$\nabla\left(\nabla \cdot \frac{\mathbf{B}}{\mu_0}\right) - \nabla^2\left(\frac{\mathbf{B}}{\mu_0}\right) = -\sigma \frac{\partial \mathbf{B}}{\partial t}$$

> From (5.32), the first term is zero, leaving
>
> $$\nabla^2 \mathbf{B} = \mu_0 \sigma \frac{\partial \mathbf{B}}{\partial t}$$
>
> This is a diffusion equation with a diffusion coefficient $D = (\mu_0 \sigma)^{-1}$. Since **B** is a vector, this corresponds to three scalar equations for the three components. The term ∇^2 is the Laplacian operator described previously.

It may appear that all of this mathematical manipulation is to show off our math skills. However, showing that we can obtain the diffusion equation from Maxwell's equations is important, because the concept of diffusion provides important physical insight into what happens to free charges in various types of materials. During further study, we will see that the diffusion process is involved in many physical phenomena involving electric and magnetic fields.

Consider the case of heat flow. We know from experience that if there is a location with a higher concentration or quantity of heat, and a second location with a lower concentration or quantity of heat, then the heat will tend to flow toward the area of lower concentration. Furthermore, the greater the temperature difference between the two locations, the faster we observe an equalization of the two quantities taking place, at least at first. As the two temperatures equalize, the rate of heat flow slows, until the two locations are at equal temperatures, and heat flow stops, except for random thermal motion. Of course, we can also change the rate of flow by changing the conditions associated with the process. For instance, we could place an insulator between the two locations, which would keep heat from flowing as quickly, as we do to prevent heat from flowing out of our homes in the winter.

In a similar fashion, we can model the flow of free charges in materials, when there is a charge differential between two locations. Also, we will see the same type of behaviors, such as rate of charge flow, and the effects of electrical insulators, in the movement and control of charge in various situations. Thus, not only charge flow in conductors can be modeled, but also flow in such diverse items as components (resistors, capacitors, cables, etc.) and materials (conductors, insulators, and semiconductors). In fact, the entire semiconductor industry and its products are based on an ability to model and control the diffusion of charge carriers in semiconducting material. Thus, the ability to relate Maxwell's equations to diffusion equations is very important and useful indeed.

EXAMPLE 5.16

Solve the diffusion equation for the case of a magnetic flux density $B_x(z, t)$ near a planar vacuum–copper interface, assuming the following values for copper: $\mu = \mu_0 = 4\pi \times 10^{-7}$ H/m and $\sigma = 5.8 \times 10^7$ S/m. Plot the solution for the spatial profile of the magnetic field, assuming a 60-Hz time-harmonic electromagnetic signal is applied.

Answer. Assuming $e^{j\omega t}$ time-variation, the diffusion equation is transformed to the following ordinary differential equation for the spatial variation of the magnetic field.

$$\frac{d^2 B_x(z)}{dz^2} = j\mu_0 \sigma \omega B_x(z)$$

where z is the coordinate normal to the vacuum–copper boundary. Assuming variation in the z direction to be $B_x(z) = B_0 e^{(-\gamma z)}$, we write

$$\gamma^2 = j\omega\mu_0\sigma \Rightarrow \gamma = \alpha + j\beta = \sqrt{j\omega\mu_0\sigma}$$

We will encounter this expression for γ^2 again in the next chapter and study it in some detail, but for now, we will accept it as part of the solution for the preceding differential equation.

The magnitude of the magnetic flux density decays exponentially in the z direction from the surface into the conductor.

$$B_x(z) = B_0 e^{-\alpha z}$$

with

$$\alpha = \sqrt{\pi f \mu_0 \sigma} = \sqrt{\pi \times 60 \times 4\pi \times 10^{-7} \times 5.8 \times 10^7} = 117.2 \text{ m}^{-1}$$

The quantity $\delta = 1/\alpha$ is called a "skin depth." Here it is $\delta = 8.5$ mm.

The plot of the spatial variation of the magnitude of the magnetic field inside the conductor is presented in the figure below.

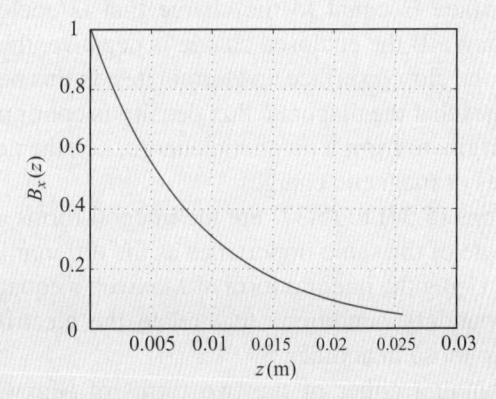

The subject of the skin depth will be encountered again in the next chapter.

As written, Maxwell's equations in (5.29) to (5.32) are partial differential equations evaluated at a particular point in space and time. A *totally equivalent* way of writing them is to write these equations as integrals. This is accomplished by integrating both sides of the first two equations over the same cross-sectional area, applying Stokes's theorem to the terms involving the curl operations, integrating both sides of the second two equations over the same volume, and applying the divergence theorem. We summarize this as follows:

$$\oint_{\Delta L} \mathbf{E} \cdot \mathbf{dl} = -\int_{\Delta s} \frac{\partial \mathbf{B}}{\partial t} \cdot \mathbf{ds} \tag{5.34}$$

$$\oint_{\Delta L} \mathbf{H} \cdot \mathbf{dl} = \int_{\Delta s} \left(\frac{\partial \mathbf{D}}{\partial t} + \mathbf{J} \right) \cdot \mathbf{ds} \tag{5.35}$$

$$\oint_{\Delta s} \mathbf{D} \cdot \mathbf{ds} = \int_{\Delta v} \rho_v dv \tag{5.36}$$

$$\oint_{\Delta s} \mathbf{B} \cdot \mathbf{ds} = 0 \tag{5.37}$$

Equation (5.34) states that the closed-line integral of the electric field around a closed loop is equal to the time rate of change of the magnetic flux that passes through the surface area defined by the closed loop. This is the meaning of Faraday's law. Equation (5.35) states that the closed-line integral of the magnetic field intensity is equal to the current that is enclosed within the loop. The current consists of the contribution due to the conduction current and the displacement current. This generalizes Ampere's circuital law, which we encountered earlier.

Equation (5.36) states that the total displacement flux Ψ_e that leaves a closed surface is equal to the charge that is enclosed within the surface (Gauss's law). If the enclosed charge is negative, then the displacement flux Ψ_e enters the closed surface and terminates on this negative charge. Equation (5.37) states that the magnetic flux density is continuous and cannot terminate nor originate from a magnetic charge, i.e., the nonexistence of magnetic monopoles or magnetic charges.

Equations (5.34) to (5.37) are the integral forms of Maxwell's equations, and they are of the same importance as the differential forms given in (5.29) to (5.32). Using the integral form of Maxwell's equations, we can derive easily the boundary conditions that relate the electromagnetic fields in one medium to those in another.

In examining either of the two forms of Maxwell's equations, we can make another useful observation. Looking at the first two equations of the differential form, (5.29) and (5.30), or the first two equations of the integral form, (5.34) and (5.35), we notice that all four of these equations have both electric field and magnetic field terms. Therefore, these are coupled equations. What this means is that when we perturb either the electric or magnetic field, we automatically affect the other field.

While this may seem like an obvious point, it has important implications both in understanding certain physical phenomena involving electromagnetic fields, as well as in the computational solutions for electromagnetic fields problems. For instance, in developing the three-dimensional version of the Finite Difference Time Domain method, we have to work with both the electric and magnetic fields, and maintain this coupled relationship. We will explore these issues in greater detail in other sections of this text.

Either of the two forms of Maxwell's equations can be used, although we will encounter the differential form more often in practice. An important derivation that describes the magnitude and direction of the flow of electromagnetic power will employ vector identities and the differential form of Maxwell's equations.

5.5 Poynting's Theorem

A frequently encountered problem in practice is to determine the direction that power is flowing if the electric and magnetic fields are measured independently in some experiment. This may not seem important in the laboratory where a signal generator can be separated from a resistive load impedance, and the direction of the flow of power can be clearly ascertained. This is not always so straight-forward in the case of electric and magnetic fields. For instance, it is clear that the sun radiates energy that is received by the Earth, and the amount of that energy can be measured. However, an investigator using a satellite floating in space may wish to determine the source of some anomalous extragalactic electromagnetic radiation in order to further map out the universe. Poynting's theorem will provide us with the method to accomplish this.

To obtain Poynting's theorem for an arbitrary volume depicted in Figure 5–9, we will require two of Maxwell's equations and a vector identity. The two equations that are required for this derivation are

$$\nabla \times \mathbf{E} = -\frac{\partial \mathbf{B}}{\partial t} \qquad (5.38)$$

and

$$\nabla \times \mathbf{H} = \frac{\partial \mathbf{D}}{\partial t} + \mathbf{J} \qquad (5.39)$$

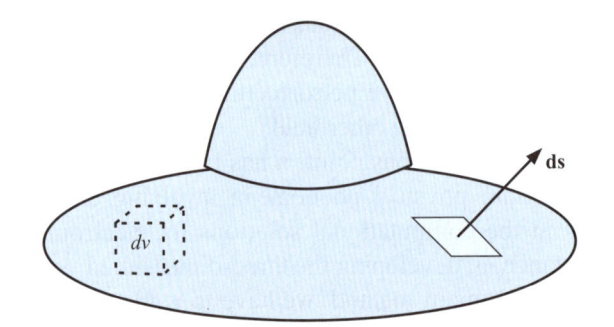

FIGURE 5-9

An arbitrarily shaped volume that contains a source of electromagnetic energy.

Let us take the scalar product of **E** with (5.39) and subtract it from the scalar product of **H** with (5.38). Performing this operation leads to

$$\mathbf{H} \cdot \nabla \times \mathbf{E} - \mathbf{E} \cdot \nabla \times \mathbf{H} = -\mathbf{H} \cdot \frac{\partial \mathbf{B}}{\partial t} - \mathbf{E} \cdot \left[\frac{\partial \mathbf{D}}{\partial t} + \mathbf{J}\right]$$

The left-hand side of this equation can be replaced using vector identity (A.9)

$$\nabla \cdot (\mathbf{A} \times \mathbf{B}) = \mathbf{B} \cdot \nabla \times \mathbf{A} - \mathbf{A} \cdot \nabla \times \mathbf{B}$$

Therefore, we obtain

$$\nabla \cdot (\mathbf{E} \times \mathbf{H}) = -\mathbf{H} \cdot \frac{\partial \mathbf{B}}{\partial t} - \mathbf{E} \cdot \frac{\partial \mathbf{D}}{\partial t} - \mathbf{E} \cdot \mathbf{J} \qquad (5.40)$$

After the introduction of the constitutive relations (5.33), the terms involving the time derivatives can be written as

$$-\mathbf{H} \cdot \frac{\partial \mathbf{B}}{\partial t} - \mathbf{E} \cdot \frac{\partial \mathbf{D}}{\partial t} = -\frac{1}{2}\frac{\partial}{\partial t}[\mu \mathbf{H} \cdot \mathbf{H} + \varepsilon \mathbf{E} \cdot \mathbf{E}] = -\frac{\partial}{\partial t}\frac{1}{2}[\mu H^2 + \varepsilon E^2] \qquad (5.41)$$

Substitute (5.41) into (5.40) and integrate both sides of the resulting equation over the same volume Δv. This volume is enclosed completely by the surface Δs. Performing this integration leads to

$$\int_{\Delta v} \nabla \cdot (\mathbf{E} \times \mathbf{H}) dv = -\frac{\partial}{\partial t}\int_{\Delta v} \frac{1}{2}[\mu H^2 + \varepsilon E^2] dv - \int_{\Delta v} \mathbf{E} \cdot \mathbf{J} dv \qquad (5.42)$$

The volume integral on the left-hand side of (5.42) can be converted to a closed-surface integral via the divergence theorem. With the substitution of Ohm's law $\mathbf{J} = \sigma \mathbf{E}$, we finally obtain

$$\oint_{\Delta s} (\mathbf{E} \times \mathbf{H}) \cdot \mathbf{ds} = -\frac{\partial}{\partial t}\int_{\Delta v} \frac{1}{2}[\mu H^2 + \varepsilon E^2] dv - \int_{\Delta v} \sigma E^2 dv \qquad (5.43)$$

5.5 Poynting's Theorem

where $\mathbf{S} = \mathbf{E} \times \mathbf{H}$ is called the ***Poynting vector***. It is the power density of the radiated electromagnetic fields in W/m^2. The direction of the radiated power is included in this vector.

Let us now give a physical interpretation to each of the three terms that appear in this equation. The units of the closed-surface integral are

$$\frac{\text{volts}}{\text{meter}} \times \frac{\text{amperes}}{\text{meter}} \times \text{meter}^2 = \text{watts}$$

or the closed-surface integral has the units of power. Using the definition of the scalar product and the fact that the notation **ds** refers to the outward normal of the surface that encloses the volume Δv, this term represents the *total power that leaves* or is radiated from the volume Δv.

The terms within the integrand of the first volume integral can be recognized as the stored magnetic energy density and the stored electric energy density that were previously described in static fields. The time derivative introduces a unit of s^{-1}. The units of this term are

$$\frac{1}{\text{second}} \times \frac{\text{joule}}{\text{meter}^3} \times \text{meter}^3 = \text{watts}$$

This term corresponds to the time derivative of the *stored electromagnetic energy* within the volume.

The units of the second volume integral correspond to Joule heating within the volume, and they are also in terms of watts.

$$\frac{1}{\text{ohms} \times \text{meter}} \times \frac{\text{volts}^2}{\text{meter}^2} \times \text{meter}^3 = \text{watts}$$

The reference to Joule heating indicates that electromagnetic power *is converted to heat* and this power cannot be recovered. A toaster uses Joule heating.

Hence Poynting's theorem states that the power that *leaves* a region is equal to the temporal decay in the energy that is stored within the volume minus the power that is dissipated as heat within it. A common-sense example will illustrate this theorem. Additional applications of this important theorem will be found in the chapter of this book that discusses radiation.

Equation (5.43), which can be considered to be a form of the conservation of energy equation, can also be written in differential form. Recalling that ***electromagnetic energy density*** is defined as

$$w = \frac{1}{2}[\mu H^2 + \varepsilon E^2] \tag{5.44}$$

and the ***power loss density*** is given by

$$p_L = \sigma E^2 \tag{5.45}$$

We can reinterpret (5.42) in the following differential form of the energy conservation of the system.

$$\nabla \cdot \mathbf{S} + \frac{\partial w}{\partial t} = -p_L \tag{5.46}$$

This equation is somewhat similar to the equation of continuity (5.18) with a "sink" term that corresponds to the Joule heating.

EXAMPLE 5.17

Using Poynting's theorem, calculate the power that is dissipated in the resistor as heat. The electric energy is supplied by the battery. Neglect the magnetic field that is confined within the resistor and calculate its value only at the surface. In addition, assume that there are conducting surfaces at the top and bottom of the resistor so they are equipotential surfaces. Also, assume that the radius of the resistor is much less than its length.

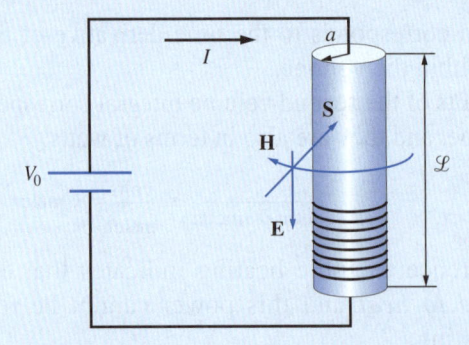

Answer. The electric field has a magnitude of $E = V_0/\mathscr{L}$ and the magnitude of the magnetic field intensity at the outer surface of the resistor is $H = I/(2\pi a)$. The direction of the Poynting vector $\mathbf{S} = \mathbf{E} \times \mathbf{H}$ is *into* the resistor. There is no energy stored in the resistor. The magnitude of the current density that is in the same direction as the electric field is $J = I/(\pi a^2)$. Therefore, the various terms in Poynting's theorem (5.43) are found to be

$$-\left(\frac{V_0}{\mathscr{L}}\right)\left(\frac{I}{2\pi a}\right)(2\pi a \mathscr{L}) = -\frac{d}{dt}\int_{\Delta v}[0 + 0]dv - \left(\frac{I}{\pi a^2}\right)\left(\frac{V_0}{\mathscr{L}}\right)(\pi a^2 \mathscr{L})$$

yielding

$$-V_0 I = -V_0 I$$

The electromagnetic energy of the battery is fully absorbed by the resistor.

5.5 Poynting's Theorem

EXAMPLE 5.18

Using Poynting's theorem, calculate the power that is flowing through the surface area at the radial edge of a capacitor. Neglect the ohmic losses in the wires connecting the capacitor with the signal generator. Also assume that the radius of the capacitor is much greater than the separation distance between the plates, or $a \gg b$.

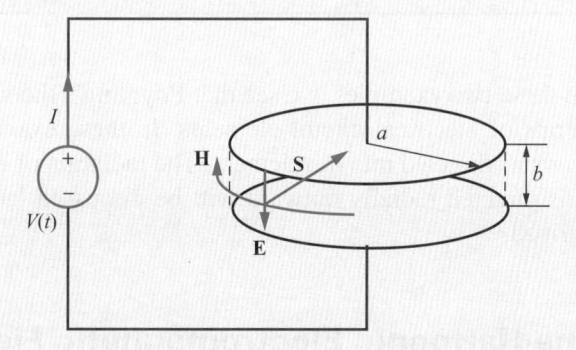

Answer. Assuming the electric field **E** is confined between the plates and is uniform, we can find the total electric energy that is stored in the capacitor to be

$$W = \left(\frac{\varepsilon E^2}{2}\right)(\pi a^2 b)$$

The total magnetic energy that is stored in the capacitor is equal to zero.

The differentiation of this electric energy with respect to time yields

$$-\frac{dW}{dt} = -\varepsilon(\pi a^2 b) E \frac{dE}{dt}$$

This is the only term that survives on the right side of (5.43) since an ideal capacitor does not dissipate energy.

The left-hand side of (5.43) requires an expression for the time-varying magnetic field intensity in terms of the displacement current. Evaluating (5.26) at the radial edge of the capacitor, we write

$$\oint \mathbf{H} \cdot d\mathbf{l} = \int_{\Delta s} \left(\varepsilon \frac{\partial \mathbf{E}}{\partial t}\right) \cdot d\mathbf{s}$$

There is no conduction current in this ideal capacitor, or $I = 0$. We obtain

$$H(2\pi a) = \varepsilon \frac{dE}{dt}(\pi a^2) \Rightarrow H = \frac{\varepsilon a}{2} \frac{dE}{dt}$$

Now we can write the Poynting vector power flow as

$$P_s = -(EH)(2\pi ab) = -\varepsilon(\pi a^2 b) E \frac{dE}{dt}$$

> The minus sign arises since the direction of the Poynting vector is radially inward. Comparing both expressions, we find that they are equal, which implies that
>
> $$P_s = -\frac{dW}{dt}$$
>
> This states that *energy is conserved* in the circuit as should be expected.

In these two examples, we see that Poynting's theorem can be interpreted in terms of electrical circuit elements. In these examples, electromagnetic power was directed into the element. The radiation of electromagnetic power that is directed radially outward will be discussed later when antennas are described.

5.6 Time-Harmonic Electromagnetic Fields

In practice, we frequently will encounter electromagnetic fields whose temporal variation is harmonic. Maxwell's equations and the Poynting vector will assume a particular form since the fields can be represented as *phasors*. In particular, we write the fields as

$$\mathbf{E}(x, y, z, t) = \text{Re}[\mathbf{E}(x, y, z)e^{j\omega t}] \tag{5.47}$$

and

$$\mathbf{H}(x, y, z, t) = \text{Re}[\mathbf{H}(x, y, z)e^{j\omega t}] \tag{5.48}$$

where Re stands for the real part. There may be a phase angle ϕ between the electric and magnetic fields that will be absorbed into the terms $\mathbf{E}(x, y, z)$ and $\mathbf{H}(x, y, z)$.

In terms of the phasors $\mathbf{E}$ and $\mathbf{H}$, we write Maxwell's equations as

$$\nabla \times \mathbf{E}(\mathbf{r}) = -j\omega\mu\mathbf{H}(\mathbf{r}) \tag{5.49}$$

$$\nabla \times \mathbf{H}(\mathbf{r}) = j\omega\varepsilon\mathbf{E}(\mathbf{r}) + \mathbf{J}(\mathbf{r}) \tag{5.50}$$

$$\nabla \cdot \mathbf{E}(\mathbf{r}) = \frac{\rho_v(\mathbf{r})}{\varepsilon} \tag{5.51}$$

$$\nabla \cdot \mathbf{B}(\mathbf{r}) = 0 \tag{5.52}$$

where the term representing the temporal variation $e^{j\omega t}$ that is common to both sides of these equations has been canceled. Hopefully, there will be little confusion in notation since we have not introduced any new symbols.

5.6 Time-Harmonic Electromagnetic Fields

EXAMPLE 5.19

Compute the frequency at which the conduction current equals the displacement current.

Answer. Using (5.33) and (5.50), we write

$$\nabla \times \mathbf{H}(\mathbf{r}) = \mathbf{J}(\mathbf{r}) + j\omega\varepsilon\mathbf{E}(\mathbf{r}) = (\sigma + j\omega\varepsilon)\mathbf{E}(\mathbf{r})$$

The frequency is given by

$$\omega = \frac{\sigma}{\varepsilon}$$

For copper, the frequency $f = \omega/2\pi$ is

$$f = \frac{\sigma}{2\pi\varepsilon_0} = \frac{5.8 \times 10^7}{2\pi \times \frac{1}{36\pi} \times 10^{-9}} \approx 1.04 \times 10^{18} \text{ Hz}$$

At frequencies much above this value, copper, which is thought to be a good conductor, acts like a dielectric.

The derivation of the Poynting vector requires some care when we are considering time-harmonic fields. This is because the Poynting vector involves the product $\mathbf{E}(\mathbf{r}) \times \mathbf{H}(\mathbf{r})$. Power is a real quantity, and we must be careful since

$$\text{Re}[\mathbf{E}(\mathbf{r})e^{j\omega t}] \times \text{Re}[\mathbf{H}(\mathbf{r})e^{j\omega t}] \neq \text{Re}[\mathbf{E}(\mathbf{r}) \times \mathbf{H}(\mathbf{r})e^{j\omega t}] \quad (5.53)$$

To effect the derivation of the Poynting vector, we make use of the following relations:

$$\text{Re}[\mathbf{E}(\mathbf{r})] = \left(\frac{\mathbf{E}(\mathbf{r}) + \mathbf{E}^*(\mathbf{r})}{2}\right) \quad \text{and} \quad \text{Re}[\mathbf{H}(\mathbf{r})] = \left(\frac{\mathbf{H}(\mathbf{r}) + \mathbf{H}^*(\mathbf{r})}{2}\right) \quad (5.54)$$

where the star indicates the complex conjugate of the function. We write

$$\text{Re}[\mathbf{E}(\mathbf{r})] \times \text{Re}[\mathbf{H}(\mathbf{r})] = \left(\frac{\mathbf{E}(\mathbf{r}) + \mathbf{E}^*(\mathbf{r})}{2}\right) \times \left(\frac{\mathbf{H}(\mathbf{r}) + \mathbf{H}^*(\mathbf{r})}{2}\right)$$

$$= \frac{\mathbf{E}(\mathbf{r}) \times \mathbf{H}^*(\mathbf{r}) + \mathbf{E}^*(\mathbf{r}) \times \mathbf{H}(\mathbf{r}) + \mathbf{E}(\mathbf{r}) \times \mathbf{H}(\mathbf{r}) + \mathbf{E}^*(\mathbf{r}) \times \mathbf{H}^*(\mathbf{r})}{4}$$

The time variation $e^{j\omega t}$ cancels in two of the terms and it introduces a factor of $e^{\pm j2\omega t}$ in the remaining two terms. After taking a time average of this power, these latter terms will contribute nothing to the result. We finally obtain the time average power to be

$$\boxed{\mathbf{S}_{av}(\mathbf{r}) = \frac{1}{2}\text{Re}[\mathbf{E}(\mathbf{r}) \times \mathbf{H}^*(\mathbf{r})] \ (\text{W/m}^2)} \quad (5.55)$$

EXAMPLE 5.20

The field vectors in free space are given by

$$\mathbf{E} = 10\cos\left(\omega t + \frac{4\pi}{3}z\right)\mathbf{u_x} \text{ (V/m)} \quad \text{and} \quad \mathbf{H} = \frac{\mathbf{u_z} \times \mathbf{E}}{120\pi} \text{ (A/m)}$$

The frequency $f = 500$ MHz. Determine the Poynting vector. The numerical value of 120π as a free-space impedance will become apparent in the next chapter.

Answer. In phasor notation, the fields are expressed as

$$\mathbf{E(r)} = 10 e^{j\left(\frac{4\pi}{3}\right)z} \mathbf{u_x} \quad \text{and} \quad \mathbf{H(r)} = \frac{10}{120\pi} e^{j\left(\frac{4\pi}{3}\right)z} \mathbf{u_y}$$

and the Poynting vector is

$$\mathbf{S_{av}(r)} = \frac{1}{2}\text{Re}[\mathbf{E(r)} \times \mathbf{H^*(r)}] = \frac{10^2}{2 \times 120\pi}\mathbf{u_z} = 0.133 \mathbf{u_z} \text{ (W/m}^2\text{)}$$

Having now manipulated the complex phasors to derive (5.55), let us apply this to the derivation of Poynting's theorem. In particular, we desire to explicitly obtain the terms $\mathbf{E(r)}$ and $\mathbf{H^*(r)}$. The procedure that we will follow is to subtract the scalar product of $\mathbf{E(r)}$ with the complex conjugate of (5.50) from the scalar product of $\mathbf{H^*(r)}$ with (5.49), resulting in

$$\mathbf{H^*(r)} \cdot \nabla \times \mathbf{E(r)} - \mathbf{E(r)} \cdot \nabla \times \mathbf{H^*(r)} = -j\omega\mu\mathbf{H(r)} \cdot \mathbf{H^*(r)}$$
$$+ j\omega\varepsilon\mathbf{E(r)} \cdot \mathbf{E^*(r)} - \mathbf{E(r)} \cdot \mathbf{J(r)}^* \quad (5.56)$$

Employing the same vector identity that we used previously to derive (5.40), we recognize that (5.56) can be written as

$$\nabla \cdot (\mathbf{E(r)} \times \mathbf{H^*(r)}) = -j\omega\mu H^2 + j\omega\varepsilon E^2 - \sigma E^2 \quad (5.57)$$

where $\mathbf{H(r)} \cdot \mathbf{H^*(r)} = H^2$, $\mathbf{E(r)} \cdot \mathbf{E^*(r)} = E^2$, and $\mathbf{E(r)} \cdot \mathbf{J(r)}^* = \sigma E^2$.

Following the procedure that has served us so well previously, we integrate the terms that appear in (5.57) over the volume of interest.

$$\int_{\Delta v} \nabla \cdot (\mathbf{E(r)} \times \mathbf{H^*(r)})dv = -j\omega \int_{\Delta v} [\mu H^2 - \varepsilon E^2]dv - \int_{\Delta v} \sigma E^2 dv \quad (5.58)$$

The volume integral is converted to a closed-surface integral that encloses the volume Δv.

$$\oint_{\Delta s} (\mathbf{E}(\mathbf{r}) \times \mathbf{H}^*(\mathbf{r})) \cdot \mathbf{ds} = -j\omega \int_{\Delta v} [\mu H^2 - \varepsilon E^2] dv - \int_{\Delta v} \sigma E^2 dv \quad (5.59)$$

The closed-surface integral represents the total power that is radiated from within the volume enclosed by this surface. The last term represents the power that is dissipated within this volume. This power could have been turned into heat and would not be recovered. The remaining two terms are the time-average energy stored within the volume. The factor *j* indicates that this is similar to the *reactive energy* stored in the capacitor or inductor in an *RLC* circuit.

5.7 Conclusion

We have now come to the end of a long journey in order to obtain the set of four Maxwell's equations that describe electromagnetic phenomena. We have demonstrated that time-varying electric and magnetic fields can be determined from each other through these equations and that they are intimately intertwined. Faraday's law of induction and Ampere's circuital law with the introduction of a displacement current relate time-varying magnetic fields to time-varying electric fields. The term that represents the displacement current arises from the requirement that the equation of continuity must be satisfied. The boundary conditions that we encountered in static fields apply equally well in time-varying fields.

There is a T-shirt that paraphrases the book of Genesis by stating that "In the beginning, God said '...' and there was light" where these equations are included within the proclamation. The goal and accomplishment of thousands of graduate students since Maxwell first inscribed these equations on paper has been to pose a new electromagnetic problem, solve it starting from these equations, and write a thesis. Even after obtaining a graduate degree, this set of equations usually appears as "Equations 1 to 4" in many of their later scholarly articles that are then stored in dusty archives. You, as a student, are not expected to write these equations on a crib sheet and bring them to an examination or even memorize them for that inquiry. *You are expected to know them!* The intellectual and even the visceral understanding of these equations is what this course and much of electrical and computer engineering is about.

5.8 Problems

5.1.1. In a source-free region, we find that $\mathbf{B} = z\mathbf{u}_y + x\mathbf{u}_z$. Does $\mathbf{E}$ vary with time?

5.1.2. A perfect conductor joins two ends of a 100 Ω resistor, and the closed loop is in a region of uniform magnetic flux density $B = 10\, e^{(-t/10)}$ T.

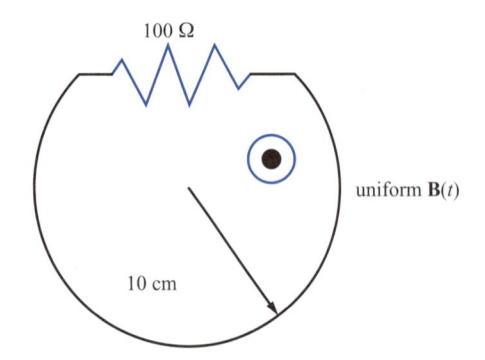

Neglecting the self-inductance of the loop, find and plot the voltage $V(t)$ that appears across the 100 Ω resistor. A device based on this principle is used to monitor time-varying magnetic fields in experiments and in biological studies.

5.1.3. A closed loop ($\Delta x = 30$ cm $\times$ $\Delta y = 20$ cm) of wire passes through a nonuniform time-independent magnetic field $\mathbf{B} = y\mathbf{u}_z$ T with a constant velocity $\mathbf{v}_0 = 5\mathbf{u}_x$ m/s. At $t = 0$, the loop's lower left corner is located at the origin. Find an expression for the voltage V, generated by the loop as a function of time. You may neglect the magnetic field created by the current in the loop.

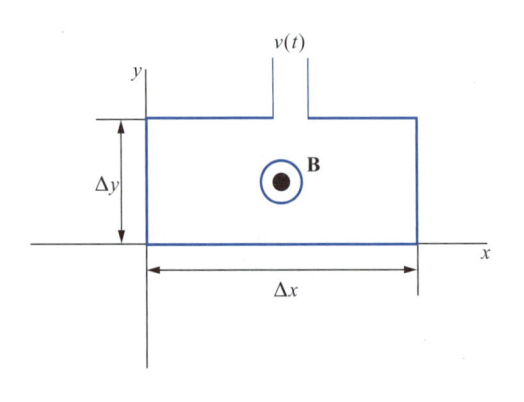

5.1.4. Repeat Problem 5.1.3 with the magnetic flux density being uniform in space $\mathbf{B} = 0.1\mathbf{u}_z$ T. Explain your result.

5.1.5. Find the generated voltage if the axle moves at a constant velocity $\mathbf{v} = v\mathbf{u}_x = 3\mathbf{u}_x$ m/s in a uniform magnetic field of $\mathbf{B} = B_0\mathbf{u}_z = 5\mathbf{u}_z$ T. At $t = 0$, the axle was at $x = 0$, $L = 40$ cm.

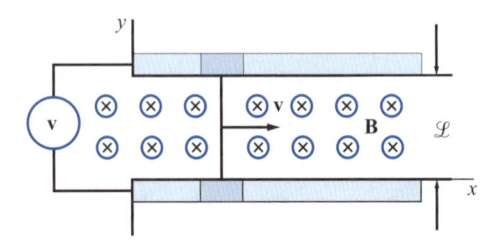

5.1.6. Repeat Problem 5.1.5 with the constraint that the rails separate with $\mathscr{L} = \mathscr{L}_0 + \mathscr{L}_1 x$. The wheels are free to slide on the "trombone-like" axle so they remain on the rails ($\mathscr{L}_0 = 0.4$ m, $\mathscr{L}_1 = 0.04$ m).

5.1.7. A tethered satellite is to be connected to the Shuttle to generate electricity as it passes through the ambient plasma. A plasma consists of a large number of positive charges and negative charges. Assuming that the Shuttle takes 1.5 hours to go around the Earth, find the expected voltage difference ΔV between the tether and the Shuttle. The Shuttle flies approximately 400 km above the earth where $\mathbf{B} \approx 10^{-5}$ T.

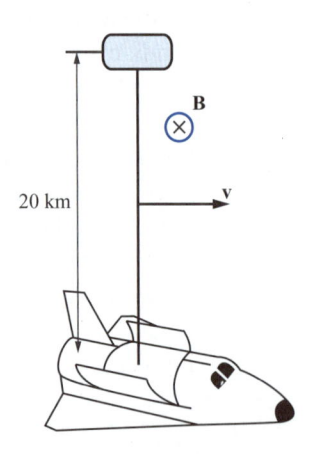

5.8 Problems

5.1.8. A conducting axle oscillates over two conducting parallel rails in a uniform magnetic field $\mathbf{B} = B_0 \mathbf{u}_z$ ($B_0 = 4$ T). The position of the axle is given by $x = (\Delta x/2)[1 - \cos\omega t]$ ($\Delta x = 0.2$ m, $\omega = 500$ s^{-1}). Find and plot the current $I(t)$ if the resistance is $R = 10 \, \Omega$ and the distance between the rails is $\Delta y = 0.1$ m.

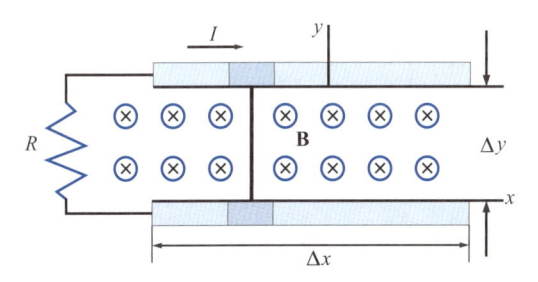

5.1.9. Repeat Problem 5.1.8 with the magnetic field also varying in time as $\mathbf{B} = B_0 \cos\omega t \, \mathbf{u}_z$ with B_0 and ω having the same values.

5.1.10. Calculate the voltage that is induced between the two nodes as the coil with dimensions 0.5 m × 0.5 m rotates in a uniform magnetic field with a flux density $B = 2$ T with a constant angular frequency $\omega = 1200$ s^{-1}.

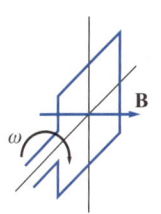

5.1.11. A square loop is adjacent to an infinite wire that carries a current I. The loop moves with a velocity $\mathbf{v} = v_0 \mathbf{u}_\rho$. The center of the loop is at ρ, and the initial position is $\rho = b$. Determine the induced voltage $V(t)$ in the loop assuming dimensions $a \times 2b$.

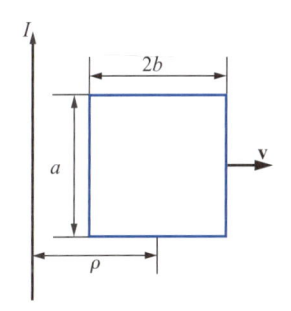

5.1.12. The 1 m long wire shown in the figure rotates with an angular frequency $\omega = 40\pi$ s^{-1} in the magnetic field $\mathbf{B} = 0.5 \cos\phi \, \mathbf{u}_\phi$ T. Find the current in the closed loop with a resistance 100 Ω.

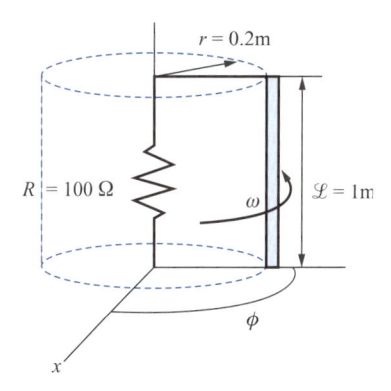

5.2.1. The current density is $\mathbf{J} = \sin(\pi x)\mathbf{u}_x$. Find the time rate of increase of the charge density $\partial \rho_v / \partial t$ at $x = 1$.

5.2.2. The current density is $\mathbf{J} = e^{(-\rho^2)} \mathbf{u}_r$ in cylindrical coordinates. Find the time rate of increase of the charge density at $\rho = 1$.

5.3.1. Compare the magnitudes of the conduction and displacement current densities in copper ($\sigma = 5.8 \times 10^7$ S/m, $\varepsilon = \varepsilon_0$), sea water ($\sigma = 4$ S/m, $\varepsilon = 81 \, \varepsilon_0$), and earth ($\sigma = 10^{-3}$ S/m, $\varepsilon = 10 \, \varepsilon_0$) at 60 Hz, 1 MHz, and at 1 GHz.

5.3.2. Given the conduction current density in a lossy dielectric as $J_c = 0.2 \sin(2\pi 10^9 t)$ A/m^2, find the displacement current density if $\sigma = 10^{-3}$ S/m and $\varepsilon_r = 6.5$.

5.4.1. Show that the fields $\mathbf{B} = B_0 \cos\omega t \, \mathbf{u}_x$ and $\mathbf{E} = E_0 \cos\omega t \, \mathbf{u}_z$ do not satisfy Maxwell's equations in air $\varepsilon_r \approx 1$. Show that the fields $\mathbf{B} = B_0 \cos(\omega t - ky)\mathbf{u}_x$ and $\mathbf{E} = E_0 \cos(\omega t - ky)\mathbf{u}_z$ satisfy these equations. What is the value of k in terms of the other stated parameters?

5.4.2. Given

$$\mathbf{E} = E_0 \cos(\omega t - ky)\mathbf{u}_z$$

and $$\mathbf{H} = \left(\frac{E_0}{Z_0}\right)\cos(\omega t - ky)\mathbf{u}_x$$

in a vacuum, find Z_0 in terms of ε_0 and μ_0 so Maxwell's equations are satisfied.

5.4.3. Do the fields
$$\mathbf{E} = E_0 \cos x \cos(\omega t) \mathbf{u}_y \text{ and } \mathbf{H} = \left(\frac{E_0}{\mu_0}\right) \sin x \sin(\omega t) \mathbf{u}_z$$
satisfy Maxwell's equations?

5.4.4. Find a charge density ρ_v that could produce an electric field in a vacuum $\mathbf{E} = E_0 \cos x \cos(\omega t) \mathbf{u}_x$.

5.4.5. Find the displacement current density flowing through the dielectric of a coaxial cable of radii a and b where $b > a$ if a voltage $V_0 \cos \omega t$ is connected between the two conducting cylinders.

5.4.6. Find the displacement current density flowing through the dielectric of two concentric spheres of radii a and b where $b > a$ if a voltage $V_0 \cos \omega t$ is connected between the two conducting spheres.

5.4.7. Starting from Maxwell's equations, derive the equation of continuity.

5.4.8. Write all of the terms that appear in Maxwell's equations in Cartesian coordinates.

5.5.1. If $\mathbf{E} = E_0 \cos(\omega t - \beta z) \mathbf{u}_y$ is a solution to Maxwell's equations, find $\mathbf{H}$. Find $\mathbf{S}_{av}$.

5.5.2. If $\mathbf{H} = H_0 \cos(\omega t - \beta z) \mathbf{u}_y$ is a solution to Maxwell's equations, find $\mathbf{E}$; find $\mathbf{S}_{av}$.

5.5.3. Compute the electric energy that is stored in a cube whose volume is 1 m^3 in which a uniform electric field of 10^4 V/m exists. Compute the stored energies if the cube is empty and if it is filled with water that has $\varepsilon = 81 \varepsilon_0$.

5.6.1. Write $\mathbf{E} = 120 \pi \cos(3 \times 10^9 t - 10z) \mathbf{u}_x$ and $\mathbf{H} = 1 \cos(3 \times 10^9 t - 10z) \mathbf{u}_y$ in phasor notation.

5.6.2. Write the phasors $\mathbf{E} = 3 e^{-j\beta z} \mathbf{u}_x$ and $\mathbf{H} = 0.4 e^{-j45°} e^{-j\beta z} \mathbf{u}_y$ in the time domain. The frequency of oscillation is ω. Find the average Poynting vector $\mathbf{S}_{av}$.

5.6.3. At a frequency of $f = 1$ MHz, verify that copper ($\sigma = 5.8 \times 10^7$ S/m, $\varepsilon_r \approx 1$) is a good conductor, and quartz ($\sigma = 10^{-17}$ S/m, $\varepsilon_r = 4$) is a good insulator.

5.6.4. Find the frequency where quartz becomes a conductor.

5.6.5. Find the frequency where copper becomes an insulator.

CHAPTER 6

Electromagnetic Wave Propagation

6.1 Wave Equation 297
6.2 One-Dimensional Wave Equation 302
6.3 Time-Harmonic Plane Waves 318
6.4 Plane-Wave Propagation in a Dielectric Medium 325
6.5 Reflection and Transmission of an Electromagnetic Wave 335
6.6 Conclusion 349
6.7 Problems 349

The study of electromagnetic wave propagation is based on the ideas of the great nineteenth-century theoretical physicist, James Clerk Maxwell. He was able to describe time-varying electromagnetic fields with four equations from which we derive the wave equation. Electromagnetic waves—which are the solutions of this equation—propagate with the velocity of light.

6.1 Wave Equation

The fact that Maxwell's equations serve as the point of embarkation for our study of electromagnetic waves should not be too surprising since the explanations for all phenomena in electromagnetic theory trace their origins to the same four equations.

The wave equation we will initially derive describes wave propagation in a homogeneous medium that could have losses. In our derivation, there will be no free charge density, hence $\rho_v = 0$, because we are interested here in electromagnetc fields outside the "source region." Therefore, Maxwell's equations are written as

$$\nabla \times \mathbf{E} = -\mu \frac{\partial \mathbf{H}}{\partial t} \tag{6.1}$$

$$\nabla \times \mathbf{H} = \varepsilon \frac{\partial \mathbf{E}}{\partial t} + \sigma \mathbf{E} \tag{6.2}$$

$$\nabla \cdot \varepsilon \mathbf{E} = 0 \tag{6.3}$$

$$\nabla \cdot \mu \mathbf{H} = 0 \tag{6.4}$$

where we have incorporated the constitutive relations

$$\left.\begin{array}{l} \mathbf{D} = \varepsilon \mathbf{E} \\ \mathbf{B} = \mu \mathbf{H} \\ \mathbf{J} = \sigma \mathbf{E} \end{array}\right\} \tag{6.5}$$

to eliminate the terms **D**, **B**, and **J**. In (6.2) we assume there are no external currents. All electromagnetic fields explicitly depend on space and time, i.e., $\mathbf{E} = \mathbf{E}(\mathbf{r}, t)$ and $\mathbf{H} = \mathbf{H}(\mathbf{r}, t)$. As we will see later, the two equations involving the divergence operation can be used to specify the value of a certain term in a vector identity, and we will initially manipulate the two equations that contain the curl operation. Also, we recall the comment we made in Chapter 4. We notice that both (6.1) and (6.2) each contain electric field and magnetic field terms. Therefore, these are coupled equations. What this means is that when we perturb either the electric or magnetic field, we automatically affect the other field.

Equations (6.1) and (6.2) are two first-order partial differential equations in the two dependent variables **E** and **H**. We can combine them into one second-order partial differential equation in terms of one of the variables. This is the same procedure we normally employ when confronted with two coupled first-order ordinary differential equations. We merely have to be careful here since we have vectors and vector operations in the equations.

For the electric field intensity **E**, we combine the two equations by taking the curl of (6.1) and inserting (6.2) for the term $\nabla \times \mathbf{H}$ that will appear on one side of the resulting equation. This operation is written in detail as

$$\nabla \times (\nabla \times \mathbf{E}) = -\mu \frac{\partial}{\partial t}(\nabla \times \mathbf{H}) = -\mu \frac{\partial}{\partial t}\left(\sigma \mathbf{E} + \varepsilon \frac{\partial \mathbf{E}}{\partial t}\right) = -\mu\sigma \frac{\partial \mathbf{E}}{\partial t} - \mu\varepsilon \frac{\partial^2 \mathbf{E}}{\partial t^2}$$

$$\tag{6.6}$$

Note that we have freely interchanged the order of differentiation of space and time in this step. The magnetic field intensity **H** can be computed later from the electric field intensity **E** using (6.1). As we will see later, there are certain advantages to solving for the electric field first, since boundary conditions on the electric field are frequently easier to specify. For example, a metal conductor has a dramatic effect on the tangential component of the electric field intensity as can be easily observed by inserting a metallic object into a microwave oven, an experiment that you are not to perform.

The next step in our journey toward understanding wave phenomena is employing the vector identity (A.15) for the first term in equation (6.6). This leads to

$$\nabla \times \nabla \times \mathbf{E} = \nabla(\nabla \cdot \mathbf{E}) - \nabla^2 \mathbf{E} = -\nabla^2 \mathbf{E} \tag{6.7}$$

In reducing this equation, we have included the fact that the charge density ρ_v is absent and the region is homogeneous. This is specified in the divergence equation (6.3). After substituting (6.7) into the vector equation (6.6), we finally obtain

$$\boxed{\nabla^2 \mathbf{E} - \mu\sigma\frac{\partial \mathbf{E}}{\partial t} - \mu\varepsilon\frac{\partial^2 \mathbf{E}}{\partial t^2} = 0} \tag{6.8}$$

This is the general homogeneous three-dimensional vector wave equation we are seeking. This equation is valid for cases that do not include external sources. Note that, as it is written, this equation does not depend on the chosen coordinate system. However, the form of Laplacian operator that should be employed in a calculation will be dictated by the coordinate system selected on the basis of any symmetry found in the problem. The polarization of the electric field **E** is determined by the polarization introduced with the excitation mechanism. This, for example, could be an antenna with a particular radiation characteristic, a laser, or a waveguide with a certain physical orientation. In the near-field region around the antenna, an inhomogeneous vector wave equation has to be solved that relates the electromagnetic field with its sources.

Polarization of the electromagnetic field is an important subject that should not be glossed over lightly. Imagine for a moment that we have two infinite, parallel metal plates that are connected to the two output terminals of a sinusoidal voltage generator as shown in Figure 6–1. The voltage applied to the top plate will be 180° out of phase with the voltage applied to the bottom plate. The resulting electric field directly between the two plates is said to be ***linearly polarized*** in one direction that we will define as the $\mathbf{u}_y$ direction.

In Figure 6–1, we have introduced a technique to create an electromagnetic field that has a ***linear polarization***. There are other polarizations such as circular and elliptical polarizations. An example will describe these polarizations after we describe the excitation of time-harmonic waves.

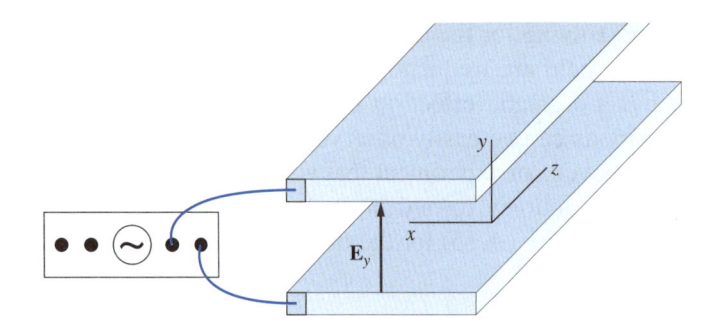

FIGURE 6–1

A technique to linearly polarize the electric field between the two infinite, parallel metal plates.

By convention, we describe the polarization of an electromagnetic wave as being determined by the electric field component rather than by the magnetic field component. In practice, we find that boundary conditions are specified by the electric field rather than the magnetic field. Ascertaining one component of an electromagnetic wave means that the other component can be determined via Maxwell's equations.

Solutions of the general three-dimensional vector wave equation (6.8) may be difficult to write down and at this stage will certainly not help further our understanding of the electromagnetic wave. We must simplify the problem.

EXAMPLE 6.1

Show that the wave equation for the magnetic field intensity **H** can be cast in the same form as (6.8).

Answer. From (6.1) and (6.2), we write

$$\nabla \times \nabla \times \mathbf{H} = \sigma(\nabla \times \mathbf{E}) + \varepsilon \frac{\partial}{\partial t}(\nabla \times \mathbf{E}) = \sigma\left(-\mu \frac{\partial \mathbf{H}}{\partial t}\right) + \varepsilon \frac{\partial}{\partial t}\left(-\mu \frac{\partial \mathbf{H}}{\partial t}\right)$$

The left-hand side of this equation is reduced via the vector identity (A.15)

$$\nabla \times \nabla \times \mathbf{H} = \nabla(\nabla \cdot \mathbf{H}) - \nabla^2 \mathbf{H} = -\nabla^2 \mathbf{H}$$

where (6.4) has been employed. Therefore, we are left with

$$\nabla^2 \mathbf{H} - \mu\sigma \frac{\partial \mathbf{H}}{\partial t} - \mu\varepsilon \frac{\partial^2 \mathbf{H}}{\partial t^2} = 0$$

Thus the wave equation for the magnetic field intensity has the same form as for the electric field intensity **E** as given in (6.8).

6.1 Wave Equation

The general wave equation written in (6.8) contains terms that make it difficult to solve, so here are a few simplifications. The first one concerns the material in which the wave is to propagate. We will initially investigate a vacuum so that the second term in (6.8) is set equal to zero since the conduction current $\mathbf{J} = \sigma \mathbf{E}$ is nonexistent. The relative permittivity ε_r and the relative permeability μ_r both $= 1$, and ε and μ can be replaced with their vacuum values ε_0 and μ_0.

The second simplification is to choose a Cartesian coordinate system with the electric field polarized in only one direction. We will choose that direction for the polarization of the electric field to be specified by the unit vector $\mathbf{u}_y$. The third simplification is to assume that the wave is a function of only one of the three variables that make up the Cartesian coordinate system, say the z coordinate. This specifies that $\partial/\partial x = \partial/\partial y = 0$. The experimental scenario depicted in Figure 6–1 indicates how this could be done. In this case, the vector wave equation (6.8) reduces to

$$\frac{\partial^2 E_y}{\partial z^2} - \mu_0 \varepsilon_0 \frac{\partial^2 E_y}{\partial t^2} = 0 \tag{6.9}$$

where the unit vector $\mathbf{u}_y$ is common to all terms in the equation. The unit vector $\mathbf{u}_y$ will not be written, but it will be understood that the electric field intensity is polarized in that particular direction. Hence, (6.9) becomes a one-dimensional scalar wave equation.

We note the appearance of the term $(\mu_0 \varepsilon_0)$ in this equation. Let us understand the meaning of this term. The dimensions of the two terms in (6.9) that involve the derivatives are respectively given by

$$\frac{\left(\frac{\text{volts}}{\text{meter}}\right)}{(\text{meter})^2} \quad \text{and} \quad \frac{\left(\frac{\text{volts}}{\text{meter}}\right)}{(\text{second})^2}$$

In order for this equation to be correct dimensionally, the term $(\mu_0 \varepsilon_0)$ must have the units of

$$\left(\frac{\text{second}}{\text{meter}}\right)^2$$

Hence $(\mu_0 \varepsilon_0)$ has the units of $(\text{velocity})^{-2}$. If we insert the numerical values for ε_0 and μ_0 and solve for this velocity, we find that it has a numerical value that has the same value as the velocity of light. Therefore, we feel comfortable substituting a symbol c into (6.9), where $c = 1/\sqrt{\mu_0 \varepsilon_0}$, without yet knowing the true meaning behind it. Such good fortune in this term's having the correct dimensions and a recognizable numerical value is not pure happenstance but is based on a firm theoretical foundation, as will be shown. The product of the velocity and the time, ct, has the dimension of distance. Hence, the vector wave equation written in (6.8) could be thought of in terms of a four-dimensional space—a concept that has occupied the time and energy of theoreticians at the graduate level and beyond.

EXAMPLE 6.2

Compute an approximate numerical value for the velocity c.

Answer. Using the numerical values for μ_0 and ε_0, we write

$$c = \frac{1}{\sqrt{\mu_0 \varepsilon_0}} \approx \frac{1}{\sqrt{(4\pi \times 10^{-7})\left(\frac{1}{36\pi} \times 10^{-9}\right)}} \approx 3 \times 10^8 \text{ m/s}$$

This is an approximate value for the speed of light. The more accurate value for the dielectric constant will slightly reduce this number.

6.2 One-Dimensional Wave Equation

In order to emphasize and understand some basic properties of waves, let us first examine waves in other disciplines before actually solving the one-dimensional wave equation. This slight diversion from our main task is to gain familiarity with the topic of waves using everyday experiences. We will look at a pulse that travels in the direction of increasing positive values of the coordinate z in a series of gedanken experiments.

6.2.1 Related Wave Experiments

The first experiment could be performed in a water tank or in a bathtub. A repetitive wave pulse is launched at a point labeled as $z = 0$ with a signal generator attached to a small plunger.[1] This plunger can move up and down as shown in Figure 6–2. The repetition frequency of the plunger motion is slow enough so the excited waves do not interfere with each other. The waves are also absorbed at the walls and at the end of the tank so no reflection of the waves occurs. The water waves propagate with a very slow velocity when

---
[1] This plunger is sometimes called a "wavemaker."

6.2 One-Dimensional Wave Equation

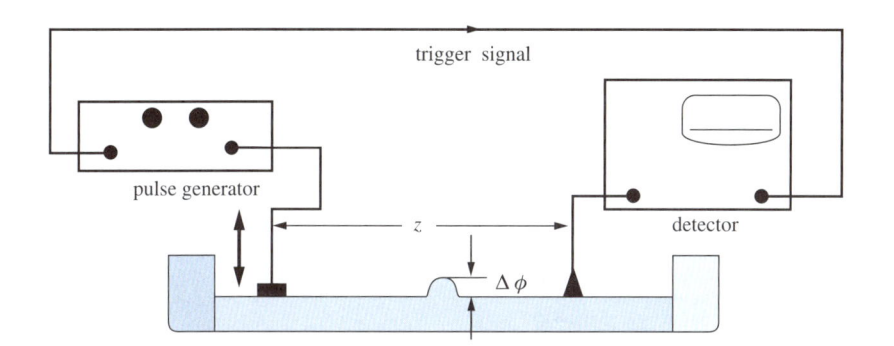

FIGURE 6–2

Wave-making experiment for water waves. The amplitude of the wave is $\Delta\phi$.

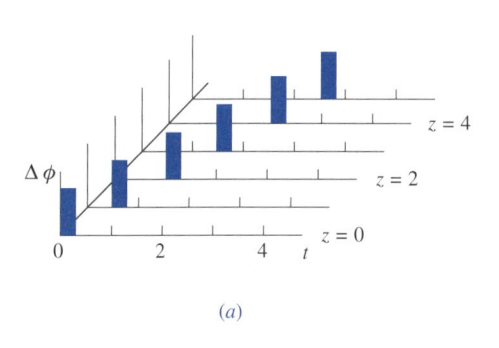

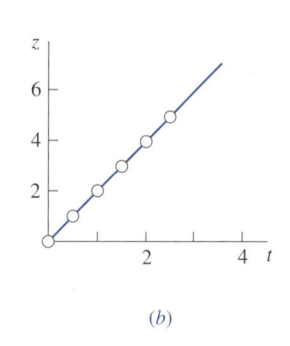

FIGURE 6–3

(*a*) Sequence of oscilloscope photographs taken at various locations in the water tank. (*b*) Trajectory of the rising edge of the wave pulse. The velocity of propagation which is determined from the slope of the line = 2.

compared with the velocity of light. Every time that the pulse is launched at $z = 0$, a trigger pulse is simultaneously sent from the signal generator to the oscilloscope. The trigger pulse will propagate at the velocity of light, so this trigger signal can be considered in the time scale of the water-wave propagation to be "instantaneous." The pulse is detected with a calibrated movable probe, and the response is displayed on an oscilloscope that is triggered from the signal generator. This detector could be a device, such as a photomultiplier, that responds to the amplitude of the reflected light from the water. If the water is uniformly illuminated, the change of curvature due to the passage of the propagating wave would alter the detected signal.

Pictures are taken from the oscilloscope at various locations z in the water tank, and a sequence of these photographs taken at equal spatial intervals is shown in Figure 6–3a. In this sequence, the $t = 0$ trigger time is lined up along one axis. Note that the pulse moves to the right with what appears to be a constant velocity.

From the sequence of photographs shown in Figure 6–3a and knowing the locations where they were taken, it is possible to obtain two numbers. These are the distance of propagation z and the time of flight t of a constant point on the pulse—say, the rising edge of the pulse. This set of numbers is plotted on the graph shown in Figure 6–3b. The experimental points appear to lie on a

FIGURE 6–4

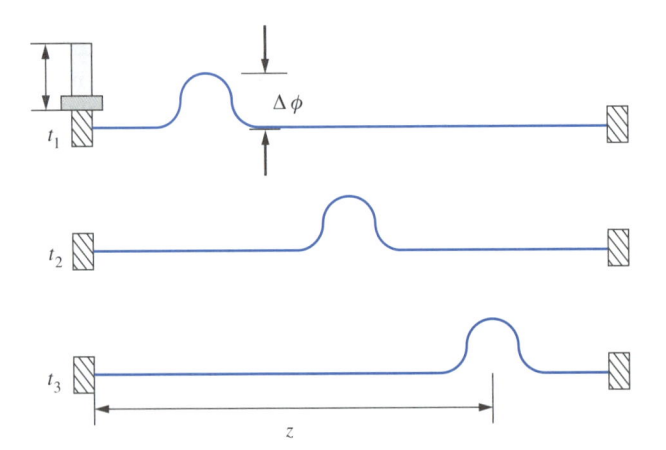

A wave can be launched on a string by "plucking" one end, and it will propagate to the other end.

straight line. The slope of this line is called the *velocity of propagation*, and this figure is called the *trajectory* of the wave propagation. The velocity of propagation for these surface waves is a function of the surface tension and mass density of the water. There are cases where this trajectory may not be a straight line, which necessitates calculating the derivative $v = \partial z / \partial t$ in order to define the slope, and hence the *local* velocity of propagation. The partial derivative notation is used in this definition because there are cases when the velocity may also depend on other quantities, such as the frequency of the wave (resulting in dispersion) or the amplitude of the wave (resulting in nonlinearity).

A second wave experiment uses a string that is stretched between two points as shown in Figure 6–4. In this case, a small perturbation is launched at one end of the string, and it propagates to the other end. As in the water tank experiment, we will neglect any reflection at the end. If a camera were available, we could take pictures of the perturbation as it moves along the string. From this sequence of pictures the trajectory could be drawn, and the velocity of the wave could be computed as shown in Figure 6–3b. In this case, the velocity of propagation is a function of the tension on the string and the mass density of the string, as is demonstrated in Example 6.3. In the case where the diameter of the string decreases as the distance increases, one would quickly encounter a nonlinear velocity of propagation. A whip makes use of this property to assist the tip of the whip to go faster than the speed of sound in air, causing the familiar "cracking" sound associated with whips.[2]

The third experiment employs a spring that is stretched between two walls. When one of the walls is suddenly moved, a resulting perturbation in

---

[2] A. Goriely & T. McMillen, "Shape of a Cracking Whip," Physical Review Letters, Vol. 88, No. 24, 17 June 2002.

6.2 One-Dimensional Wave Equation

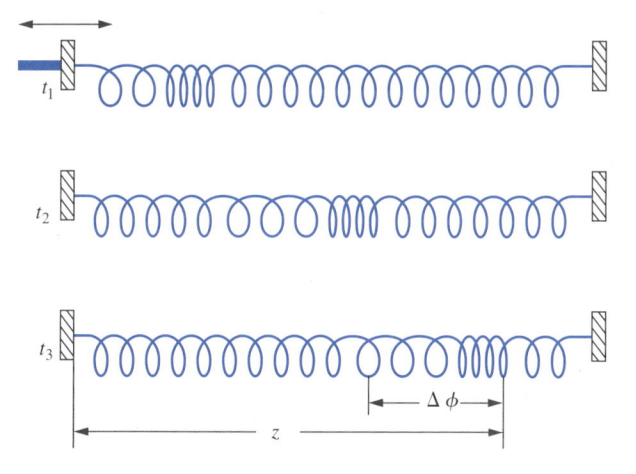

FIGURE 6–5

A sudden compression–rarefaction in the spring causes a perturbation whose amplitude will propagate on the spring.

the spring propagates to the other end, as shown in Figure 6–5. Once again, the trajectory can be drawn as shown in Figure 6–3b, and the velocity of propagation can be computed. This velocity will depend on the elasticity and mass density of the spring.

In all three examples, some general conclusions can be drawn about the nature of wave propagation. The host medium does not propagate. Only the perturbation propagates, and it propagates with a definite velocity that is determined by the properties of the medium in which it propagates. After the perturbation passes, the host medium *returns* to its original unperturbed state since we are assuming that the perturbations are small. The host medium is not affected by the passage of the wave—no energy remains at any spot in the medium. There are no local hot spots caused by a large-amplitude wave locally heating the medium. Except for a later comment, we will not further examine this important class of wave-propagation problems, which involves an energy transfer from one form to another, where it is either locally absorbed or dissipated.

There is a major difference between these three experiments. In the first two cases, the perturbation $\Delta\phi$ was transverse to the direction of the wave's propagation. In the third case, the perturbation $\Delta\phi$ was in the direction of propagation. The first two cases are classified as ***transverse waves*** and the third case is known as a ***longitudinal wave***. The electromagnetic waves we will study in this text are transverse waves. Sound waves, in which there are sequences of local compressions and rarefactions that propagate in the air, are longitudinal waves. Transverse electromagnetic waves can propagate in a vacuum; longitudinal sound waves require something to be compressed and, therefore, cannot propagate in a vacuum.

EXAMPLE 6.3

Consider a wave that propagates in a non-electromagnetic milieu. Derive the wave equation for the *transverse* waves that propagate along a string. The string has a mass density ρ_m and the string is under a tension force T. Neglect the weight of the string.

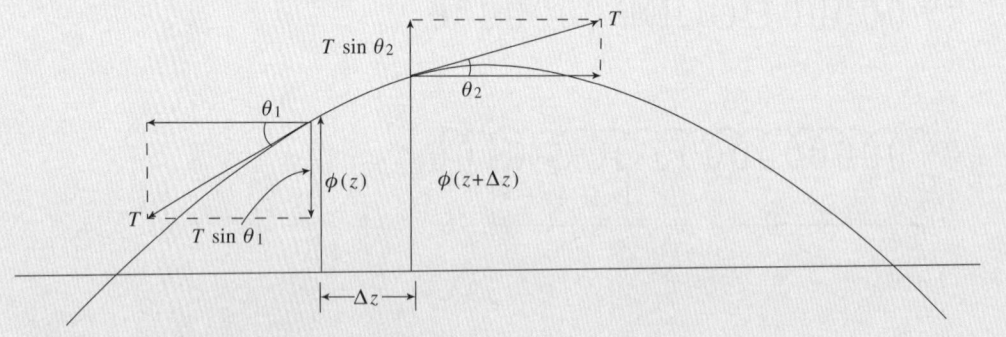

Answer. A segment Δz of the string that experiences a displacement $\Delta\phi$ normal to the string is shown in the figure. The tension T is constant along the string. If the tension were not a constant, the string would break. For small amplitudes $\Delta\phi$, we can write

$$\sin\theta \approx \tan\theta \approx \frac{\partial(\Delta\phi)}{\partial z}$$

Hence the vertical force F on the string can be written as

$$F = T\sin\theta_2 - T\sin\theta_1 = T\left\{\left.\frac{\partial(\Delta\phi)}{\partial z}\right|_{z+\Delta z} - \left.\frac{\partial(\Delta\phi)}{\partial z}\right|_z\right\} \approx T\Delta z\frac{\partial^2(\Delta\phi)}{\partial z^2}$$

Newton's equation of motion states that

$$F = m\frac{\partial^2(\Delta\phi)}{\partial t^2} = \rho_m \Delta z\frac{\partial^2(\Delta\phi)}{\partial t^2}$$

Equating these two expressions, we obtain the wave equation

$$\frac{T}{\rho_m}\frac{\partial^2(\Delta\phi)}{\partial z^2} - \frac{\partial^2(\Delta\phi)}{\partial t^2} = 0$$

The velocity of the wave along the string is given by $v = \sqrt{T/\rho_m}$.

Waves are omnipresent in the universe, and the example of the waves that propagate along the string is an easily visualized example of a transverse wave. The electromagnetic waves that we encounter will also be transverse waves. We could have derived a wave equation and determined the propagation velocity for all three of the waves shown earlier. This was done here only for the transverse wave that propagates along the string in Example 6.3. Suffice it to say that a standard wave equation for the amplitude $\Delta\phi$ would result in all cases, and the wave would have its own unique velocity of propagation.[3] It would have the same mathematical form as the wave equation for the electromagnetic waves, as will be shown in the following discussion. Experiments for other waves of the type indicated here can also be performed with electromagnetic waves, and the velocity of propagation can be appropriately measured.

6.2.2 Analytical Solution of One-Dimensional Equation—Traveling Waves

Equation (6.9) can now be rewritten as

$$\frac{\partial^2 E_y}{\partial z^2} - \frac{1}{c^2}\frac{\partial^2 E_y}{\partial t^2} = 0 \tag{6.10}$$

The *most general* nontrivial solution[4] of this equation is given by

$$E_y(z, t) = F(z - ct) + G(z + ct) \tag{6.11}$$

where F and G are *arbitrary* functions that are determined by the function generator that excites the wave. This can be checked by substituting (6.11) into (6.10). The excitation could be a pulse, a step function, a continuous time-harmonic wave, or any other function. The solution $F(z - ct)$ is a **traveling wave** in the $+z$ direction, while the solution $G(z + ct)$ is a traveling wave in the $-z$ direction. There are two possible scenarios that would create both waves simultaneously. The first assumes that there are two function generators, one at $z = -\infty$ and the other at $z = +\infty$. The second assumes that there is a source at $z = -\infty$ and that there is a reflecting boundary at some location—say at $z = 0$. In this case, there will be an **incident** wave and a **reflected** wave. Reflection will be discussed later.

In order to show that the general solution presented in (6.11) is, indeed, the mathematical solution for the scalar wave equation (6.10), we need only substitute the solution into this partial differential equation. Let us do this operation very methodically since this is a crucial point in our argument. It is easiest to first define two new independent variables ς and ψ that include the independent variables of space z and time t in a particular format as

$$\varsigma = z - ct$$
$$\psi = z + ct \tag{6.12}$$

and then use the chain rule for differentiation.

---

[3] These equations are derived in several texts. It is also shown here that these waves carry both energy and momentum as they propagate. The authors are familiar with A. Hirose and K. E. Lonngren, *Introduction to Wave Phenomena* (New York: Wiley-Interscience, 1985), reprinted by Malabar, FL: Krieger, 1999, 2001.
[4] An electric field that is a constant, such as zero, is also a solution to this equation.

Therefore, the first derivative of our general solution can be written as

$$\frac{\partial E_y}{\partial t} = \frac{dF}{d\varsigma}\left(\frac{\partial \varsigma}{\partial t}\right) + \frac{dG}{d\psi}\left(\frac{\partial \psi}{\partial t}\right) = \frac{dF}{d\varsigma}(-c) + \frac{dG}{d\psi}(c)$$

$$\frac{\partial E_y}{\partial z} = \frac{dF}{d\varsigma}\left(\frac{\partial \varsigma}{\partial z}\right) + \frac{dG}{d\psi}\left(\frac{\partial \psi}{\partial z}\right) = \frac{dF}{d\varsigma}(1) + \frac{dG}{d\psi}(1)$$

where the last derivatives are computed from (6.12). The second derivatives lead to

$$-\frac{1}{c^2}\frac{\partial^2 E_y}{\partial t^2} = -\frac{1}{c^2}\left[\frac{d^2F}{d\varsigma^2}(-c)^2 + \frac{d^2G}{d\psi^2}(c)^2\right]$$

$$\frac{\partial^2 E_y}{\partial z^2} = \frac{d^2F}{d\varsigma^2}(1)^2 + \frac{d^2G}{d\psi^2}(1)^2$$

Hence we have verified that (6.11) is indeed the most general solution of (6.10), since the functions $F(\varsigma) = F(z - ct)$ and $G(\psi) = G(z + ct)$ are completely arbitrary—such as a pulse or a sine wave that depends only on the excitation. In order to actually draw the functions, *numerical values* must be specified for the arguments of the functions. Now we will invoke one of the truisms of life: "We are all getting older!" This truth states that time is always increasing. It manifests itself in the following way with respect to the solution of (6.10). The variables ς and ψ were chosen to have particular numerical values, say ς_0 and ψ_0, where

$$\varsigma_0 = z_0 - ct_0$$

$$\psi_0 = z_0 + ct_0$$

We assume that the functions $F(\varsigma)$ and $G(\psi)$ at a time $t = t_0$ are located at $z = -z_0$ and at $z = +z_0$, respectively. The independent variables ς and ψ must also have the same numerical values so that the shape of the signal is not altered at a later time, $t > t_0$. This implies that the location z of the pulse must change correspondingly. The signal $F(\varsigma) = F(z - ct)$ will pass to increasing values of z. The signal $G(\psi) = G(z + ct)$ will pass to decreasing values of z. This is shown in Figure 6–6 for two pulses of particular shape at two values of time.

We call this effect **wave propagation** (or just propagation). Assuming that there is no distortion in the wave, we can follow the same point of the perturbation as it propagates. This implies that $d\varsigma = 0$ and $d\psi = 0$. The velocity of

6.2 One-Dimensional Wave Equation

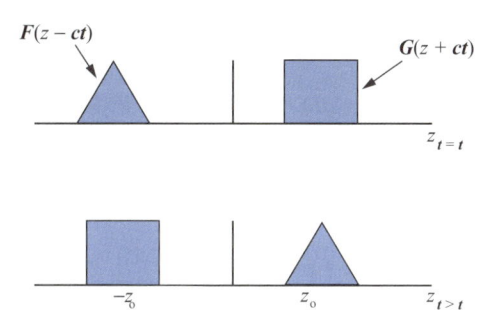

FIGURE 6-6

Solutions of the one-dimensional wave equation at two different times. Each pulse has a particular shape that is determined by the function generator.

wave propagation is then computed from the relations

$$d\varsigma = dz - c\,dt = 0$$
$$d\psi = dz + c\,dt = 0 \tag{6.13}$$

from which we determine the propagation velocity of the two functions to be

$$\frac{dz}{dt} = \pm c \tag{6.14}$$

The sign difference indicates that $F(z - ct)$ propagates to increasing values of the coordinate z, and $G(z + ct)$ propagates to decreasing values of the coordinate z.

Having just solved for the electric field, we could compute the magnetic field intensity **H** from Maxwell's equations. While we will defer this computation for a little while, rest assured that both components are required for the propagation of an electromagnetic wave.

In order to obtain a particular solution of (6.10) rather than the general solution (6.11), we will add two more initial conditions:

1. We assume that the solution is a known function $P(z)$ at the time $t = 0$.
2. We assume that the derivative of the solution is a known function $Q(z)$ at the time $t = 0$.

$$\left. \begin{aligned} E_y(z, 0) &= P(z) \\ \frac{\partial E_y}{\partial t}(z, 0) &= Q(z) \end{aligned} \right\} \tag{6.15}$$

A *particular solution* of the wave equation that satisfies (6.15) is given by

$$E_y(z, t) = \frac{1}{2}[P(z - ct) + P(z + ct)] + \frac{1}{2c}\int_{z-ct}^{z+ct} Q(\xi)d\xi$$
$$= \frac{1}{2}[P(z - ct) + P(z + ct)] + \frac{1}{2}[R(z + ct) - R(z - ct)]$$
(6.16)

where the new auxiliary function $R(z) = \frac{1}{c}\int_0^z Q(z')dz'$ is introduced.

EXAMPLE 6.4

Using a technique developed by D'Alambert, verify that (6.16) is a solution of (6.10) that satisfies the initial conditions given in (6.15).

Answer. We assume the general solution is given by (6.11)

$$\boxed{E_y(z, t) = F(z - ct) + G(z + ct)}$$

After taking the derivative with respect to the variable t and evaluating the result at the time $t = 0$, the following system of two functional equations is obtained.

$$F(z) + G(z) = P(z)$$

$$c\left(-\frac{dF}{dz} + \frac{dG}{dz}\right) = Q(z)$$

where $\varsigma = \psi = z$ at the time $t = 0$ is used. The integration of the second equation yields

$$F(z) + G(z) = P(z)$$

$$-F(z) + G(z) = \frac{1}{c}\int_0^z Q(z')dz'$$

The addition and subtraction of these two equations leads to the following set of simultaneous equations

$$F(z) = \frac{1}{2}P(z) - \frac{1}{2c}\int_0^z Q(\xi)d\xi$$

$$G(z) = \frac{1}{2}P(z) + \frac{1}{2c}\int_0^z Q(\xi)d\xi$$

The substitution of both functions into (6.11) along with appropriate replacements of the variables $z \to \varsigma = z - ct$ in the first function and $z \to \psi = z + ct$ in the second function

6.2 One-Dimensional Wave Equation

yields the following solution for the initial value problem

$$E_y(z,t) = \frac{1}{2}[P(z-ct) + P(z+ct)] + \frac{1}{2c}\int_{z-ct}^{z+ct} Q(\xi)d\xi$$

$$= \frac{1}{2}[P(z-ct) + P(z+ct)] + \frac{1}{2}[R(z+ct) - R(z-ct)]$$

This is (6.16).

EXAMPLE 6.5

Show that the function $F(z-ct) = F_0 e^{-(z-ct)^2}$ is a solution of the wave equation (6.10). This is a so-called Gaussian-pulse traveling wave. In addition, relate the solution to Example 6.4.

Answer. Let $\varsigma = z - ct$. Therefore $F(z-ct) = F_0 e^{-\varsigma^2}$ and $G(z+ct) = 0$. Then we write using the chain rule

$$\frac{\partial F}{\partial z} = \frac{dF}{d\varsigma}\frac{\partial \varsigma}{\partial z} = F_0[-2\varsigma e^{-\varsigma^2}](1)$$

$$\frac{\partial^2 F}{\partial z^2} = F_0[(-2 + 4\varsigma^2)e^{-\varsigma^2}](1)^2 \tag{1}$$

$$\frac{\partial F}{\partial t} = \frac{dF}{d\varsigma}\frac{\partial \varsigma}{\partial t} = F_0[-2\varsigma e^{-\varsigma^2}](-c)$$

$$-\frac{1}{c^2}\frac{\partial^2 F}{\partial t^2} = -\frac{1}{c^2}F_0[(-2 + 4\varsigma^2)e^{-\varsigma^2}](-c)^2 \tag{2}$$

Adding equations (1) and (2) yields equation (6.10)

$$F_0[(-2+4\varsigma^2)e^{-\varsigma^2}](1)^2 - \frac{1}{c^2}F_0[(-2+4\varsigma^2)e^{-\varsigma^2}](-c)^2 = 0$$

A sequence of pulses taken at successive times illustrates the propagation of the pulses. The velocity of propagation is c.

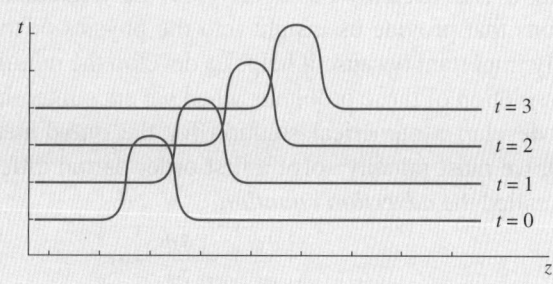

We can define the propagation of the Gaussian pulse as an initial value problem that was discussed in Example 6.4. In this case,

$$P(z) = F(\varsigma)|_{t=0} = F_0 e^{-z^2}$$

Now

$$\frac{\partial F}{\partial t} = \frac{dF}{d\varsigma}\frac{\partial \varsigma}{\partial t} = 2F_0 c\varsigma e^{-\varsigma^2}$$

$$Q(z) = \frac{\partial F}{\partial t}(\varsigma)\bigg|_{t=0} = 2F_0 c z e^{-z^2}$$

A simple integration yields the auxiliary function

$$R(z) = \frac{1}{c}\int_0^z Q(\xi)d\xi = \frac{1}{c}2cF_0\int_0^z \xi e^{-\xi^2}d\xi = F_0[1 - e^{-z^2}] = F_0 - F(z)$$

After substituting this result into (6.16), we obtain

$$E_y(z, t) = \frac{1}{2}F(z - ct) + \frac{1}{2}F(z + ct) + \frac{1}{2}[F_0 - F(z + ct)] - \frac{1}{2}[F_0 - F(z - ct)]$$

$$= F(z - ct)$$

which is a traveling Gaussian wave towards $+z$ with a speed c.

6.2.3 MATLAB Solution of One-Dimensional Equation— Finite Difference in Time-Domain Method

The numerical solution of the general wave equation is a formidable task, and one quickly encounters difficulties that are beyond the scope of this text. The major strength of numerical methods is that you can find a numerical solution to the problem. Unfortunately, this is also a weakness, because generating a number doesn't provide a lot of insight into the underlying physics or hint at a general solution to the problem. Fortunately, the use of modern software tools, such as MATLAB, allows us to carry out these calculations and create plots of solutions that provide us insight into the physics of the problem. This is extremely important because it helps us develop the intuition to "know" the form of the solution of those problems when we encounter them in the future.

To develop a numerical solution for the one-dimensional wave equation (6.10), we must initially solve a first-order partial differential equation sometimes called the *advection equation*.

$$\boxed{\frac{\partial \phi}{\partial z} + \frac{1}{c}\frac{\partial \phi}{\partial t} = 0} \tag{6.17}$$

6.2 One-Dimensional Wave Equation

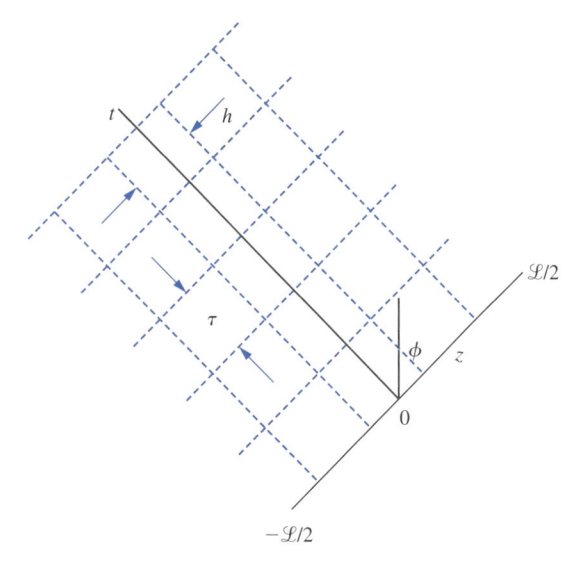

FIGURE 6-7

Numerical grid that uses periodic boundary conditions.

The advection equation is an equation that governs the transport of a conserved scalar quantity in a vector field. An example would be sugar dissolving in water that is being heated, or a pollutant spreading through a flowing stream. This equation is known to be difficult to solve numerically in the general case.

For the initial condition

$$\phi(z, t = 0) = F(z) \tag{6.18}$$

the analytical solution of the advection equation is given by

$$\phi(z, t) = F(z - ct) \tag{6.19}$$

which is also a solution of the wave equation (6.10). This can easily be checked by substituting (6.19) into (6.17).

Both the wave equation and the advection equation belong to the same family of equations called **hyperbolic equations**. The diffusion equation is in the **parabolic equation** family, and Laplace's and Poisson's equations are in the **elliptic equation** family. We will focus our attention here on the advection equation as it is simpler.

As shown in Figure 6–7, we consider that the space z and time t can be drawn in a three-dimensional figure. The amplitude ϕ of the wave is specified by the third coordinate. In Figure 6–7, we set up a numerical grid. First, we have divided up the region $\mathcal{L}$ in which the wave propagates into N sections. In the figure, we have chosen $N = 4$. Hence we write

$$h \equiv \frac{\mathcal{L}}{N} \tag{6.20}$$

We assume that the velocity of propagation is c and that it takes a time τ for the wave to propagate a distance h. Therefore

$$h = c\tau \tag{6.21}$$

With these restrictions, we will jump over numerical stability reservations that were originally noted by Courant–Fredrichs–Lewy (CFL). We will leave them as exercises to examine the cases where $h \neq c\tau$.

In addition to stability restrictions, we have also invoked **periodic boundary conditions**. This states that once a numerically calculated wave reaches the boundary at $z = +\mathscr{L}/2$, it reappears at the same time at the boundary $z = -\mathscr{L}/2$ and continues to propagate in the region $-\mathscr{L}/2 \leq z \leq +\mathscr{L}/2$. As shown in Figure 6–7, we do not actually evaluate the wave at these two edges but rather at one-half of a spatial increment $h/2$ removed from them at $z = -\mathscr{L}/2 + h/2$ and at $z = +\mathscr{L}/2 - h/2$.

Let us now convert the advection equation (6.17) to the finite difference form that can be handled by the computer. The **time derivative** is replaced using the **forward difference method** that was introduced in Chapter 4

$$\frac{\partial \phi}{\partial t} \Rightarrow \frac{\phi(z_i, t_n + \tau) - \phi(z_i, t_n)}{\tau} \tag{6.22}$$

In this notation with reference to Figure 6–7, we have

$$z_i = \left(i - \frac{1}{2}\right)h - \frac{\mathscr{L}}{2} \tag{6.23}$$

$$t_n = (n - 1)\tau$$

The **space derivative** is replaced using the **central difference method**

$$\frac{\partial \phi}{\partial z} \Rightarrow \frac{\phi(z_i + h, t_n) - \phi(z_i - h, t_n)}{2h} \tag{6.24}$$

Substitute (6.22) and (6.24) into the advection equation (6.17) and obtain

$$\frac{\phi(z_i + h, t_n) - \phi(z_i - h, t_n)}{2h} + \frac{1}{c}\frac{\phi(z_i, t_n + \tau) - \phi(z_i, t_n)}{\tau} = 0 \tag{6.25}$$

In (6.25), three of the four terms are evaluated at the same time t_n and one term is evaluated at the next increment in time, $t_n + \tau$. From (6.25), we obtain this term

$$\phi(z_i, t_n + \tau) = \phi(z_i, t_n) - \frac{c\tau}{2h}[\phi(z_i + h, t_n) - \phi(z_i - h, t_n)] \tag{6.26}$$

The finite difference method based on this equation is called the **Finite Difference Time Domain** (FDTD) method. This is valid in the interior range $2 \leq n \leq N - 1$. In (6.26), we note that all values are known initially at the time

6.2 One-Dimensional Wave Equation

$t_0 = 0$. Hence, we use (6.26) to evaluate the values at the next increment in time; this is also called a "leapfrog" scheme. With the imposition of periodic boundary conditions, we must use (6.26) carefully in order to find the values at the boundaries. This manifests itself with the requirement that

$$\phi(z_1, t_n + \tau) = \phi(z_1, t_n) - \frac{c\tau}{2h}[\phi(z_2, t_n) - \phi(z_N, t_n)]$$

$$\phi(z_N, t_n + \tau) = \phi(z_N, t_n) - \frac{c\tau}{2h}[\phi(z_1, t_n) - \phi(z_{N-1}, t_n)]$$

(6.27)

which makes the iteration process consistent.

EXAMPLE 6.6

Use (6.26) and (6.27) to find the evolution of a rectangular pulse whose initial shape is defined by

$$\phi\left(-\frac{h}{2} \leq z \leq \frac{h}{2}, t = 0\right) = 1$$

$$\phi\left(|z| > \frac{h}{2}, t = 0\right) = 0$$

Use the grid depicted in Figure 6–7. The stability requirement $h = c\tau$ is also to be invoked in this calculation.

Answer. We take a grid with a small number of points $N = 4$. The number of the calculated time steps is taken to be nstep = 5. We tabulate the computed values to be

				t			
		0	τ	2τ	3τ	4τ	5τ
	$-3\mathcal{L}/8$	0	$-1/2$	$-1/2$	$1/2$	$5/2$	$9/2$
z	$-\mathcal{L}/8$	1	$1/2$	$-1/2$	$-3/2$	$-3/2$	$1/2$
	$\mathcal{L}/8$	1	$3/2$	$3/2$	$1/2$	$-3/2$	$-7/2$
	$3\mathcal{L}/8$	0	$1/2$	$3/2$	$5/2$	$5/2$	$1/2$

Note that the signal becomes distorted and increases in value as it propagates. It is unstable for every value of time! Our imposition of the stability requirement $h = c\tau$ did not ensure stability in this case!

Fortunately for us, there is a simple solution to the instability problem that we encountered in Example 6.6. This is the **Lax method**. It replaces (6.26) with the slightly different iteration equation

$$\phi(z_i, t_n + \tau) = \frac{1}{2}[\phi(z_i + h, t_n) + \phi(z_i - h, t_n)] - \frac{c\tau}{2h}[\phi(z_i + h, t_n) - \phi(z_i - h, t_n)] \quad (6.28)$$

The first term on the right side is the average of the two neighboring terms. Similarly, the two equations that represent the periodic boundary conditions are modified to

$$\phi(z_1, t_n + \tau) = \frac{1}{2}[\phi(z_2, t_n) + \phi(z_N, t_n)] - \frac{c\tau}{2h}[\phi(z_2, t_n) - \phi(z_N, t_n)]$$

$$\phi(z_N, t_n + \tau) = \frac{1}{2}[\phi(z_1, t_n) + \phi(z_{N-1}, t_n)] - \frac{c\tau}{2h}[\phi(z_1, t_n) - \phi(z_{N-1}, t_n)] \quad (6.29)$$

EXAMPLE 6.7

Repeat Example 6.6 using the Lax method, and sketch the solution with $h = c\tau$ again.

Answer. In MATLAB language, we write (6.28) and (6.29) in three steps. The first step finds the new interior values of ϕ in terms of the previous interior values. These iterations are specified to have iterations in the range $2 \leq i \leq (N-1)$. The remaining two steps take care of the periodic boundary conditions at $i = 1$ and at $i = N$. The results for the case of $N = 4$ for the first five time steps are given in the table below.

		\		t			
		0	τ	2τ	3τ	4τ	5τ
	$3\mathscr{L}/8$	0	0	1	1	0	0
z	$-\mathscr{L}/8$	1	0	0	1	1	0
	$\mathscr{L}/8$	1	1	0	0	1	1
	$3\mathscr{L}/8$	0	1	1	0	0	1

6.2 One-Dimensional Wave Equation

Note that, in this case, we have **stability**. In addition, the pulse is not distorted as it propagates. We plot the solution below.

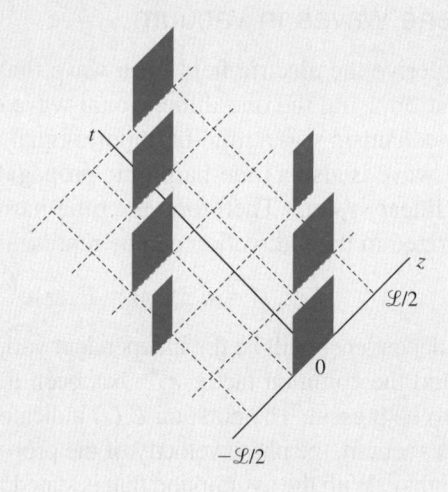

EXAMPLE 6.8

Calculate the temporal and spatial evolution of a narrow pulse. Use 50 grid points and 50 time steps.

Answer. The solution is shown in the following two figures. The initial pulse and the final pulse are shown in (*a*), and the propagation is shown in (*b*).

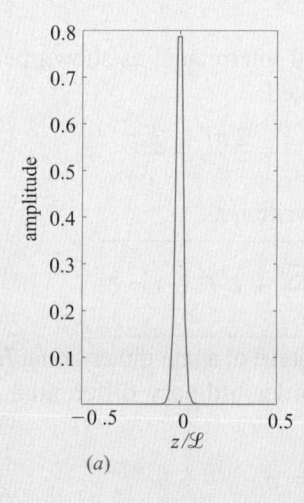

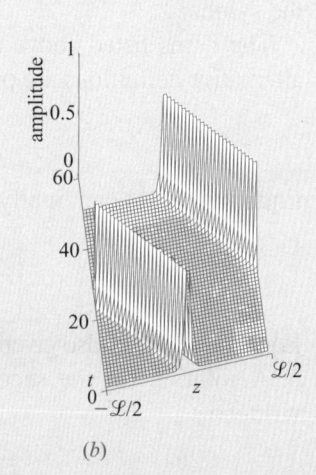

6.3 Time-Harmonic Plane Waves

6.3.1 Plane Waves in Vacuum

Let us now derive the electric field for a wave that is generated with a sinusoidal excitation using the one-dimensional wave equation (6.10). Since the excitation mechanism was a time-harmonic signal, we should expect that the propagating wave is also a time-harmonic propagating wave, because we are assuming a linear system. Therefore, the time-harmonic time variation of an E field polarized in the y direction can be represented by

$$E(z, t) = E_y(z, t) = E_y(z)e^{j\omega t} \tag{6.30}$$

where the z dependence will be the independent variable and will not be explicitly stated, and the common factor $e^{j\omega t}$ has been deleted to save space, but is understood to be present. The notation $E_y(z)$ indicates that this term is a **phasor** quantity. In a vacuum, the phase velocity of the propagating wave is equal to the velocity of light c. With the assumption that is stated in (6.30), we write (6.10) as

$$\frac{\partial^2 E_y(z)}{\partial z^2} - \frac{1}{c^2}\frac{\partial^2 E_y(z)}{\partial t^2} = \left[\frac{d^2 E_y(z)}{dz^2} - \frac{(j\omega)^2}{c^2}E_y(z)\right]e^{j\omega t} = 0$$

or

$$\frac{d^2 E_y(z)}{dz^2} + \left(\frac{\omega}{c}\right)^2 E_y(z) = 0 \tag{6.31}$$

where the exponential term $e^{j\omega t}$ has been removed. The ratio (ω/c) has the dimensions m^{-1}. This ratio is called the **wave number**, and it is usually given the symbol k.

The terms listed above are all interrelated as shown below, where several alternative definitions are presented.

$$k = \frac{\omega}{c} = \frac{2\pi f}{c} = \frac{2\pi}{\lambda} \tag{6.32}$$

Equation (6.31) can finally be written as

$$\boxed{\frac{d^2 E_y(z)}{dz^2} + k^2 E_y(z) = 0} \tag{6.33}$$

Equation (6.33) is also given the name of a one-dimensional **Helmholtz equation**.

A solution for the second-order ordinary differential equation (6.33) is written as

$$E_y(z) = ae^{-jkz} + be^{+jkz} \tag{6.34}$$

6.3 Time-Harmonic Plane Waves

where we have let the two constants of integration a and b be real quantities. A substitution into (6.30) yields

$$E_y = ae^{j(\omega t - kz)} + be^{j(\omega t + kz)} \tag{6.35}$$

Taking the *real part* of (6.35), we find the following expression for the electric field

$$E_y = a\cos(\omega t - kz) + b\cos(\omega t + kz) \tag{6.36}$$

Either form of the electric field can be used, since trigonometric functions are merely linear combinations of exponential functions. We could have equivalently chosen the imaginary part of (6.35) and obtained the sine functions. It is most common to use the real part assumption. In doing this, we have actually applied Euler's identity

$$e^{j\theta} = \cos\theta + j\sin\theta \tag{6.37}$$

The first term in either expression (6.35) or (6.36) corresponds to a wave propagating to increasing values of the spatial coordinate z, and the second term corresponds to a wave propagating to decreasing values of z. Both terms are particular forms of the general solution (6.11). The constants a and b are specified by the initial excitation signal and the known direction of propagation.

We have already discovered in (6.14) that the ***phase velocity*** $v_\phi = dz/dt$ of the propagation of these two traveling waves is determined from the equation

$$\boxed{v_\phi \equiv \pm\frac{\omega}{k} = \pm c \text{ (m/s)}} \tag{6.38}$$

because the medium in which the wave is propagating is a vacuum.

In general, the phase velocity is a vector quantity. This is because the phase velocity has both a magnitude and a direction associated with it. It can have a value that is *greater* than the velocity of light. Even water waves breaking on the beach can have an infinite phase velocity *along* the beach if the water waves are propagating exactly perpendicular to the beach, as shown in Figure 6–8b. In this case, the wave would hit two separate points on the beach at exactly the same time. This leads to an infinite phase velocity along the beach. There is, however, *no energy transported* along the beach. Suffice it to say that Einstein's insight has yet to be disproved when he said that particles and energy can go no faster than the speed of light. In fact, effects of gravity are not instantaneous but have recently been observed to obey this upper limit for their velocity of propagation.

We can also use the wave number to tell us in which direction the wave is propagating if not directed along the z axis. In this case, the wave number will

FIGURE 6–8

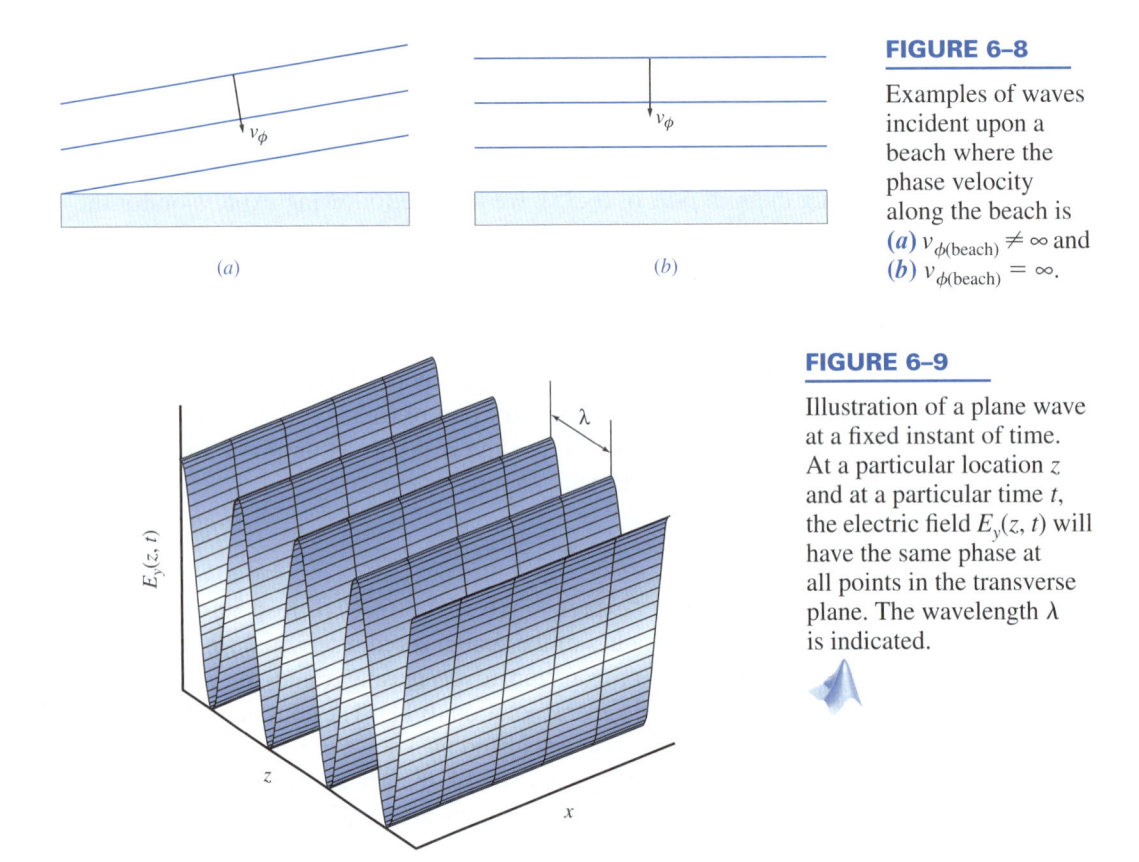

Examples of waves incident upon a beach where the phase velocity along the beach is (**a**) $v_{\phi\text{(beach)}} \neq \infty$ and (**b**) $v_{\phi\text{(beach)}} = \infty$.

FIGURE 6–9

Illustration of a plane wave at a fixed instant of time. At a particular location z and at a particular time t, the electric field $E_y(z, t)$ will have the same phase at all points in the transverse plane. The wavelength λ is indicated.

be a vector—(so indicated with the vector notation, **k**). It is frequently called the **wave vector**. If the position at which the wave is to be determined is indicated with the position vector **r**, then the term kz is replaced with the scalar product $\mathbf{k} \cdot \mathbf{r}$. The magnitude of the wave vector is $2\pi/\lambda$, where λ is the wavelength of the wave. We will encounter this notation later.

The previous discussion asserts that the wave depends on only one spatial coordinate. This was chosen to be the z coordinate, and we have assumed that $\partial E_y(z, t)/\partial x = \partial E_y(z, t)/\partial y = 0$. A two-dimensional picture of a segment of this one-dimensional wave is shown in Figure 6–9. The transverse coordinate could have been either x or y. There is no change in the value of the perturbation of the electric field in a plane that its transverse to the z axis. In this plane, all of the field components have the same phase. This is defined as being a **plane wave**. Plane waves are friendly from a theoretical and pedagogical point of view, since the vector wave equation has been reduced to a one-dimensional scalar wave equation in Cartesian coordinates. This equation is the easiest to solve. There are few cases where plane waves do actually exist.[5] However, in free space such waves would have to be launched either at $z = \pm\infty$ (so any spherical radiation effects originating from a finite-sized antenna would have decayed to zero), or the antenna would have to be infinite in its transverse dimension.

EXAMPLE 6.9

In an experiment performed in a vacuum, we simultaneously measure the electric field to be polarized in the y direction at $z = 0$ and also one wavelength away at $z = 2$ cm. The amplitude is 2 μV/m. Find the frequency of excitation, and write an expression that describes the wave if the wave is moving in the direction of increasing values of the coordinate z.

Answer. Since the wavelength was specified to be $\lambda = 0.02$ m, we can compute the wave number

$$k = \frac{2\pi}{\lambda} = \frac{2\pi}{0.02} = 100\pi \ (\text{m}^{-1})$$

Since the wave is propagating in a vacuum, the frequency of oscillation can be found from

$$f = \frac{\omega}{2\pi} = \frac{kc}{2\pi} = \frac{100\pi \times 3 \times 10^8}{2\pi} = 15 \times 10^9 \ \text{Hz} = 15 \ \text{GHz}$$

where G is the abbreviation for "giga" and stands for 10^9. Finally, the wave can be expressed as

$$\mathbf{E}(z, t) = 2 \times 10^{-6} \cos[2\pi(15 \times 10^9 t - 50z)]\mathbf{u}_y \ \text{V/m}$$

The polarization of the electric field and the direction of wave propagation are clearly indicated in this solution.

EXAMPLE 6.10

Show that a linearly polarized plane wave can be resolved into two equal amplitude waves that rotate about the z axis.

Answer. The linearly polarized wave

$$\mathbf{E}_y(z, t) = E_0 \mathbf{u}_y e^{j(\omega t - kz)}$$

---

[5] Plane waves are also encountered in ideal lossless transmission lines or coaxial cables.

can be written as a sum of two components $\mathbf{E}_y(z, t) = \mathbf{E}_y^+(z, t) + \mathbf{E}_y^-(z, t)$, where

$$\mathbf{E}_y^+(z, t) = \frac{E_0}{2}(\mathbf{u}_y - j\mathbf{u}_x)e^{j(\omega t - kz)}$$

$$\mathbf{E}_y^-(z, t) = \frac{E_0}{2}(\mathbf{u}_y + j\mathbf{u}_x)e^{j(\omega t - kz)}$$

We will show that these two components represent, respectively, a right-hand rotating wave and a left-hand rotating wave, each with an amplitude $E_0/2$. Taking into account that

$$\mathbf{u}_x = -\mathbf{u}_\phi \sin\phi$$
$$\mathbf{u}_y = \mathbf{u}_\phi \cos\phi$$

we obtain

$$\mathbf{u}_y - j\mathbf{u}_x = \mathbf{u}_\phi e^{j\phi}$$
$$\mathbf{u}_y + j\mathbf{u}_x = \mathbf{u}_\phi e^{-j\phi}$$

which demonstrates that the first and second waves are rotating in opposite directions. They are *circularly polarized* waves. The polarization of the wave is a subject in the next section. These are sometimes called *symmetrical components* in the study of electrical machinery.

6.3.2 Magnetic Field Intensity and Characteristic Impedance

The preceding discussion emphasized the propagation characteristics of only the electric field component of the electromagnetic wave. We will now compute the magnetic field component of the plane wave. As the reader might expect, it will follow directly from Maxwell's equations. For the time-harmonic wave studied in the previous section, the appropriate equation is

$$\nabla \times \mathbf{E}(z, t) = -j\omega \mathbf{B}(z, t) = -j\omega\mu_0 \mathbf{H}(z, t) \qquad (6.39)$$

Let us explicitly find the magnetic field component for the wave that is propagating in the $+z$ direction and is defined by the first term in (6.35). We write

$$\mathbf{H}(z, t) = -\frac{1}{j\omega\mu_0} \begin{vmatrix} \mathbf{u}_x & \mathbf{u}_y & \mathbf{u}_z \\ \frac{\partial}{\partial x} & \frac{\partial}{\partial y} & \frac{\partial}{\partial z} \\ 0 & ae^{j(\omega t - kz)} & 0 \end{vmatrix}$$

6.3 Time-Harmonic Plane Waves

FIGURE 6–10

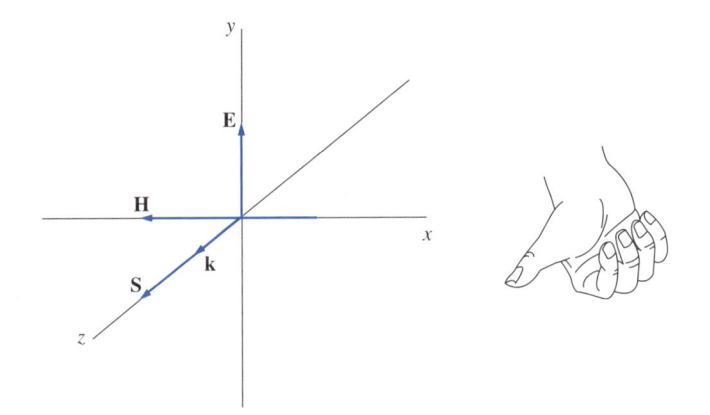

Orientation of the electric and magnetic fields at an instant in time that are propagating in the $+z$ direction. The right-hand rule indicates the direction of propagation from Poynting's theorem. Only two of the three vectors and the right hand are required to determine the third vector.

or after expanding the determinant, we write

$$\mathbf{H}(z, t) = -\frac{1}{j\omega\mu_0} \frac{\partial (ae^{j(\omega t - kz)})}{\partial z} (-\mathbf{u_x}) = -\left[\frac{k}{\omega\mu_0}\right] ae^{j(\omega t - kz)} \mathbf{u_x} \quad (6.40)$$

Figure 6–10 illustrates the field components for the electric field intensity $\mathbf{E}$ given in (6.35) and the magnetic field intensity $\mathbf{H}(z, t)$ given in (6.40) for an electromagnetic wave propagating in the $+z$ direction. Also shown is the direction of propagation, which is indicated by the wave vector $\mathbf{k} = k\mathbf{u_z}$. The constant in the denominator is a quantity that has the dimension of an impedance, since the ratio of

$$\frac{E}{H} = \frac{\text{V/m}}{\text{A/m}} = \Omega$$

Therefore,

$$Z_c = \frac{\omega\mu_0}{k} = \frac{kc\mu_0}{k} = c\mu_0 = \frac{\mu_0}{\sqrt{\mu_0\varepsilon_0}} = \sqrt{\frac{\mu_0}{\varepsilon_0}} \quad (\Omega)$$

$$\boxed{Z_c = \sqrt{\frac{\mu_0}{\varepsilon_0}} \quad (\Omega)} \quad (6.41)$$

This ratio is called the ***characteristic impedance*** or the ***intrinsic wave impedance*** of the medium. Substituting the values for a vacuum, we find the characteristic impedance equal to $120\pi\ \Omega$. We can then rewrite

(6.40) to be

$$\mathbf{H}_x(z,t) = \frac{-1}{Z_c} ae^{j(\omega t - kz)} \mathbf{u_x} = \frac{1}{Z_c}(\mathbf{u_z} \times \mathbf{E}_y(z,t)) \text{ (A/m)} \quad (6.42)$$

Looking now at various properties of this wave—particularly the propagation direction of the electromagnetic power—we invoke Poynting's theorem and find that the power will flow in the direction of the **Poynting vector S** given by $\mathbf{S} = \mathbf{E}(z,t) \times \mathbf{H}(z,t) = S\mathbf{u_z}$. From Figure 6–10, we see that this will be in the $+z$ direction—the same direction as the wave vector **k**. This is consistent with our initial specification to examine only the electromagnetic wave component that is propagating in the $+z$ direction. Since both field components are propagating in the $+z$ direction, we naturally expect that the electromagnetic power flows in the same direction.[6] By measuring the electric and magnetic field components at the same instant of time to determine the polarization of the two components and by computing the Poynting vector, space scientists and radio astronomers can determine the actual origin of radiation detected by satellites or large ground-based antennas. The actual power that flows through a given surface area can be computed by integrating the Poynting vector over that area.

EXAMPLE 6.11

Compute the characteristic impedance of a vacuum Z_0.

Answer. Equation (6.41) yields

$$Z_0 = \frac{|\mathbf{E}(z,t)|}{|\mathbf{H}(z,t)|} = \sqrt{\frac{\mu_0}{\varepsilon_0}} \approx \sqrt{\frac{4\pi \times 10^{-7}}{\frac{1}{36\pi} \times 10^{-9}}} = 120\pi \approx 377 \, \Omega$$

If we know the value of one of the field components, we also know the other field component. We need only multiply or divide it by the characteristic impedance Z_c. However, readers should be reminded that this impedance is just a ratio of two field quantities and cannot be measured by running outside with the two leads of an ohmmeter held up in either hand.

---

[6] Certain waves will have their phase velocity and the direction of energy propagation in the opposite directions. Such "backward" waves need not concern us here since we are examining vacuum conditions. They will be described in reference to a particular dispersive transmission line.

EXAMPLE 6.12

For the electric field described in Example 6.9, find the magnetic field intensity.

Answer. We can compute $\mathbf{H}(z, t)$ from (6.42) to be

$$\mathbf{H}(z, t) = \frac{2 \times 10^{-6}}{377} \cos[2\pi(15 \times 10^9 t - 50z)](\mathbf{u}_z \times \mathbf{u}_y)$$

$$\mathbf{H}(z, t) = 5.3 \times 10^{-9} \cos[2\pi(15 \times 10^9 t - 50z)](-\mathbf{u}_x) \text{ A/m}$$

Note the direction of the magnetic field intensity. This direction is required so the power will flow in the $+z$ direction.

EXAMPLE 6.13

The electric field in a vacuum is given by

$$\mathbf{E}(z, t) = 10 \cos(\omega t - kz)\mathbf{u}_x \text{ V/m}$$

Find the average power in a circular area in a plane defined by $z = $ constant and whose radius is $R = 3$ m.

Answer. We first write the electric field in complex form as

$$\mathbf{E}(z, t) = 10 e^{j(\omega t - kz)} \mathbf{u}_x$$

Since $Z_0 = 120\pi$, we write

$$\mathbf{H}(z, t) = \frac{10}{120\pi} e^{j(\omega t - kz)} \mathbf{u}_y \text{ A/m}$$

The average power through this plane is

$$P_{av} = \frac{1}{2} \text{Re} \left\{ \int_{\Delta s} \mathbf{E}(z, t) \times \mathbf{H}(z, t)^* \cdot \mathbf{ds} \right\} = \frac{1}{2}(10)\left(\frac{10}{120\pi}\right)(\pi 3^2) = 3.75 \text{ W}$$

6.4 Plane Wave Propagation in a Dielectric Medium

6.4.1 Plane Wave Propagation in a Lossless Dielectric Medium

In the definition of both the wave number that was applicable for a vacuum (6.32), we showed that it was a function of both the permeability of free space μ_0 and the permittivity of free space ε_0 via the relation $k = 2\pi f \sqrt{\varepsilon_0 \mu_0}$. If the

FIGURE 6–11

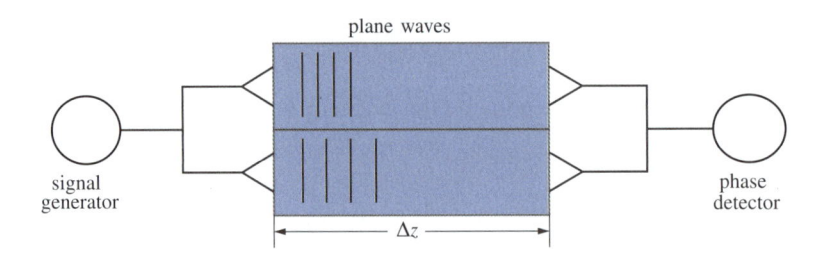

Plane waves are launched from the signal generator. The phase of detected signals can be monitored.

space were now uniformly filled with a gas or a linear homogeneous dielectric (such as a plastic whose dielectric constant $\varepsilon = \varepsilon_r \varepsilon_0$, where $\varepsilon_r > 1$), the wave number k would have to be modified by replacing ε_0 with ε. We can usually assume the vacuum value μ_0 for the permeability μ since it differs significantly only for wave propagation in iron, nickel, and cobalt—and we usually do not consider electromagnetic wave propagation in these materials.

As we will emphasize later, this effect can be used to determine the properties of the material. It also plays an important role in the refraction of light that allows eye glasses to work as vision-correcting devices. To introduce this technique here and to describe the wave propagation in a dielectric, let us assume that two materials are juxtaposed as shown in Figure 6–11.

In this figure, we have assumed that plane waves are launched in two adjacent media and that they propagate to the end, where any phase difference that may exist between the two signals can be detected. The wave numbers in the two media are

$$\left. \begin{array}{l} k_1 = 2\pi f \sqrt{\varepsilon_1 \mu_0} \\ k_2 = 2\pi f \sqrt{\varepsilon_2 \mu_0} \end{array} \right\} \tag{6.43}$$

Both signals will travel the same distance Δz, although with different phase velocities $v_1 = \omega/k_1$ and $v_2 = \omega/k_2$. This difference delays the arrival of one signal with respect to the other and creates a detectable phase difference $\delta\theta$. The phase difference $\delta\theta$ is given by

$$\delta\theta = k_1 \Delta z - k_2 \Delta z = 2\pi f (\sqrt{\varepsilon_1 \mu_0} - \sqrt{\varepsilon_2 \mu_0}) \Delta z \tag{6.44}$$

If we know the total phase change in the signal passing through one of the paths, say $\Delta\theta_1$ where

$$\Delta\theta_1 = 2\pi f \sqrt{\varepsilon_1 \mu_0} \Delta z \tag{6.45}$$

we can then calculate the phase difference $\delta\theta$ to be

$$\frac{\delta\theta}{\Delta\theta_1} = \frac{2\pi f (\sqrt{\varepsilon_1 \mu_0} - \sqrt{\varepsilon_2 \mu_0}) \Delta z}{2\pi f \sqrt{\varepsilon_1 \mu_0} \Delta z} = 1 - \frac{\sqrt{\varepsilon_2}}{\sqrt{\varepsilon_1}} \tag{6.46}$$

6.4 Plane Wave Propagation in a Dielectric Medium

Hence, if the material in one of the regions is known (say, a vacuum), then one can identify the material in the other region and determine the relative dielectric constant of that material. Modified versions of this idea have been used to determine plastic or paper properties in the manufacturing plant, and this is a standard tool in such industries.

The ratio of the phase velocity of light in a vacuum c to the phase velocity in the dielectric is called the ***index of refraction*** for the material (here, medium #1 is a vacuum and medium #2 is the unknown material). The index of refraction is usually given the symbol n and is defined as

$$n \equiv \frac{v_{\text{vacuum}}}{v_{\text{dielectric}}} = \sqrt{\varepsilon_r} \tag{6.47}$$

Optical materials are typically specified in terms of their index of refraction.

EXAMPLE 6.14

Using the setup shown in Figure 6–11, calculate the phase difference $\delta\theta$ if one region is filled with a gas with $\varepsilon_r = 1.0005$ and the other region is a vacuum. The frequency of oscillation is 10 GHz, and the length $\Delta z = 1$ m.

Answer. The phase difference $\delta\theta$ is given by (6.44)

$$\delta\theta = 2\pi f(\sqrt{\varepsilon_0\mu_0} - \sqrt{\varepsilon_r\varepsilon_0\mu_0})\Delta z = 2\pi f\sqrt{\varepsilon_0\mu_0}(1 - \sqrt{\varepsilon_r})\Delta z$$

$$= \frac{2\pi \times 10^{10}}{3 \times 10^8}(1 - \sqrt{1.0005}) = \frac{2\pi \times 10^{10}}{3 \times 10^8}(-0.00025)$$

$$= -0.052 \text{ radians} \approx -3°$$

This number is small but detectable. We can increase the resolution by increasing the length Δz, as long as Δz is less than a wavelength, or by passing the wave through the regions several times.

6.4.2 Plane Wave Propagation in a Lossy Dielectric Medium

Electromagnetic waves can propagate in a material that has a nonzero conductivity σ. In order to show this, we return to the time-harmonic form of Maxwell's equations with the conduction current $\mathbf{J}(\mathbf{r}) = \sigma\mathbf{E}(\mathbf{r})$ included. We write the two curl equations for phasors as

$$\nabla \times \mathbf{E}(\mathbf{r}) = -j\omega\mu\mathbf{H}(\mathbf{r}) \tag{6.48}$$

$$\nabla \times \mathbf{H}(\mathbf{r}) = \mathbf{J}(\mathbf{r}) + j\omega\varepsilon\mathbf{E}(\mathbf{r}) = (\sigma + j\omega\varepsilon)\mathbf{E}(\mathbf{r}) \tag{6.49}$$

As before, we will assume that there is no free charge density ($\rho_v = 0$). Following the same procedure we have used before, we combine these two equations to yield a vector wave equation.

$$\nabla^2 \mathbf{E}(\mathbf{r}) - j\omega\mu(\sigma + j\omega\varepsilon)\mathbf{E}(\mathbf{r}) = 0 \tag{6.50}$$

It can easily be shown that the magnetic field intensity $\mathbf{H}(\mathbf{r})$ satisfies the same equation. Again, we will assume that the electric field is linearly polarized in the $\mathbf{u}_y$ direction and the wave propagates only in the $\mathbf{u}_z$ direction. With these simplifications, the resulting wave equation in phasor notation becomes

$$\frac{d^2 E_y(z)}{dz^2} - j\omega\mu(\sigma + j\omega\varepsilon)E_y(z) = 0 \tag{6.51}$$

This equation can be written as

$$\frac{d^2 E_y(z)}{dz^2} - \gamma^2 E_y(z) = 0 \tag{6.52}$$

where

$$\gamma^2 = j\omega\mu(\sigma + j\omega\varepsilon) = (j\omega)^2 \mu\varepsilon\left(1 + \frac{\sigma}{j\omega\varepsilon}\right) \tag{6.53}$$

Equation (6.52) is similar to (6.33) except that the coefficient γ is complex. We write γ as

$$\gamma = \alpha + j\beta \tag{6.54}$$

It consists of a real part α and an imaginary part $j\beta$. In a vacuum, $\alpha = 0$ and $\beta = k = \omega\sqrt{\mu_0\varepsilon_0}$, or it is directly proportional to the frequency ω. As a result, the phase velocity $v_\phi = \omega/\beta = c$ is a constant. In the general case, the real and the imaginary parts of the *propagation constant* γ are nonlinear functions of the frequency ω. After a little algebra, we obtain

$$\alpha(\omega) = \sigma\sqrt{\frac{\mu}{2\varepsilon\left[1 + \sqrt{1 + \left(\frac{\sigma}{\omega\varepsilon}\right)^2}\right]}}$$

$$\beta(\omega) = \omega\sqrt{\frac{\mu\varepsilon}{2}\left[1 + \sqrt{1 + \left(\frac{\sigma}{\omega\varepsilon}\right)^2}\right]} \tag{6.55}$$

The substitution of (6.55) into (6.54) leads to (6.53) for the term γ^2. Therefore, the phase velocity depends on the frequency. This is called **dispersion**, and the

6.4 Plane Wave Propagation in a Dielectric Medium

medium in which the wave is propagating is a ***dispersive medium***.

$$v_\phi(\omega) = \frac{\omega}{\beta} = \frac{1}{\sqrt{\frac{\mu\varepsilon}{2}\left[1 + \sqrt{1 + \left(\frac{\sigma}{\omega\varepsilon}\right)^2}\right]}} \quad (6.56)$$

The ***group velocity*** v_g is defined as

$$v_g \equiv \frac{\partial \omega}{\partial \beta} = \frac{1}{\partial \beta / \partial \omega} \text{ (m/s)} \quad (6.57)$$

We also encounter the group velocity when we describe wave propagation on a transmission line. Here, the differentiation of (6.55) using the definition given in (6.57) leads to the following group velocity:

$$v_g(\omega) = \sqrt{\frac{2}{\mu\varepsilon}} \frac{\sqrt{\left[1 + \sqrt{1 + \left(\frac{\sigma}{\omega\varepsilon}\right)^2}\right]\sqrt{1 + \left(\frac{\sigma}{\omega\varepsilon}\right)^2}}}{\left\{\left[1 + \sqrt{1 + \left(\frac{\sigma}{\omega\varepsilon}\right)^2}\right]\sqrt{1 + \left(\frac{\sigma}{\omega\varepsilon}\right)^2} - \frac{1}{2}\left(\frac{\sigma}{\omega\varepsilon}\right)^2\right\}} \quad (6.58)$$

Group velocity is different from phase velocity (6.56). The group velocity may equal the velocity of energy transport in certain dispersive media, while in other cases, the energy transport velocity may be altogether different.

The electric field component for a wave that propagates to increasing values of the coordinate z can finally be written as

$$E_y(z, t) = E_{y0}e^{(j\omega t - \gamma z)} = E_{y0}e^{-\alpha z}e^{j(\omega t - \beta z)} \quad (6.59)$$

This implies that an electromagnetic wave will propagate with a ***phase constant*** β in the conducting medium, but the wave amplitude will be decreased with an ***attenuation constant*** α as it propagates in space. The units of the attenuation constant α are given in nepers/meter. If $\alpha = 1$ Np/m, the amplitude of the wave will decrease to $e^{-1} \approx 0.368$ of its original value at a distance of 1 m. An attenuation of 1 Np/m equals $20\log_{10}(e) = 8.686$ dB/m. Although this decay of the wave amplitude occurs in space, we recall that it is similar to a time constant in circuits where the decay was in time.

Once again, the magnetic field intensity has only an $H_x(z)$ component, which can be computed from (6.48). The ratio of these two terms yields the characteristic impedance of the conductor—which in this case is complex

$$Z_c(\omega) = \frac{E_y(z)}{H_x(z)} = \frac{j\omega\mu}{\alpha(\omega) + j\beta(\omega)} \quad (6.60)$$

where both parts of the propagation constant have to be replaced with the expressions (6.55). For a vacuum, $\alpha = 0$ and $\beta = k$ and equation (6.60) reduces to (6.41).

EXAMPLE 6.15

A 10 V/m plane wave whose frequency is 300 MHz propagates in the +z direction in an infinite medium. The electric field is polarized in the $\mathbf{u}_x$ direction. The parameters of the medium are $\varepsilon_r = 9$, $\mu_r = 1$, and $\sigma = 10$ S/m. Write a complete time-domain expression for the electric field.

Answer. The attenuation constant is obtained from (6.55), and we write

$$\alpha = 10 \sqrt{\frac{(4\pi \times 10^{-7}) \times 36\pi}{2 \times (9 \times 10^{-9}) \left[1 + \sqrt{1 + \left(\frac{10 \times 36\pi}{2\pi \times 300 \times 10^6 \times 9 \times 10^{-9}}\right)^2}\right]}}$$

This gives a numerical value $\alpha = 108.01$ Np/m. Similarly, equation (6.55) yields the phase constant

$$\beta = 2\pi \times 300 \times 10^6$$

$$\times \sqrt{\frac{(4\pi \times 10^{-7})}{2}\left(\frac{9}{36\pi} \times 10^{-9}\right)\left[1 + \sqrt{1 + \left(\frac{10 \times 36\pi}{2\pi \times 300 \times 10^6 \times 9 \times 10^{-9}}\right)^2}\right]}$$

This gives a numerical value $\beta = 109.65$ rad/m. The complex propagation constant is $\gamma \approx 108 + j110$ m^{-1}. Hence the electric field is

$$\mathbf{E}(z,t) = 10 e^{-108z} \cos[(2\pi \times 300 \times 10^6)t - 110z]\mathbf{u}_x \text{ V/m}$$

EXAMPLE 6.16

For the medium described in Example 6.15, plot the phase velocity and the group velocity as functions of frequency in the frequency range $30 < f < 3000$ MHz. Recall that 1 GHz equals 1000 MHz.

Answer. The phase velocity v_ϕ and the group velocity v_g based on equations (6.56) and (6.58) are shown in the following figure. Both velocities are normalized by the vacuum value of the velocity of light $c = 3 \times 10^8$ m/s.

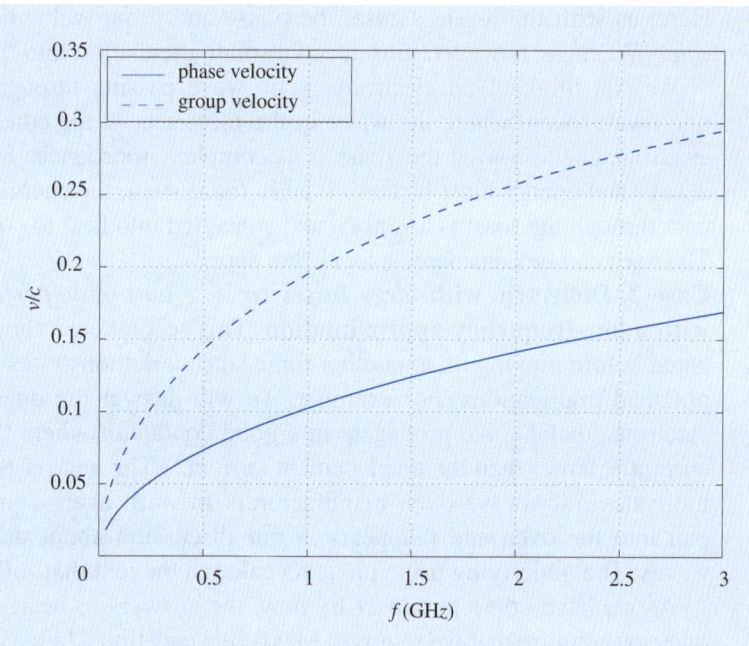

6.4 Plane Wave Propagation in a Dielectric Medium

It is evident that both velocities increase with increasing frequency. The theoretical limit of the two velocities when $f \to \infty$ is the same.

$$v_\infty = \frac{c}{\sqrt{\varepsilon_r}} = \frac{3 \times 10^8}{\sqrt{9}} = 10^8 \text{ m/s}$$

There are two particular cases that should be investigated further.

Case 1. Dielectric with *small losses* ($\sigma \ll \omega\varepsilon$) with a high-frequency approximation. In this case, we can replace the complex factor in the brackets in (6.55) with

$$1 + \sqrt{1 + \left(\frac{\sigma}{\omega\varepsilon}\right)^2} \approx 2 \tag{6.61}$$

From (6.55), (6.56), and (6.58), we obtain the approximate attenuation and propagation constants to be

$$\alpha \approx \frac{\sigma}{2}\sqrt{\frac{\mu}{\varepsilon}}$$
$$\beta \approx \omega\sqrt{\mu\varepsilon} \tag{6.62}$$
$$v_\phi = v_g \approx \frac{1}{\sqrt{\mu\varepsilon}}$$

Here, as with the lossless case, the phase and group velocities remain the *same*. We have, however, introduced a small attenuation into the wave.

We can think of an electromagnetic wave passing through a roast in a microwave oven, where the water in the meat acts as the conductor. For the electromagnetic wave, the roast is a complex impedance. Since the field decays and energy must be conserved in the system, the energy that does not pass through the roast is absorbed and converted into heat to cook our dinner. The roast can be considered a local "hot spot."

Case 2. Dielectric with *large losses* ($\sigma \gg \omega\varepsilon$) or *imperfect conductor* with a low-frequency approximation. This second case should be investigated before moving on to another topic since it demonstrates a multitude of practical implications. For example, we will answer the question "Can an electromagnetic wave propagate in a good conductor where the conduction current is larger than the displacement current?" The answer to this question motivates microwave oven manufacturers to warn against putting a metal pan into the oven and reappears in our discussion about the reflection of waves. The underlying principle is revealed in the text that follows.

As we have come to expect by now, the answers to nearly all important questions require that we return to Maxwell's equations. In the case where the conduction current is much larger than the displacement current, we rewrite the second curl equation (6.49) as

$$\nabla \times \mathbf{H}(\mathbf{r}) = \mathbf{J}(\mathbf{r}) + j\omega\varepsilon\mathbf{E}(\mathbf{r}) \approx \sigma\mathbf{E}(\mathbf{r})$$

We again obtain (6.52) where

$$\gamma = \sqrt{j\omega\mu\sigma} \qquad (6.63)$$

which is a consequence of (6.53). From (6.55) and taking into account that

$$1 + \sqrt{1 + \left(\frac{\sigma}{\omega\varepsilon}\right)^2} \approx \left(\frac{\sigma}{\omega\varepsilon}\right) \qquad (6.64)$$

we obtain the following asymptotic terms:

$$\alpha \approx \beta \approx \sqrt{\frac{\omega\mu\sigma}{2}} \qquad (6.65)$$

Because $\sqrt{j} = e^{j45°} = (1 + j)/\sqrt{2}$, (6.65) follows directly from (6.63). The phase and group velocities will actually be different.

Equation (6.59) also gives the solution in this particular case—spatial propagating waves with an exponential decay in the amplitude. We can write the attenuation constant as $\alpha = 1/\delta$, where δ is a constant with a dimension m. This length is called the ***skin depth*** of the material. The explicit

6.4 Plane Wave Propagation in a Dielectric Medium

expression for it is

$$\delta = \sqrt{\frac{2}{\omega\mu\sigma}} = \frac{1}{\sqrt{\pi f \mu \sigma}} \quad (\text{m}) \tag{6.66}$$

The phase constant has the same value as the attenuation constant $\alpha = \beta = 1/\delta$. For a given medium ($\mu$, σ = constant), the skin depth decreases with increasing frequency—this is known as the **skin effect**. For the particular case of a perfect conductor (or **superconductor** when $\sigma \to \infty$), the skin depth is zero, and it is independent of the frequency. The electric and magnetic fields *do not penetrate* into the medium at all.

Imagine the problem that would arise in a microwave oven if the skin depth were much less than the size of the roast. The edge would become charred and the central part of the roast would remain uncooked. The wavelength at the standard microwave frequency of $f = 2.45$ GHz is

$$\lambda = \frac{c}{f} = \frac{3 \times 10^8}{2.45 \times 10^9} \approx 0.12 \text{ m}$$

Since the roast is heated from all directions, this is not a major problem.

EXAMPLE 6.17

Calculate the skin depth of copper at a frequency of 3 GHz. The conductivity of copper is $\sigma = 5.8 \times 10^7$ S/m, and $\mu \approx \mu_0$.

Answer. Equation (6.66) yields

$$\delta = \frac{1}{\sqrt{\pi f \mu \sigma}} = \frac{1}{\sqrt{\pi (3 \times 10^9)(4\pi \times 10^{-7})(5.8 \times 10^7)}}$$

$$= 1.21 \times 10^{-6} \text{ m} = 1.21 \text{ } \mu\text{m}$$

This is a very small value because copper is a good conductor and the frequency is sufficiently high.

EXAMPLE 6.18

Plot the spatial variation of the real part of the electric field $E_y(z)$ at a fixed time ($t = 0$) of a plane wave propagating into a copper conductor. The frequency of the wave is 3 GHz, and it has an initial amplitude of $E_{y0} = 10$ V/m. Assume the material parameters are $\varepsilon_r = 1$, $\mu_r = 1$, and $\sigma = 5.8 \times 10^7$ S/m.

Answer. First, we check the validity of the approximation that the conduction current is larger than the displacement current. This is satisfied if $\sigma \gg \omega\varepsilon$. We compute

$$\omega\varepsilon = (2\pi \times 3 \times 10^9) \times \left(\frac{1}{36\pi} \times 10^{-9}\right) = 0.1667 \ll \sigma = 5.8 \times 10^7 \text{ S/m}$$

From (6.59), it follows that the amplitude of the wave $E_y(z)$ decays to e^{-1} of its initial value E_{y0} in one skin depth δ. Copper is a good conductor, and this reduction in the amplitude occurs within the first wavelength. We have $\lambda = 2\pi/\beta = 2\pi\delta$ or $\delta/\lambda = 1/2\pi < 1$. The spatial response for the electric field $E_y(z)$ at one instant in time is shown in the figure below.

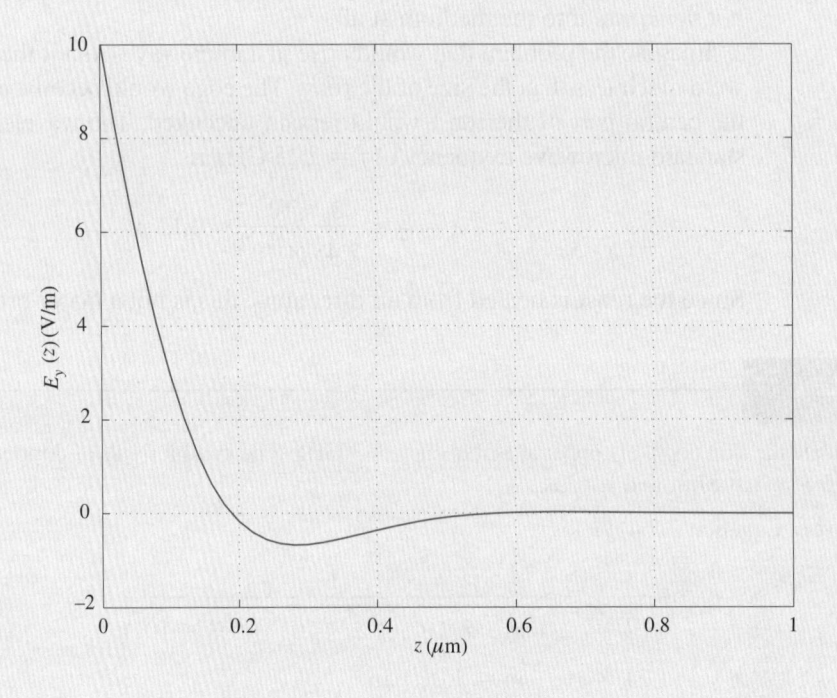

The skin depth for a particular material *depends on the frequency* of the electromagnetic wave and the conductivity of the material. There are cases where a metal is gold plated in order to decrease the skin depth by an additional small percentage since the conductivity of gold is slightly higher than aluminum or even brass. Electromagnetic energy that propagates into a conductor will be converted into heat, and hence will be lost. This implies that electromagnetic waves that are guided by a conducting surface such as a stripline will attenuate as they propagate. This will have practical importance in modern communications systems.

Reflection and Transmission of an Electromagnetic Wave

From (6.60), one can find the characteristic complex impedance of the conductor by taking into account (6.65)

$$Z_{conductor} = (1 + j)\sqrt{\frac{\omega\mu}{2\sigma}} \qquad (6.67)$$

For large values of conductivity σ that are found in many metallic materials, this value is approximately *zero* (perfect conductor). This will be important when we discuss the next topic—reflection of electromagnetic waves.

6.5 Reflection and Transmission of an Electromagnetic Wave

6.5.1 Normal Incidence—Propagating Waves

The subject of electromagnetic waves could now be concluded if we were just interested in studying wave propagation in an infinite medium. We are confronted, however, with a universe with various objects in it. What happens if a plane wave hits one of these objects? The object could be either another dielectric or a conductor. The answer to this question requires that we invoke the boundary conditions derived in Chapter 4. The formal procedure can be illustrated with the plane wave as normally incident upon the interface. Plane waves that are incident at other angles are described on the CD that is included with the book.

Let us assume that a plane wave is launched in the region $z < 0$ in a lossless material whose dielectric constant is ε_1 and that the wave is *perpendicularly* incident on a second lossless material located in the region $z \geq 0$ whose dielectric constant is ε_2. The case of perpendicular incidence is depicted in Figure 6–12. The permeabilities of both materials are both equal to their vacuum values μ_0.

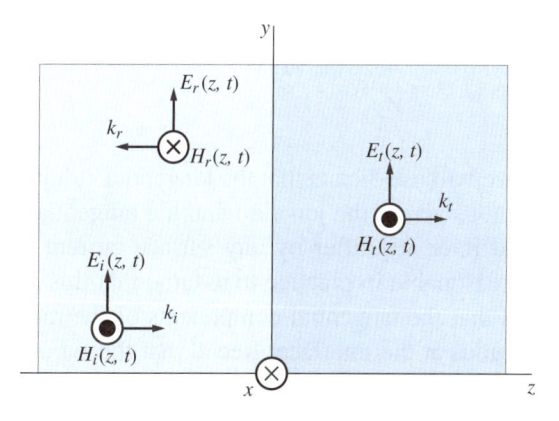

FIGURE 6–12

An electromagnetic wave $[E_i(z, t) \times H_i(z, t)]$ is normally incident upon an interface at $z = 0$. This results in an electromagnetic wave that is transmitted into the region $z > 0$ $[E_t(z, t) \times H_t(z, t)]$ and an electromagnetic wave that is reflected back into the region $z < 0$ $[E_r(z, t) \times H_r(z, t)]$. The x axis is into the paper.

Using the convention that arises from Poynting's theorem, we define the direction of propagation via the right-hand rule. This suggests that the polarization of one of the electromagnetic field components of the reflected field would have to be altered after the incident wave strikes the interface. Let us choose that component to be the magnetic field and assume that the electric field is unchanged.

The electric field components of the incident (the reflected and the transmitted electromagnetic wave) are respectively written from (6.35) as

$$E_{y,i}(z, t) = A_i e^{j(\omega t - k_1 z)}$$
$$E_{y,r}(z, t) = B_r e^{j(\omega t + k_1 z)} \quad (6.68)$$
$$E_{y,t}(z, t) = A_t e^{j(\omega t - k_2 z)}$$

where k_1 is the wave number in region 1 and k_2 is the wave number in region 2. We employ the constants A and B respectively to indicate the terms that propagate to increasing and decreasing values of the coordinate z that was introduced in (6.11). The additional subscripts i, r, and t indicate the incident, the reflected, and the transmitted terms, respectively. Since the materials were assumed to be lossless, the waves will not attenuate as they propagate ($\alpha = 0$ and $\beta = k$). The magnetic field intensities $H_x(z, t)$ for the three field components can be computed directly from Maxwell's equations or by using the characteristic impedances found in the two regions. We will use the simpler approach here. We write that

$$H_{x,i}(z, t) = -\frac{A_i}{Z_{c,1}} e^{j(\omega t - k_1 z)}$$
$$H_{x,r}(z, t) = \frac{B_r}{Z_{c,1}} e^{j(\omega t + k_1 z)} \quad (6.69)$$
$$H_{x,t}(z, t) = -\frac{A_t}{Z_{c,2}} e^{j(\omega t - k_2 z)}$$

The boundary conditions we will use dictate that the tangential components of the electric field be continuous across the junction and the tangential components of the magnetic field intensity differ by any surface current that is located at the interface. It is reasonable in practice to assume that this current is equal to zero. This implies that the tangential components of the magnetic field intensity are also continuous at the interface. Recall that these boundary conditions were discussed previously.

Reflection and Transmission of an Electromagnetic Wave

Before applying these boundary conditions, we should be aware of a trick that can be used in the calculation at this stage. By choosing that the interface between the two media should be located at $z = 0$, the exponential factor $e^{j(\omega t \pm kz)}\big|_{z=0} = e^{j\omega t}$ will cancel in all of the terms. This is nothing profound, but making this choice does lighten the burden. We use the same trick in our study of transmission lines.

At the boundary at $z = 0$, we therefore write that

$$E_{y,i}(z = 0, t) + E_{y,r}(z = 0, t) = E_{y,t}(z = 0, t)$$
$$H_{x,i}(z = 0, t) + H_{x,r}(z = 0, t) = H_{x,t}(z = 0, t) \quad (6.70)$$

Evaluating (6.68) and (6.69) at $z = 0$ and substituting them into these expressions, we obtain

$$A_i + B_r = A_t$$
$$\frac{A_i}{Z_{c,1}} - \frac{B_r}{Z_{c,1}} = \frac{A_t}{Z_{c,2}} \quad (6.71)$$

In order to proceed, we must assume that something is known—for instance, the amplitude of the incident electric field intensity A_i. In terms of this known quantity, we can find the other two terms from (6.71) to be

$$\boxed{\Gamma \equiv \frac{B_r}{A_i} = \frac{Z_{c,2} - Z_{c,1}}{Z_{c,1} + Z_{c,2}}} \quad (6.72)$$

and

$$\boxed{\mathrm{T} \equiv \frac{A_t}{A_i} = \frac{2 Z_{c,2}}{Z_{c,1} + Z_{c,2}}} \quad (6.73)$$

The symbols Γ and T are called the **reflection coefficient** and the **transmission coefficient**, respectively. The symbol T is also used to indicate a period of time. However, the difference should be clear from the context where the symbol is being used.

Hence, knowing the characteristic impedance of the materials lets us determine the propagation characteristics and amplitudes of both the wave that is transmitted into the second material and the wave that is reflected at the interface and propagates in the first material. If the characteristic impedances on both sides of the interface are equal, all of the incident electromagnetic energy is transmitted into region 2 and none is reflected back into region 1. This is called **matching** the media, and it has many practical applications.

Since the characteristic impedance $Z_c = \sqrt{\mu/\varepsilon}$, we can write the reflection and transmission coefficients in terms of the relative dielectric constants of the two dielectric materials. We find

$$\Gamma = \frac{\sqrt{\varepsilon_1} - \sqrt{\varepsilon_2}}{\sqrt{\varepsilon_1} + \sqrt{\varepsilon_2}} \qquad (6.74)$$

and

$$T = 1 + \Gamma = \frac{2\sqrt{\varepsilon_1}}{\sqrt{\varepsilon_1} + \sqrt{\varepsilon_2}} \qquad (6.75)$$

EXAMPLE 6.19

Describe the expected reflection-transmission characteristics of a time harmonic electromagnetic wave normally incident at a layered dielectric. Find the total reflection and transmission coefficients of a single layer with known thickness d and dielectric constant $\varepsilon_2 = \varepsilon_r \varepsilon_0$ ($\varepsilon_1 = \varepsilon_3 = \varepsilon_0$). Plot the frequency dependence of these coefficients assuming $d = 0.1$ m and $\varepsilon_r = 4$.

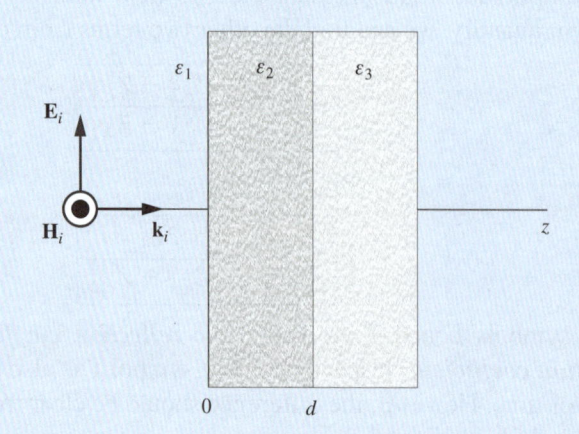

Answer. We will use the notation "+" to indicate a wave moving to increasing values of z and "−" to indicate a wave moving to decreasing values of z. The additional numerical subscript specifies the region in which the wave propagates. The superscript indicates the numerical iteration of reflection transmission. A portion of the electromagnetic wave that is incident from material 1 is transmitted into material 2, and a portion is reflected. The wave now propagating in material 2 encounters the interface separating materials 2 and 3. A portion of this wave is transmitted into material 3, and a portion is reflected back to the original interface. Once the wave enters material 3, it propagates to $z \to \infty$ since there are no other boundaries. The amplitudes of the total transmitted and reflected fields are given

Reflection and Transmission of an Electromagnetic Wave

by

$$E_t(z) = \sum_{n=1}^{\infty} E(z)_{+(3)}^{(n)}$$

$$E_r(z) = \sum_{n=1}^{\infty} E(z)_{-(1)}^{(n)}$$

where n is the index of the particular wave that traverses region 2. The amplitude of each individual component is determined from repeated applications of (6.74) and (6.75). The wave as it propagates is depicted below.

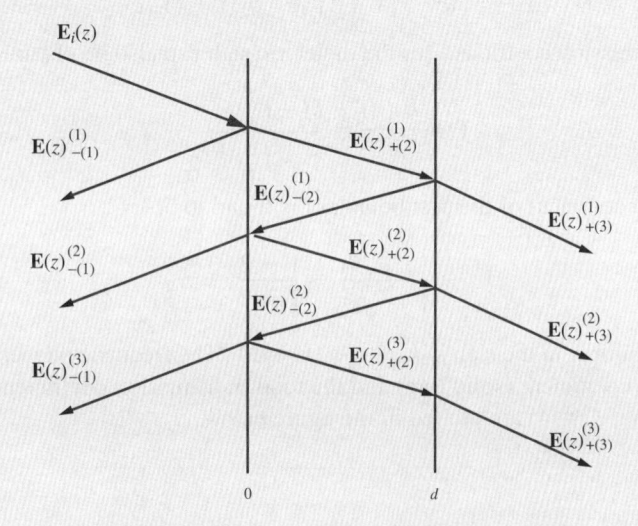

However, the phase of the individual terms in the summation is different— after every crossing of the dielectric slab, an additional phase difference ($k_2 d$) appears. Suppose the reflection coefficient from the first interface is Γ_1 in the $+z$ direction and $-\Gamma_1$ in the opposite direction, then $\Gamma_2 = -\Gamma_1$ in the $+z$ direction and $-\Gamma_2 = \Gamma_1$ in the opposite direction (for the particular case where medium 3 is identical to medium 1). Then the transmission coefficient through the first interface in the $+z$ direction is $T_1^+ = 1 + \Gamma_1$, while in the $-z$ direction, it is $T_1^- = 1 - \Gamma_1$. We write the following expression for the total reflected field:

$$E_r(z) = E_i \{ \Gamma_1 + T_1^+ \Gamma_2 T_1^- e^{j2k_2 d} + T_1^+ (-\Gamma_2 \Gamma_1) \Gamma_2 T_1^- e^{j4k_2 d}$$
$$+ T_1^+ (-\Gamma_2 \Gamma_1)^2 \Gamma_2 T_1^- e^{j6k_2 d} + \cdots \}$$
$$= E_i \{ \Gamma_1 - (1 - \Gamma_1^2) \Gamma_1 e^{j2k_2 d} [1 + (\Gamma_1^2 e^{j2k_2 d}) + (\Gamma_1^2 e^{j2k_2 d})^2 + \cdots] \}$$

The total reflection coefficient for the dielectric slab can be calculated by realizing that this equation can be written as

$$\Gamma \equiv \frac{E_r(z)}{E_i} = \Gamma_1\left[1 - \frac{(1-\Gamma_1^2)e^{j2k_2d}}{1-\Gamma_1^2 e^{j2k_2d}}\right] = \frac{\Gamma_1(1-e^{j2k_2d})}{1-\Gamma_1^2 e^{j2k_2d}}$$

Similarly, we obtain the following expression for the transmitted field.

$$E_t(z) = E_i\{T_1^+ T_2^+ e^{jk_2d} + T_1^+(-\Gamma_2\Gamma_1)T_2^+ e^{j3k_2d}$$
$$+ T_1^+(-\Gamma_2\Gamma_1)^2 T_2^+ e^{j5k_2d} + \cdots\}$$
$$= E_i\{(1-\Gamma_1^2)e^{jk_2d}[1 + (\Gamma_1^2 e^{j2k_2d}) + (\Gamma_1^2 e^{j2k_2d})^2 + \ldots]\}$$

The total transmission coefficient for the dielectric slab can also be obtained as

$$T \equiv \frac{E_t(z)}{E_i} = \frac{(1-\Gamma_1^2)e^{jk_2d}}{1-\Gamma_1^2 e^{j2k_2d}}$$

The reflection coefficient of the first boundary is found to be

$$\Gamma_1 = \frac{1-\sqrt{\varepsilon_r}}{1+\sqrt{\varepsilon_r}} = \frac{1-2}{1+2} = -\frac{1}{3}$$

and the wave number in the slab is $k_2 = \sqrt{\varepsilon_r}k = 2k$. The frequency dependence of the total reflection coefficient (solid line) and the total transmission coefficient (dashed line) ($d = 0.1$ m, $\varepsilon_r = 4$) are plotted in the figure below.

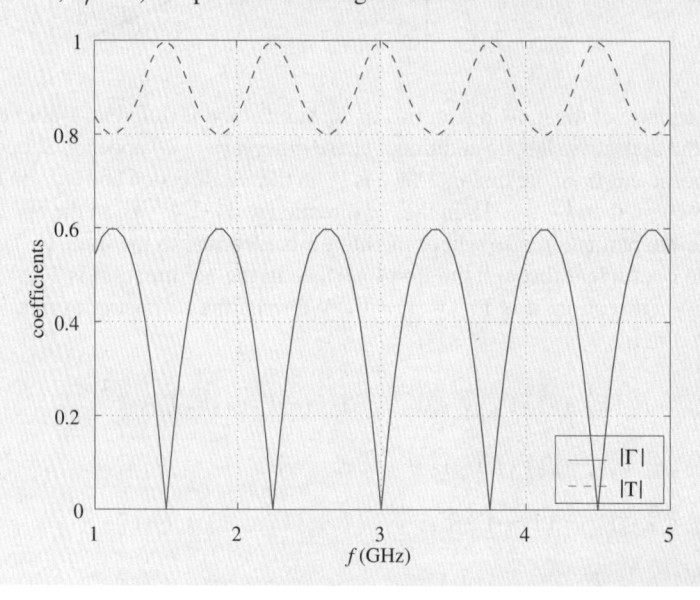

Reflection and Transmission of an Electromagnetic Wave

Choosing the frequency $f = 3$ GHz that corresponds to a free-space wavelength $\lambda = d = 0.1$ m, where the coefficients are $\Gamma = 0$ and $|T| = 1$ (a *transparent* slab), it is easy to check whether the following condition is fulfilled for every frequency.

$$|\Gamma|^2 + |T|^2 = 1$$

This results from the fact that the dielectric constant is a real value and no energy is absorbed in the slab—the entire incident energy is either reflected or transmitted.

If material 2 were a good conductor, a peculiar response would result. The characteristic impedance of the conductor would be very small, and it would approach zero as the conductivity approached infinity ($Z_{c,2} \to 0$). In this case, there would be *no* transmission of electromagnetic energy *into* the conductor or $T = 0$, which directly follows from (6.73). It would all be reflected. From (6.72) we compute that $\Gamma = -1$, which states that the polarization of the reflected electric field would be *opposite* of the incident one. Recall the warning on your microwave oven "Do not place a metal pan inside!"

This can best be demonstrated if we return to the most general solution of the wave equation given in (6.11). We will assume that the wave that arrives from $z = -\infty$ is a pulse instead of a time-harmonic wave and that it propagates toward the interface as shown in Figure 6–13(a) and (b). At $t = t + 3\Delta t$, the tangential electric field must be equal to zero as shown in Figure 6–13(c). For times $t > t + 3\Delta t$, a reflected pulse will propagate back to $z = -\infty$ as shown in Figure 6–13(d) and (e).

One could think of this reflection in the following terms. A ***virtual pulse*** with a negative amplitude is launched at $z = +\infty$ by a virtual pulse generator at

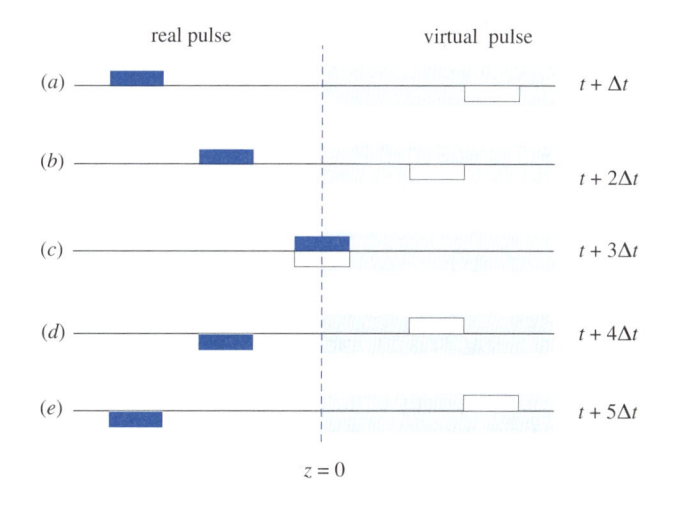

FIGURE 6–13

Propagation of a pulse from $z = -\infty$ and its location at five times.

the *same time* that the real pulse is launched from $z = -\infty$. This virtual pulse propagates to decreasing values of z with the same speed as the real pulse. At the time $t = t + 3\Delta t$, both pulses meet at $z = 0$. At $z = 0$, they *interchange* natures in that the virtual pulse now becomes a **real pulse** continuing its propagation to decreasing values of the coordinate z. The real pulse transforms into a virtual pulse that propagates into the conductor. At the time $t = t + 3\Delta t$, the amplitude of the two pulses add up to equal zero at the interface—in order to satisfy the requirement that the tangential component of the electric field must be equal to zero at a perfect conductor.

EXAMPLE 6.20

Pulse radars can be used to determine the velocity of speeding automobiles. Show how such a radar might work.

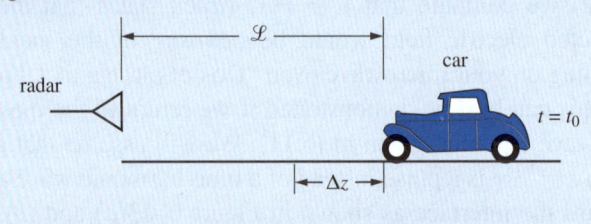

Answer. A repetitive electromagnetic pulse from the radar is incident upon the automobile. Due to the high conductivity of the car, it is reflected back to the radar, and the total time of flight $(\Delta t)_i$ of the ith pulse can be measured. The repetition frequency of the radar pulses is $(\delta t)^{-1}$. During the time δt, the car travels a distance Δz. The velocity of the automobile v_{car} can be computed from

$$(\Delta t)_i - (\Delta t)_{i+1} = \frac{2\mathscr{L}}{c} - \frac{2(\mathscr{L} - \Delta z)}{c} = \frac{2\Delta z}{c} = \frac{2v_{car}\delta t}{c}$$

or

$$v_{car} = \frac{c[(\Delta t)_i - (\Delta t)_{i+1}]}{2\delta t}$$

The actual distance of the car from the radar ($\mathscr{L}$) is not important. Since the total time of flight Δt and the intervals between the pulses δt are known (as they are parameters involving the radar set and the velocity of light c), the velocity of the car v_{car} can be computed. If the computed velocity is below the speed limit, no ticket needs to be issued.

6.5.2 Fabry–Perot Resonator—Standing Waves

Now let us return to investigating the behavior of a time-harmonic signal that is incident upon a perfectly conducting wall with a reflection coefficient $\Gamma = -1$. The total electric field in the region $z < 0$ will be equal to the sum of the incident and the reflected components, which we write as

$$E_y(z, t) = \text{Re}\{B[e^{j(\omega t - kz)} - e^{j(\omega t + kz)}]\} = \text{Re}\{-Be^{j\omega t}[e^{jkz} - e^{-jkz}]\}$$
$$= \text{Re}\{-j2Be^{j\omega t} \sin kz\}$$

or

$$E_y(z, t) = A \sin \omega t \sin kz \qquad (6.76)$$

where $A = 2B$. The tangential electric field $E_y(z, t) = 0$ at $z = 0$. In this case, the signal that consists of two oppositely propagating waves appears to be stationary in space and merely oscillating in time. This is called a **standing wave**. This standing wave results from the constructive and destructive interference of the two counter-propagating waves. This is shown in Figure 6–14 at equal intervals of time ($\Delta t = T/12$, $T = 2\pi/\omega$). The separation distance between successive null points called **nodes** is equal to one-half of the wavelength λ. The point where the electric field is a maximum is called an **antinode**.

If we moved a small detecting probe that responded only to the magnitude of the electric field and whose response time was slow with respect to the

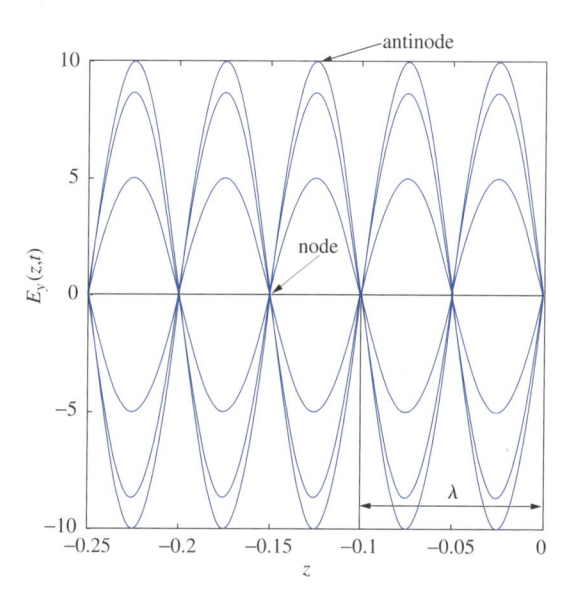

FIGURE 6–14

Standing wave depicted at equal temporal intervals in the period of oscillation ($A = 10$ V/m, $\lambda = 0.1$ m).

sinusoidal oscillation along the z axis, the response of the slowly moving probe would spatially alternate between zero and a maximum value and back to zero as the probe moved over a distance equal to one-half of a wavelength. In practice, a probe that would respond to the magnitude of the electric field would be calibrated; then, we could *measure* the wavelength with a high degree of accuracy. It would also be connected to a diode detector. Such a detector is sometimes is called a "square-law" detector.

EXAMPLE 6.21

An electromagnetic wave propagating in a vacuum in the region $z < 0$ is normally incident upon a perfect conductor located at $z = 0$. The frequency of the wave is 3 GHz. The amplitude of the incident electric field is 10 V/m, and it is polarized in the $\mathbf{u_y}$ direction. Determine the phasor and the instantaneous expressions for the incident and the reflected field components.

Answer. At the frequency $f = 3$ GHz, we compute

$$\omega = 2\pi f = 6\pi \times 10^9 \text{ rad/s}$$

$$k = \frac{\omega}{c} = \frac{6\pi \times 10^9}{3 \times 10^8} = 20\pi \text{ 1/m}, \quad \mathbf{k} = k\mathbf{u_z}$$

In *phasor* notation, the incident wave is expressed as

$$\mathbf{E_i}(z) = 10 e^{-j20\pi z} \mathbf{u_y} \text{ V/m}$$

$$\mathbf{H_i}(z) = \frac{1}{Z_c}(\mathbf{u_z} \times \mathbf{E_i}(z)) = -\frac{10}{120\pi} e^{-j20\pi z} \mathbf{u_x} \text{ A/m}$$

where the characteristic impedance of free space is $Z_0 = 120\pi\ \Omega$. The fields can also be written as

$$\mathbf{E_i}(z, t) = \text{Re}\{\mathbf{E_i}(z)e^{j\omega t}\} = 10 \cos(6\pi \times 10^9 t - 20\pi z)\mathbf{u_y} \text{ V/m}$$

$$\mathbf{H_i}(z, t) = \text{Re}\{\mathbf{H_i}(z)e^{j\omega t}\} = -\frac{10}{120\pi} \cos(6\pi \times 10^9 t - 20\pi z)\mathbf{u_x} \text{ A/m}$$

In phasor notation, the reflected wave is expressed as

$$\mathbf{E_r}(z) = -10 e^{+j20\pi z} \mathbf{u_y} \text{ V/m}$$

$$\mathbf{H_r}(z) = \frac{1}{Z_c}(-\mathbf{u_z} \times \mathbf{E_r}(z)) = -\frac{10}{120\pi} e^{+j20\pi z} \mathbf{u_x} \text{ A/m}$$

Reflection and Transmission of an Electromagnetic Wave

The reflected fields can also be written as

$$\mathbf{E_r}(z, t) = \text{Re}\{\mathbf{E_r}(z)e^{j\omega t}\} = -10\cos(6\pi \times 10^9 t + 20\pi z)\mathbf{u_y} \text{ V/m}$$

$$\mathbf{H_r}(z, t) = \text{Re}\{\mathbf{H_r}(z)e^{j\omega t}\} = -\frac{10}{120\pi}\cos(6\pi \times 10^9 t + 20\pi z)\mathbf{u_x} \text{ A/m}$$

After the constructive and destructive interference occurs, standing waves appear as shown in Figure 6–14. The separation between the nodes in the standing wave is $\lambda/2 = 5$ cm ($\lambda = 2\pi/k = 2\pi/20\pi = 0.1$ m).

An examination of the standing wave depicted in Figure 6–14 leads us to conjecture that it should be possible to insert another high-conductivity metal wall at any of the nodes where the tangential electric field is equal to zero *without* altering the remaining electric field structure. The applicable boundary condition is that the tangential electric field must be *zero* at a conducting surface. This conjecture is depicted in Figure 6–15, where plates have been inserted at two of the many possible locations. For the moment, we will assume that the plates, which are infinite in transverse extent, are instantaneously inserted at the nodes such that the electromagnetic energy is "trapped" between the plates and nothing else is disturbed. This energy is actually "coupled" between the plates with an antenna structure, a topic to be discussed later.

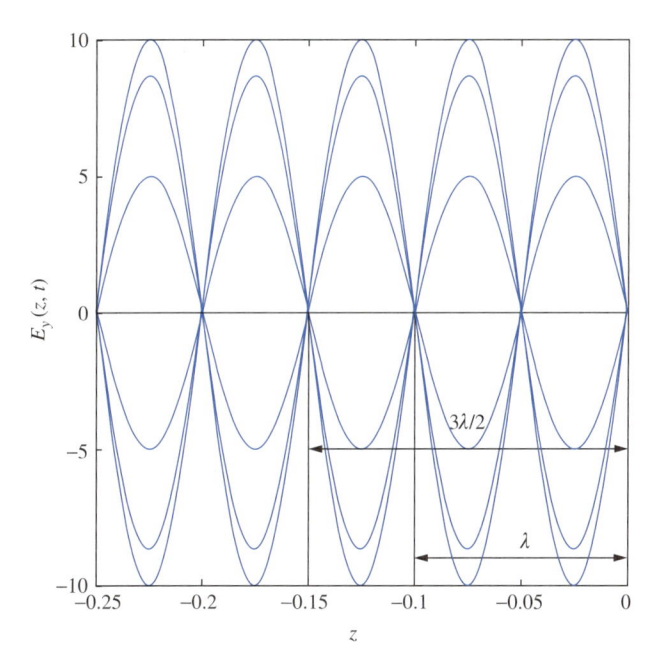

FIGURE 6–15

By inserting thin conducting plates separated by $(n\lambda/2)$ at the locations where the standing wave is zero, the electromagnetic field structure will not be altered. Two of many possible locations are indicated in the figure.

Let us now formally derive this result using the one-dimensional Helmholtz equation (6.33), which we rewrite as

$$\frac{d^2 E_y(z)}{dz^2} + k^2 E_y(z) = 0 \tag{6.77}$$

Recall that we have assumed a time-harmonic signal. The solution of this equation is given by

$$E_y(z) = A \sin kz + B \cos kz \tag{6.78}$$

The constants of integration A and B are specified by the boundary condition that the tangential electric field must be equal to zero at a metal wall. These determine the constant $B = 0$ and $k = (n\pi/\mathscr{L})$, where n is an integer and $\mathscr{L}$ is the distance between the metal walls. If the maximum electric field has a magnitude E_{y0}, then the spatial distribution of the electric field is given by

$$E_y(z) = E_{y0} \sin\left(\frac{n\pi z}{\mathscr{L}}\right) \tag{6.79}$$

The parallel plate cavity depicted in Figure 6–15 is called a **Fabry–Perot resonator or Fabry–Perot cavity**. This cavity has a very high Q that could approach one million. Remember that the Q of an ordinary electrical circuit is of the order of ten.[7] Since it is very frequency selective, it has received wide application as the cavity that encloses various "lasing" materials. The total lasing-material-cavity entity carries the acronym *laser* for "light amplification by stimulated emission of radiation." At light frequencies, it is fair to assume that the transverse dimension measures so large a number of wavelengths that it can be approximated as infinity.

We recall that the wave number k is a function of the frequency of oscillation ω and the velocity of light in the region between the two parallel plates $c = 1/\sqrt{\varepsilon \mu_0}$ where $\varepsilon = \varepsilon_r \varepsilon_0$. For the cavities depicted in Figure 6–15, this resonant frequency $\omega = \omega_r$ will be given by

$$\frac{\omega_r}{c} = k = \frac{n\pi}{\mathscr{L}}$$

or

$$\omega_r = \frac{n\pi}{\mathscr{L}} \frac{1}{\sqrt{\varepsilon \mu_0}} \tag{6.80}$$

---

[7] The Q of a circuit or a cavity is defined as $Q = 2\pi$ (energy stored)/(power dissipated per cycle).

Reflection and Transmission of an Electromagnetic Wave

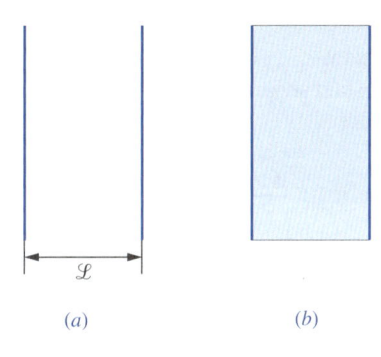

FIGURE 6–16

(*a*) An empty Fabry–Perot cavity.
(*b*) A Fabry–Perot cavity filled with a dielectric $\varepsilon = \varepsilon_r \varepsilon_0$.

(*a*)   (*b*)

For the two cavities depicted in Figure 6–16, which are either empty or filled with a dielectric, we find that the two Fabry–Perot cavities will resonate with slightly different frequencies. The difference between these two frequencies $\Delta \omega$ is given by

$$\Delta \omega = \omega_{ra} - \omega_{rb} = \frac{n\pi}{\mathscr{L}} \frac{1}{\sqrt{\varepsilon_0 \mu_0}} - \frac{n\pi}{\mathscr{L}} \frac{1}{\sqrt{\varepsilon_r \varepsilon_0 \mu_0}}$$

If the resonant frequency for the vacuum case (Figure 6–16*a*) can be computed or measured, this frequency difference can be written as

$$\frac{\Delta \omega_r}{\omega_{ra}} = 1 - \frac{\omega_{rb}}{\omega_{ra}} = 1 - \frac{1}{\sqrt{\varepsilon_r}} \tag{6.81}$$

EXAMPLE 6.22

An empty microwave Fabry–Perot cavity has a resonant frequency of 35 GHz. Determine the thickness $\Delta \mathscr{L}$ of a sheet of paper that is then inserted between the plates if the resonant frequency changes to 34.99 GHz. The separation $\mathscr{L}$ between the parallel plates is 50 cm. Assume that the integer n that specifies the mode does not change. You may ignore any reflection at the paper interface.

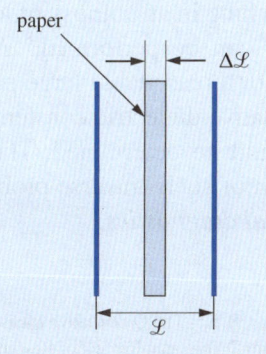

Answer. The relative dielectric constant ε_{paper} of paper as determined from Appendix B is $\varepsilon_{paper} \approx 3$. The relative dielectric constant separating the plates with the paper inserted can be approximated as

$$\varepsilon_r \approx \frac{(\mathscr{L} - \Delta\mathscr{L})}{\mathscr{L}} + \varepsilon_{paper}\frac{\Delta\mathscr{L}}{\mathscr{L}} = 1 + (\varepsilon_{paper} - 1)\frac{\Delta\mathscr{L}}{\mathscr{L}}$$

Therefore, we write

$$\frac{\omega_{vacuum} - \omega_{paper\ inserted}}{\omega_{vacuum}} = 1 - \frac{1}{\sqrt{\varepsilon_r}}$$

$$= 1 - \frac{1}{\sqrt{1 + (\varepsilon_{paper} - 1)\frac{\Delta\mathscr{L}}{\mathscr{L}}}} \approx \frac{\Delta\mathscr{L}}{2\mathscr{L}}(\varepsilon_{paper} - 1)$$

Inserting the values, we compute

$$\frac{35 - 34.99}{35} = \frac{0.01}{35} \approx \frac{\Delta\mathscr{L}}{2 \times 50}(3 - 1)$$

or $\Delta\mathscr{L} \approx 0.014$ cm.

From this example and the example mentioned earlier, we can discern that high-frequency electromagnetic waves can be used in the diagnostics of various materials. This is a practical technique that has received wide currency in manufacturing paper where the ratio of less expensive water to the more costly wood pulp determines the ultimate grade of the paper. The relative dielectric constants of wood pulp and water are different.

Medical diagnostics to determine the ratio of the diseased to the undiseased portion of a lung in an autopsy of a patient who died—say, of pulmonary emphysema—can be performed.[8] Assuming that one of the lungs is sufficiently dried or a reasonably large portion of one could be used to yield a value for the relative dielectric constant for the lung, the percentage of the diseased lung could be determined. The disease has "eaten" holes in the lung. The solution of such "inverse problems" is part of an interesting new area called ***medical diagnostics***.

---

[8] Jutabha, O. and Lonngren, K.E., "The Quantitative Measurement of Emphysema Using a Microwave Technique—A Feasibility Study," American Review of Respiratory Diseases, Vol. 99, 1969, pp. 101–103.

6.6 Conclusion

The propagation of plane electromagnetic waves encountered in this chapter has provided us with the first application of Maxwell's equations. Electromagnetic waves are transverse waves in that the electric and magnetic fields are in a plane that is perpendicular to the direction of propagation. This is similar to waves that propagate on the surface of water or along a string and contrasts with sound waves, which are longitudinal. The velocity of propagation c is determined by the materials in which the wave is propagating, with the highest velocity being in a vacuum $c = 1/\sqrt{\varepsilon_0 \mu_0} \approx 3 \times 10^8$ m/s. The ratio of the electric field to the magnetic field intensity is given by the characteristic impedance, which in a vacuum has a value of $Z_0 = 120\pi \approx 377\ \Omega$. Time-harmonic waves have certain unique features such as wavelengths, frequencies of oscillation, and phase velocities, but the general wave properties remain. Lossy materials will attenuate the wave as it propagates.

If the wave propagates from one material to another, a portion of the incident wave will be reflected back into the first material, and a portion will be transmitted into the second material. Good conductors will reflect most of the incident wave. The boundary conditions will determine the amplitude of each of these terms. If the wave reaches a knife-edge boundary of a conducting screen, the wave will change its direction of propagation or it will be diffracted by the knife edge. The diffracted wave will then cause an oscillation of the amplitude of the wave in the illuminated region near the line of sight. However, a portion of this wave will also appear in the shadow region behind the screen.

6.7 Problems

6.1.1. In terms of the standard units mass kg, length m, time s, and charge C, show that $(\varepsilon_0 \mu_0)^{-1/2}$ has the units of a velocity.

6.1.2. Prove that $\mathbf{E}(z, t)$ and $\mathbf{H}(z, t)$ are orthogonal in a vacuum for an arbitrary function of $(z - ct)$.

6.2.1. Let $F(z - ct) = 1$ and $G(z + ct) = -2$ for $|z - ct| \leq 1$ and $|z + ct| \leq 1$, respectively, and $F(z - ct) = G(z + ct) = 0$ elsewhere. Accurately sketch the pulse with the velocity $c = 2$ at three times: $t = 0, t = 1,$ and $t = 3$.
Note: Normalized variables are used.

6.2.2. Define the functions $F(z - v_1 t)$ and $G(z + v_2 t)$ from the following sketch, which was drawn at the times $t = 0$ and $t = 2$.

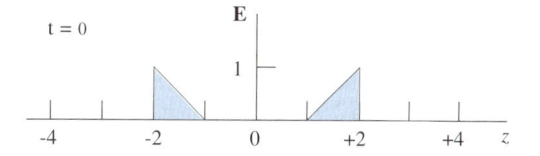

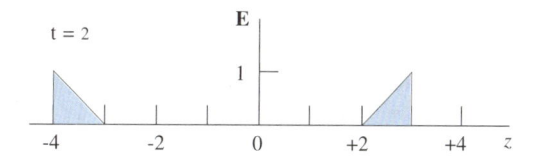

6.2.3. If the waves in Problem 6.2.2 were electromagnetic waves, find the ratio of the dielectric constants ε_2 and ε_1 for the two regions ($z < 0$)/($z > 0$) if the relative permeabilities were equal to one in the two regions.

6.2.4. A displacement wave on a string is described by a harmonic wave $\psi(z, t) = 0.02\sin[2\pi(10t - 0.5z)]$ m where z is in meters and t is in seconds. Find
(a) Propagation velocity v
(b) Wavelength λ and wave number k
(c) Frequency f and angular frequency ω
(d) Period T
(e) Direction of propagation
(f) Amplitude of the wave A

6.2.5. Plot the wave given in Problem 6.2.4 as a function of z at $t =$ (a) 0 s, (b) 0.025 s, (c) 0.05 s, and (d) 0.075 s. Convince yourself that the wave pattern progresses in the $+z$ direction as time increases.

6.2.6. Show that the Gaussian pulse defined by $\psi(z, t) = 0.5e^{-(z-5t)^2}$ satisfies a wave equation (6.10). Plot this function as a function of z for the three times: $t = 0$, 0.5 s, and 1 s.

6.2.7. Repeat Problem 6.2.6 with $h/c\tau = 1/2$.

6.2.8. Repeat Problem 6.2.6 with $h/c\tau = 3/2$.

6.2.9. Show that Maxwell's equations can be cast in the form of a one-dimensional diffusion equation (provided the direction of propagation is z)

$$\frac{\partial^2 E_y(z, t)}{\partial z^2} - \mu\sigma\frac{\partial E_y(z, t)}{\partial t} = 0$$

Describe when this derivation might be valid.

6.2.10. Using the substitution $\xi = z/\sqrt{Dt}$ and the chain rule, show that the partial differential equation given in Problem 6.2.9 will transform into an ordinary differential equation. Find the units of the diffusion coefficient $D = 1/\mu\sigma$.

6.2.11. Solve the ordinary differential equation obtained in Problem 6.2.10 with the boundary conditions that the electric field goes to 0 as $z \to \infty$ and the integral over all space of the electric field is a constant. In addition, the electric field in the region external to the point where it is excited is equal to zero just after the excitation. This can be interpreted as the diffusion of a "pulse."

6.2.12. The initial condition at $t = 0$ for a wave is

$$\psi(z, 0) = 5e^{-10z}$$

Write a MATLAB program to show the propagation of this wave in the region $0 \leq z \leq 1$ if $c = 1$.

6.2.13. The initial condition at $t = 0$ for a wave is

$$\psi(z, 0) = 5e^{-10z}$$

Write a MATLAB program to show the propagation of this wave in the region $0 \leq z \leq 2$ if $c = 2$.

6.2.14. The initial condition at $t = 0$ for a wave is

$$\psi(z, 0) = 5e^{-10z}$$

Write a MATLAB program to show the propagation of this wave in the region $0 \leq z \leq 0.5$ if $c = 0.5$.

6.3.1. Snapshots of two cycles of the electromagnetic wave propagating in a vaccum are taken at three locations: $z = 0$, $z = 1$, and $z = 2$ m. Find the wavelength and the frequency of the wave. Write the equation that describes the electric field.

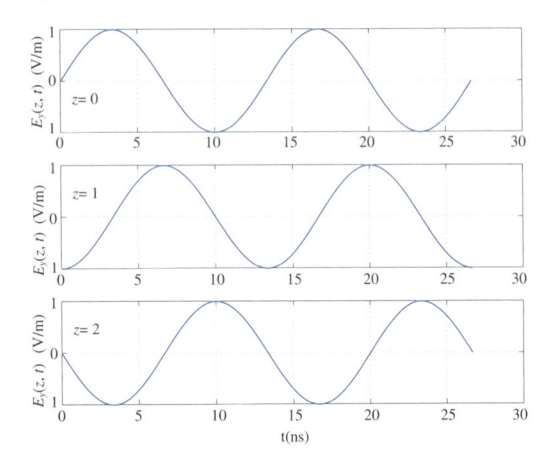

6.3.2. The electric field of a uniform plane wave propagating in air is given by $\mathbf{E}(z, t) = E_0[\sin(\omega t - kz)\mathbf{u_x} + \cos(\omega t - kz)\mathbf{u_y}]$ V/m. Using a sketch, show that it is justified to call this wave circularly polarized.

6.3.3. The electric field of a uniform plane wave propagating in air is given by $\mathbf{E}(z, t) = E_0[\sin(\omega t - kz)\mathbf{u_x} + a\sin(\omega t - kz + \delta)\mathbf{u_y}]$ V/m ($a \neq 1$). Using an accurately drawn sketch with $E_0 = 1$, $a = 2$, and $d = \pi/4$,

6.7 Problems

show that it is justified to call this wave elliptically polarized.

6.3.4. Show that the circular polarization and the linear polarization are special cases of an elliptical polarization.

6.3.5. If we know that the magnetic field intensity of an electromagnetic wave is $\mathbf{H}(z, t) = H_0 e^{j(\omega t + kz)} \mathbf{u}_y$ A/m, find the direction of power flow, the electric field $\mathbf{E}$, and the time-average Poynting vector $\mathbf{S}_{av}$.

6.3.6. The electric field of an electromagnetic wave is $\mathbf{E}(z, t) = -10 e^{j(\omega t - kz)} \mathbf{u}_x$ V/m. Find the magnetic field intensity $\mathbf{H}(z, t)$. Compute the time-average Poynting vector $\mathbf{S}_{av}$.

6.4.1. An electromagnetic wave with a frequency $f = 1$ MHz propagates in a dielectric material ($\varepsilon_r = 4$, $\mu_r = 1$) and has an electric field component $\mathbf{E}_y(z, t) = 1.3 \cos(\omega t - kz)$ V/m. Find the velocity of the wave $\mathbf{v}$, the wave vector $\mathbf{k}$, the characteristic impedance of the material Z_c, and the magnetic field $\mathbf{H}(z, t)$.

6.4.2. In free space, a signal generator launches an electromagnetic wave that has a wavelength of 10 cm. As the same wave propagates in a material, its wavelength is reduced to 8 cm. In the material, the amplitude of the electric field $\mathbf{E}(z)$ and the magnetic field intensity $\mathbf{H}(z)$ are measured to be 50 V/m and 0.1 A/m, respectively. Find the generator frequency f, ε_r, and μ_r for the material.

6.4.3. In free space, a signal generator launches an electromagnetic wave that has a wavelength of 3 cm. As the same wave propagates in a material, its wavelength is reduced to 1.5 cm. In the material, the amplitude of the electric field $\mathbf{E}(z)$ and the magnetic field intensity $\mathbf{H}(z)$ are measured to be 60 V/m and 0.1 A/m, respectively. Find the generator frequency f, ε_r, and μ_r for the material.

6.4.4. Find the attenuation constant α (Np/m), the phase constant β (rad/m) and the phase velocity v_ϕ if the conductivity σ of the material is such that $\sigma = \omega\varepsilon$ (the material parameters are $\mu_r = 1$ and $\varepsilon_r = 2.5$). The wavelength in free space is $\lambda_0 = 30$ cm.

6.4.5. Convert the phase constant β into °/m and the attenuation constant α into dB/m for Problem 6.4.4.

6.4.6. At what frequencies can Earth be considered a perfect dielectric if $\sigma = 5 \times 10^{-3}$ S/m, $\mu_r = 1$, and $\varepsilon_r = 8$? Can α be neglected at these frequencies? Find the characteristic impedance Z_c?

6.4.7. Find the skin depth δ at a frequency 4 MHz in aluminum, where $\sigma = 3.82 \times 10^7$ S/m and $\mu_r = \varepsilon_r = 1$. Also find the phase velocity v_ϕ.

6.4.8. Compute the skin depth of copper, graphite, and seawater at $f = 2.45$ GHz.

6.5.1. A fisherman in the sea detects a fish at a depth d with a radar operating at a frequency f. Find d if the delay time of the reflected signal is $\tau = 20$ ns. Assume that fish scales are perfect conductors and the conductivity of the water σ satisfies the condition $\sigma \ll \omega\varepsilon$.

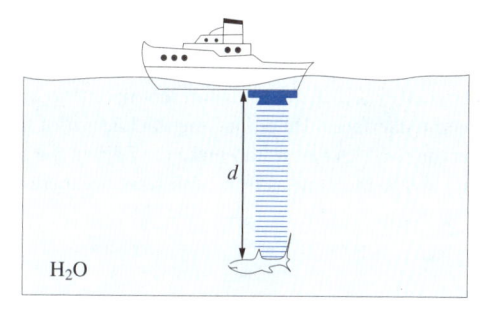

6.5.2. Assume that a wave reflector were installed at $z = 5$ in Problem 6.2.5. This reflector causes a positive amplitude pulse to be reflected as a negative amplitude pulse (reflection coefficient $\Gamma = -1$). Reflection implies that a wave traveling to increasing values of z would start traveling to decreasing values of z after reflection. Accurately sketch the expected oscilloscope pictures.

6.5.3. An electromagnetic wave with an amplitude of 1 V/m is normally incident from a vacuum into a dielectric having a relative dielectric constant $\varepsilon_r = 4$. Find the amplitude of the reflected and the transmitted electric fields and the incident, reflected, and transmitted powers.

6.5.4. An electromagnetic wave with an amplitude of 1 V/m is normally incident from a vacuum into a dielectric having a relative dielectric constant $\varepsilon_r = 4$. Find the amplitude of the reflected and the transmitted magnetic field intensities.

6.5.5. An electromagnetic wave with an amplitude of 1 V/m is normally incident from a dielectric having a relative dielectric constant $\varepsilon_r = 4$ into a vacuum. Find the amplitude of the reflected and the transmitted electric fields and the incident, reflected, and transmitted powers.

6.5.6. An electromagnetic wave with an amplitude of 1 V/m is normally incident from a dielectric having a relative dielectric constant $\varepsilon_r = 4$ into a vacuum. Find the amplitude of the reflected and the transmitted magnetic field intensities.

6.5.7. In Example 6.20, a speeder pleads to the judge that because of inclement weather when the radar was tested and calibrated, the calibration was incorrect. If the radar assumed a calibration in a vacuum and said the speeder was traveling at 25% over the speed limit, what would the dielectric constant of ambient space have to be in order that the defendant would go free?

6.5.8. A plane electromagnetic wave $\mathbf{E}(z, t) = E_0 \cos(\omega t - kz)\mathbf{u}_y$ V/m is incident upon two air–dielectric interfaces. Determine the thickness d of the dielectric slab (ε_r) that would make the field in the region $z < 0$ be the same as if the slab were not there.

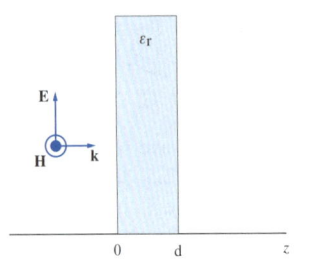

6.5.9. A dielectric slab (ε_r) is inserted between two plane wave launching horns. Waves will be reflected and transmitted at each interface. Determine the ratio of $E_B(z)/E_A(z)$ for a wave that passes through the region B.

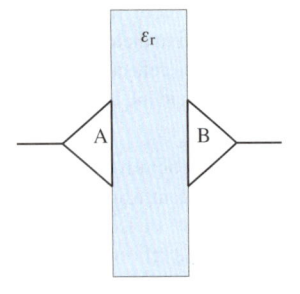

6.5.10. A dielectric that is $\lambda/4$ thick separates two dielectrics. Find the values of ε_r so that none of the power launched from A will be reflected back to A.

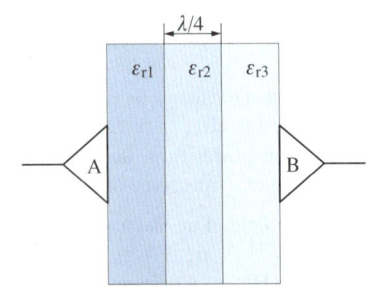

6.5.11. A time-harmonic electromagnetic wave in a vacuum is incident upon an ideal conductor located at $z = 0$, and a standing wave is created in the region $z < 0$. With a crystal detector connected to a volt meter, we measure a null voltage at equal increments of 10 cm in the region $z < 0$. Find the frequency of oscillation of the electromagnetic wave.

6.5.12. A helium–neon laser emits light at a wavelength of 6328 Å $= 6.328 \times 10^{-7}$ m in air. Calculate the frequency of oscillation of the laser, the period of the oscillation, and the wave number. The symbol Å is called an Ångstrom where 1 Å $= 10^{-10}$ m.

6.5.13. Estimate the number of wavelengths of helium–neon laser light ($\lambda = 6328$ Å where 1 Å $= 10^{-10}$ m) that can be found between the two parallel end plates that are separated by 1 m. You may assume $\varepsilon_r = 1$ between the end plates.

6.5.14. The resonant frequency of a Fabry–Perot cavity caused by the introduction of a dielectric is changed from its vacuum value of 10 GHz to 9.9 GHz. Calculate the relative dielectric constant ε_r of the perturbing material.

6.5.15. The relative dielectric constant of a slice of lung of thickness $\Delta \mathscr{L}$ is found to be 1.5. A diseased lung of thickness $\Delta \mathscr{L}$ as shown in Figure 6–16 is inserted between the plates of a Fabry–Perot cavity. The cavity has a resonant frequency of 9.9 GHz for the undiseased lung and 9.95 GHz for the diseased lung. Find the percentage of the diseased lung that has been eaten away by emphysema.

CHAPTER 7

Transmission Lines

7.1	Equivalent Electrical Circuits	354
7.2	Transmission Line Equations	357
7.3	Sinusoidal Waves	362
7.4	Terminations	367
7.5	Impedance and Matching of a Transmission Line	373
7.6	Smith Chart	381
7.7	Transient Effects and the Bounce Diagram	390
7.8	Pulse Propagation	397
7.9	Lossy Transmission Lines	402
7.10	Dispersion and Group Velocity	406
7.11	Conclusion	414
7.12	Problems	414

In this chapter we model three electrical transmission systems: a coaxial cable, a strip line, and two parallel wires, which can be used to transfer both signals and power. Our models will employ familiar electrical circuit elements: inductors, capacitors, and resistors. The equations that describe the propagation of the voltage and current signals along these *transmission lines* are called the ***telegraphers' equations***. These equations were obtained in the mid-to-late 1800's to describe the propagation of telegraph signals along the long copper cables connecting the United States with Europe under the Atlantic Ocean, as well as spanning the United States on poles from coast to coast. From the telegraphers' equations we can derive another important equation called the ***wave equation***. This equation describes how

---

Note to Students: Your instructor may have chosen to introduce this chapter very early in your course. No need to be anxious! It is very common for this approach to be taken as a logical extension of what you have recently learned in your circuit theory course. Every precaution has been taken so that the material in the chapter can be studied without reference to and assumed knowledge of the previous chapters. For more information about this approach, see the detailed comments in the Preface.

waves propagate from one point to another, and will turn out to be valid whether the waves move along a transmission line or are radiated into space.

However, although we are using familiar circuit components, under certain circumstances we require several extensions to the circuit theory we have previously studied. First, the energy transfer associated with the signals moving through our circuit model takes a small, but measurable, amount of time. Thus, we need to incorporate the propagation time into our equations explicitly to account for this effect. Second, instead of using the common discrete or "lumped" electrical components that we are familiar with from circuit theory, we will be using distributed circuit elements to explain why the signals require a finite amount of time to move through the circuit and the impact of that finite propagation time on circuit operation. Additionally, these finite propagation delays will help us to understand such concepts as characteristic impedance, signal or wave reflections, and matching techniques. These concepts are very important in understanding the proper operation of modern integrated circuits, especially in the case of VLSI circuits or microprocessors, where we want to move signals through very complicated circuits in a very short time and must compensate for the physical effects of these delays.

In our initial discussion of transmission lines we assume that the wave suffers no loss of energy as it propagates and, therefore, do not include any resistors in the introductory model. Loss terms (resistances and conductances) will be introduced after we develop and explore the lossless transmission line model.

7.1 Equivalent Electrical Circuits

The three common types of transmission lines in use are depicted in Figure 7–1. These are (*a*) the coaxial cable—this structure usually has a dielectric that separates the inner conductor from the outer one, (*b*) the strip line or microstrip line—this structure has an insulator, or dielectric, that separates the two flat conductors, and (*c*) two parallel conducting wires or the ***twin lead***—this structure may have air or a dielectric that separates the two wires. The dielectric in these three

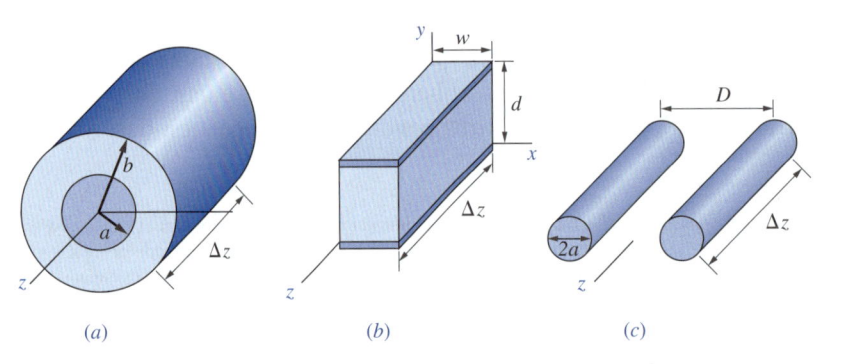

FIGURE 7–1

Three common transmission lines. (*a*) Coaxial cable. (*b*) Microstrip line. (*c*) Two-wire line (sometimes called a twin lead).

7.1 Equivalent Electrical Circuits

types of transmission lines is used to maintain a constant separation between the metallic elements so that the electrical properties of the transmission lines are constant. In our everyday experiences, we recognize these lines, respectively, as being transmission lines that (a) connect the cable TV into the house, (b) connect two components in an integrated circuit within the TV, and (c) provide the connection between a TV set and the external antenna.

Rather than fully examining the electromagnetic field distribution within these transmission lines we will, for now, simplify our discussion by using a simple model consisting of distributed inductors and capacitors. This model will be valid if any dimension that is transverse to the direction of propagation is much less than the free space wavelength. If the dimension is comparable with the wavelength, then a more complicated analysis will be required that may involve various numerical tools.

In the three transmission lines depicted in Figure 7–1, such an analysis would show that the three lines support the propagation of a wave that has both the electric field intensity and the magnetic field intensity in a plane that is transverse to the direction of propagation. This is sometimes called a ***transverse electromagnetic*** (TEM) mode of propagation. The analysis would require that there be no losses in either the dielectric separating the conductors or in the conductors themselves. To make this simplification in the mode structure, we first have to define the capacitance C and inductance L in terms of electromagnetic fields. The detailed calculation of the equivalent electrical circuit elements for these three transmission lines using electromagnetic field theory is presented in Chapters 2 and 3 and Appendix D. At the present time we will use the results of these calculations and our knowledge of circuit theory to understand the implications of these results. The parameters for the three transmission lines are summarized in Table 7–1. We note that

TABLE 7–1 Electrical Circuit Elements for Transmission Lines in Figure 7–1*

Line	Inductance	Capacitance
Coaxial cable	$L = \dfrac{\mu}{2\pi} \ln\left(\dfrac{b}{a}\right) \Delta z$	$C = \dfrac{2\pi\varepsilon}{\ln(b/a)} \Delta z$
Microstrip line	$L = \dfrac{\mu d}{w} \Delta z$	$C = \dfrac{\varepsilon w}{d} \Delta z$
Twin lead	$L = \dfrac{\mu}{\pi} \cosh^{-1}\left(\dfrac{D}{2a}\right) \Delta z$	$C = \dfrac{\pi\varepsilon}{\cosh^{-1}(D/2a)} \Delta z$

*For the microstrip line, we are using simplified expressions. A more accurate expression could be obtained using numerical techniques that would include the effects of fringing fields. The parameters for the material between the two conductors are the permeability $\mu = \mu_r \mu_0$ and the permittivity $\varepsilon = \varepsilon_r \varepsilon_0$.

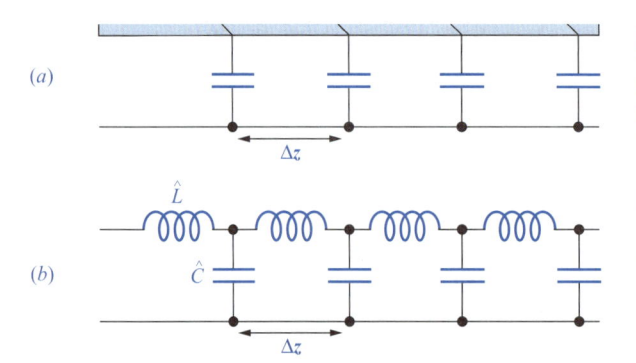

FIGURE 7–2

(*a*) Distributed transmission line.
(*b*) Equivalent circuit of this transmission line. The circuit elements are given in their per-unit-length values. In each section of length Δz, the values are $L = \hat{L}\,\Delta z$ and $C = \hat{C}\,\Delta z$, respectively. Hence transmission line models can be easily constructed in the laboratory.

the factor Δz, which represents a short distance containing the distributed circuit parameter, can be separated from the other terms for the three structures.

We can model, therefore, the three transmission lines depicted in Figure 7–1 with an equivalent circuit consisting of an infinite number of distributed inductors and capacitors as shown in Figure 7–2a. We could think of constructing such a distributed line by wrapping wire uniformly around a broomstick and locating it at a constant distance above a ground plane. In addition to the uniformly distributed inductance, there would be a uniformly distributed capacitance between the wires of the coil and the ground plane. If the spirit moved us, we could actually carry out the tedious task of soldering a very large number of uniformly distributed capacitors from the coil to the ground plane.

The distributed transmission line model depicted in Figure 7–2 has incorporated some obvious simplifications. In particular, there are no elements that would describe any loss of energy as the wave propagates through the transmission line. Loss can be incorporated with a resistor in series with the inductor or with a conductance in parallel with the capacitor. The effects of these additional elements are described later. In addition, parasitic capacitances exist between the wires that constitute the distributed inductance in Figure 7–2a. We assume these capacitances initially will be very small, and we can neglect them at this stage of our discussion. We will later see that the inclusion of these elements gives rise to a wave number, and hence a phase velocity, that is dependent upon the frequency of the wave. This phenomenon is called ***dispersion***, and it is discussed at the end of this chapter. The equivalent circuit that we will now use is depicted in Figure 7–2b. The reader who is comfortable with circuit theory could consider the transmission line as a large number of distributed two-port networks and would be guided by well-worn and understood techniques. We choose, however, not to follow this path, preferring to interpret the signals in terms of waves. In particular, we choose the path that leads to a coupled

7.2 Transmission Line Equations

set of first-order partial differential equations that are called the telegraphers' equations. This set of equations describes the temporal and spatial evolution of voltage and current signals along this transmission line. In addition, we can manipulate the set and obtain the wave equation.

7.2 Transmission Line Equations

To analyze the equivalent circuit of the lossless transmission line, it is simpler to use Kirchhoff's laws rather than Maxwell's equations at this stage. The various currents and voltages are shown in Figure 7–3. To simplify the notation, we define the inductance and capacitance per unit length,

$$\hat{L} = \frac{L}{\Delta z} \quad \text{and} \quad \hat{C} = \frac{C}{\Delta z}$$

which have the units of **henries per unit length** and **farads per unit length**, respectively. Readers may encounter alternative notations in other books.

The current entering the node at the location z is $I(z)$. From Kirchhoff's current law, part of this current flows through the capacitor, and the rest flows into section. Thus,

$$I(z, t) = \hat{C}\Delta z \frac{\partial V(z, t)}{\partial t} + I(z + \Delta z, t) \tag{7.1}$$

This can be rewritten as

$$\frac{I(z + \Delta z, t) - I(z, t)}{\Delta z} = -\hat{C}\frac{\partial V(z, t)}{\partial t} \tag{7.2}$$

In the limit as $\Delta z \to 0$, the term on the left-hand side of (7.2) can be recognized as a spatial derivative. Therefore, (7.2) becomes

$$\boxed{\frac{\partial I(z, t)}{\partial z} = -\hat{C}\frac{\partial V(z, t)}{\partial t}} \tag{7.3}$$

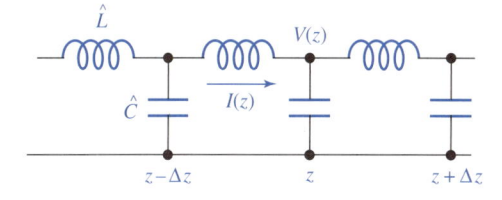

FIGURE 7–3

The lossless transmission line model comprises a number of identical sections. The length of each section is Δz, and each section contains an inductance and a capacitance. The values are $\hat{L} = L/\Delta z$ and $\hat{C} = C/\Delta z$.

Similarly, the sum of the voltage drops in this section can also be calculated using Kirchhoff's voltage law, and we find

$$V(z - \Delta z, t) = \hat{L}\Delta z \frac{\partial I(z, t)}{\partial t} + V(z, t) \tag{7.4}$$

Rewriting (7.4), we obtain

$$\frac{V(z, t) - V(z - \Delta z, t)}{\Delta z} = -\hat{L}\frac{\partial I(z, t)}{\partial t} \tag{7.5}$$

Again, the left-hand side of (7.5) is recognized as a spatial derivative in the limit as $\Delta z \to 0$, and we have

$$\boxed{\frac{\partial V(z, t)}{\partial z} = -\hat{L}\frac{\partial I(z, t)}{\partial t}} \tag{7.6}$$

The two linear coupled first-order partial differential equations (7.3) and (7.6) are the telegraphers' equations. They are sometimes also referred to as the **Heaviside equations** in honor of the physicist Oliver Heaviside (1850–1925), who successfully employed new mathematical tools in their solution. In fact, he developed his own form of calculus called "operational calculus," which can be used to simplify the carrying out of complicated mathematical procedures. Although the approach did not initially have the same rigorous foundation as the traditional form of the calculus, it was eventually put on a firm footing.[1]

We can eliminate one of the functions $V(z, t)$ or $I(z, t)$ from these two coupled first-order equations to obtain a second-order partial differential equation. This can be done for either the voltage $V(z, t)$ or the current $I(z, t)$. The resulting equations are

$$\frac{\partial^2 V(z, t)}{\partial z^2} - \hat{L}\hat{C}\frac{\partial^2 V(z, t)}{\partial t^2} = 0 \tag{7.7}$$

$$\frac{\partial^2 I(z, t)}{\partial z^2} - \hat{L}\hat{C}\frac{\partial^2 I(z, t)}{\partial t^2} = 0 \tag{7.8}$$

---

[1] Another example of a mathematical technique that was successfully used prior to its rigorous establishment is use of the Dirac "delta" function. Dirac and others made considerable use of it during the period that led to the development of quantum mechanics. This was a decade before mathematicians established the basis for its use.

7.2 Transmission Line Equations

Both (7.7) and (7.8) are in the form of the well-known wave equation, as found in standard mathematics and physics texts. In the present case, the velocity of propagation v is defined as

$$\boxed{v = \frac{1}{\sqrt{\hat{L}\hat{C}}} \text{ (m/s)}} \quad (7.9)$$

Recall that the units for $\hat{L}$ and $\hat{C}$ are henries per meter and farads per meter. This implies that (7.9) does indeed have the proper units of the velocity, meters per second (m/s). The choice of the symbol v in (7.9) to represent the velocity of propagation in a transmission line is reasonable since this is also the velocity of a plane electromagnetic wave propagating in the material that separates the conductors in a coaxial cable or between the two parallel wires. We reserve the symbol c for the velocity of propagation in free space.

EXAMPLE 7.1

Show that a transmission line consisting of distributed linear resistors and capacitors in the configuration below can be used to model diffusion.

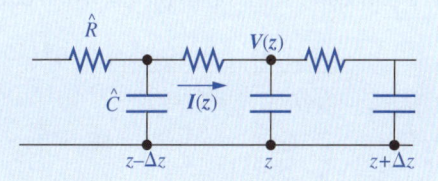

Answer. Assume that the resistance and the capacitance per unit length are defined as $\hat{R} = R/\Delta z$ and $\hat{C} = C/\Delta z$, respectively, where Δz is the length of a section. The potential drop ΔV across the resistor $R = \hat{R}\Delta z$ and the current ΔI through the capacitor $C = \hat{C}\Delta z$ can be written as

$$\Delta V(z, t) = I(z, t)\hat{R}\Delta z$$

$$\Delta I(z, t) = \hat{C}\Delta z \frac{\partial V(z, t)}{\partial t}$$

In the limit of $\Delta z \to 0$, this reduces to the following set of equations:

$$\frac{\partial V(z, t)}{\partial z} = I(z, t)\hat{R}$$

$$\frac{\partial I(z, t)}{\partial z} = \hat{C} \frac{\partial V(z, t)}{\partial t}$$

A second-order partial differential equation for the potential $V(z, t)$ follows:

$$\frac{\partial^2 V(z, t)}{\partial z^2} = \hat{R}\frac{\partial I(z, t)}{\partial z} = \hat{R}\left(\hat{C}\frac{\partial V(z, t)}{\partial t}\right)$$

$$\frac{\partial^2 V(z, t)}{\partial z^2} = \hat{R}\hat{C}\frac{\partial V(z, t)}{\partial t}$$

This equation is in the form of a **diffusion equation**, with a diffusion coefficient given by

$$D = \frac{1}{\hat{R}\hat{C}}$$

The dimensions of the diffusion coefficient D are square meters per second (m^2/s).

EXAMPLE 7.2

Define the diffusion constant for the distributed "RC" transmission line of Example 7.1 as $D = 1/\hat{R}\hat{C}$. Show that a particular solution to the diffusion equation, repeated here as

$$\frac{\partial^2 V(z, t)}{\partial z^2} = \frac{1}{D}\frac{\partial V(z, t)}{\partial t}$$

is given by

$$V(z, t) = \frac{1}{2\sqrt{D\pi t}} e^{\left(-\frac{z^2}{4Dt}\right)}$$

Since the capacitors are linear, this voltage will also correspond to the charge on a particular capacitor. This follows from $Q(z, t) = \hat{C}V(z, t)$ with linear capacitors.

Answer. Differentiating the solution with respect to z, we obtain

$$\frac{\partial V(z, t)}{\partial z} = \frac{1}{2\sqrt{D\pi}}\left(-\frac{z}{2Dt^{3/2}}\right) e^{\left(-\frac{z^2}{4Dt}\right)}$$

and

$$\frac{\partial^2 V(z, t)}{\partial z^2} = \frac{1}{2\sqrt{D\pi}}\left(-\frac{1}{2Dt^{3/2}} + \frac{z^2}{4D^2 t^{5/2}}\right) e^{\left(-\frac{z^2}{4Dt}\right)}$$

7.2 Transmission Line Equations

Differentiating the solution with respect to t, we obtain

$$\frac{1}{D}\frac{\partial V(z,t)}{\partial t} = \frac{1}{D}\frac{1}{2\sqrt{D\pi}}\left(-\frac{1}{2t^{3/2}} + \frac{z^2}{4Dt^{5/2}}\right)e^{\left(-\frac{z^2}{4Dt}\right)}$$

Hence, the equation is satisfied. The voltages at the normalized times 1 through 4 are shown below. Note that the peak remains at $z = 0$ as time increases. The total area under the curve at each of the times remains equal to one and is independent of time.

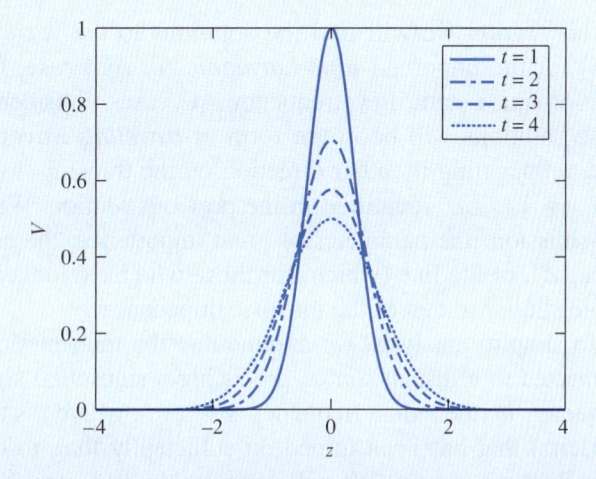

The solution shown in this figure is valid if a certain amount of charge is placed at the location $z = 0$. There is a decrease in amplitude at the center by 50% as time increases by a factor of 4. The area under every curve, equal to the total charge, is a constant.

ANIMATIONS

Diffusion analyzed in the above example would be similar to filling a balloon with helium, popping it at $t = 0$, and then monitoring the helium density in space a later time. One can think of this source also in terms of a ***Dirac delta function*** since the area under this function is a constant. A second boundary condition would be to fix the voltage at $z = 0$ with a battery and a switch that was closed at $t = 0$. Diffusion differs from waves in that the solution predicts that the voltage could change at $z = \pm\infty$ at a time $t = 0^+$. The diffusion equation is also called the ***heat equation*** from thermodynamics. A calculation associated with the second boundary condition would be to compute the temporal and spatial evolution of the temperature in an object if one end touches a hot plate located at $z = 0$ and whose temperature remains constant for all times.

The reader might find it instructive to analyze the diffusion of a step voltage excited at a location $z = 0$ at a time $t = 0$ of a distributed *RC* transmission line containing several sections using a computer circuit analysis program such as SPICE. The process of diffusion is common in semiconductors. Diffusion is significantly different from the wave propagation that we encounter in normal transmission lines.

7.3 Sinusoidal Waves

In this section we will find the solutions to the wave equations (7.7) and (7.8) for the important *time-harmonic* (or AC) case. Unlike the static DC case or quasi-static low frequency AC cases considered in circuit theory, these solutions will be in the form of *traveling waves* of voltage and current, propagating in either direction on the transmission line with the velocity $v = 1/\sqrt{\hat{L}\hat{C}}$ identified in the previous section. We will identify a new transmission line parameter of great importance, the *characteristic impedance*, Z_c, of the line (which, for those who have looked at Chapter 6, plays a role similar to that of the intrinsic impedance).

To simplify analysis, we assume that the transmission line in question is connected to a distant source producing a sinusoidal signal at a single fixed frequency ω (the radian frequency $\omega = 2\pi f$, where f is the angular frequency in Hertz) that has been turned on sufficiently long to ensure that transients have decayed to zero. We will consider the important case of transient excitation of transmission lines in a later section (this case is important due to the use of high-speed digital circuitry).

The primary simplification that occurs in the time-harmonic case results from the use of *phasors* to represent time-varying quantities. As usual, we define the phasor voltage and current, $V(z)$ and $I(z)$, to be the complex functions of position that satisfy the equations

$$V(z, t) = \text{Re}\{V(z)e^{j\omega t}\} \tag{7.10}$$

$$I(z, t) = \text{Re}\{I(z)e^{j\omega t}\} \tag{7.11}$$

Note that, unlike phasors in low frequency AC circuits that are simply complex numbers, for transmission lines, the phasor voltage and current are complex functions of position z on the line. As noted in Chapter 1, when using phasors, time-differentiation becomes multiplication by the factor $j\omega$. Then, the phasor forms of the wave equations (7.7) and (7.8) become with the help of (7.9)

7.3 Sinusoidal Waves

$$\frac{d^2V(z)}{dz^2} + k^2 V(z) = 0 \tag{7.12}$$

$$\frac{d^2I(z)}{dz^2} + k^2 I(z) = 0 \tag{7.13}$$

We have defined a new constant k, known as the **wave number**, given by $k = \omega/v$, with units of radians per meter. It can be shown that the wave number is related to the wavelength λ of the voltage or current wave as $k = 2\pi/\lambda$. Equations (7.12) and (7.13) have general solutions in terms of either trigonometric functions or complex exponentials. For instance, the solution of (7.12) can be written in either of the following forms:

$$V(z) = A_1 \cos kz + B_1 \sin kz \tag{7.14}$$

$$V(z) = A_2 e^{-jkz} + B_2 e^{+jkz} \tag{7.15}$$

We will choose the exponential form (7.15) of the solution, since it is easier to interpret in terms of propagating waves of voltage on the transmission line (see Example 7.3).

EXAMPLE 7.3

The voltage wave that propagates along a transmission line is detected at the indicated points. From this data, write an expression for the wave. Note that there is a propagation of the sinusoidal signal to increasing values of the coordinate z.

Answer. From the data shown in (*a*), we can obtain the following information. The peak-to-peak amplitude of the wave is $2V_0$. The wave is propagating to increasing values of the coordinate z. The period of the wave is 2 s, hence the frequency of oscillation of the wave is $f = 1/2 = 0.5$ Hz. The velocity of propagation v is obtained from the slope of the trajectory shown in (*b*). The value is found to be $v = (5-1)/(1-0) = 4$ m/s. The wave number k is computed to be

$$k = \frac{\omega}{v} = \frac{1}{4}\left(2\pi \frac{1}{2}\right) = \frac{\pi}{4} \text{ m}^{-1}$$

The wavelength λ of the wave is equal to

$$\lambda = \frac{2\pi}{k} = \frac{4 \times 2\pi}{\pi} = 8 \text{ m}$$

The wave is given by

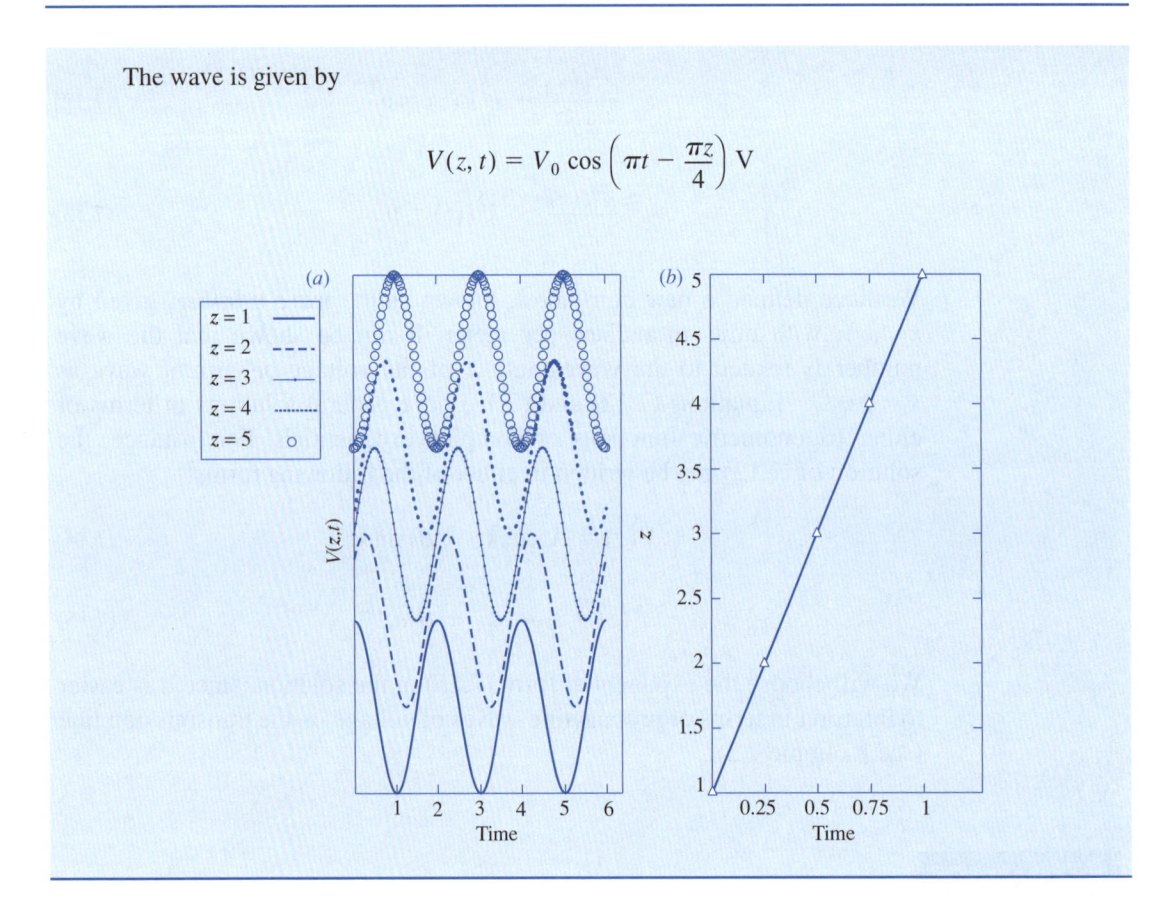

$$V(z, t) = V_0 \cos\left(\pi t - \frac{\pi z}{4}\right) \text{ V}$$

Now assume that the source is located far from the point of interest (at location $z = -\infty$) and that the transmission line is of infinite extent. In this case, there would only be a forward traveling wave of voltage on the transmission line. This corresponds to setting the constant B_2 to zero in (7.15). Denote the constant A_2 by V_0. Then the transmission line voltage is given by $V(z) = V_0 e^{-jkz}$. Now, the phasor form of equation (7.6) states

$$\frac{dV(z)}{dz} = -jkV(z) = -j\omega \hat{L} I(z) \tag{7.16}$$

We may solve this to obtain

$$I(z) = \frac{k}{\omega \hat{L}} V(z) = \frac{k}{\omega \hat{L}} V_0 e^{-jkz} \tag{7.17}$$

7.3 Sinusoidal Waves

We are interested in the ratio of the voltage to the current. It has units of impedance, and by (7.16), we have the following value, a very important transmission line parameter known as the ***characteristic impedance***, Z_c:

$$Z_c = \frac{V(z)}{I(z)} = \frac{\omega \hat{L}}{k} \tag{7.18}$$

An alternative, simpler expression for Z_c may be obtained by recalling that $k = \omega/v$ and $v = 1/\sqrt{\hat{L}\hat{C}}$ to obtain

$$\boxed{Z_c = \sqrt{\frac{\hat{L}}{\hat{C}}} \ (\Omega)} \tag{7.19}$$

EXAMPLE 7.4

Calculate the velocity of propagation and the characteristic impedance of a coaxial cable. The radius of the inner conductor is 3 mm and the radius of the outer conductor is 6 mm. Assume free space between these two conductors.

Answer. From Table 7–1, the inductance per unit length is given by

$$\hat{L} = \frac{\mu_0}{2\pi} \ln\left(\frac{b}{a}\right) = \frac{4\pi \times 10^{-7}}{2\pi} \ln\left(\frac{6}{3}\right) = 0.14 \ \mu\text{H/m}$$

The capacitance per unit length is given by

$$\hat{C} = \frac{2\pi \varepsilon_0}{\ln\left(\frac{b}{a}\right)} = \frac{2\pi \left(\frac{1}{36\pi} \times 10^{-9}\right)}{\ln\left(\frac{6}{3}\right)} = 80 \ \text{pF/m}$$

The velocity of propagation is computed from (7.9) to be

$$v = \frac{1}{\sqrt{\hat{L}\hat{C}}} = \frac{1}{\sqrt{(0.14 \times 10^{-6})(80 \times 10^{-12})}} = 3 \times 10^8 \ \text{m/s}$$

The characteristic impedance of the coaxial cable is computed from (7.19) to be

$$Z_c = \sqrt{\frac{\hat{L}}{\hat{C}}} = \sqrt{\frac{0.14 \times 10^{-6}}{80 \times 10^{-12}}} \approx 42 \ \Omega$$

The velocity of propagation and the characteristic impedance can be decreased if a dielectric is inserted between the two conductors.

The characteristic impedance is equal to the ratio of the phasor voltage to current on the line in the case where only one wave, traveling either forward or backward, exists on the line. In the general case, waves traveling in both directions may exist, and the ratio of total voltage to total current takes a more complex form (more on this later). It is also worth noting the analogy with the intrinsic impedance for a plane wave traveling in a region, which we found as $\eta = \sqrt{\mu/\epsilon}$. As with intrinsic impedance, if we know the forward voltage wave, we may find the forward current wave by dividing by the characteristic impedance. Another important parameter is the length $\mathscr{L}$ of the transmission line. This length is usually normalized by the wavelength λ of the propagating wave. If the transmission line were lossy, the characteristic impedance would be a complex impedance instead of the real number obtained here. Since the transmission line has been assumed to be lossless, we could let $Z_c = R_c + jX_c$, where R_c is a real number and $X_c = 0$ with no loss of generality. We will keep, however, the more general notation of defining the characteristic impedance as Z_c even for the lossless transmission line.

Table 7–1 summarizes the inductance per unit length and the capacitance per unit length for the transmission lines most commonly used. Table 7–2 summarizes the velocity of propagation and the characteristic impedance for the same transmission lines.

TABLE 7–2 Velocity of Propagation and Characteristic Impedance of Transmission Lines in Figure 7–1*

Line	Velocity of propagation	Characteristic impedance
Coaxial cable	$v = \dfrac{1}{\sqrt{\hat{L}\hat{C}}} = \dfrac{1}{\sqrt{\mu\varepsilon}}$	$Z_c = \sqrt{\dfrac{\hat{L}}{\hat{C}}} = \sqrt{\dfrac{\mu}{\varepsilon}}\left(\dfrac{\ln(b/a)}{2\pi}\right)$
Microstrip line	$v = \dfrac{1}{\sqrt{\hat{L}\hat{C}}} = \dfrac{1}{\sqrt{\mu\varepsilon}}$	$Z_c = \sqrt{\dfrac{\hat{L}}{\hat{C}}} = \sqrt{\dfrac{\mu}{\varepsilon}}\left(\dfrac{d}{w}\right)$
Twin lead	$v = \dfrac{1}{\sqrt{\hat{L}\hat{C}}} = \dfrac{1}{\sqrt{\mu\varepsilon}}$	$Z_c = \sqrt{\dfrac{\hat{L}}{\hat{C}}} = \sqrt{\dfrac{\mu}{\varepsilon}}\left(\dfrac{\cosh^{-1}(D/2a)}{\pi}\right)$

*The parameters for the material between the two conductors are the permeability $\mu = \mu_r\mu_0$ and the permittivity $\varepsilon = \varepsilon_r\varepsilon_0$.

The velocity of propagation of the wave is independent of the dimensions of the transmission line, and it is only a function of the electrical parameters of the material that separates the two conductors. However, the characteristic

7.4 Terminations

impedance depends upon the geometry and physical dimensions of the transmission line along with the electrical parameters. This fact will be important when two transmission lines are connected together and a signal is launched on one of them.

We have assumed so far that the transmission line was of infinite extent. A practical transmission line will have both a beginning and an end positioned at finite locations in space. We, therefore, must incorporate this subject into our discussion and will examine terminations in this section.

A finite length of a lossless transmission line is illustrated in Figure 7–4. The transmission line has characteristic impedance Z_c that is a real quantity. It is convenient to assume that the source of the time-harmonic wave is at $z = -\infty$ and the termination is located at $z = 0$. This termination could be either an impedance or another transmission line with a different characteristic impedance. We will also assume that the signal generator was also turned on sufficiently long ago that all transient effects will have disappeared. These choices are predicated on our desire to simplify our discussion as much as possible.

The phasor voltage at any point on the line is given by (7.15) as

$$V(z) = A_2 e^{-jkz} + B_2 e^{+jkz} \tag{7.20}$$

The current is found by substitution of (7.20) into (7.17):

$$I(z) = \frac{1}{Z_c}[A_2 e^{-jkz} - B_2 e^{+jkz}] \tag{7.21}$$

Here we have used the definition (7.18) of the characteristic impedance. At the location of the load, the ratio of voltage to current must be equal to Z_L. According to the coordinates shown in Figure 7–4, the load is at location $z = 0$. Hence, we divide (7.20) by (7.21) and set $z = 0$ in the result to obtain the equality

$$Z_L = \frac{V(z=0)}{I(z=0)} = Z_c \frac{A_2 + B_2}{A_2 - B_2} \tag{7.22}$$

We are interested in finding the ratio of B_2 to A_2, since the latter represents the magnitude of the wave incident on the load Z_L (traveling in the +z direction),

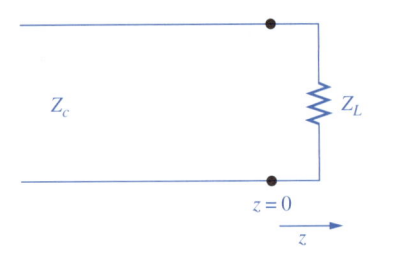

FIGURE 7-4

A semi-infinite transmission line that is terminated in a load impedance Z_L.

whereas the former represents the magnitude of the wave reflected off the load (traveling in the $-z$ direction). After some algebra, we obtain the result, giving a formula for the **reflection coefficient**,

$$\Gamma \equiv \frac{B_2}{A_2} = \frac{Z_L - Z_c}{Z_L + Z_c} \qquad (7.23)$$

It is customary to define the normalized load impedance, z_L, as the ratio $z_L = Z_L/Z_c$. Then we have the alternative form for the reflection coefficient:

$$\Gamma \equiv \frac{B_2}{A_2} = \frac{z_L - 1}{z_L + 1} \qquad (7.24)$$

It is useful to express the total phasor voltage and current on the transmission line in terms of the wave number, characteristic impedance, and reflection coefficient. Using (7.23) in (7.20) and (7.21) gives

$$V(z) = V_0[e^{-jkz} + \Gamma e^{+jkz}] \qquad (7.25)$$

$$I(z) = \frac{V_0}{Z_c}[e^{-jkz} - \Gamma e^{+jkz}] \qquad (7.26)$$

We have denoted the remaining arbitrary amplitude coefficient A_2 by V_0.

The ratio of these expressions for total line voltage and current has the units of impedance, and is denoted $Z(z)$, the total impedance at location z on the line. Note that, as mentioned earlier, $Z(z)$ is a complicated function of position on the line, and is not simply equal to the characteristic impedance Z_c. However, there is on very important case where $Z(z)$ is equal to Z_c, when the load is *matched* to the transmission line. If we choose a load with the matched impedance $Z_L = Z_c$, then equation (7.23) gives a zero reflection coefficient, and the ratio of voltage to current is simply Z_c.

7.4 Terminations

EXAMPLE 7.5

The dielectric in an infinitely long coaxial cable has a value for its relative dielectric constant of $\varepsilon_r = 2$ for $z < 0$ and $\varepsilon_r = 3$ for $z > 0$. Calculate the reflection coefficient Γ for a wave that is incident from $z = -\infty$.

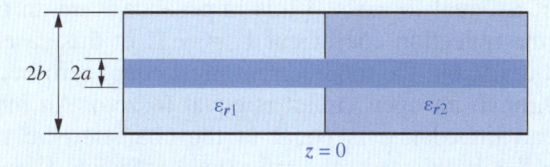

Answer. The characteristic impedance of a coaxial line is given in Table 7–2. The load impedance Z_L that appears in (7.24) will be the characteristic impedance of the coaxial cable in the region $z > 0$ since it acts as a load for the coaxial cable in the region $z < 0$. Hence, we write the reflection coefficient in terms of the characteristic impedance of the two transmission lines in unnormalized form.

$$\Gamma = \frac{Z_2 - Z_1}{Z_2 + Z_1} = \frac{\sqrt{\frac{\mu_0}{\varepsilon_{r2}\varepsilon_0}}\left(\frac{\ln(b/a)}{2\pi}\right) - \sqrt{\frac{\mu_0}{\varepsilon_{r1}\varepsilon_0}}\left(\frac{\ln(b/a)}{2\pi}\right)}{\sqrt{\frac{\mu_0}{\varepsilon_{r2}\varepsilon_0}}\left(\frac{\ln(b/a)}{2\pi}\right) + \sqrt{\frac{\mu_0}{\varepsilon_{r1}\varepsilon_0}}\left(\frac{\ln(b/a)}{2\pi}\right)}$$

$$= \frac{\frac{1}{\sqrt{\varepsilon_{r2}}} - \frac{1}{\sqrt{\varepsilon_{r1}}}}{\frac{1}{\sqrt{\varepsilon_{r2}}} + \frac{1}{\sqrt{\varepsilon_{r1}}}} = \frac{\frac{1}{\sqrt{3}} - \frac{1}{\sqrt{2}}}{\frac{1}{\sqrt{3}} + \frac{1}{\sqrt{2}}} \approx -0.1$$

The reflection coefficient Γ is determined entirely by the value of the impedance of the load and the characteristic impedance of the transmission line. If the line is homogeneous and it has no discontinuities, this is a good assumption. It implies that the voltages and the currents that appear at any point along the transmission line are determined by the signal generator and the load impedance—which may be several wavelengths apart. Remember that our signal generator was turned on a long time ago, and no transients are relevant in this discussion. We will examine transients later, however, as they are very important.

The reflection coefficient Γ for a lossless transmission line can have any complex value with magnitude less than or equal to one. The load impedance may have any complex value, but for the moment, we restrict our discussion to lossless transmission lines with real characteristic impedances.

If the load impedance were a short circuit ($Z_L = 0$), then the reflection coefficient Γ that we compute from (7.24) would yield $\Gamma = -1$. At the load, the total voltage at $z = 0$ that consists of the sum of the incident and the reflected components must equal zero because the voltage across a short-circuit is equal to zero. If this impedance were an open circuit ($Z_L = \infty$), then the reflection coefficient $\Gamma = +1$. In this case, the total voltage can be arbitrary, but the total current must equal zero because the current flowing through an open circuit is equal to zero. An interesting case arises if the load impedance is equal to the characteristic impedance of the line ($Z_L = Z_c$). In this case, the reflection coefficient $\Gamma = 0$, and we say that the line is **matched**. This matching is very important in practice since all of the energy that is transported down the line is absorbed at the load impedance, and none will be reflected back toward the signal generator. All of the energy will be "gainfully employed" in the load impedance, and there will be none to come back. We will find that techniques can be used to achieve this desirable state of operation, even if the load impedance has a value that differs from the value of the characteristic impedance of the transmission line.

Let us for the moment examine the voltage waveform that arises when a sinusoidal voltage wave is incident upon a short circuit ($Z_L = 0$) or an open circuit ($Z_L = \infty$). In the first case, the reflection coefficient as computed from (7.24) is $\Gamma = -1$; for the second case, $\Gamma = +1$. From (7.25) and the definition of phasors, we have

$$V(z,t) = \text{Re}\{V_0[e^{-jkz} - e^{+jkz}]e^{+j\omega t}\} = 2V_0 \sin kz \cos(\omega t - \pi/2) \quad (7.27)$$

$$V(z,t) = \text{Re}\{V_0[e^{-jkz} + e^{+jkz}]e^{+j\omega t}\} = 2V_0 \cos kz \cos \omega t \quad (7.28)$$

In writing these equations we have used the equations $e^{-jkz} - e^{+jkz} = -2j \sin kz$, $e^{-jkz} + e^{+jkz} = 2 \cos kz$, and Euler's identity $e^{+j\omega t} = \cos \omega t + j \sin \omega t$.

The addition of the two waves that individually propagate to increasing values of the spatial coordinate z and to decreasing values of z creates a signal that appears to be stationary in space but whose magnitude oscillates in time from zero to twice the value of the incident wave. This effect is called a **standing wave** since the resulting signal does not appear to propagate. We illustrate the measured voltages as a function of space at various times for the two values of load impedance in Figure 7–5. Standing waves are also

7.4 Terminations

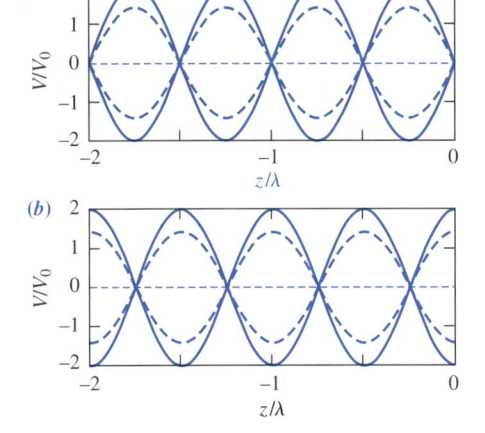

FIGURE 7–5

Transmission line that has a characteristic impedance Z_c. The line is excited at $z = -\infty$ with a sinusoidal voltage and the voltage is depicted at various times.
(a) Standing voltage wave if $Z_L = 0$.
(b) Standing voltage wave if $Z_L = \infty$. The maximum amplitude of the voltage standing wave is $2V_0$.

EXAMPLE 7.6

Determine the standing current waves that correspond to the standing voltage waves depicted in Figure 7–5.

Answer. The current wave is equal to the voltage wave divided by the characteristic impedance of the transmission line. In this case, $I_0 = V_0 / Z_c$. The standing current wave is depicted below for two cases: (a) $Z_L = 0$ and (b) $Z_L = \infty$. Note that the standing current waves differ in phase by 90° with respect to the standing voltage waves. From Figure 7–5 (reproduced below, left), we note that the maximum voltage that occurs along the line is equal to $2V_0$ and the minimum voltage is 0. The magnitude of the voltage repeats itself every *half* wavelength. This is a crucial observation since a typical voltage detector used in practice is a "square-law device" that responds to the magnitude of the voltage and cannot distinguish between plus or minus voltages.

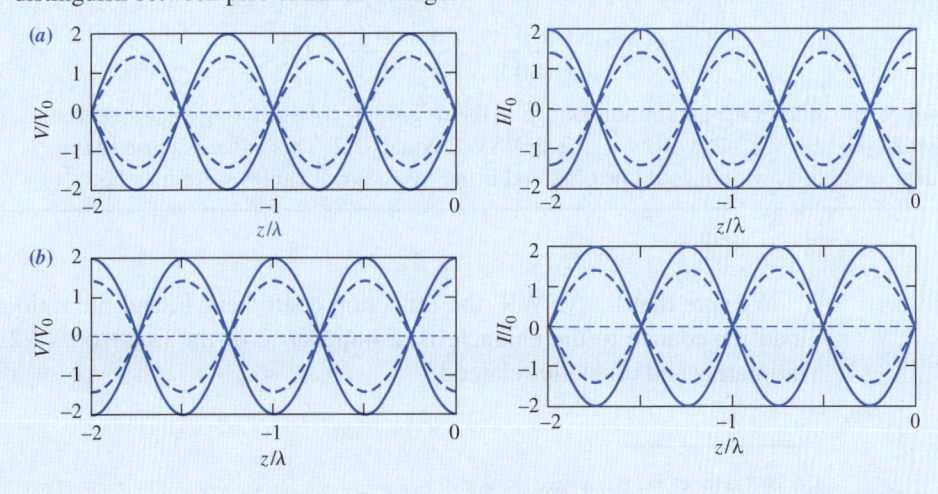

found on the strings of a violin as the force of the fingers on the player's left hand create a local "short circuit" for the waves that are excited by the bow or by the plucking of the right hand. The waves propagate along the string and are trapped by the fingers of the left hand and the bridge of the violin.

The ratio of the maximum voltage to the minimum voltage that appears along this transmission line, which in this case is ∞, is an important quantity. The maximum value of the voltage is equal to the magnitude of the incident voltage plus the magnitude of the reflected voltage. The minimum value equals the difference of these two quantities. The ratio is called the *voltage standing wave ratio*. The abbreviation VSWR[2] is frequently used. From (7.25), it may be shown that for an arbitrary load

$$\boxed{\text{VSWR} = \frac{V_{max}}{V_{min}} = \frac{1 + |\Gamma|}{1 - |\Gamma|}} \qquad (7.29)$$

We can solve this equation for $|\Gamma|$ and obtain

$$\boxed{|\Gamma| = \frac{\text{VSWR} - 1}{\text{VSWR} + 1}} \qquad (7.30)$$

EXAMPLE 7.7

Compute the ratio of the maximum voltage to the minimum voltage of a wave propagating in the coaxial cable described in Example 7–5 in the region $z < 0$.

Answer. The reflection coefficient Γ was computed in Example 7–5 to be $\Gamma = -0.1$. From (7.29), we write

$$\text{VSWR} = \frac{1 + |\Gamma|}{1 - |\Gamma|} = \frac{1 + |-0.1|}{1 - |-0.1|} = 1.2$$

Hence, the ratio of the maximum voltage to the minimum voltage along the transmission line in the region $z < 0$ will be 1.2. The VSWR equals 1.2. This value is close to the ideal value of 1, which would be obtained if the two coaxial cables were matched.

We note that the VSWR, the reflection coefficient Γ, and the ratio of the load impedance to the characteristic impedance of the line from (7.23) are intimately and crucially related.

---

[2] VSWR is sometimes pronounced "vizwar".

If the amplitudes of the voltages are not very large, this may be a moot point. However, if the voltages [$V_0(1 + |\Gamma|)$] are large—say, above the breakdown conditions of electronic components—serious problems may occur even if the amplitude of the incident wave V_0 is beneath this critical value. The magnitude of the reflection coefficient Γ must be reduced. From (7.23), this implies that the load impedance Z_L should have a value that approaches the characteristic impedance Z_c of the line. The line must be matched!

7.5 Impedance and Matching of a Transmission Line

The ratio of total phasor voltage to total phasor current on a transmission line has units of impedance. However, as we have seen, this impedance is not a constant, and varies with location on the line. This behavior is due to fact that the line voltage and current consists of incident and reflected waves traveling in opposite directions on the line.

From equations (7.25) and (7.26), we have

$$Z(z) = \frac{V(z)}{I(z)} = \frac{V_0[e^{-jkz} + \Gamma e^{+jkz}]}{\frac{V_0}{Z_c}[e^{-jkz} - \Gamma e^{+jkz}]} = Z_c \frac{e^{-jkz} + \Gamma e^{+jkz}}{e^{-jkz} - \Gamma e^{+jkz}} \quad (7.31)$$

Now make use of equation (7.23) for Γ. Substitution gives

$$Z(z) = Z_c \frac{e^{-jkz} + \frac{Z_L - Z_c}{Z_L + Z_c} e^{+jkz}}{e^{-jkz} - \frac{Z_L - Z_c}{Z_L + Z_c} e^{+jkz}}$$

After some manipulation, we obtain

$$Z(z) = Z_c \frac{2Z_L \cos kz - j2Z_c \sin kz}{2Z_c \cos kz - j2Z_L \sin kz} = Z_c \frac{Z_L - jZ_c \tan kz}{Z_c - jZ_L \tan kz} \quad (7.32)$$

This formula is most often used to find the impedance at one particular location of importance on a transmission line: its input terminals. Consider the transmission line circuit shown in 7.5(a). A load with given impedance Z_L is connected to a transmission line segment of length ℓ. We assume that the wave number k and the characteristic impedance Z_c of the transmission line are

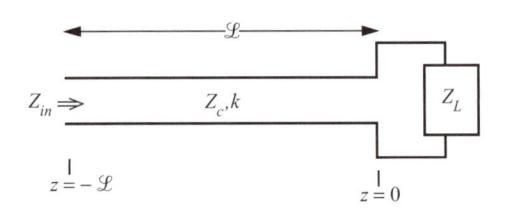

FIGURE 7–5A

The input impedance, $Z_{in}(\ell)$, of a transmission line segment of length ℓ, wave number k and the characteristic impedance Z_c terminated by a load impedance Z_L.

known. The quantity to be computed is the **input impedance** of the loaded transmission line segment; that is, the impedance seen looking into the input terminals of the transmission line, as depicted in the figure.

What we are asked for is simply the impedance $Z(z)$ on the line at the location of the input terminals. Referring to the figure, we see that this impedance, denoted $Z_{in}(\mathcal{L})$, is given by evaluating equation (7.32) at location $z = -\mathcal{L}$:

$$Z_{in}(\mathcal{L}) = Z(z = -\mathcal{L}) = Z_c \frac{Z_L + jZ_c \tan k\mathcal{L}}{Z_c + jZ_L \tan k\mathcal{L}} \qquad (7.33)$$

This can also be written as a normalized impedance $z_{in} = Z_{in}/Z_c$, where we have followed the convention of using a lower case letter to indicate a normalized impedance.

$$z_{in}(\mathcal{L}) = \frac{z_L + j \tan(k\mathcal{L})}{1 + jz_L \tan(k\mathcal{L})} \qquad (7.34)$$

The impedance z_{in} at this location will be called the normalized **input impedance** of the transmission line.

Recall that the wave number can be defined in terms of the wavelength λ as $k = 2\pi/\lambda$. Therefore, the length of the transmission line $\mathcal{L}$ can also be normalized by the wavelength $k\mathcal{L} = (2\pi/\lambda)\mathcal{L}$. The implication of this normalization is that the value of the load impedance will repeat itself every half wavelength, since $\tan(k\mathcal{L}) = \tan(k\mathcal{L} + n\pi)$, where n is an integer. If the length of the transmission line is one-quarter of a wavelength, we obtain

$$k\mathcal{L} = \frac{2\pi}{\lambda}\left(\frac{\lambda}{4}\right) = \frac{\pi}{2}$$

The trigonometric function $\tan(\pi/2) \longrightarrow \infty$, and the input impedance is given by

7.5 Impedance and Matching of a Transmission Line

$$Z_{in}\left(z = -\frac{\lambda}{4}\right) = Z_c \frac{Z_L + jZ_c\infty}{Z_c + jZ_L\infty} \longrightarrow Z_{in(\lambda/4)} = \frac{(Z_c)^2}{Z_L} \qquad (7.35)$$

This implies that the normalized input impedance z_{in} of a one-quarter wavelength transmission line that is terminated with a load impedance Z_L will have a numerical value that is equal to the normalized load admittance $y_L = 1/z_L$.

EXAMPLE 7.8

A signal generator whose frequency $f = 100$ MHz is connected to a coaxial cable of characteristic impedance $100 \, \Omega$ and length of 100 m. The velocity of propagation is equal to 2×10^8 m/s. The transmission line is terminated with a load impedance of $50 \, \Omega$. Calculate the impedance at a distance of 50 m from the load.

Answer. The normalized load impedance z_L is equal to 1/2. The wavelength λ is calculated from

$$\lambda = \frac{v}{f} = \frac{2 \times 10^8}{1 \times 10^8} = 2 \text{ m}$$

The wave number k is calculated to be

$$k = \frac{2\pi}{\lambda} = \frac{2\pi}{2} = \pi \text{ m}^{-1}$$

The normalized input impedance is calculated using (7.34):

$$z_{in}(z = -50 \text{ m}) = \frac{z_L + j \tan(k\mathcal{L})}{1 + jz_L \tan(k\mathcal{L})} = \frac{\frac{1}{2} + j \tan(50\pi)}{1 + j\frac{1}{2} \tan(50\pi)} = \frac{1}{2}$$

Therefore, the input impedance at this location is equal to

$$Z_{in} = z_{in} Z_c = 50 \, \Omega$$

This *one-quarter-wavelength* transmission line will be useful in joining two transmission lines that have different characteristic impedances or in matching a load, as will be demonstrated now. One of the simplest matching techniques is to use a quarter-wave transformer. A *quarter-wave*

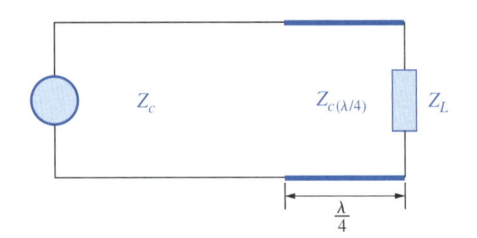

FIGURE 7-6

A transmission line with a characteristic impedance $Z_c \neq Z_L$ is joined to the load with a quarter-wave transformer.

transformer is a section of transmission line that has a particular characteristic impedance that is identified as $Z_{c(\lambda/4)}$. In addition, the length of this matching transmission line will be specified in terms of the length of the wave as it propagates in this quarter-wavelength transmission line. We're assuming that the value of the characteristic impedance $Z_{c(\lambda/4)}$ of this particular transmission line can be specified by the user. The value of this characteristic impedance will be chosen such that the reflection coefficient Γ at the input of the matching transmission line's section is equal to zero. This is shown in Figure 7–6.

In specifying the value of this characteristic impedance $Z_{c(\lambda/4)}$ of this quarter-wavelength transmission line, we recall the reflection coefficient Γ that was given in (7.23). In the present application, we assume that the load impedance is the input impedance of the quarter-wavelength transmission line:

$$\Gamma \equiv \frac{Z_{in} - Z_c}{Z_{in} + Z_c} \tag{7.36}$$

where using (7.23),

$$Z_{in}\left(z = -\frac{\lambda}{4}\right) = \frac{\left(Z_{c(\lambda/4)}\right)^2}{Z_L} \tag{7.37}$$

In order to minimize the reflection coefficient, this input impedance should be chosen to have a value that is equal to the characteristic impedance Z_c of the transmission line connected to the signal generator. From (7.37), we obtain the characteristic impedance of the *matching* transmission line to be

$$\boxed{Z_{c(\lambda/4)} = \sqrt{Z_c Z_L} \ \ (\Omega)} \tag{7.38}$$

7.5 Impedance and Matching of a Transmission Line

The characteristic impedance of this matching transmission line is chosen to have a value that is equal to the geometric mean of the load impedance and the characteristic impedance of the transmission line that is connected to the signal generator.

This technique has certain disadvantages in that it is frequency sensitive, since the velocity of propagation (and, therefore, the wavelength) is determined by the material parameters in the matching transmission line. Techniques from modern filter theory can, to a certain extent, be employed to desensitize this restriction.

Three particular values of the load impedances will be examined at this point. In the first case, the load impedance Z_L equals the characteristic impedance of the transmission line Z_c. In this case, the transmission line is **matched**. There will be no reflected component of the incident wave, and we find from (7.32) that the impedance $Z(z)$ will always be equal to the characteristic impedance Z_c. For the other two cases, the load impedance is either a short circuit ($Z_L = 0$) or an open circuit ($Z_L = \infty$). For two cases, the impedance is found from (7.33) to be

$$Z_{in}(z = -\mathcal{L})|_{Z_L = 0} = jZ_c \tan(k\mathcal{L})$$

$$Z_{in}(z = -\mathcal{L})|_{Z_L = \infty} = \frac{Z_c}{j \tan(k\mathcal{L})} = -jZ_c \cot(k\mathcal{L}) \quad (7.39)$$

In practice, it is easier to make a terminating load impedance that is a short circuit rather than an open circuit because of the fringing fields that could exist at an open circuit. In both cases, the input impedance will be a reactance, $Z_{in} = jX_{in}$. The value of this reactance as a function of line length is depicted in Figure 7–7. The value will range from $-j\infty < Z_{in} < +j\infty$, and

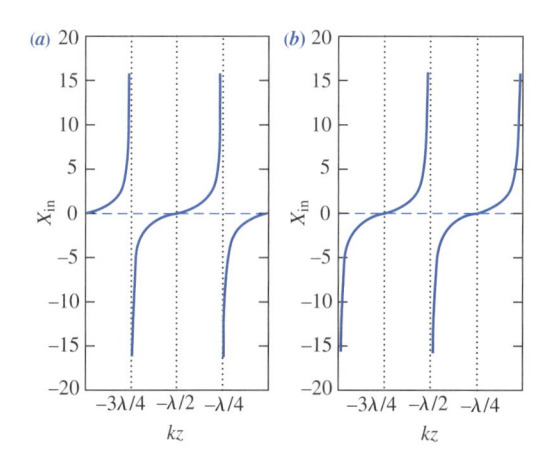

FIGURE 7–7

Input reactance of (*a*) a short-circuited transmission line and (*b*) an open-circuited transmission line. The vertical lines are repeated for equal intervals of $\lambda/4$.

TABLE 7–3 Input Impedance of Short-Circuited or Open-Circuited Lossless Transmission Lines

Length	Short circuit	Open circuit
$0 < \mathscr{L} < \dfrac{\lambda}{4}$	Inductive	Capacitive
$\dfrac{\lambda}{4} < \mathscr{L} < \dfrac{\lambda}{2}$	Capacitive	Inductive

is specified by the length of this transmission line. This implies that we can have every possible value of reactance that is either capacitive or inductive.

The reactance along a lossless transmission line of both short and open circuits can vary from $-\infty < X_{in} < +\infty$. The precise value depends on the length of the line. We summarize the input impedance of the two lines in Table 7–3. From (7.39), this also implies that the input susceptance $B_{in} = -1/X_{in}$ can also change and have any value from $-\infty < B_{in} < +\infty$ along this lossless transmission line.

There are some practical consequences to the fact that the input reactance or susceptance of a transmission line that terminates in either a short or an open can have any value from $-\infty$ to $+\infty$. From a practical point of view, it is better at this stage to think in terms of admittances rather than impedances, as will presently become clear.

EXAMPLE 7.9

A lossless transmission line is terminated with an impedance whose value is half the characteristic impedance on the line. What impedance should be inserted in parallel with the line $\lambda/4$ in front of the load to minimize the reflection of the wave back toward the signal generator?

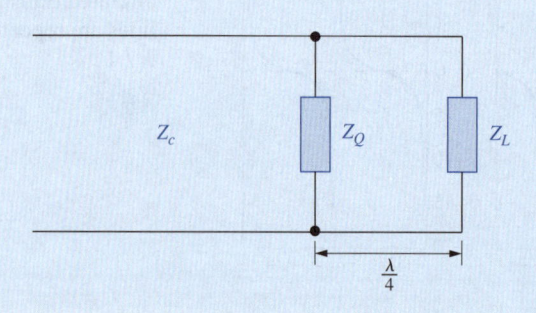

7.5 Impedance and Matching of a Transmission Line

Answer. In order to reduce the reflection, the parallel combination of Z_Q and the input impedance at this location should be equal to the characteristic impedance of the transmission line. Therefore, we write

$$Z_c = \frac{Z_Q Z_{in(\lambda/4)}}{Z_Q + Z_{in(\lambda/4)}} = \frac{Z_Q\left(\frac{Z_c^2}{Z_L}\right)}{Z_Q + \left(\frac{Z_c^2}{Z_L}\right)} = \frac{Z_Q\left(\frac{Z_c^2}{\left(\frac{Z_c}{2}\right)}\right)}{Z_Q + \left(\frac{Z_c^2}{\left(\frac{Z_c}{2}\right)}\right)}$$

$$1 = \frac{Z_Q 2}{Z_Q + 2Z_c} \Rightarrow Z_Q = 2Z_c$$

Let us assume that we have a transmission line terminated with a load impedance Z_L or a load admittance Y_L that are not equal to the line's characteristic impedance Z_c or characteristic admittance Y_c. At some distance d_1 from the load, the input admittance of the line will have a value that is equal to $Y_c + jB$ as shown in Figure 7–8a. The characteristic admittance Y_c of the transmission line is defined as $Y_c = 1/Z_c$.

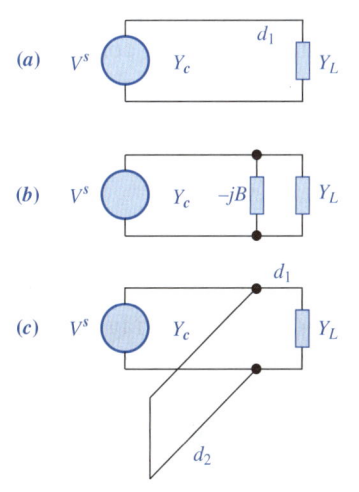

FIGURE 7–8

(a) The input admittance of a transmission line at a distance d_1 from the load is equal to $Y_{in} = Y_c + jB$.
(b) The addition of a susceptance whose value is equal to $-jB$ at a distance d_1 from the load admittance causes the input admittance at that point to be equal to Y_c.
(c) The addition of a short-circuited transmission line whose length is d_2 at the location d_1 will match the parallel combination of the transformed load admittance and the matching transmission line.

At this distance d_1 from the load, a susceptance $-jB$ is added in *parallel* with the transmission line, causing the total admittance to the left of this point to be equal to Y_c as shown in Figure 7–8b. If the transmission line were a coaxial cable, this could be accomplished by connecting a circuit element from the center conductor to the outer conductor. For the strip line, this element would be between the top and the bottom conductors. A series connection of the additional matching element is more difficult to achieve in practice since it would involve the separation of the transmission line into two sections and the subsequent insertion of the matching impedance.

The addition of the matching element at this location implies that the transmission line will be matched from this point back to the signal generator. A transmission line terminated in a short circuit can have any value of susceptance or reactance. Therefore, a shorted transmission line of the appropriate length should be connected at the location d_1 as shown in Figure 7–8c. Since it is difficult to construct an open circuit because there are fringing fields and leakage currents, one normally uses a short circuit as a load. Remember that an open circuit can be located one-quarter wavelength from a short circuit.

The length of this transmission line, which is called a *stub*, is chosen so that its input admittance will equal $-jB$. The load impedance that includes the added short-circuited transmission line now is matched to the rest of the transmission line. This process of matching is called *single-stub matching*. This adjustable-length transmission line is sometimes called a *trombone line*. In the next section, we will examine the matching in more detail after the Smith chart is introduced.

Single-stub matching requires that there be two adjustable distances—the location of the stub d_1 and the length of the stub d_2. There are several cases where it is not practical to make the distance d_1 adjustable, since it requires the milling of a narrow slit in the outer conductor of the coaxial cable or in one of the conductors of the strip line. In these cases, one may have to resort to the addition of a second or a third stub at additional fixed distances from the first one in order to match the load impedance to the transmission line.

All distances mentioned in this process are normalized by the wavelength of the wave. This implies that one can only match a load admittance at certain discrete frequencies. To improve upon this situation so that a band of frequencies can be matched, we bring into play the capacity developed in circuit theory for designing flat pass-band filters. While broadband operation can be achieved by a designer using these concepts, the appropriate techniques are better left to advanced courses in microwave design.

7.6 Smith Chart

In the previous section, we learned that the input impedance of a transmission line depends upon the impedance of the load, the characteristic impedance of the transmission line, and the distance between the load impedance and the point of observation. In addition, the value of the input impedance periodically varies in space. This was demonstrated in Figure 7–7 for the case of a load that was either a short circuit or an open circuit. Since the input impedance (7.33) and the normalized input impedance (7.34) involve a trigonometric function, we can infer that this periodicity will also be true for an arbitrary load impedance. Rather than always running to a calculator or a computer whenever an equation such as the one for normalized input impedance (7.34) is presented, methods have been developed to solve these equations graphically. This graphical approach is based on a tool called a **Smith chart**, after its developer, Bell Labs engineer P. H. Smith.

The equation that describes the normalized impedance at any location (7.34) is a complex equation, which is rewritten here

$$z_{in}(z = -\mathcal{L}) = \frac{z_L + j\tan(k\mathcal{L})}{1 + jz_L\tan(k\mathcal{L})} \tag{7.40}$$

An arbitrary normalized load impedance will also be a complex function. We can write it as

$$z_L = r + jx \tag{7.41}$$

where $r = R_L/Z_c$ and $x = X_L/Z_c$. The lowercase notation implies a normalized impedance that has been divided by the characteristic impedance of the lossless transmission line Z_c. The subscript L has also been dropped from r and x in order to later conform to the Smith chart notation. Remember that we have assumed that the characteristic impedance Z_c is a real quantity since we are presently considering only lossless transmission lines.

The reflection coefficient Γ given in (7.24) is also a complex quantity that can be written as

$$\Gamma = \Gamma_r + j\Gamma_i = \frac{z_L - 1}{z_L + 1} \tag{7.42}$$

In this equation, Γ_r is the real part and Γ_i is the imaginary part of the reflection coefficient Γ. Solving this equation for z_L, we obtain

$$z_L = \frac{1 + \Gamma}{1 - \Gamma} \tag{7.43}$$

or

$$r + jx = \frac{1 + \Gamma_r + j\Gamma_i}{1 - \Gamma_r - j\Gamma_i} \tag{7.44}$$

This equation is simplified by multiplying the numerator and the denominator by the complex conjugate of the denominator, and we find

$$r + jx = \frac{1 - \Gamma_r^2 - \Gamma_i^2}{(1 - \Gamma_r)^2 + \Gamma_i^2} + j\frac{2\Gamma_i}{(1 - \Gamma_r)^2 + \Gamma_i^2} \quad (7.45)$$

Equating the real and the imaginary parts of (7.45) yields two equations that can, after some algebra, be written in the following form:

$$\left(\Gamma_r - \frac{r}{r+1}\right)^2 + \Gamma_i^2 = \left(\frac{1}{r+1}\right)^2 \quad (7.46)$$

and

$$(\Gamma_r - 1)^2 + \left(\Gamma_i - \frac{1}{x}\right)^2 = \left(\frac{1}{x}\right)^2 \quad (7.47)$$

Writing the equations for the real and imaginary terms in this format allows us to recognize both as equations for a family of circles in a plane whose axes are labeled as Γ_r and Γ_i.[3] The center and radius of each circle are determined by the value of the normalized resistance r and the normalized reactance x. The maximum magnitude of the reflection coefficient Γ is equal to 1. Therefore, all of these complete circles or portions of various circles should reside within a large circle whose radius is equal to 1.

Before describing the application of the Smith chart, let us summarize its properties with reference to Figure 7–9. For the case of the constant r circles, we find that

1. The centers of all of the constant r circles lie on the horizontal axis, which is the real part of the reflection coefficient Γ_r axis.
2. As the value of r increases from $r = 0$ to $r = \infty$, the circles become progressively smaller.
3. All constant r circles pass through the point $\Gamma_r = 1, \Gamma_i = 0$.
4. The normalized resistance $r = \infty$ is at the point $\Gamma_r = 1, \Gamma_i = 0$.

For the constant x circles, which in actuality are portions of complete circles, we conclude the following:

1. The centers of all of the constant x circles lie on the $\Gamma_r = 1$ line. The circles with $x > 0$ (inductive reactance) lie above the Γ_r axis, and the circles with $x < 0$ (capacitive reactance) lie below the Γ_r axis.
2. As the values of x change from $x = 0$ to $x = +\infty$ or $x = -\infty$, the circles become progressively smaller.
3. The normalized reactances $x = \pm\infty$ are at the point $\Gamma_r = 1, \Gamma_i = 0$.

---

[3] The general form of the equation for a circle in the $x - y$ plane centered at the point (x_0, y_0) with radius r is $(x - x_0)^2 + (y - y_0)^2 = r^2$.

7.6 Smith Chart

The Smith chart has the property that the constant r circles are orthogonal to the constant x circles at every intersection. The actual load impedance that is connected to a transmission line whose characteristic impedance is Z_c is given by $Z_L = Z_c(r + jx)$. The results are depicted in Figure 7–9.

The Smith chart's fine-scale gradation is determined by the user. The normalized values of the resistance and reactance (which range from $0 < r < \infty$ and $-\infty < x < +\infty$) have also been included in this chart. The intersection of an r circle and an x circle specifies the normalized impedance, and the intersection of the two circles is orthogonal at that point. The evenly spaced marks on the circumference of the Smith chart indicate the fraction of a half-wavelength, since the impedance repeats itself every half-wavelength. Since we are examining a lossless transmission line, the magnitude of the reflection coefficient Γ is a constant at every point between the load and the signal generator. This is shown by carefully locating a circle that is centered on the origin of this coordinate system and whose radius is equal to the magnitude of the reflection coefficient Γ. A clockwise rotation of this circle will be in the direction toward the signal generator, and a counterclockwise rotation will be in the direction toward the load impedance. The horizontal Γ_r axis and the vertical Γ_i axis have been removed from the Smith chart. Although the Smith chart has been derived in terms of an impedance, it works equally well for admittances.

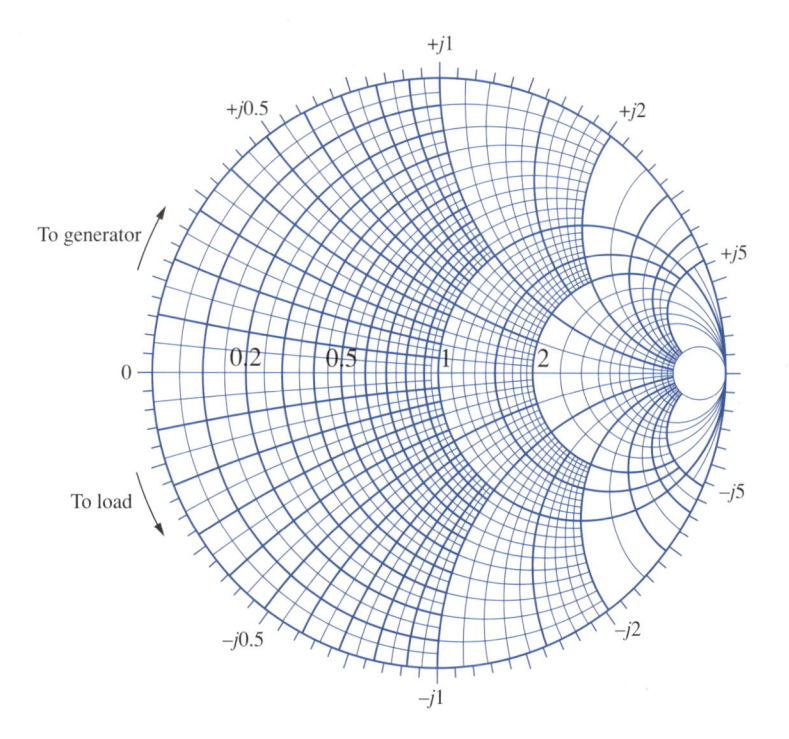

FIGURE 7–9

A Smith chart created with MATLAB.

Why have we spent all this effort and energy to obtain a chart in some strange coordinate system? The answer may seem opaque at this stage, but several examples will be used to illustrate a few of the many applications of the Smith chart. There are many laboratory instruments and software packages that use Smith charts as their display format. High-frequency circuit designers prefer this format, as it allows them to easily visualize the circuit response as a function of both the load impedance and the frequency.

EXAMPLE 7.10

On the simplified Smith chart, locate the normalized impedances.

(a) $z = 1 + j0$
(b) $z = 0.5 - j0.5$
(c) $z = 0 + j0$
(d) $z = 0 - j1$
(e) $z = 1 + j2$
(f) $z = \infty$

Answer.

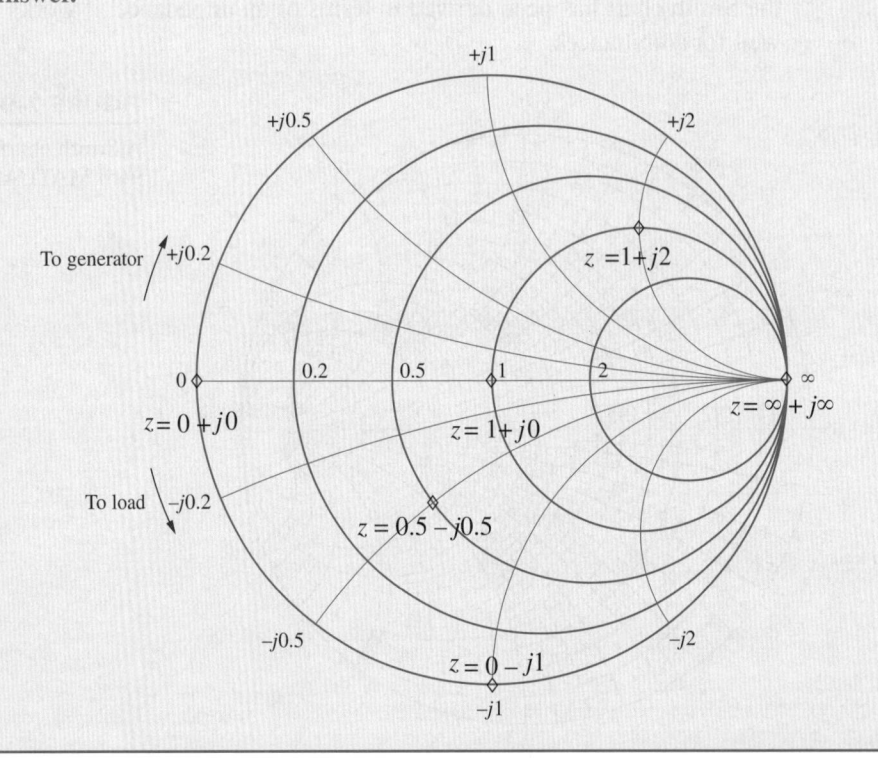

7.6 Smith Chart

From (7.42) we realize that the reflection coefficient is a complex quantity that can also be expressed in polar coordinates:

$$\Gamma = |\Gamma|e^{j\theta_L} = \frac{z_L - 1}{z_L + 1} \quad (7.48)$$

The magnitude of the reflection coefficient $|\Gamma|$ can have any value in the range $0 \leq |\Gamma| \leq 1$. This value is determined by the value of the normalized load impedance z_L and will not change regardless of location on the lossless transmission line. Since the load impedance can be complex, there will be a phase angle θ_L associated with the reflection coefficient Γ.

The normalized input impedance at any point $z = -z'$ on the transmission line can be written from (7.31) as

$$Z(z = -z') = \frac{V(z = -z')}{I(z = -z')} = Z_c \frac{\left(e^{jkz'} + \Gamma e^{-jkz'}\right)}{\left(e^{jkz'} - \Gamma e^{-jkz'}\right)} = Z_c \frac{1 + \Gamma e^{-j2kz'}}{1 - \Gamma e^{-j2kz'}} \quad (7.49)$$

If we substitute (7.48) into (7.49), we obtain

$$\frac{Z_{in}}{Z_c} = \frac{1 + \Gamma e^{-j2kz'}}{1 - \Gamma e^{-j2kz'}} = \frac{1 + |\Gamma|e^{j\phi}}{1 - |\Gamma|e^{j\phi}} \quad (7.50)$$

where

$$\phi = \theta_L - 2kz' \quad (7.51)$$

Remember that the magnitude of the reflection coefficient remains a constant as we move along the transmission line back toward the signal generator. In comparing (7.43) and (7.50), we note that the only difference is a phase shift ϕ that is linearly proportional to the distance z'. This implies that we can easily make this translation on the Smith chart by rotating the initial value of the load impedance along a circle whose radius is equal to the magnitude of the reflection coefficient $|\Gamma|$. A clockwise rotation is directed toward the signal generator, and a counterclockwise rotation is directed toward the load impedance. The amount of rotation depends upon the distance $2kz' = 4\pi z'/\lambda$, where this phase angle must be subtracted from the initial value used to locate the load impedance on the Smith chart. This impedance repeats when the argument changes by a factor of 2π, which occurs every half wavelength on the line.

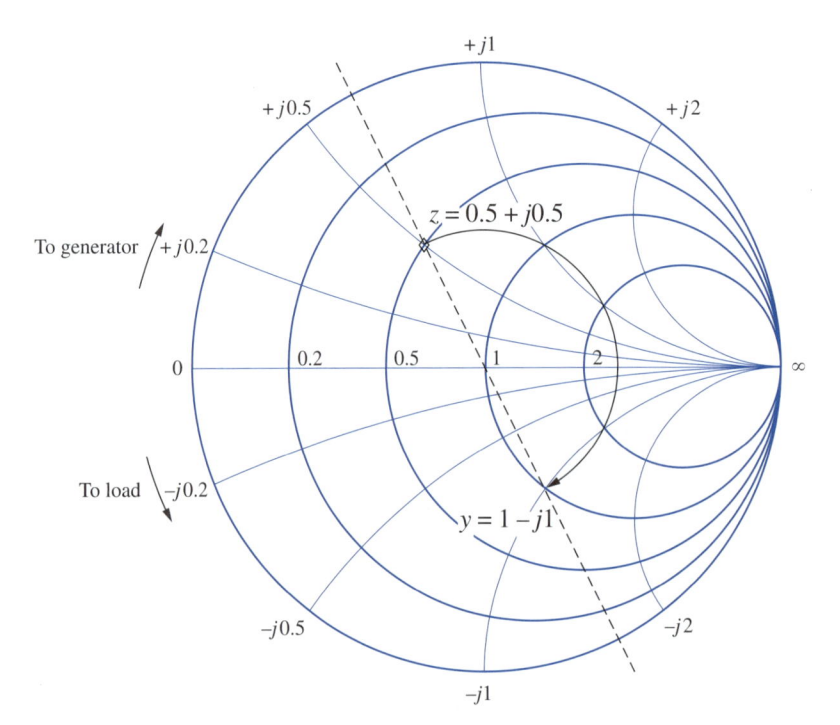

FIGURE 7–10

The transformation of an impedance to an admittance using the Smith chart. A semicircle whose radius is equal to the magnitude of the reflection coefficient is drawn. This corresponds to motion of a distance of $\lambda/4$ along the transmission line.

If we choose the distance to be equal to be $z' = \lambda/4$, the rotation will be equal to π radians. Referring back to (7.35), remember that this distance converts the numerical value of an impedance into the numerical value of an admittance. For example, the impedance $z = 0.5 + j0.5$ corresponds to an admittance of

$$y = \frac{1}{0.5 + j0.5} = 1 - j1$$

Rather than performing this computation with a calculator, it can also be found directly from the Smith chart. The normalized impedance is first located on the chart. A semicircle that has a radius equal to the reflection coefficient is drawn on the chart along with a straight line that passes through the center of the Smith chart and this impedance. The intersection of the line with the semicircle will yield directly the proper value of the admittance as illustrated in Figure 7–10.

Figure 7–10 further shows that a normalized load impedance equal to 0 will yield a normalized load admittance that is equal to $\pm j\infty$. This implies that the input impedance that is $\lambda/4$ from a short circuit will be an input impedance equal to that of an open circuit. In practice, it is difficult to make an open circuit in the transmission line due to the fringing fields, but a quarter-wavelength shorted transmission line can be used to create an open circuit.

EXAMPLE 7.11

A load impedance $Z_L = 50 + j50$ Ω terminates a transmission line that is 5 m long and has a characteristic impedance of $Z_c = 25$ Ω. Using the Smith chart, find the impedance at the signal generator if the frequency of oscillation $f = 1 \times 10^5$ Hz. The phase velocity for this transmission line is $v = 2 \times 10^6$ m/s.

Answer. The wavelength λ is

$$\lambda = \frac{v}{f} = \frac{2 \times 10^6}{1 \times 10^5} = 20 \text{ m}$$

The distance between the load and the generator is $\lambda/4$. The normalized load impedance is

$$z_L = \frac{50 + j50}{25} = 2 + j2$$

The normalized load impedance is first located on the Smith chart.

Locate the center of a compass at the center of the Smith chart and draw an arc a distance of $\lambda/4$ in the clockwise direction which is toward the generator. The normalized load impedance at the generator z_{in} as read from the Smith chart is $z_{in} = 0.25 - j0.25$. Therefore, the input impedance that is connected to the signal generator is

$$Z_{in} = Z_c z_{in} = 25 (0.25 - j0.25) = 6.25 - j6.25 \text{ Ω}$$

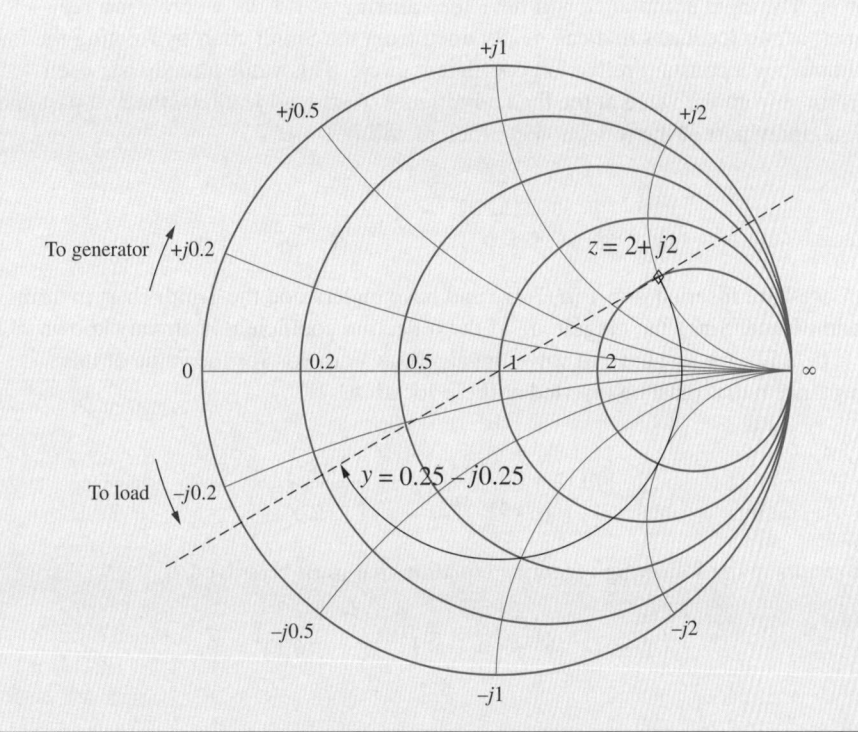

Figure 7–10 also reveals an important potential application of the Smith chart. This chart can be interpreted equally well in terms of an impedance or an admittance, and we note that a constant coefficient circle will pass through the "real part equals 1" circle at two locations. Let us now interpret this as an admittance chart. At either of the locations where it has passed through the $g = 1$ circle, an admittance can be added in parallel with the transmission line. Rather than separating a transmission line and inserting a matching element in series, it is better just to insert the matching element in parallel. It should have a value that will cause the input admittance from that location back to the signal generator to have a normalized value of 1. This is called *matching* a transmission line. Let us illustrate this with an example.

EXAMPLE 7.12

A load admittance has the value $y_L = 0.2 - j0.5$. Find the locations where a matching admittance should be placed. In addition, find the value for the matching admittance.

Answer. The input admittance will have the value $y_{in} = 1 \pm jb$, where b is a real number, at two locations that can be obtained from the Smith chart by rotating the load admittance on a constant reflection coefficient circle. This value already has been determined from the value at the load admittance. At these locations, the real part and the imaginary part of the reflection coefficient will be

$$\Gamma_r = \frac{b^2}{4 + b^2} \qquad \Gamma_i = \frac{2b}{4 + b^2}$$

where we have inserted $g = 1$ in (7.42) and have understood the Smith chart in terms of the admittance. Since the magnitude of the reflection coefficient is already known ($|\Gamma| = 0.7257$), we just have to solve the algebraic equation for the value of the susceptance b that must be inserted at these locations.

$$|\Gamma|^2 = \left(\frac{b^2}{4 + b^2}\right)^2 + \left(\frac{2b}{4 + b^2}\right)^2$$

This results in the following algebraic equation that must be solved.

$$\left(1 - |\Gamma|^2\right)b^4 + \left(4 - 8|\Gamma|^2\right)b^2 - 16|\Gamma|^2 = 0$$

7.6 Smith Chart

The roots of this polynomial are calculated to be $b^2 = 4.45$ and $b^2 = -4$, and because b must be a real number, we neglect the second solution. Thus, we find two real solutions ($b_1 \approx 2.11$ and $b_2 \approx -2.11$), which are in good agreement with the two graphical solutions obtained from the Smith chart. The foregoing calculation using the Smith chart is shown below.

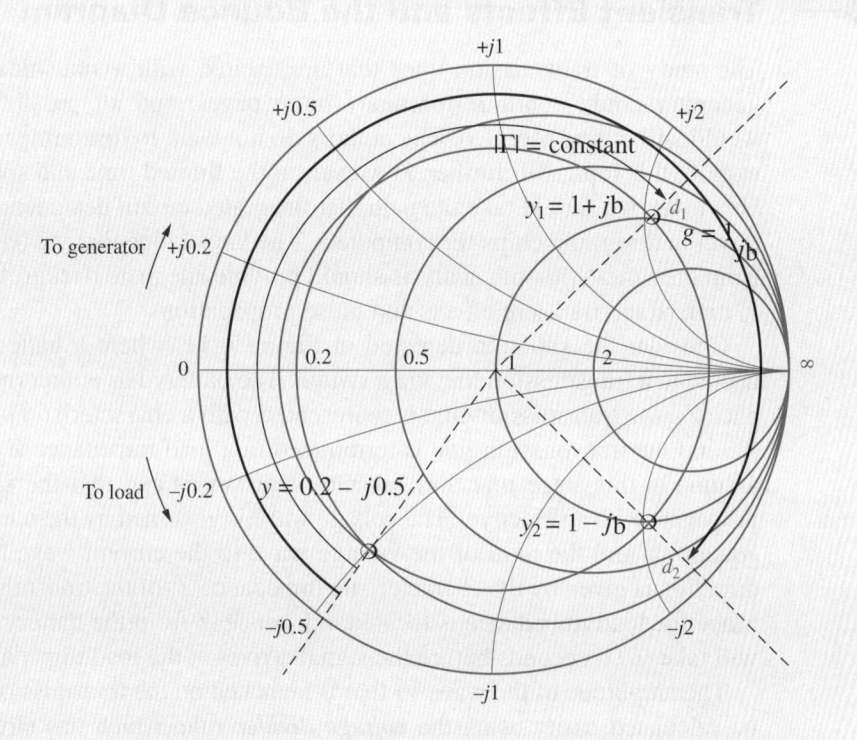

The location of the load admittance is indicated with a ◯, and the radius of the circle that passes through this point is equal to the magnitude of the reflection coefficient circle. The circle passes through the $g = 1$ circle at two locations. The closest one is at a distance d_1 from the load admittance. One should insert an admittance equal to $-jb$ at that location. The transmission line will be matched from that point back to the signal generator. There is a second location d_2 that is further from the load admittance where one could insert an admittance that is equal to $+jb$, and it could be used if it is inconvenient to choose the first location.

The matching elements that were employed in Example 7.12 could be created easily by using a section of a transmission line that is terminated in a short circuit. The disadvantage of using the **single-stub** matching technique is that the distance d must be adjustable. In practice, this would typically be accomplished with a narrow slit inserted into the coaxial cable or the strip line. There are techniques that can be used to minimize the reflection coefficient. The techniques, however, are frequency sensitive and may even be based on electrical circuit filter theory.

7.7 Transient Effects and the Bounce Diagram

The study of transmission lines that are excited with a sinusoidal voltage generator could continue for many more pages, and all possible aspects would still not be revealed. The authors do not want to discourage efforts to explore this topic still further. However, in the limited time and space available to us, we should take the path that integrated circuit designers regularly travel in designing chips for computers. The knowledge that we have gained from the time-harmonic analysis should provide adequate background for us to understand transient effects and pulse propagation.

Consider the situation depicted in Figure 7–11, where a battery is connected to a transmission line via a switch. The battery has an internal impedance Z_b, the transmission line is represented with a characteristic impedance Z_c, and the transmission line is terminated in a load impedance Z_L. We will assume at this stage that they are pure resistances and that the signal will propagate with a velocity v. The voltage wave is governed by the telegraphers' equations, and the ratio of the voltage wave to the current wave in a given direction is given by the characteristic impedance Z_c of the transmission line. Since the load impedance is located a distance $\mathscr{L}$ from the battery switch, it will take $\mathscr{L}/v$ seconds before the signal arrives at the load impedance.

The amplitude of the wave V_1 that is launched on the transmission line can be calculated easily using the **voltage divider** rule, which has already been encountered in earlier courses. This amplitude is dictated by the two impedances at the input of the transmission line—namely, the battery impedance and the characteristic impedance of the transmission line. It is given by

$$V_1 = \frac{Z_c}{Z_b + Z_c} V_b \tag{7.52}$$

After a time $\tau = \mathscr{L}/v$, the front of this propagating voltage step arrives at the load impedance Z_L. At this time, a portion of the incident voltage is reflected from the load impedance, and a portion of the incident voltage is "transmitted" or absorbed by the load impedance. The reflection coefficient at the load Γ_L is given by

$$\Gamma_L = \frac{Z_L - Z_c}{Z_L + Z_c} = \frac{V_2}{V_1} \tag{7.53}$$

7.7 Transient Effects and the Bounce Diagram

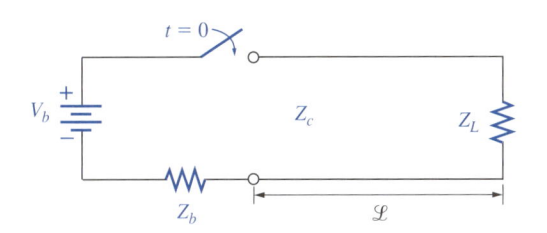

FIGURE 7-11

The battery is connected to the transmission line along with a switch that is closed at $t = 0$.

The amplitude of the reflected voltage step V_2 can be positive or negative, depending on the relative values of the impedances that appear in (7.53).

Eventually, the front of this propagating reflected voltage step V_2 reaches the battery impedance at a time $2\tau = 2(\mathcal{L}/v)$. This front will be reflected from the battery impedance with a reflection coefficient Γ_b, where

$$\Gamma_b = \frac{Z_b - Z_c}{Z_b + Z_c} = \frac{V_3}{V_2} \qquad (7.54)$$

The battery impedance is essentially a "load impedance" for the incident wave V_2, and a voltage step V_3 will be reflected toward the load impedance.

The front of this reflected propagating voltage step V_3 reaches the load impedance, where a portion of this voltage step will be reflected by the load impedance. This process may continue indefinitely.

The front of the propagating voltage step "bounces" back and forth between the load impedance and the battery impedance. There is a graphical technique to evaluate the voltage at any location on the line as a function of the time. This technique makes it possible to predict the response seen on an oscilloscope that is connected to a certain location on the transmission line. It is formed by plotting the trajectory of this wave front as shown in Figure 7–12, and is called a **bounce diagram**. It is convenient that the vertical and horizontal axes—which are, respectively, normalized time and position—both be dimensionless. The prediction of the temporal response at a given location is obtained by inserting a vertical line on the bounce diagram at that location. The intersection of the trajectory with this line indicates that the voltage at that location will change its value by the amplitude of that particular component of the wave.

Recall that a battery is the source of the voltage, and it has a constant value. Therefore, it is reasonable to assume that the voltage behind the propagating front of all the components will also be a constant and will have the same value as that of the front. The voltage at any location along the transmission line is just the summation of the individual components.

ANIMATIONS

$$V = V_1 + V_2 + V_3 + \cdots = \frac{Z_c}{Z_b + Z_c}(1 + \Gamma_L + \Gamma_L\Gamma_b + \cdots)V_b \qquad (7.55)$$

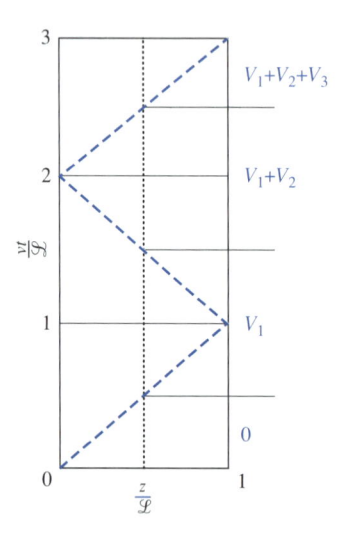

FIGURE 7-12

The bounce diagram. The magnitude of the slope of each line is equal to 1. The amplitude of each individual component is usually specified. A vertical line at a certain position on the transmission line, which in the figure is at the midpoint of the transmission line, indicates the location of an oscilloscope probe. The intersection of this vertical line and the trajectory marks the times when the voltage will change. These points are indicated with the short horizontal lines, and the subsequent voltage during that interval is given.

The reader might suspect that, as time passes, an asymptotic value might be reached as the voltage step bounces back and forth between the battery and the load impedance in Figure 7–11. Such suspicions will be well rewarded. Every voltage step that is incident on the load impedance will be reflected with a reflection coefficient Γ_L, and every voltage step incident on the battery will be reflected with a reflection coefficient Γ_b. Adding up all the individual contributions and regrouping the terms will lead to the following steady state value:

$$V = V_1 + V_2 + V_3 + V_4 + \cdots$$

$$= \frac{Z_c}{Z_b + Z_c}\left(1 + \Gamma_L + \Gamma_L\Gamma_b + \Gamma_L\Gamma_b\Gamma_L + \cdots\right)V_b$$

$$= \frac{Z_c}{Z_b + Z_c}\left\{\left[1 + \left(\Gamma_L\Gamma_b\right) + \left(\Gamma_L\Gamma_b\right)^2 + \cdots\right]\right.$$

$$\left. + \Gamma_L\left[1 + \left(\Gamma_L\Gamma_b\right) + \left(\Gamma_L\Gamma_b\right)^2 + \cdots\right]\right\}V_b \qquad (7.56)$$

The terms within the square brackets can be written using the closed form summation:

$$1 + \xi + \xi^2 + \cdots = \frac{1}{1-\xi} \qquad \text{for } |\xi| < 1 \qquad (7.57)$$

7.7 Transient Effects and the Bounce Diagram

Since the magnitude of a reflection coefficient is always less than or equal to one, so is the product of two of them, $|\Gamma_L \Gamma_b| < 1$, so we can employ this summation relation and obtain

$$V = \frac{Z_c}{Z_b + Z_c}\left(\frac{1 + \Gamma_L}{1 - \Gamma_L \Gamma_b}\right) V_b \qquad (7.58)$$

Substituting the definitions for the reflection coefficient at the load impedance and at the battery impedance, (7.58) becomes

$$V = \frac{Z_c}{Z_b + Z_c}\left\{\frac{1 + \left[(Z_L - Z_c)/(Z_L + Z_c)\right]}{1 - \left[(Z_L - Z_c)/(Z_L + Z_c)\right]\left[(Z_b - Z_c)/(Z_b + Z_c)\right]}\right\} V_b$$

This simplifies to

$$V = \frac{Z_L}{Z_b + Z_L} V_b \qquad (7.59)$$

Based on our previous experience in circuit theory, this value for the steady state load voltage should not be too surprising.

The current that flows through the load impedance is equal to the voltage at the load divided by the load impedance. The long-time asymptotic value is given by

$$I = \frac{V}{Z_L} = \frac{1}{Z_b + Z_L} V_b \qquad (7.60)$$

EXAMPLE 7.13

A 12-V battery is connected via a switch to a transmission line that is 6 m long. The characteristic impedance of the transmission line is 50 Ω, the battery impedance is 25 Ω, and the transmission line is terminated in a load impedance of 25 Ω. The velocity of propagation along this transmission line is 2×10^6 m/s. Find and sketch the voltage at the midpoint of this transmission line during the time interval $0 < t < 9$ μs.

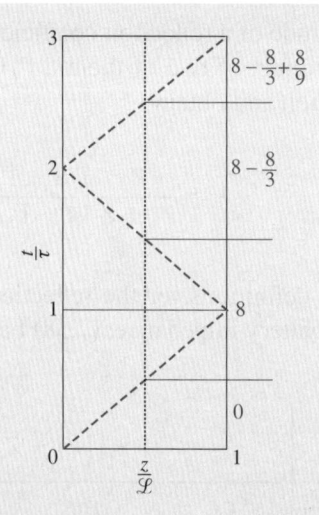

Answer. The amplitude of the wave that is launched on the transmission line is calculated from

$$V_1 = \frac{Z_c}{Z_b + Z_c} V_b = \frac{50}{25 + 50} 12 = 8 \text{ V}$$

The reflection coefficient at the load is equal to

$$\Gamma_L = \frac{Z_L - Z_c}{Z_L + Z_c} = \frac{25 - 50}{25 + 50} = -\frac{1}{3}$$

The reflection coefficient at the battery is equal to

$$\Gamma_b = \frac{Z_b - Z_c}{Z_b + Z_c} = \frac{25 - 50}{25 + 50} = -\frac{1}{3}$$

In order to calculate the voltage at the midpoint of the transmission line, we make use of the bounce diagram. In this case, we clearly identify the amplitudes of the waves. The normalized time is t/τ, where $\tau = \mathcal{L}/v = 3\,\mu\text{s}$. The bounce diagram is obtained first.

Using the bounce diagram, the voltage at the midpoint of the transmission line is equal to 0 until the front of the wave arrives. The voltage increases to the amplitude of the wave, and it remains at that value until the wave that is reflected from the load impedance passes the midpoint. This reflected wave is reflected again at the battery and arrives at the midpoint. The following figure depicts the expected response of the oscilloscope that is located at the midpoint of the transmission line. The final value of this voltage is calculated from (7.59) to be

$$V = \frac{Z_L}{Z_b + Z_L} V_b = \frac{25}{25 + 25} 12 = 6 \text{ V}$$

7.7 Transient Effects and the Bounce Diagram

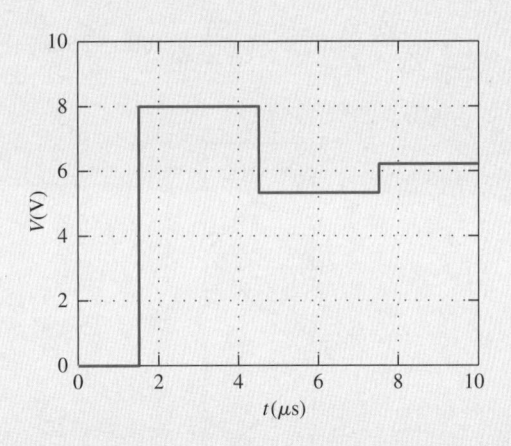

EXAMPLE 7.14

A battery with 0 internal impedance has an open circuit voltage of 100 volts. At a time $t = 0$, this battery is switched into a 50-Ω air–dielectric coaxial cable via a 150-Ω resistor. The cable is 300 m long and is terminated in a load of 33.3 Ω.

(a) Sketch a bounce diagram for the first 4 μs after the switch is closed.
(b) Draw a graph of the voltage that appears across the load impedance as a function of time.
(c) Find the asymptotic values of V_L and I_L as $t \to \infty$.

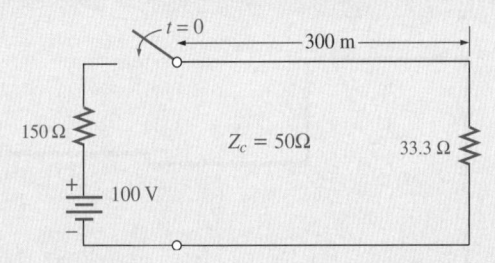

Answer.

(a) The two reflection coefficients and the incident voltage step that propagates on the line are given by

$$\Gamma_b = \frac{Z_b - Z_c}{Z_b + Z_c} = \frac{150 - 50}{150 + 50} = \frac{1}{2}$$

$$\Gamma_L = \frac{Z_L - Z_c}{Z_L + Z_c} = \frac{33.3 - 50}{33.3 + 50} = -\frac{1}{5}$$

$$V_1 = \frac{Z_c}{Z_b + Z_c} V_b = \frac{50}{150 + 50} 100 = 25 \text{ V}$$

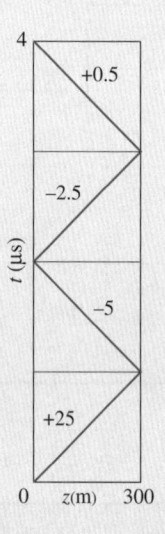

Since the coaxial cable is filled with air, the velocity of propagation equals 3×10^8 m/s. The voltage signal takes 1 μs to travel from one end to the other.

(b) The asymptotic voltages and currents as computed from (7.59) and (7.60) are for $V_{t \to \infty} = 18.2$ V and $I_{t \to \infty} = 0.55$ A.

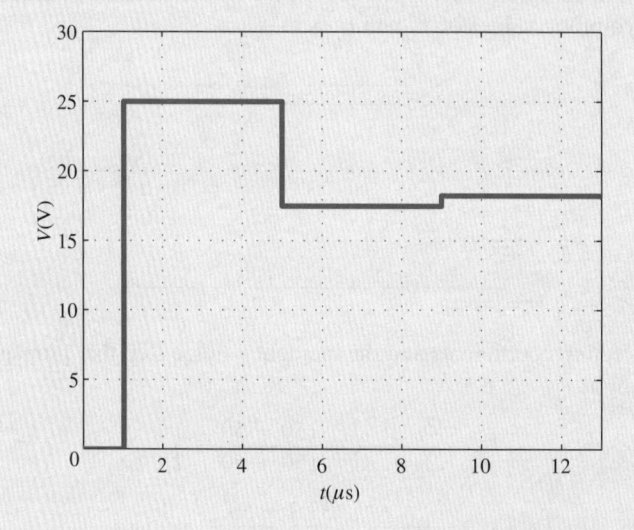

7.8 Pulse Propagation

In the previous section, we examined the transient characteristics of a step voltage as it propagated along a transmission line. We observed that the front propagated with a definite velocity, and it took a nonzero time for the signal to pass from one point to another. This time was found to be of the order of $\mathscr{L}/v$, where $\mathscr{L}$ is the distance to be traveled and v is the velocity of propagation of the signal. In this section, we will devote our energies to the study of the propagation of a voltage pulse along a transmission line. An emphasis on this particular topic is certainly justified, owing to its practical importance in digital integrated circuits and in practical laboratory measurements.

Let us consider a transmission line that connects a pulse generator to a load impedance as shown in Figure 7–13. As in the previous section describing transient effects, we can calculate the amplitude V_1 of the pulse launched on the transmission line using a voltage divider rule

$$V_1 = \frac{Z_c}{Z_g + Z_c} V_g \tag{7.61}$$

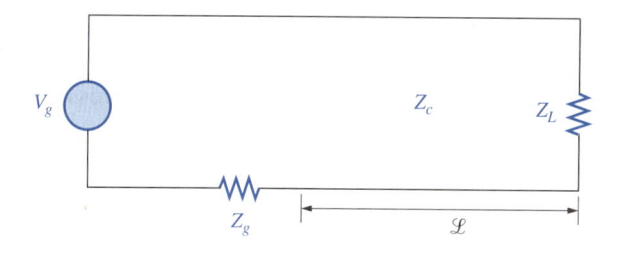

FIGURE 7–13

A pulse generator V_g that has an internal impedance of Z_g is connected to a transmission line that has a characteristic impedance Z_c and is terminated in a load impedance Z_L.

We will assume that the temporal width Δt of the pulse is much less than the time it takes for the pulse to travel from one end of the transmission line to the other; that is, $\Delta t \ll \mathscr{L}/v$. For instance, if the transmission line has length 3 meters and we assume propagation at the velocity of light, this means the pulse width is much less than 10 nanoseconds.

$$\frac{\mathscr{L}}{v} \approx \frac{3 \text{ m}}{3 \times 10^8 \text{ m/s}} \approx 10 \times 10^{-9} \text{ s}$$

We have used the velocity of light in this estimate, although in actuality the velocity of propagation will be decreased by the square root of the dielectric constant of the meduim that exists between the two conductors.

The pulse V_1 that is launched from the signal generator along the transmission line propagates toward the load impedance. A portion of the pulse is absorbed by the load impedance, and a portion of the pulse is reflected back toward the signal generator. The amount that is absorbed or the amount that is reflected is determined by the ratio of the load impedance to the characteristic impedance of the transmission line. We can compute the value of the absorbed signal from a consideration of the flow of energy along the transmission line.

The energy of the incident pulse $P_{inc}\Delta t$ is divided into the energy in the reflected wave $P_{ref}\Delta t$ and the energy that is absorbed in the load impedance $P_{abs}\Delta t$. The energy must be conserved in this junction. We write

$$P_{inc}\Delta t = P_{ref}\Delta t + P_{abs}\Delta t \tag{7.62}$$

The common factor Δt will cancel in this expression, and the various terms for the power can be written in terms of the impedances and the reflection coefficient

$$\frac{V_1^2}{Z_c} = \frac{(\Gamma V_1)^2}{Z_c} + \frac{V_L^2}{Z_L} \tag{7.63}$$

where V_1 is the incident voltage, Γ is the reflection coefficient, and V_L is the voltage that appears across the load impedance. Inserting the expression for the reflection coefficient (7.24), we write

$$\frac{V_1^2}{Z_c} = \frac{\left(\left[(Z_L - Z_c)/(Z_L + Z_c)\right]V_1\right)^2}{Z_c} + \frac{V_L^2}{Z_L} \tag{7.64}$$

Solving for the ratio of this voltage divided by the incident voltage V_1, we obtain

$$V_L = V_1\sqrt{\frac{Z_L}{Z_c} - \frac{\left[(Z_L - Z_c)/(Z_L + Z_c)\right]^2 Z_L}{Z_c}} = V_1\frac{2Z_L}{Z_L + Z_c}$$

or

$$\boxed{T = \frac{V_L}{V_1} = \frac{2Z_L}{Z_L + Z_c}} \tag{7.65}$$

where T is defined as the *transmission coefficient*. It obeys the important relation to the reflection coefficient: $T = 1 + \Gamma$.

7.8 Pulse Propagation

EXAMPLE 7.15

Calculate the transmission coefficient T for a wave that is propagating in the $+z$ direction in a coaxial cable. The relative dielectric constant of the separating dielectric in the region $z < 0$ is 2 and in the region $z > 0$ is 3. The physical dimensions of the cable are the same in all regions.

Answer. Using (7.65) and the results given in Table 7–2, we write

$$T = \frac{2\sqrt{(\mu/3\varepsilon_0)}[\ln(b/a)/2\pi]}{\sqrt{(\mu/3\varepsilon_0)}[\ln(b/a)/2\pi] + \sqrt{(\mu/2\varepsilon_0)}[\ln(b/a)/2\pi]}$$

$$= \frac{2/\sqrt{3}}{(1/\sqrt{3}) + (1/\sqrt{2})} = 0.9$$

This point can be emphasized if we consider the joining of two transmission lines that may have different dimensions, as shown in Figure 7–14. The characteristic impedance of the transmission line between the two electrodes is different in the two regions of the integrated circuit. A portion of the signal launched from one electrode will reach the second electrode, and a portion will be reflected back to the original electrode.

A sequence of pictures of an incident pulse propagating from one transmission line into another one is shown in Figure 7–15. Note that the signals on the two lines may propagate with differing velocities. As usual, $v = 1/\sqrt{\hat{L}\hat{C}}$, but the capacitance and inductance per unit length may differ on the lines.

Knowing the velocity of propagation v on a transmission line has some very practical consequences. Let us assume that we can launch a pulse on a transmission line and measure the time ΔT that it takes for a reflected pulse to

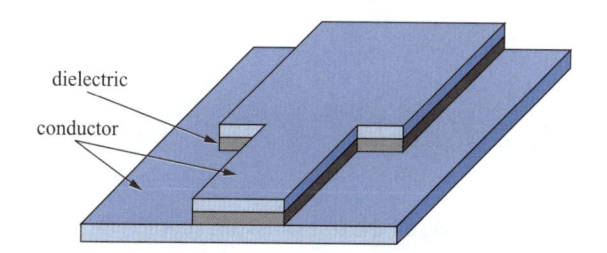

FIGURE 7–14

An integrated circuit transmission line element, called microstrip transmission line.

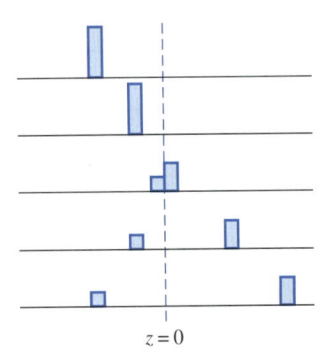

FIGURE 7-15

Snapshots of a voltage pulse crossing a discontinuity in a transmission line. The pictures are taken at equal intervals in time, and the velocity of propagation has $v_{z>0} > v_{z<0}$. In addition, the characteristic impedances of the two lines have the relative values $Z_{c(z>0)} > Z_{c(z<0)}$.

return. From these data, we can exactly compute the unknown distance d from the pulse generator to where the reflection took place, since $\Delta T = 2(d/v)$. Imagine trying to locate a fault in an integrated circuit or a short circuit in a cable that is buried underground. Knowing in advance where to probe or dig might save many hours of frustration. This practical technique is called *time domain reflectometry*.

EXAMPLE 7.16

Using the reflection coefficient Γ and the transmission coefficient T, show that energy is conserved at the junction between two lossless transmission lines.

Answer. Conservation of energy implies that (7.62) must be satisfied. This implies that

$$\frac{V_{inc}^2}{Z_{c1}} = \frac{(\Gamma V_{inc})^2}{Z_{c1}} + \frac{(TV_{inc})^2}{Z_{c2}}$$

Substituting the values for the reflection coefficient and the transmission coefficient, we write (after canceling the value of the incident voltage wave)

7.8 Pulse Propagation

$$\frac{1}{Z_{c1}} = \frac{|\Gamma|^2}{Z_{c1}} + \frac{|T|^2}{Z_{c2}} = \frac{\left[(Z_{c2} - Z_{c1})/(Z_{c2} + Z_{c1})\right]^2}{Z_{c1}} + \frac{\left[2Z_{c2}/(Z_{c2} + Z_{c1})\right]^2}{Z_{c2}} = \frac{1}{Z_{c1}}$$

EXAMPLE 7.17

A 1-V pulse propagates from $z < 0$ on a transmission line. The line is terminated in an open circuit at $z = 0$. Four oscilloscopes are triggered by the same pulse generator and are located at $z_a = -6$, $z_b = -4$, $z_c = -2$, and $z_d = 0$ m. Find the velocity of propagation and interpret the voltage signals on the oscilloscopes. Sketch the corresponding voltage signals if the transmission line is terminated in a short circuit.

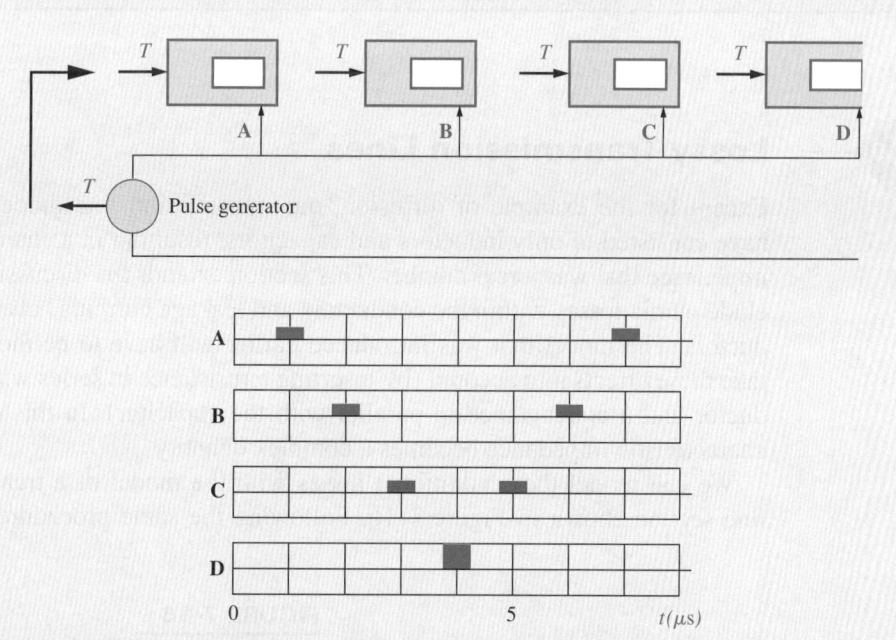

Answer. From the traces on Oscilloscopes A and B, we find the velocity of propagation to be $v = \Delta z / \Delta t = 2 \text{ m}/1\,\mu\text{s} = 2 \times 10^6 \text{ m/s}$. Oscilloscope D is at the location of the open circuit, and the incident and the reflected pulses add together. The signals that are detected after $t = 4$ s are the reflected pulses that propagate toward the pulse generator. The voltage signals detected by the oscilloscopes if the transmission line is terminated in a short circuit are depicted below. The voltage across the short circuit must be zero; hence, the signal at Oscilloscope D is zero.

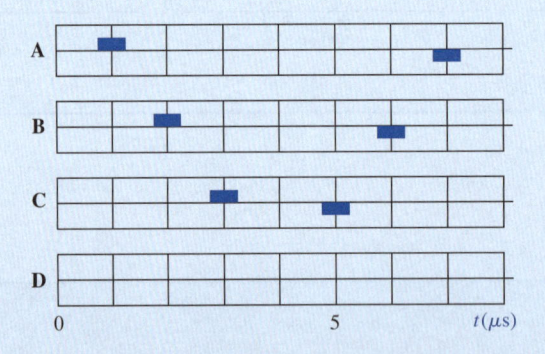

It should be noted that the two oscilloscope pictures correspond to the voltage pulses. If the oscilloscope pictures corresponded to the current pulses, the first picture would correspond to a load impedance that was a short circuit, and the second picture would correspond to an open circuit.

7.9 Lossy Transmission Lines

Except for the example of diffusion, our transmission line models so far have consisted of only inductors and capacitors, resulting in a characteristic impedance that was a real number. This section extends the discussion to include ohmic losses within the conductors and leakage currents between conductors. The model that was introduced earlier will have to be modified to take these effects into account (by inserting a resistance in series with the inductor and a conductance in parallel with the capacitor). In this case, the characteristic impedance becomes a complex quantity.

We can model these additional losses with the model of a transmission line section shown in Figure 7–16. Following the same procedure that we

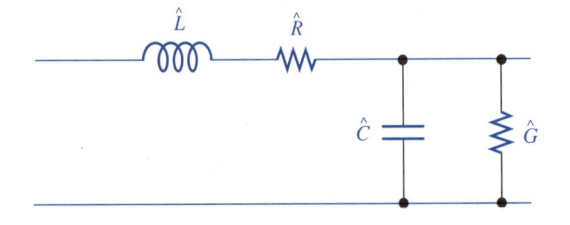

FIGURE 7–16

Model of a section whose length is Δz of a transmission line that includes loss terms. The units of all of the elements are per unit length.

7.9 Lossy Transmission Lines

employed to write the first-order partial differential equations for the lossless transmission line (7.3) and (7.6), we obtain

$$\frac{\partial I(z,t)}{\partial z} = -\hat{C}\frac{\partial V(z,t)}{\partial t} - \hat{G}V(z,t) \quad (7.66)$$

$$\frac{\partial V(z,t)}{\partial z} = -\hat{L}\frac{\partial I(z,t)}{\partial t} - \hat{R}I(z,t) \quad (7.67)$$

where the circuit elements are defined as

$$\hat{L} = \frac{L}{\Delta z} \quad \hat{C} = \frac{C}{\Delta z} \quad \hat{R} = \frac{R}{\Delta z} \quad \hat{G} = \frac{G}{\Delta z} \quad (7.68)$$

The set of the first-order partial differential equations (7.66) and (7.67) are also known as the telegraphers' equations. The wires that crossed the United States were lossy.

In order to factor in the loss, it is convenient to assume that there is a time-harmonic excitation of the transmission line. If we make this assumption, then (7.66) and (7.67) become

$$\frac{\partial I(z)}{\partial z} = -\left[\hat{G} + j\omega\hat{C}\right]V(z) \quad (7.69)$$

$$\frac{\partial V(z)}{\partial z} = -\left[\hat{R} + j\omega\hat{L}\right]I(z) \quad (7.70)$$

where we introduce the phasor notation for the currents and voltages. The terms within the square brackets are denoted by distributed admittance $\hat{Y}$ and distributed impedance $\hat{Z}$ quantities, respectively. Hence we can rewrite these two equations as

$$\frac{dI(z)}{dz} = -\hat{Y}V(z) \quad (7.71)$$

$$\frac{dV(z)}{dz} = -\hat{Z}I(z) \quad (7.72)$$

The coupled first-order ordinary differential equations can be used to obtain a pair of second-order ordinary differential equations for each of the dependent variables. We write them as

$$\frac{d^2 I(z)}{dz^2} = \hat{Z}\hat{Y}I(z) \quad (7.73)$$

$$\frac{d^2V(z)}{dz^2} = \hat{Z}\hat{Y}V(z) \tag{7.74}$$

The solutions of the phasor line voltage and current are

$$V(z) = V_1 e^{-\gamma z} + V_2 e^{+\gamma z}$$
$$I(z) = I_1 e^{-\gamma z} + I_2 e^{+\gamma z} \tag{7.75}$$

We have introduced the **complex propagation constant**

$$\boxed{\gamma = \alpha + j\beta = \sqrt{\hat{Z}\hat{Y}} = \sqrt{(\hat{R} + j\omega\hat{L})(\hat{G} + j\omega\hat{C})}} \tag{7.76}$$

As previously, the terms V_1 and I_1 correspond to the amplitudes of the forward propagating voltage and current waves, and the terms V_2 and I_2 correspond to the amplitudes of the backward propagating waves, respectively. The time varying waves are recovered from the phasors in the usual fashion. Using (7.76) in (7.75) gives

$$V(z, t) = V_1 e^{-\alpha z}\cos(\omega t - \beta z) + V_2 e^{+\alpha z}\cos(\omega t + \beta z)$$
$$I(z, t) = I_1 e^{-\alpha z}\cos(\omega t - \beta z) + I_2 e^{+\alpha z}\cos(\omega t + \beta z) \tag{7.77}$$

We recognize that (7.77) represents waves that decay exponentially as they propagate (with the terms V_1 and I_1 decaying for increasing values of z and the terms V_2 and I_2 decaying for decreasing values of z). A "snapshot" of a decaying wave propagating in the forward direction is shown in Figure 7–17. It is possible to determine the values of α and β from the figure.

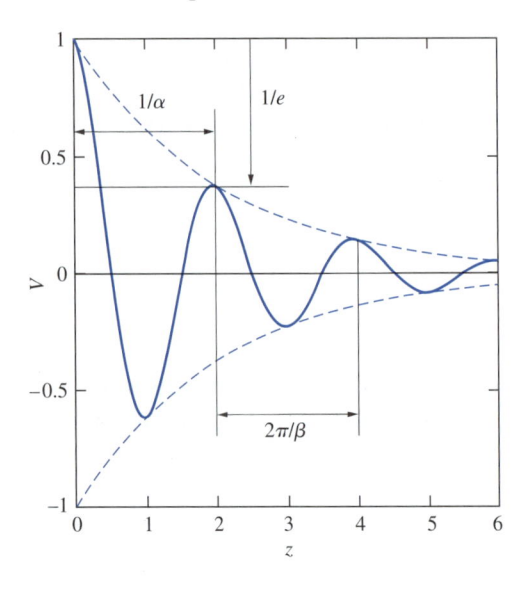

FIGURE 7–17

A plot of a time-harmonic voltage signal at an instant in time as a function of position. From this figure, one can determine the complex propagation constant

$$\gamma = \alpha + j\beta = -\frac{1}{2} - j\frac{2\pi}{2} = -\frac{1}{2} - j\pi$$

7.9 Lossy Transmission Lines

For the case where the loss terms are small, we can approximate this complex propagation constant (7.76) as

$$\gamma = \alpha + j\beta = \sqrt{\left(j\omega\hat{L}\right)\left(j\omega\hat{C}\right)}\sqrt{\left(1 + \frac{\hat{R}}{j\omega\hat{L}}\right)\left(1 + \frac{\hat{G}}{j\omega\hat{C}}\right)} \qquad (7.78)$$

$$\approx j\omega\sqrt{\hat{L}\hat{C}}\left[1 - j\left(\frac{\hat{R}}{2\omega\hat{L}} + \frac{\hat{G}}{2\omega\hat{C}}\right)\right]$$

Here we have used the binomial approximation $(1 - x)^n \simeq 1 - nx$ for $x \ll 1$ to simplify the result. From the approximation (7.78), we see that the wave will propagate in the $\pm z$ directions, but the amplitude will attenuate as it propagates. The attenuation constant α will be given approximately by

$$\alpha \approx \omega\sqrt{\hat{L}\hat{C}}\left(\frac{\hat{R}}{2\omega\hat{L}} + \frac{\hat{G}}{2\omega\hat{C}}\right) = \frac{1}{2}\sqrt{\hat{L}\hat{C}}\left(\frac{\hat{R}}{\hat{L}} + \frac{\hat{G}}{\hat{C}}\right) \qquad (7.79)$$

Note that the attenuation constant is independent of frequency.

EXAMPLE 7.18

Find the complex propagation constant if the circuit elements satisfy the ratio $\hat{R}/\hat{L} = \hat{G}/\hat{C}$. Interpret the propagation of such a signal that propagates on this line.

Answer. The complex propagation constant (7.77) can be written as

$$\gamma = \alpha + j\beta = \sqrt{\left(\hat{R} + j\omega\hat{L}\right)\left(\hat{G} + j\omega\hat{C}\right)}$$

$$= \sqrt{(\hat{R} + j\omega\hat{L})\left(\frac{\hat{R}\hat{C}}{\hat{L}} + j\omega\hat{C}\right)} = \left(\hat{R} + j\omega\hat{L}\right)\sqrt{\frac{\hat{C}}{\hat{L}}}$$

In this case, the attenuation constant α and the phase velocity ω/β are independent of frequency. This implies that there will be no distortion of a signal as it propagates on this transmission line. There will only be a constant attenuation of the signal. The characteristic impedance of this transmission line

$$Z_c = \sqrt{\frac{\hat{Z}}{\hat{Y}}} = \sqrt{\frac{\hat{R} + j\omega\hat{L}}{\hat{G} + j\omega\hat{C}}} = \sqrt{\frac{\hat{R} + j\omega\hat{L}}{\left(\hat{R}\hat{C}/\hat{L}\right) + j\omega\hat{C}}} = \sqrt{\frac{\hat{L}}{\hat{C}}}$$

is also independent of frequency. This transmission line is called a *distortionless line.*

EXAMPLE 7.19

The attenuation on a 50-Ω distortionless transmission line is 0.01 dB/m. The line has a capacitance of 0.1×10^{-9} F/m.

(a) Find the values of the transmission line parameters $\hat{L}$, $\hat{R}$, and $\hat{G}$.
(b) Find the velocity of wave propagation.

Answer.

(a) Since this is a distortionless transmission line, we write

$$Z_c = \sqrt{\frac{\hat{L}}{\hat{C}}} = 50 \ \Omega \Rightarrow \hat{L} = (50)^2 \left(0.1 \times 10^{-9}\right) = 2.5 \times 10^{-7} \ \text{H/m}$$

$$\alpha = \hat{R}\sqrt{\frac{\hat{C}}{\hat{L}}} = 0.01 \ \text{dB/m} = \frac{0.01}{8.69} \ \text{Np/m} = 0.0012 \ \text{Np/m}$$

From this, we write

$$\hat{R}\sqrt{\frac{\hat{C}}{\hat{L}}} = \frac{\hat{R}}{50} = 0.0012 \Rightarrow \hat{R} = 0.0575 \ \Omega/\text{m}$$

We also have the distortionless line criteria

$$\frac{\hat{R}}{\hat{L}} = \frac{\hat{G}}{\hat{C}} \Rightarrow \hat{G} = \frac{0.0575}{50^2} = 2.3 \times 10^{-5} \ \text{S/m}$$

(b) The phase velocity is

$$v = \frac{1}{\sqrt{\hat{L}\hat{C}}} = \frac{1}{\sqrt{\left(2.5 \times 10^{-7}\right)\left(0.1 \times 10^{-9}\right)}} = 2 \times 10^8 \ \text{m/s}$$

7.10 Dispersion and Group Velocity

So far, we have generalized our model of a real transmission line to take into account the effect of losses due to the finite conductivity of real conductors, leading to the inclusion of the resistive term $\hat{R}$, and losses due to the non-zero conductivity of the dielectric, leading to the inclusion of the conductive term $\hat{G}$. These effects, the finite conductivity of the wire and the leakage currents through the dielectric introduced loss mechanisms, which lead to

7.10 Dispersion and Group Velocity

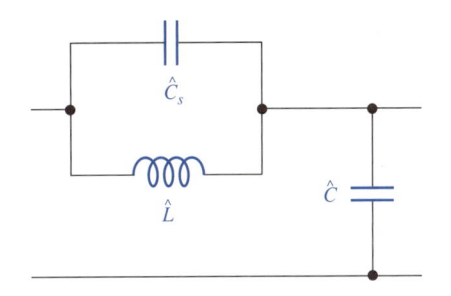

FIGURE 7-18

A model of a section of a transmission line, whose length is Δz, that includes dispersion.

waves whose amplitude will decrease as they propagate on the transmission line. In this section, we will examine another physical effect, known as dispersion, which is found in cases where the wavelength is comparable with the physical dimensions of the transmission line or in cases where the permittivity of the separating dielectric depends upon the frequency. As we will find, dispersion limits the frequency response of the transmission line, leads to signal distortion for non-sinusoidal signals, and prompts the introduction of a new velocity of propagation that is called the ***group velocity***, denoted v_g. All real transmission lines exhibit this effect. We will discuss a simple circuit model for a dispersive transmission line.

A simple model consisting of linear elements that we will use to introduce the concept of dispersion is shown in Figure 7–18. Following the same procedure that we have used previously, we can write down the telegraphers' equations that are applicable for a transmission line consisting of a large number of these sections. One caveat is immediately encountered—the current that enters the node at the left will subdivide into one current I_L that passes through the inductor and another current that passes through the capacitor I_c that is in parallel with this inductor:

$$I(z, t) = I_L(z, t) + I_c(z, t) \tag{7.80}$$

The appropriate equation that describes the voltage drop across the inductor is

$$\frac{\partial V(z, t)}{\partial z} = -\hat{L} \frac{\partial I_L(z, t)}{\partial t} \tag{7.81}$$

The equation that describes the voltage drop across the capacitor that is in parallel with this inductor is

$$\frac{\partial V(z, t)}{\partial z} = -\frac{1}{\hat{C}_s} \int I_c \, dt \tag{7.82}$$

The units of this additional capacitor are F·m rather than F/m. The equation that describes the current that passes through the shunt capacitor is given by

$$\frac{\partial I(z,t)}{\partial z} = -\hat{C}\frac{\partial V(z,t)}{\partial t} \tag{7.83}$$

From the set of equations (7.80) through (7.83), we derive the following wave equation:

$$\frac{\partial^2 V(z,t)}{\partial z^2} - \hat{L}\hat{C}\frac{\partial^2 V(z,t)}{\partial t^2} + \hat{L}\hat{C}_s\frac{\partial^4 V(z,t)}{\partial z^2 \partial t^2} = 0 \tag{7.84}$$

Let us assume that there is a time-harmonic signal generator connected to this transmission line, which is infinitely long. The complex time-varying wave that is excited and propagates on this line is of the form

$$V(z,t) = V_0 e^{j(\omega t - \beta z)} \tag{7.85}$$

The substitution of (7.85) into (7.84) leads to the **dispersion relation** that relates the propagation constant β to the frequency of the wave ω. We obtain

$$\left[(-j\beta)^2 - \hat{L}\hat{C}(j\omega)^2 + \hat{L}\hat{C}_s(j\omega)^2(-j\beta)^2\right]V_0 e^{j(\omega t - \beta z)} = 0 \tag{7.86}$$

where the terms within the square brackets yield the dispersion relation. We write this as

$$\beta = \pm \frac{\omega\sqrt{\hat{L}\hat{C}}}{\sqrt{1 - \omega^2 \hat{L}\hat{C}_s}} \tag{7.87}$$

The propagation constant is a nonlinear function of frequency as shown in Figure 7–19a.

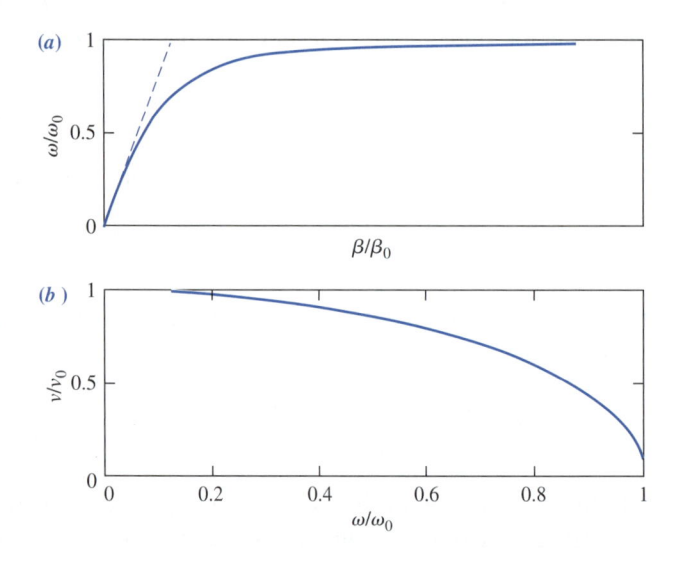

FIGURE 7–19

The normalized propagation characteristics of a dispersive transmission line.
(a) The solid line is the dispersion relation (7.87). The dotted line is a nondispersive propagation constant.
(b) The phase velocity is a function of frequency.

7.10 Dispersion and Group Velocity

We have found that in this case the propagation constant depends on frequency. This phenomenon is called **dispersion**. The consequences of dispersion will dramatically influence the propagation of waves. One could think of the action of a low-pass filter in describing the effect of this line. The propagation constant will be a real number for frequencies from 0 up to a value that is called the cutoff frequency ω_0,

EXAMPLE 7.20

Derive the dispersion relation (7.87) using (7.76).

Answer. From (7.76), we write

$$\gamma = \alpha + j\beta = \sqrt{\hat{Z}\hat{Y}} = \sqrt{\left(\frac{(j\omega\hat{L})(1/j\omega\hat{C}_s)}{j\omega\hat{L} + (1/j\omega\hat{C}_s)}\right)(j\omega\hat{C})}$$

$$= \sqrt{\frac{j\omega \hat{L}\hat{C}/\hat{C}_s}{j\left(\omega\hat{L} - (1/\omega\hat{C}_s)\right)}} = j\frac{\omega\sqrt{\hat{L}\hat{C}}}{\sqrt{1 - \omega^2 \hat{L}\hat{C}_s}}$$

Below a certain frequency called the **cutoff frequency** this expression is purely imaginary. At these frequencies, we can set $\alpha = 0$ and the wave is not attenuated as it propagates.

where

$$\omega_0 = \frac{1}{\sqrt{\hat{L}\hat{C}_s}} \text{ (rad/s)} \quad (7.88)$$

This cutoff frequency is equal to the resonant frequency of the "tank" circuit in the series arm. At this resonant frequency, the tank circuit will appear to be an infinite impedance. Above this frequency, the propagation constant will be imaginary, and the wave will not propagate. In addition, the velocity of propagation v_0 in the nondispersive frequency range is given by

$$v_0 = \frac{1}{\sqrt{\hat{L}\hat{C}}} \text{ (m/s)} \quad (7.89)$$

The wave number β_0 in this region is

$$\beta_0 = \frac{\omega_0}{v_0} \text{ (rad/m)} \quad (7.90)$$

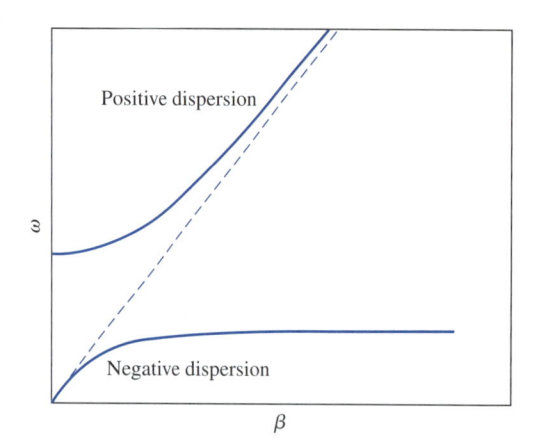

FIGURE 7-20

Dispersion curves.

Dispersive transmission lines and dispersive media in general are very important and common in practice. Dispersion can be created by finite transverse dimensions as in waveguides or by interatomic dimensions as in materials.

Dispersion implies that the propagation constant depends on the frequency of oscillation. The relationship between these two quantities is often depicted as shown in Figure 7–20. We find that the curvature of the resulting dispersion curve can have different slopes, and these are referred to as *positive dispersion* and *negative dispersion*, respectively.

Electromagnetic waves that are confined to propagate within metallic structures can only propagate if the frequency is above some cutoff frequency that could be determined by the physical dimensions of the structure. Recall that we encountered this topic in our discussion of plane waves. In the case of negative dispersion, a wave with a frequency from 0 to some cutoff frequency will propagate. Above this frequency, the wave will not propagate. There are various longitudinal waves that exist in ionized gases called *plasmas*, which exhibit this characteristic.

The question then arises, "What will happen if there are two signals that are propagating in the same linear medium but with slightly different frequencies?" The answer to this question will suggest that there is another velocity, called the *group velocity*, that exists in a dispersive medium. We have already learned that a narrow pulse can be examined using a Fourier analysis, and the pulse would consist of a number of high frequency components. If this pulse propagates in a dispersive region it will be difficult, if not impossible, to reconstruct the original pulse at a later stage, because the higher frequency components will be severely attenuated and portions of the signal at different frequencies will propagate down the line at different pahse velocities.

7.10 Dispersion and Group Velocity

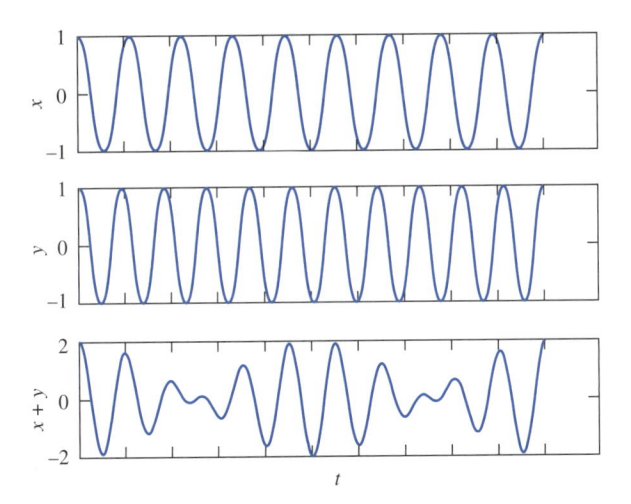

FIGURE 7–21

The linear summation of two cosine waves with slightly different frequencies of oscillation is depicted in the figure. There is constructive and destructive interference between these two signals that is indicated in the bottom figure.

In answering the question that was just posed, we consider two waves that each have the same amplitude V_0 and the following frequencies of oscillation:

$$\omega_1 = \omega_0 + \Delta\omega \quad \text{and} \quad \omega_2 = \omega_0 - \Delta\omega \tag{7.91}$$

Corresponding to each of these frequencies, a signal that propagates in the positive z direction is excited. From either of the dispersion curves in Figure 7–20, the corresponding propagation constants are obtained as

$$\beta_1 = \beta_0 + \Delta\beta \quad \text{and} \quad \beta_2 = \beta_0 - \Delta\beta \tag{7.92}$$

Since superposition will apply, the two waves can be added together in order to find the sum total of the response.

$$V(z, t) = V_0[\cos(\omega_1 t - \beta_1 z) + \cos(\omega_2 t - \beta_2 z)] \tag{7.93}$$

Let us apply a trigonometric identity to (7.93) and obtain

$$V(z, t) = 2V_0 \cos(\Delta\omega t - \Delta\beta z)\cos(\omega_0 t - \beta_0 z) \tag{7.94}$$

In Figure 7–21, we illustrate the summation procedure by just adding together two cosine signals with slightly different frequencies.

If we examine these two signals at various locations that are equally spaced as shown in Figure 7–22, we will be able to ascertain both the

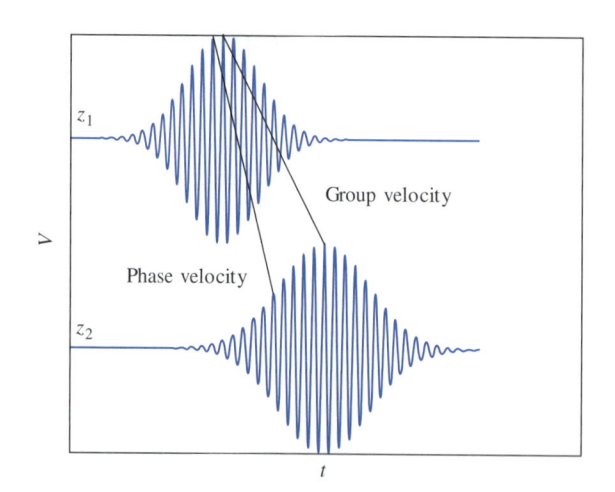

FIGURE 7–22

The propagation of a signal in a dispersive medium. The signals are detected at two locations. A point of constant phase and the peak of the envelope are followed. The point of constant phase propagates with the phase velocity, and the modulation envelope propagates with the group velocity. In this figure, the phase velocity is greater than the group velocity.

propagation velocity of a point of constant phase and the velocity of this modulation. The points of constant phase yield the phase velocity $v_p = \omega_0 / \beta_0$. The velocity of the amplitude modulation is $v_g \approx \Delta\omega / \Delta\beta$. The velocity of the modulation in the limit $v_g = \Delta\omega / \Delta\beta \longrightarrow \partial\omega / \partial\beta$ is called the ***group velocity*** of the wave. In a nondispersive medium, the two velocities are identical. However, in a dispersive medium they can be vastly different and can even have the opposite sign.

An illustration of signals that are detected at increasing values of the distance is shown in Figure 7–22. A point of constant phase can be followed, and it will yield data to compute the phase velocity. The envelope of the modulating signal will propagate with the group velocity. Frequently, this modulating signal spreads as it propagates, causing the detected signal to become vastly distorted. For example, a very narrow pulse excitation that contains a very large number of frequency components and is propagating in a dispersive medium could appear as a time-harmonic signal with only one frequency component at distances far from the point of excitation.

Suffice it to say, dispersion has very dramatic effects on the propagation of electromagnetic waves. If the medium, in addition to being dispersive, were also nonlinear (in that the velocity of propagation depended upon the amplitude of the propagating wave), then it might be possible to have nonlinear waves, which are called ***solitons***, propagating in this medium. A nonlinear dispersive transmission line can be constructed by replacing the linear shunt capacitor shown in Figure 7–18

7.10 Dispersion and Group Velocity

EXAMPLE 7.21

(a) Find the phase and group velocities for a normal transmission line depicted below.
(b) Find the phase and group velocities for a transmission line in which the circuit elements are interchanged.

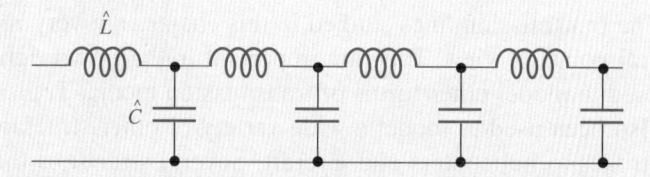

Answer.

(a) From (7.76), we write

$$\gamma = \alpha + j\beta = \sqrt{\hat{Y}\hat{Z}} = \sqrt{(j\omega\hat{L})(j\omega\hat{C})} = j\omega\sqrt{\hat{L}\hat{C}}$$

The phase velocity v_p is computed to be

$$v_p = \frac{\omega}{\beta} = \frac{1}{\sqrt{\hat{L}\hat{C}}}$$

The group velocity v_g is computed to be

$$v_g = \frac{\partial \omega}{\partial \beta} = \frac{1}{\left(\frac{\partial \beta}{\partial \omega}\right)} = \frac{1}{\sqrt{\hat{L}\hat{C}}}$$

The two velocities are equal in this case and independent of frequency.

(b) From (7.76), we write

$$\gamma = \alpha + j\beta = \sqrt{\hat{Z}\hat{Y}} = \sqrt{\left(\frac{1}{j\omega\hat{C}}\right)\left(\frac{1}{j\omega\hat{L}}\right)} = \frac{1}{j\omega\sqrt{\hat{L}\hat{C}}}$$

The phase velocity v_p is computed to be

$$v_p = \frac{\omega}{\beta} = -\omega^2\sqrt{\hat{L}\hat{C}}$$

The group velocity v_g is computed to be

$$v_g = \frac{\partial \omega}{\partial \beta} = \frac{1}{\left(\frac{\partial \beta}{\partial \omega}\right)} = \omega^2\sqrt{\hat{L}\hat{C}}$$

In this case, the phase and the group velocities are in the opposite direction, and both of them depend on frequency.

with a capacitance whose value depends upon the local value of the voltage of the wave. This is also called a *nonlinear varactor diode*. This is a topic of current research interest in several scientific and engineering communities.

7.11 Conclusion

The transmission lines studied in this chapter are very important from several points of view. They are important in their own right, and they can be used to model other forms of transmission media. Transmission lines have also been used to model a wide variety of other structures, from lightning strokes to helicopters and aircraft. Several structures that are in wide use (such as a coaxial cable, a strip line, and two parallel wires) can be modeled with a structure that consists of distributed inductors and capacitors. The direct application of Kirchhoff's laws leads to two first-order partial differential equations known as the telegraphers' equations. Eliminating one of the dependent variables between these two equations leads to a wave equation.

A sinusoidal signal generator will launch sinusoidal waves to increasing or decreasing values of the spatial coordinate assuming that the initial transient effects can be neglected. The amplitude of the wave will repeat itself every *half wavelength*. The ratio of the voltage wave propagating in one direction to the current wave propagating in the same direction is the characteristic impedance of the transmission line. Terminating the transmission line with either a load impedance or another transmission line introduced the concepts of reflection and transmission coefficients, standing waves, the VSWR, and of impedance matching. The Smith chart is used to facilitate this matching. Transient effects and their subsequent propagation, along with pulse propagation, were analyzed with a bounce diagram. The final asymptotic state of a transmission line excited by a step voltage was found. Finally, the effects of loss and dispersion were analyzed.

7.12 Problems

7.1.1. Show that the equivalent circuit element parameters for the coaxial cable and the strip line are correct representations.

7.1.2. Derive the formulas for the capacitance and inductance of a coaxial transmission line presented in Table 7–1.

7.2.1. Show that the quantity $v = 1/\sqrt{\hat{L}\hat{C}}$ does indeed have the units of a velocity.

7.2.2. Show that the units of the diffusion coefficient $D = 1/\hat{R}\hat{C}$ do indeed have the units of (length)^2/time.

7.2.3. Show that a function which represents a wave that propagates to decreasing values of z satisfies the wave equation (7.7).

7.2.4. Let us replace the linear capacitors in Figure 7–3 with nonlinear varactor diodes whose capacitance depends

7.12 Problems

on the voltage applied across them. In this case, the current ΔI into the diode can be written as $\Delta I = \partial Q(V)/\partial t$. Derive the resulting wave equation for this transmission line.

7.2.5. Demonstrate that both forms of the general solution, (7.15) and (7.16), satisfy the phasor wave equation (7.13). Show that the two terms in each of these solutions are independent functions, so that a general solution must be written as a linear combination of these functions with arbitrary coefficients as given. (Hint: The Wronskian can be used to show that functions are independent.)

7.3.1. Find an expression for the characteristic impedance of the strip line.

7.3.2. Find an expression for the characteristic impedance of a twin lead.

7.3.3. In an integrated circuit, a dielectric with $\varepsilon_r = 2$ is inserted between two metal conductors. The width of the top metal strip is 10 μm, and its separation from the bottom-grounded metal plane is 5 μm. Find the characteristic impedance of this transmission line and the velocity of a signal.

7.3.4. Design a strip line with a glass insulator that will have a characteristic impedance of 5 Ω. You will have some freedom in this design, but there is one constraint—it is to be used in an integrated circuit.

7.3.5. A TV twin lead consists of two parallel 1-mm diameter copper wires separated by 1 cm of a rubber dielectric with $\varepsilon_r = 3$. What is the capacitance per meter of this twin lead? What is its characteristic impedance?

7.3.6. Prove that the voltage that appears across a load impedance will be less than the incident wave if $Z_L < Z_c$.

7.4.1. A VSWR is measured along a transmission line to be 2. Find two values for the reflection coefficient Γ. Which of these values will correspond to $Z_L < Z_c$ and which to $Z_L > Z_c$, given that Z_L and Z_c are real?

7.4.2. For Problem 7.4.1, find the two values of Z_L if $Z_c = 50$ Ω.

7.4.3. A load impedance $Z_L = 25$ Ω is connected to a transmission line whose characteristic impedance is 50 Ω. Using (7.33), plot the impedance as a function of a distance from the load to a total distance of 2λ.

7.5.1. Using (7.33), prove that the load impedance will repeat itself every $\lambda/2$.

7.5.2. Using (7.33), prove that the input impedance of a transmission line terminated in a short circuit with a length $\mathcal{L}$, where $\lambda/4 < \mathcal{L} < \lambda/2$ is capacitive.

7.5.3. Using (7.33), prove that the input impedance of a transmission line terminated in an open circuit with a length $\mathcal{L}$, where $\lambda/4 < \mathcal{L} < \lambda/2$ is inductive.

7.5.4. An air-filled 50-Ω coaxial cable that is 1 m long is excited with a 300-MHz signal generator. The line is terminated with a load impedance $Z_L = (25 + j25)$ Ω. What is the input impedance of this line?

7.5.5. A shorted 50-Ω transmission line of length $\mathcal{L}$ has an input admittance of $-j\,0.01$ S. Find the length of the line in λ.

7.5.6. Using (7.46) and (7.47), draw the circles for the values of $r = 0, 1$, and ∞ and $x = -1, 0$, and 1 to convince yourself that Figure 7–9 is correct.

7.6.1. Using a Smith chart, find the impedance Z_{in} of a 50-Ω coaxial cable that is terminated in a load $Z_L = (25 + j25)\Omega$. The coaxial cable has a length of $3\lambda/8$.

7.6.2. Using a Smith chart, find the admittance Y_{in} of a 50-Ω coaxial cable that is terminated in a load $Z_L = (25 + j25)$ Ω. The coaxial cable has a length of $\lambda/8$.

7.6.3. Using a Smith chart, find the distance from a load impedance $Z_L = (25 + j25)$ Ω that is connected to a 50-Ω coaxial cable where the normalized input admittance $Y_{in} = 1 + jB_{in}$. How long should a transmission line that is terminated in a short circuit be in order to match the transmission?

7.6.4. A load impedance $Z_L = (100 - j100)$ Ω terminates a 50-Ω transmission line. Find the characteristic impedance of the quarter-wavelength matching transmission line.

7.7.1. A lossless battery is connected to an ideal transmission line with a characteristic impedance Z_c of length $\mathcal{L}$ that is terminated in a short circuit. Sketch the potential at $z = \mathcal{L}/2$ as a function of time $0 < t < 4(\mathcal{L}/v)$. The switch is closed at $t = 0$.

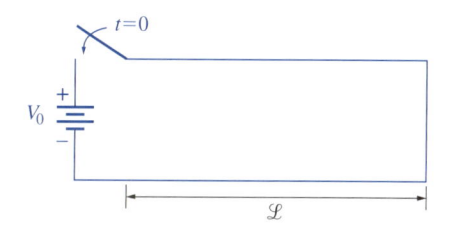

7.7.2. Sketch the current profile at $z = \mathcal{L}/2$ as a function of time $0 < t < 4(\mathcal{L}/v)$ at $z = \mathcal{L}/2$ for the transmission line stated in Problem 7.7.1.

7.7.3. A lossless battery is connected to an ideal transmission line with a characteristic impedance Z_c of length $\mathcal{L}$ that is terminated in an open circuit. Sketch the potential at $z = \mathcal{L}/2$ as a function of time $0 < t < 4(\mathcal{L}/v)$. The switch is closed at $t = 0$.

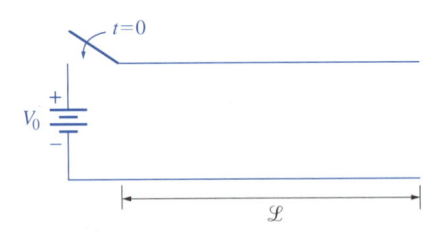

7.7.4. Sketch the current profile at $z = \mathcal{L}/2$ as a function of time $0 < t < 4(\mathcal{L}/v)$ for the transmission line stated in Problem 7.7.2.

7.7.5. Two transmission lines are joined with a resistor R_L.

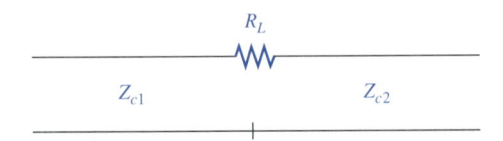

Show that the transmitted voltage V_T in Line 2 can be written as

$$V_T = \frac{Z_{c2}}{R_L + Z_{c2}}(1 + \Gamma)V_{inc}$$

where

$\Gamma = (R_L + Z_{c2} - Z_{c1})/(R_L + Z_{c2} + Z_{c1})$

if the voltage incident from $z = -\infty$ is V_{inc}.

7.7.6. At $t = 0$, the switch located at the load is closed. Sketch the voltage and the current at the load as a function of time $\mathcal{L}/v$. The impedance at the load R_L = the characteristic impedance of the line Z_c.

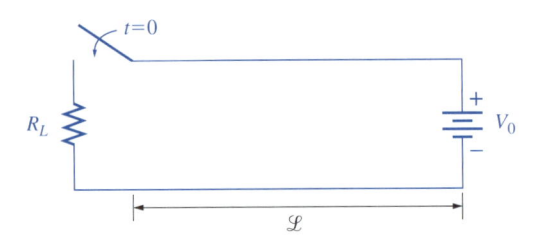

7.7.7. A transmission line with two switches, one at the battery and one at the load, is shown below. Initially, switch S_1 is closed and switch S_2 is open. At $t = 0$, S_1 is opened and S_2 is closed. Sketch the voltage $V_{aa'}$ as a function of time.

If there were a load impedance located at the midpoint of this particular transmission line, it would be possible to generate a large voltage pulse across this load impedance. This is called a **Blumlein transmission line**.

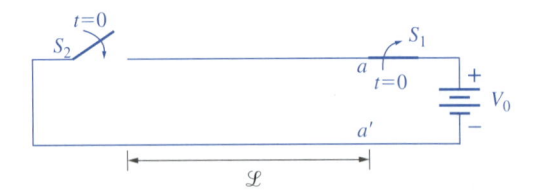

7.7.8. A pulse generator is connected to a transmission line of length $\mathcal{L} = 2$ m having $Z_c = 50\,\Omega$, $R_L = 20\,\Omega$, and $R_g = 30\,\Omega$. The propagation velocity in this transmission line is equal to 10^8 m/s. The amplitude of the pulse is 1 V and its width is 10^{-9} s. Plot the voltage at $z = \mathcal{L}/2$ as a function of time, $0 \le t \le 100$ ns.

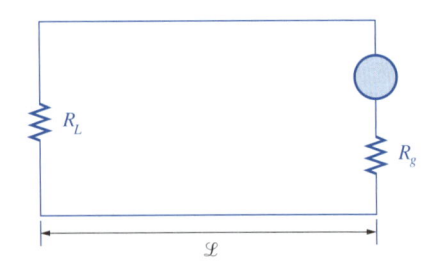

7.8.1. In a digital computer we want to transmit a sequence of binary pulses from a pulse generator to another point where they are to be sampled. Let us assume that the pulse sequence shown below is launched by the pulse generator shown in Problem 7.7.7, and we want to detect the sequence at a time $T = 3\mathcal{L}/2v$ later. Describe any limitations that might be imposed on the speed of this computer.

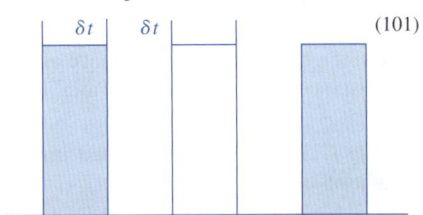

(101)

7.12 Problems

7.8.2. Generalize the results of Problem 7.8.1 to 32-bit and 64-bit machines.

7.9.1. Repeat Problem 7.7.8 if the transmission line is lossy and the signal decays as $e^{-\alpha z}$ as it propagates, where $\alpha = 0.01, 0.1,$ and 1.

7.9.2. Sketch the dispersion relation for a transmission line consisting of a series-resonant circuit in the series branch and a capacitor in the shunt branch. Calculate the propagation constant using (7.77).

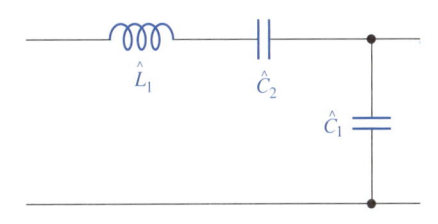

7.9.3. Describe the dispersion relation for a transmission line consisting of an inductor in the series branch and a tank circuit in the shunt branch.

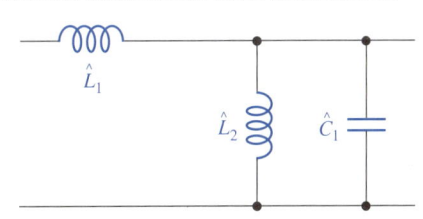

7.10.1. Repeat Problem 7.9.2 with the addition of a series resistor $\hat{R}_s$ added in series with the inductor. The dispersion relation will be complex.

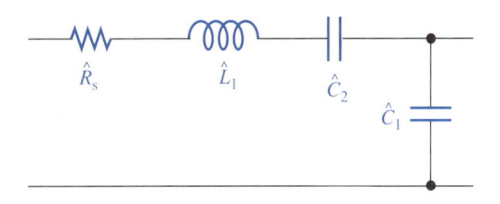

7.10.2. Repeat Problem 7.9.3 with the addition of a series resistor $\hat{R}_s$ added in series with the series inductor. The dispersion relation will be complex.

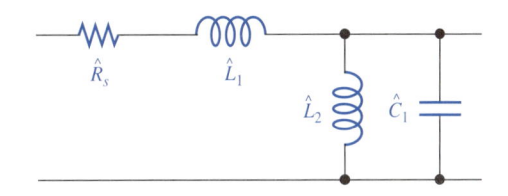

CHAPTER 8

Radiation of Electromagnetic Waves

8.1	Radiation Fundamentals	419
8.2	Infinitesimal Electric Dipole Antenna	427
8.3	Finite Electric Dipole Antenna	434
8.4	Loop Antennas	440
8.5	Antenna Parameters	443
8.6	Antenna Arrays	455
8.7	Conclusion	466
8.8	Problems	467

In the previous chapters, we learned that electromagnetic waves can propagate in free space of an infinite extent and that these same waves can also propagate along a common transmission line. A question that remains to be answered is whether the same electromagnetic waves can be excited in a finite region and then be launched or radiated into infinite space. In this chapter, we first examine the fundamentals of the radiation of electromagnetic waves, which provides a natural lead-in to the important topic of antennas.

8.1 Radiation Fundamentals

Before examining the radiation properties of an antenna, we should first understand the physical process that causes the radiation of electromagnetic waves. This means that we have to examine possible radiation characteristics of an electric charge from a fundamental viewpoint. There are certain requirements that an electric charge must meet in order for it to radiate electromagnetic waves. These requirements will be presented from an intuitive point of view. Once we understand this general argument, it is only a short step further in the same direction toward developing an understanding of antenna radiation theory. The principle of superposition applies in the linear medium being considered in this text, and the antenna can be considered to comprise a large number of charges. The argument also illustrates the type of calculation that can be written on the backs of old envelopes.

Radiation of Electromagnetic Waves

We can understand radiation of electromagnetic waves using Poynting's theorem that the total power P_{rad} radiated from a source is given by the following closed surface integral:

$$P_{rad} = \oint_{\Delta s} \mathbf{E} \times \mathbf{H} \cdot \mathbf{ds} \tag{8.1}$$

Poynting's theorem tells us that the radiation of electromagnetic waves from a source located within a volume completely enclosed by a surface requires *both* an electric field and a magnetic field—the two fields being coupled together via Maxwell's equations.

A *stationary* charge (discussed in Chapter 2) will *not* radiate electromagnetic waves. This can be easily understood. Since a stationary charge will cause no current to flow, there can be no magnetic field associated with it. From (8.1), the total radiated power is therefore equal to zero—from which we can conclude that there will be no radiation of electromagnetic waves from a stationary charge.

We can also come to this conclusion from another point of view. If the point where the power is to be detected is far from the source and there is a spherically radiating wave, it would almost appear to be a plane wave at large distances from the charge. We can make use of the fact that the electric and magnetic field intensities of propagating waves are related through the wave impedance of free space Z_0, as described in Chapter 6. The magnitude of the magnetic field intensity H can be found from the electric field intensity E via $H = E/Z_0$.

Therefore, a source of electromagnetic power located at the center of a sphere whose radius is R (shown in Figure 8–1) would radiate a total power

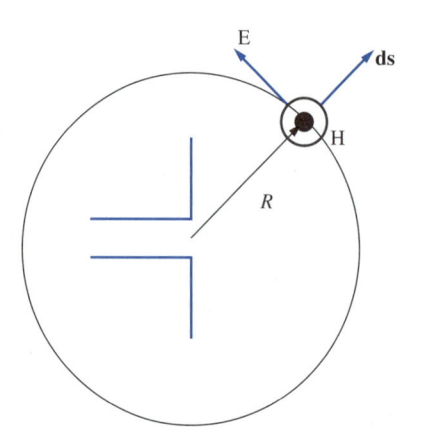

FIGURE 8–1

Antenna radiation of electromagnetic waves. For a stationary charge, **H** will be equal to zero.

8.1 Radiation Fundamentals

whose value can be written as

$$P_{rad} \approx \frac{E^2}{Z_0}(4\pi R^2) \tag{8.2}$$

Let us assume at this stage that the antenna is an isotropic radiator and has no directional characteristics. The total radiated power is equal to that which is delivered from the source, which we will assume to be a constant. Hence the total radiated power is independent of the distance R. Therefore, we would conclude that the electric field E of an electromagnetic wave must decrease with increasing distance as R^{-1}. However, we find that the electric field from a static charge varies as R^{-2}. Hence we come to the same conclusion—that stationary charges cannot radiate electromagnetic waves.

We can also argue that a stationary charge will not radiate by examining Figure 8–1. The electric field associated with radiation is in the surface of the sphere, while the electric field associated with a stationary charge is entirely in the radial direction.

Our next question is whether a charge that is in motion with a *constant velocity* $v \ll c$ can radiate electromagnetic waves. We know that a charge in motion constitutes a current, and currents cause magnetic fields. We are not able to invoke the previous argument (based on the radiated power) that we used to show the lack of radiation from a static charge, since both an electric field and a magnetic field are now present. Instead, we will use a slightly different argument that is still based on the Poynting vector.

Let us assume that a positive charge Q is moving in the positive $\mathbf{u_x}$ direction with a constant velocity $\mathbf{v}$ as shown in Figure 8–2. This velocity will be chosen to be much less than the velocity of light c so it is *nonrelativistic*. We do not want to wade into the deep waters of relativity or advanced topics in physics at this time.

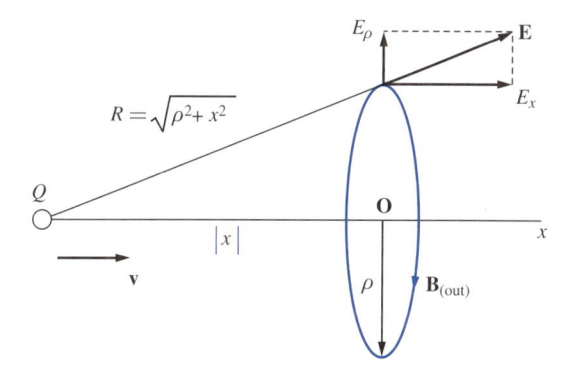

FIGURE 8–2

Electric and magnetic fields due to a moving charge. The velocity $\mathbf{v}$ is a constant and $v \ll c$.

FIGURE 8-3

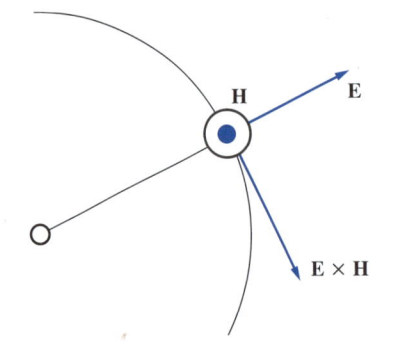

The Poynting vector associated with a charge moving with a constant velocity **v**.

The static electric field **E** from the charge Q is computed to be

$$\mathbf{E} = \frac{Q}{4\pi\varepsilon_0} \frac{1}{\rho^2 + x^2} \mathbf{u}_\rho \qquad (8.3)$$

where $R = \sqrt{\rho^2 + x^2}$. The magnetic field can be computed from the Biot–Savart law. This leads to

$$\mathbf{H} = \frac{1}{4\pi} \frac{Q(\mathbf{v} \times \mathbf{u}_\rho)}{\rho^2 + x^2} \qquad (8.4)$$

Let us compute the direction of Poynting's vector associated with these two fields. This is facilitated by examining a sphere centered on the charge at a certain instant in time, as shown in Figure 8–3. The electric field caused by a charge moving with a uniform velocity is entirely normal to the spherical surface, and the magnetic field is tangent to the surface. Hence, the Poynting vector $\mathbf{S} = \mathbf{E} \times \mathbf{H}$ is completely confined within the spherical surface, and it does *not* radiate in the radial direction away from the charge. Is there any hope for radiation?

EXAMPLE 8.1

Calculate the component of the Poynting vector in the $\mathbf{u}_x$ direction in Figure 8–2 and the total energy flow rate through an infinitely large plane placed normal to the x axis. Discuss the meaning of this result.

Answer. The magnitude of the x component of the Poynting vector is computed from $|\mathbf{E}_\rho \times \mathbf{B}|/\mu_0$. This leads to

$$S_x = \left\{\frac{Q\rho}{4\pi\varepsilon_0(\rho^2 + x^2)^{3/2}}\right\}\left\{\frac{Qv\rho}{4\pi(\rho^2 + x^2)^{3/2}}\right\} = \frac{Q^2 v \rho^2}{16\pi^2 \varepsilon_0 (\rho^2 + x^2)^3}$$

8.1 Radiation Fundamentals

The total energy flow rate becomes

$$P_{rad} = \int_0^\infty S_x 2\pi\rho d\rho = \frac{Q^2 v}{8\pi\varepsilon_0}\int_0^\infty \frac{\rho^3 d\rho}{(\rho^2+x^2)^3}$$

Using the results of Problem 8.1.1 to evaluate the integral, we obtain

$$P_{rad} = \frac{Q^2 v}{32\pi\varepsilon_0}\frac{1}{x^2}\ (W)$$

The distance $|x|$ is the instantaneous separation between the charge and the plane. In the one-dimensional system being considered here ($\mathbf{v} \rightarrow v\mathbf{u}_x$), the velocity $\mathbf{v}$ can be written as $\mathbf{v} = dx/dt\,\mathbf{u}_x$. The power can be rewritten in the form

$$P_{rad} = -\frac{d}{dt}\left[\frac{Q^2}{32\pi\varepsilon_0 x}\right]$$

The quantity

$$W = \frac{Q^2}{32\pi\varepsilon_0|x_0|}$$

is the electrostatic energy stored in the region $x > x_0$ as in Problem 8.1.2. Therefore, the power that is calculated using Poynting's theorem can be interpreted as the flow rate of electrostatic energy stored in space and *it has nothing to do with radiation*. The magnetic energy will be of the order of $(v/c)^2$ times the electric energy and will be very small in nonrelativistic cases.

In order to answer the question about whether there can be any radiation at all, let us consider a charge initially at rest at point A, which is accelerated in the x direction as shown in Figure 8–4. The acceleration lasts for a duration of Δt seconds until it reaches point B, after which the charge moves with a constant velocity $v \ll c$ to point C and beyond. Remember that a signal cannot propagate faster than the velocity of light.

We know that stationary charges and charges moving with a constant velocity do not radiate electromagnetic waves and have an electric field that is radially directed with respect to the charge. Thus, the electric field lines when the charge is at point A (stationary charge) and at point C (charge moving at constant velocity) are entirely radial. These electric field lines must be continuous since they are caused by the same charge. They are connected with "kinked" lines which are disturbances in the electric field lines caused by acceleration of the charge and propagate with the speed of light.

FIGURE 8-4

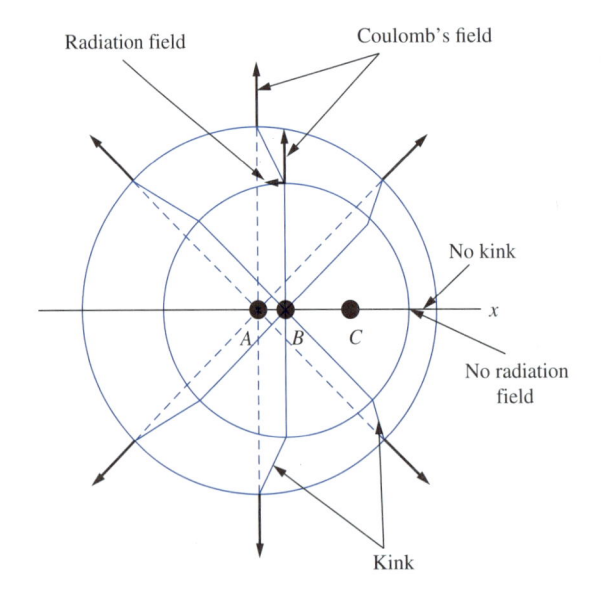

A charge that is accelerated does radiate electromagnetic waves. The dark lines are electric field lines **E**.

It takes a time Δt for the charge to move from point A to point B; therefore, the separation between the two circles is approximately $c\Delta t$. In the kinks there are components of electric field that are perpendicular to the Coulomb field. These transverse components are responsible for the radiation. Note that in this argument, there are directions where there are no radiated electric fields and only the static Coulomb field exists. The maximum radiated electric field will occur along the line that is perpendicular to the charge's acceleration.

Consider a point D in Figure 8–5 that is normal to the direction of the charge's velocity at a certain instant. Let t be the time after the charge is accelerated from a stationary point A to point B where it has a velocity $v = a\Delta t$, where a is the acceleration. We will assume that $\Delta t \ll t$, so the distance $AB + BC \approx BC = vt$.

At point D, there will be two components of an electric field. The first is the radial Coulomb field that is given by

$$E_0 = \frac{Q}{4\pi\varepsilon_0 R^2} = \frac{Q}{4\pi\varepsilon_0 (ct)^2} \tag{8.5}$$

The radiation field E_t can be computed from the triangle JKD.

$$\frac{JK}{KD} = \frac{JK}{AB+BC} \approx \frac{JK}{BC} = \frac{c\Delta t}{vt} = \frac{E_0}{E_t} \tag{8.6}$$

Solving (8.6) for E_t, we obtain

$$E_t = \frac{vt}{c\Delta t}E_0 = \frac{vt}{c\Delta t}\frac{Q}{4\pi\varepsilon_0(ct)^2} = \frac{Q}{4\pi\varepsilon_0 c^2}\frac{v}{\Delta t}\frac{1}{R} \tag{8.7}$$

8.1 Radiation Fundamentals

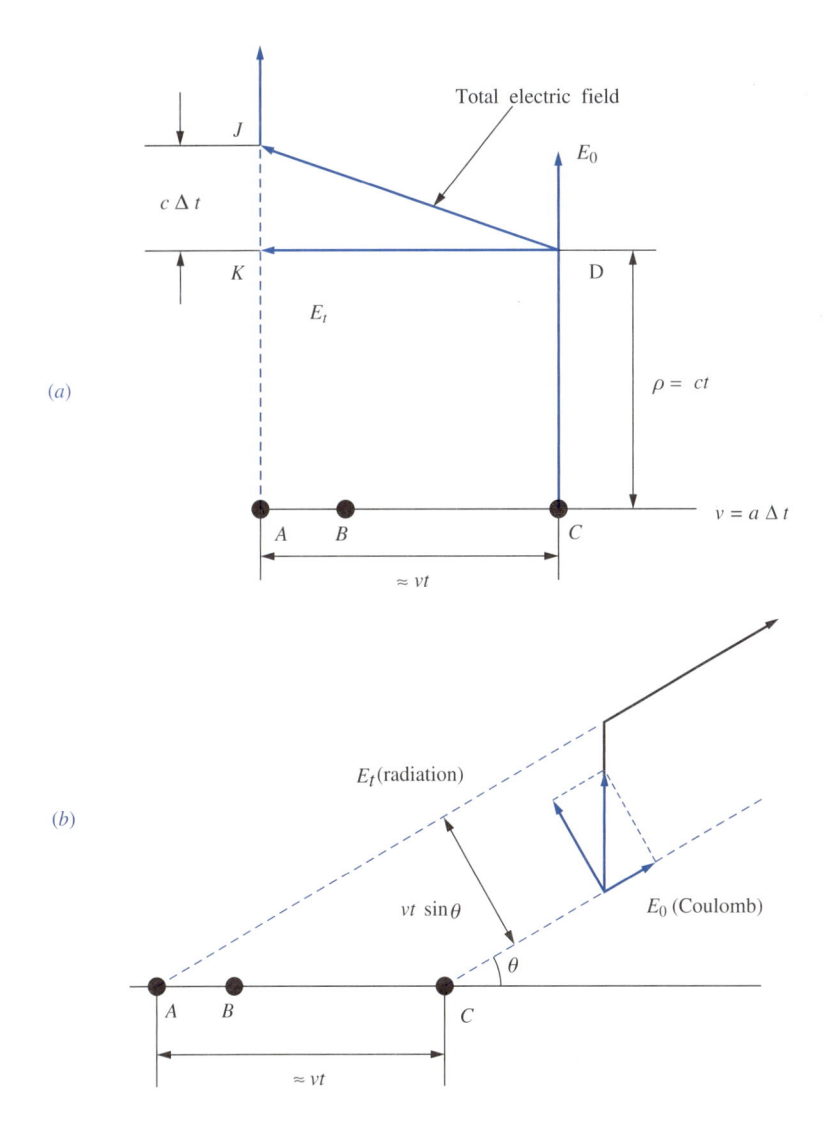

FIGURE 8-5

(*a*) The components of the electric field caused by a charge Q that is accelerated during a time Δt from points A to B. (*b*) The electric fields at an arbitrary angle θ.

Eureka! This is what we were looking for! There is a transverse component of the electric field that is proportional to the *acceleration* $v/\Delta t$, and it has the proper spatial variation $1/R$, which is required in the Poynting vector (8.2). From Figure 8–5, we note that there is a preferred direction for this radiation. If we define the angle θ as being the angle between the point of observation and the velocity of the accelerated charge, (8.7) can be written, using $\mu_0 = 1/(\varepsilon_0 c^2)$, as

$$E_t = \frac{Q\mu_0 [a] \sin \theta}{4\pi \quad R} \qquad (8.8)$$

where $[a] = a(t' = t - R/c)$ is the acceleration at an earlier time—or, as it is frequently called, a ***retarded acceleration*** of the charge taken in a previous moment t'. As it can be seen from Figure 8–5, the signal needs time $\tau = R/c$ to travel between points C and D. This is the "propagation delay" due to the finite speed of light that we encountered in our earlier discussion of plane waves and transmission lines.

The magnetic field intensity H_t associated with E_t can be computed using the wave impedance of the vacuum Z_0 and the fact that the electric and magnetic field intensities are related by this wave impedance. Including $\mu_0/Z_0 = 1/c$, we write

$$H_t = \frac{Q}{4\pi c} \frac{[a]\sin\theta}{R} \qquad (8.9)$$

The Poynting vector is directed *radially outward*, and its magnitude S is given by

$$S = \frac{Q^2[a]^2 \mu_0}{16\pi^2 c} \frac{\sin^2\theta}{R^2} \qquad (8.10)$$

If we insert numerical values for the coefficient in (8.10), we find that a very small number will be obtained.

Therefore, the requirement that must be satisfied for electromagnetic waves to be radiated from a source is that there be charged particles that are either *accelerated* or *decelerated*. Since acceleration is a complex quantity, it has both an associated magnitude and direction. Therefore, if charged particles change their speed or their direction, or both, it is equivalent to an acceleration. Thus, we will see accelerated or decelerated charge associated with any charge motion other than a stationary charge or a charge experiencing uniform motion.

A current with a time-harmonic variation certainly satisfies this requirement, and this will be the source used for the antennas. Once this current is specified, the electromagnetic fields can be computed. Of course we must keep in mind that the electric field at a distance R from the current element will be *retarded* in time from this oscillating current. The retardation is given by Einstein's requirement that nothing can go faster than the velocity of light. Any effect will appear at a time R/c after the cause.

Now that a basic physical mechanism for radiation has been determined, we can describe the practical radiation characteristics of antennas. This chapter will examine several antenna structures and define important antenna parameters.

EXAMPLE 8.2

Assume that an antenna could be described as being an ensemble of N oscillating electrons with a frequency ω in a plane that is orthogonal to the distance R. Find an expression for the electric field $E_\perp$ that would be detected at that location.

Answer. The maximum electric field is computed from (8.8) with $\theta = 90°$. We obtain

$$E_\perp = \frac{NQ\mu_0}{4\pi}\frac{1}{R}\left[\frac{dv}{dt}\right] = \frac{\mu_0}{4\pi R}\frac{1}{dt}\left[\frac{dJ}{dt}\right]$$

where the electric current density $J = NQv$ is introduced. This equation shows that radiation occurs when there is an ***oscillating current*** because the derivative dJ/dt must be different from zero. If we assume that the direction of oscillation in the orthogonal plane is x, then $x(t) = x_m \sin \omega t$ and $v(t) = dx(t)/dt = \omega x_m \cos \omega t$. The substitution of these terms in the equation for the current density $J(t) = \omega NQ x_m \cos \omega t$. Finally, the expression for the transverse electric field becomes

$$E_\perp(R, t) = \omega^2 \frac{NQx_m \mu_0}{4\pi}\frac{1}{R}\sin \omega t'$$

This expression shows that the electric field is proportional to the square of the frequency, which implies that radiation of electromagnetic waves is essentially a high-frequency phenomenon rather than a static occurrence.

8.2 Infinitesimal Electric Dipole Antenna

In applying the basic principles presented in the previous section, we will start with a very elementary antenna—the infinitesimal electric dipole that is shown in Figure 8–6. We assume that the excitation is a time-harmonic signal whose frequency is ω. This will result in electromagnetic radiation that has a time-harmonic frequency of oscillation and a wavelength λ. The length of the antenna, $\mathscr{L}$, is assumed to be short in comparison with the length of the excitation signal wavelength, λ, or $\mathscr{L} \ll \lambda$. Typically, this means that $\mathscr{L} < \lambda/50$. Also, for this and other examples to follow, we will assume that the antennas are very thin. That is, the radius of the antenna element, r_a, is much smaller than a wavelength, or $r_a \ll \lambda$.

The small plates that are placed at the ends of the dipole provide capacitive loading. The short length and the presence of these plates result in a uniform current I along the dipole length. The dipole may be excited by a balanced transmission line, such as a coaxial line. The radiation from the transmission line connection and the end plates of the dipole is considered to be negligible. The

FIGURE 8-6

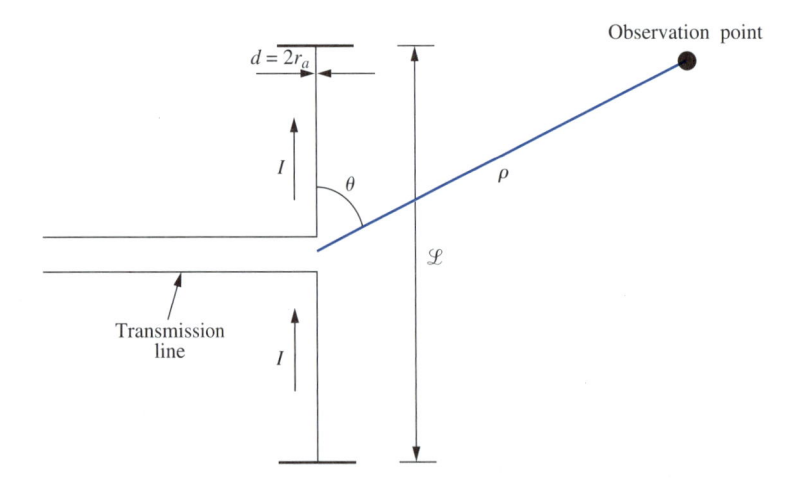

An oscillating short electric dipole radiates electromagnetic waves.

diameter d of the dipole is also small in comparison with its length ($d << \mathcal{L}$). The current on this antenna follows directly from the equation of continuity

$$I = \frac{dQ}{dt} \tag{8.11}$$

For a time-harmonic excitation with a frequency ω, (8.11) reduces to $I(\mathbf{r}) = j\omega Q(\mathbf{r})$.

The vector potential $\mathbf{A}(\mathbf{r}, t)$ can be calculated from the current density $\mathbf{J}(\mathbf{r}, t)$ in the wire using a three-dimensional generalization of the wave equation, which we will demonstrate in example 8.3 below

$$\boxed{\nabla^2 \mathbf{A}(\mathbf{r}, t) - \frac{1}{c^2}\frac{\partial^2 \mathbf{A}(\mathbf{r}, t)}{\partial t^2} = -\mu_0 \mathbf{J}(\mathbf{r}, t)} \tag{8.12}$$

EXAMPLE 8.3

Show that the vector-potential $\mathbf{A}$ satisfies a three-dimensional inhomogeneous wave equation (8.12).

Answer. Recall Maxwell's equations from Chapter 5, rewritten here in a slightly different form as

$$\nabla \times \mathbf{E} = -\frac{\partial \mathbf{B}}{\partial t}$$

$$\nabla \times \mathbf{B} = \frac{1}{c^2}\frac{\partial \mathbf{E}}{\partial t} + \mu_0 \mathbf{J}$$

$$\nabla \cdot \mathbf{E} = \frac{\rho_v}{\varepsilon_0}$$

$$\nabla \cdot \mathbf{B} = 0$$

8.2 Infinitesimal Electric Dipole Antenna

The magnetic flux density is found from the vector potential using the definition $\mathbf{B} = \nabla \times \mathbf{A}$. Substituting this relation into the second equation yields

$$\nabla \times \nabla \times \mathbf{A} = \frac{1}{c^2} \frac{\partial \mathbf{E}}{\partial t} + \mu_0 \mathbf{J}$$

Using the vector identity for the repeated vector operation given in Appendix A, $\nabla \times \nabla \times \mathbf{A} = \nabla \nabla \cdot \mathbf{A} - \nabla^2 \mathbf{A}$, this equation becomes

$$\nabla \nabla \cdot \mathbf{A} - \nabla^2 \mathbf{A} = \frac{1}{c^2} \frac{\partial \mathbf{E}}{\partial t} + \mu_0 \mathbf{J}$$

Substituting

$$\mathbf{B} = \nabla \times \mathbf{A}$$

into the first Maxwell equation yields

$$\nabla \times \left(\mathbf{E} + \frac{\partial \mathbf{A}}{\partial t} \right) = 0 \text{ or } \mathbf{E} + \frac{\partial \mathbf{A}}{\partial t} = -\nabla V$$

This leads to

$$\mathbf{E} = -\nabla V - \frac{\partial \mathbf{A}}{\partial t}$$

where V is the scalar electric potential discussed in Chapter 2. Substituting this expression for $\mathbf{E}$ into the equation for the vector potential yields the following result:

$$\nabla \nabla \cdot \mathbf{A} - \nabla^2 \mathbf{A} = \nabla \left(-\frac{1}{c^2} \frac{\partial V}{\partial t} \right) - \frac{1}{c^2} \frac{\partial^2 \mathbf{A}}{\partial t^2} + \mu_0 \mathbf{J}$$

This equation can be simplified by including one more constraint that relates the vector potential and the scalar electric potential. This is called the **Lorenz Gauge**.

$$\nabla \cdot \mathbf{A} + \frac{1}{c^2} \frac{\partial V}{\partial t} = 0$$

The application of this constraint leads to the vector wave equation for the vector potential.

$$\nabla^2 \mathbf{A} - \frac{1}{c^2} \frac{\partial^2 \mathbf{A}}{\partial t^2} = -\mu_0 \mathbf{J}$$

A similar wave equation can also be derived for the scalar potential. This follows directly from the third Maxwell equation and application of the Lorenz Gauge. We obtain the scalar wave equation.

$$\nabla^2 V - \frac{1}{c^2} \frac{\partial^2 V}{\partial t^2} = -\frac{\rho_v}{\varepsilon_0}$$

In the case of a static field, the second term on the left-hand side of (8.12) disappears and the equation reduces to Poisson's equation. It can be shown that (8.12) has a solution similar to the result that we obtained there. The only difference is that we must be cognizant of the finite velocity of light c. With this in mind we realize that there are two distinct times that must be incorporated into the solution. The current density $\mathbf{J}(\mathbf{r}', t)$ must be replaced with the *retarded* current density $\mathbf{J} \equiv \mathbf{J}(\mathbf{r}', t - R/c)$ in order to find the vector potential $\mathbf{A}(\mathbf{r}, t)$.

$$\mathbf{A}(\mathbf{r}, t) = \frac{\mu_0}{4\pi} \int_{\Delta v} \frac{\mathbf{J}(\mathbf{r}', t - R/c)}{R} dv' \tag{8.13}$$

We leave it to the reader as an exercise to check that (8.13) is actually a solution of the inhomogeneous wave equation (8.12). This equation takes into account the finite velocity of propagation for the electromagnetic waves c. The electric and magnetic fields radiated from the antenna can be found in terms of the vector potential $\mathbf{A}(\mathbf{r}, t)$ as shown in Example 8.3.

We will now be able to obtain the general solution for the case of a infinitesimal electric dipole antenna. A time-harmonic current density can be written as

$$\mathbf{J}\left(\mathbf{r}', t - \frac{R}{c}\right) = \mathbf{J}(\mathbf{r}') e^{-j(\omega t - \mathbf{k} \cdot \mathbf{R})} \tag{8.14}$$

In writing this equation, we have defined the distance in the spherical coordinates. The distance from the center of the dipole $R = r$ and $k = \omega/c$ is the **wave number**. The volume of the dipole antenna can be approximated as $dv' = \mathscr{L} ds'$. The current becomes $\mathbf{J}(\mathbf{r}') ds' = I(z)\mathbf{u}_z$. Therefore, the vector potential can be calculated directly from (8.13).

$$\mathbf{A}(r) = \mathbf{u}_z \frac{\mu_0 I \mathscr{L}}{4\pi} \left(\frac{e^{-jkr}}{r} \right) \tag{8.15}$$

where $I\mathscr{L}$ is the current element of the radiating dipole. This infinitesimal dipole antenna is also known as a **Hertzian dipole**.

Let us now find the field components at large distances from the antenna. In particular, we will assume that these distances are much greater than the wavelength of the wave ($r \gg \lambda$). This is called the *far field* of the antenna. It is also known as the **radiation field** or **Fraunhofer field** of the antenna. In order to do this, we have to express the unit vector $\mathbf{u}_z$ in spherical coordinates as

$$\mathbf{u}_z = \cos\theta\, \mathbf{u}_r - \sin\theta\, \mathbf{u}_\theta \tag{8.16}$$

8.2 Infinitesimal Electric Dipole Antenna

The components of the vector potential $\mathbf{A}(\mathbf{r}) = A_z(z)\mathbf{u}_z$ in spherical coordinates are given by

$$A_r = A_z(r)\cos\theta = \frac{\mu_0 I \mathscr{L}}{4\pi}\left(\frac{e^{-jkr}}{r}\right)\cos\theta$$

$$A_\theta = -A_z(r)\sin\theta = -\frac{\mu_0 I \mathscr{L}}{4\pi}\left(\frac{e^{-jkr}}{r}\right)\sin\theta \quad (8.17)$$

$$A_\phi = 0$$

The magnetic field intensity is computed from the vector potential using the definition of the curl operation in spherical coordinates. We find that

$$\mathbf{H}(\mathbf{r}) = \frac{1}{\mu_0}\nabla \times \mathbf{A} = \frac{1}{\mu_0 r}\left[\frac{\partial(rA_\theta)}{\partial r} - \frac{\partial A_r}{\partial \theta}\right]\mathbf{u}_\phi$$

$$= -\frac{I(z)}{4\pi}k^2 \sin\theta \left[\frac{1}{jkr} + \frac{1}{(jkr)^2}\right]e^{-jkr}\mathbf{u}_\phi \quad (8.18)$$

or

$$\boxed{\begin{array}{c} H_r = 0 \\ H_\phi \approx \dfrac{jkI\mathscr{L}e^{-jkr}}{4\pi r}\sin\theta \\ H_\theta = 0 \end{array}} \quad (8.19)$$

where only the term in the far field is retained, since $kr = (2\pi/\lambda r) \gg 1$. We will provide a more explicit derivation of the far-field requirements shortly. The electric field is computed from Maxwell's equations, where we have set the current density J equal to zero for now. We then write

$$\mathbf{\mathcal{E}}(\mathbf{r}) = \frac{1}{j\omega\varepsilon_0}\nabla \times \mathbf{H}(\mathbf{r}) = \frac{1}{j\omega\varepsilon_0}\left[\frac{1}{r\sin\theta}\frac{\partial[(\sin\theta)H_\phi]}{\partial\theta}\mathbf{u}_r - \frac{1}{r}\frac{\partial(rH_\phi)}{\partial r}\mathbf{u}_\theta\right] \quad (8.20)$$

The components of the electric field are calculated from (8.20). In the far-field region, we find them to be

$$\boxed{\begin{array}{c} E_r \approx 0 \\ E_\theta \approx \dfrac{jZ_0 kI\mathscr{L}e^{-jkr}}{4\pi r}\sin\theta \\ E_\phi = 0 \end{array}} \quad (8.21)$$

where

$$Z_0 = \frac{E_\theta(r)}{H_\phi(r)} = \sqrt{\frac{\mu_0}{\varepsilon_0}} \approx 377\,\Omega$$

is the wave impedance of vacuum. Equation (8.21) agrees with the result found in Example 8.2, which was obtained using a different point of view. This result could also be obtained by replacing the term $dJ/dt \rightarrow j\omega J$, including the retarded time (8.14), and making the substitutions $\omega\mu_0 \rightarrow kZ_0$ and $J(r)\Delta v \rightarrow I\mathscr{L}$ (where Δv is the antenna volume).

Let us now consider the angular distribution of the radiated fields, also known as the radiation pattern of the antenna. For instance, the radiation pattern of an infinitesimal dipole antenna is shown in Figure 8–7. How do we obtain this pattern?

Consider equations (8.21) and (8.19). Both $E_\theta(r)$ and $H_\phi(r)$ are proportional to $\sin\theta$, which implies that there will be an angular variation for both fields. They both have a maximum value when $\theta = 90°$, the direction perpendicular to the dipole axis. Additionally, they both have a minimum value when $\theta = 0°$, which is the direction off the ends of the dipole. Conceptually, this makes sense, since in the direction perpendicular to the antenna there is more area for the energy to radiate away from. In a similar fashion, we can expect little radiation off the ends of the dipole, since there is little area from which the energy can radiate.

The contours depicted in Figure 8–7 in blue are called *lobes*. Although this particular pattern has only two lobes, it is very common to have multiple lobes radiated by an antenna. The lobe in the direction of the maximum is usually called the *main lobe*, while any others are called *side lobes*. Sometimes the pattern has a main lobe, several side lobes, and one lobe located 180° from the main lobe. This specific lobe is called a back lobe. Also associated with an antenna pattern is the idea of a null. A null is a minimum value that occurs between two lobes. Thus, for Figure 8–7, we have main lobes at 90 and 270° and nulls at 0 and 180°.

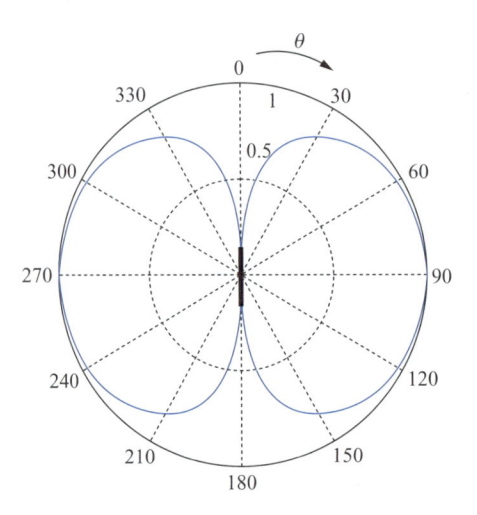

FIGURE 8–7

Radiation pattern for an infinitesimal dipole antenna.

8.2 Infinitesimal Electric Dipole Antenna

If we were to circle the antenna at a constant radius and monitor the received signal with a detector that was sensitive to the phase of the detected signal, we would note a phase shift of 180° as we move from one lobe to the adjacent one. The lobe structure is another example of constructive and destructive interference. Much of the effort in antenna analysis in more advanced study is oriented toward placing these maxima and minima in specific locations.

EXAMPLE 8.4

A small antenna that is 1 cm in length and 1 mm in diameter is designed to transmit a signal at 1 GHz inside the human body in a medical experiment. Assuming the dielectric constant of the body is similar to that of distilled water ($\varepsilon_r = 80$) and that the conductivity σ can be neglected, compute the maximum electric field at the surface of the body that is approximately 20 cm from the antenna. The maximum current that can be applied to the antenna is 10 μA. Find the distance from the antenna r_1 where the signal will be attenuated by 3 dB.

Answer. The wavelength of the electromagnetic wave within the body is computed to be

$$\lambda = \frac{c}{f\sqrt{\varepsilon_r}} = \frac{3 \times 10^8}{10^9 \sqrt{80}} \approx 3.3 \text{ cm}$$

The characteristic impedance of the body is

$$Z_c = \sqrt{\frac{\mu_0}{\varepsilon_0 \varepsilon_r}} = \frac{377}{\sqrt{80}} \approx 42 \text{ }\Omega$$

Since the dimensions of the antenna are significantly less than the wavelength, we can apply (8.20) for $\theta = 90°$ and replace $Z_0 \to Z_c$. Therefore,

$$|E_\theta| = \frac{I\mathcal{L}}{4\pi} Z_c k \frac{1}{r} = \frac{10^{-5} \times 10^{-2}}{4\pi} \times 42 \times \frac{2\pi}{0.033} \times \frac{1}{0.2} \approx 320 \text{ }\mu\text{V/m}$$

An attenuation of 3 dB means that the power is reduced by a factor of 2, or is one-half as large as the original amount. Since the power is related to the square of the electric field, the equivalent reduction in the electric field is related to the square root of the power reduction. Thus, a reduction in power by a factor of two is the same as a reduction in field by the square root of two, or a factor of 1.414. The distance is found to be

$$r_1 = \sqrt{2} r \approx 1.41 \times 0.2 = 0.28 \text{ m}$$

For most common antenna applications, the far-field region is the region of greatest interest, although there are antenna applications where the near-field region is of great interest. In the far-field region, both of the electromagnetic field components are transverse to the direction of propagation. This is clearly seen from equations (8.19) and (8.21). In order to calculate the radiated power from this antenna, one need only perform the surface integration of the time-average Poynting vector.

$$P_{rad} = \frac{1}{2}\text{Re}\left[\oint_{\Delta s} (\mathbf{E}(\mathbf{r}) \times \mathbf{H}^*(\mathbf{r}))_{av} \cdot \mathbf{ds}\right] = \frac{1}{2}Z_0 \int_{\phi=0}^{2\pi}\int_{\theta=0}^{\pi} |H_\phi(r)|_{av}^2 r^2 \sin\theta d\theta d\phi$$

(8.22)

After substituting (8.18) into (8.22) and performing the integration over the variable ϕ, (8.22) becomes

$$P_{rad} = \frac{Z_0 k^2 (I_{av}\mathcal{L})^2}{16\pi}\int_0^\pi \sin^3\theta d\theta = -\frac{Z_0 k^2 (I_{av}\mathcal{L})^2}{16\pi}-\int_0^\pi (1-\cos^2\theta)d(\cos\theta)$$

The last integral gives a factor of $-4/3$. The radiated power from this antenna is

$$P_{rad} = \frac{Z_0 (k\mathcal{L})^2 I_{av}^2}{12\pi} \qquad (8.23)$$

Here the constant current I can be replaced with its average value I_{av}, assuming that there is a slow variation in space.

8.3 Finite Electric Dipole Antenna

We have now covered the basic idea of the radiation of electromagnetic waves from an infinitesimal electric dipole antenna. If we should venture into the hinterlands, we might see some tall structures that reach into the heavens with flashing red lights at the top to warn passing airplanes. Since these antennas certainly do not seem to fall into the "small" class, we will describe a technique to generalize our treatment of antennas so that more realistically-sized antennas can be studied. The generalization is based on superposition—we can consider the radiation from a realistically sized antenna as being the linear summation of the radiation from an infinite number of Hertzian dipoles.

Let us consider two thin metallic rods having a total length $\mathcal{L}$. The length of the *finite dipole (linear antenna)* may be of the order of the free space wavelength λ of the electromagnetic wave that is to be radiated. A sinusoidal voltage generator whose frequency of oscillation is ω is connected between

8.3 Finite Electric Dipole Antenna

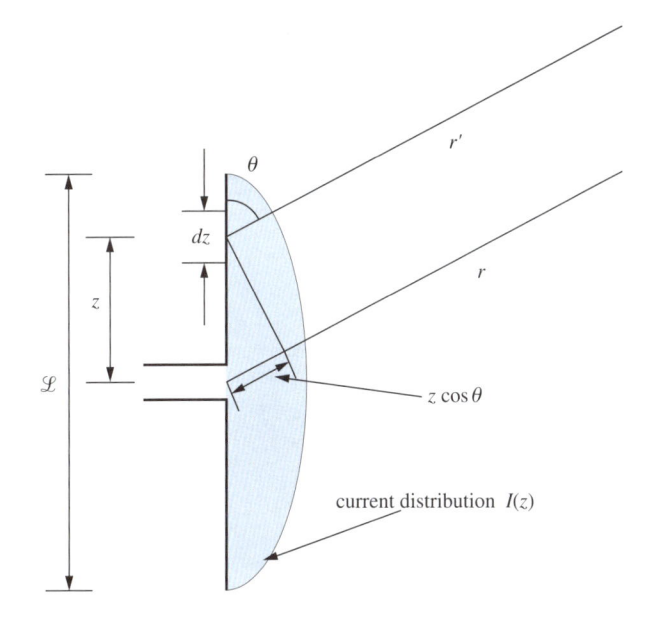

FIGURE 8–8

A center-fed dipole with an arbitrary current distribution $I(z)$.

the two rods as shown in Figure 8–8. This voltage generator induces a current in the rods that can have a distribution $I(z)$, which is governed by the shape and length of the conductor.

For a first approximation, it is reasonable to assume that the current distribution at the ends of the antenna ($z = \pm \mathcal{L}/2$) is equal to zero and that the current distribution is symmetrical about the center ($z = 0$). The first assumption is predicated on the idea that no conduction current could extend beyond the metallic surface. Since the antenna is "center fed," symmetry arguments can be applied. The assumption for the approximated current distribution that is to be used in various integral expressions, such as the finite dipole equivalent to (8.22), requires some ingenuity since it is not a quantity that is measured in the laboratory. Additionally, the actual distribution will depend upon the antenna's boundary conditions, such as length and shape. A typical requirement is that the current distribution be selected so certain integrals can actually be performed. Computers have now alleviated this restriction, and more realistic distributions can be employed. However, finding the actual current distribution $I(z)$ is a difficult task—an *integral equation* must be solved. The iterative solution of this integral equation will require that the boundary conditions are still satisfied for all of the current distributions that are chosen. The final assumption

FIGURE 8-9

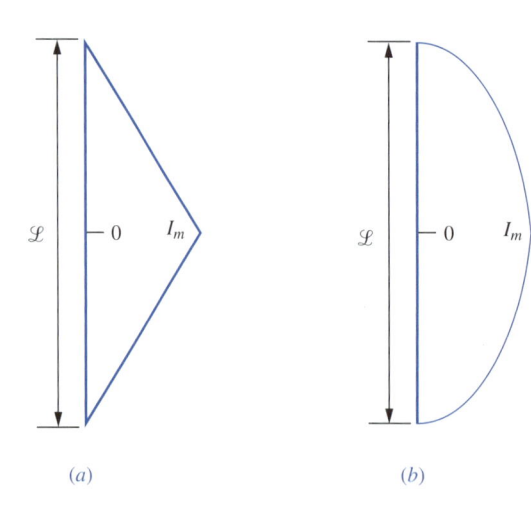

Two possible distributions of current on an antenna.
(*a*) Initial triangular distribution.
(*b*) Actual distribution ($\mathscr{L} = \lambda/2$).

includes the boundary conditions, and the investigator can iterate the solution starting from a very simple function that satisfies these conditions. The iteration procedure is used to initially assume a function for the current distribution, calculate the electric field resulting from this distribution, measure the electric field in the laboratory, modify the initial choice for the current distribution, and repeat the procedure again. For example, starting from a triangular current distribution shown in Figure 8–9*a*, we continue on with a more realistic solution that is shown in Figure 8–9*b*. This problem is a subject for an advanced course in electromagnetics, and we will skip it here.

One simple and quite reasonable approximation for the actual current distribution is the following sinusoidal function:

$$I(z) = I_m \sin\left[k\left(\frac{\mathscr{L}}{2} - |z|\right)\right] \tag{8.24}$$

The far-field radiation properties such as the radiated power, power density and radiation pattern are not very sensitive to the specific choice for the current distribution, since they have a $1/r$ dependency, and do not vary much in the far field. However, near-field properties, such as input impedance, or near-field radiation patterns are very sensitive to the actual current distribution. Many of these terms will be defined in the next few sections.

8.3 Finite Electric Dipole Antoine

> **EXAMPLE 8.5**
>
> Describe the excitation of a center-fed dipole antenna using a transmission line model.
>
> **Answer.** The current distribution of both the incident and the reflected components of the current on an ***open-circuited transmission line*** as discussed in the previous chapter are depicted in the figure. Its spatial distribution is sinusoidal as shown in (*a*). By bending the transmission line at $\lambda/4$ from the end, we form a half-wave dipole ($\mathscr{L} = \lambda/2$) with the proper current distribution. This model has assumed that the terminating $\lambda/4$ of the transmission line is unaffected by the bending of the transmission line. The actual distribution of the current on the line will be *altered*, since the load is not infinite due to the fringing effects.

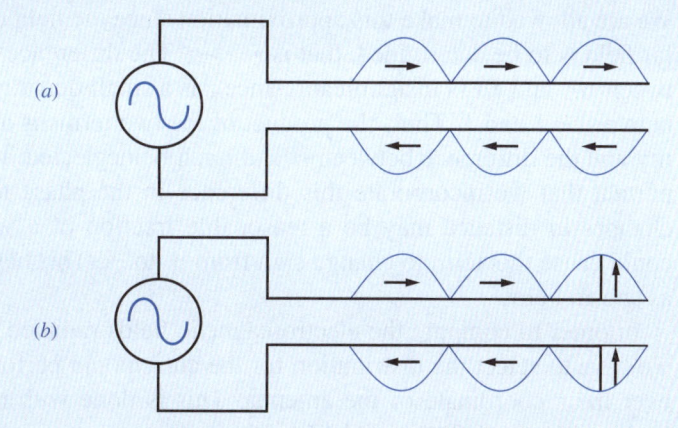

To show the development of the integration procedure, we will calculate the radiation pattern of a finite dipole antenna, where the finite dipole antenna can be represented as a linear combination or collection of infinitesimal electric dipole current elements. We will see later that this is the same basic procedure we will follow in calculating the radiation pattern of an antenna array, or a collection of antenna elements. We will discuss antenna arrays in greater detail in Section 8.6 of this chapter. Returning to the development of the integration procedure, we will use the results for an infinitesimal electric dipole as given in equation (8.21), where we replace the current element $I(z)$ with the differential current element $I(z)dz$

$$dE_\theta = j\frac{Z_0 k}{4\pi}(I(z)dz)\frac{e^{-jkr'}}{r'}\sin\theta \qquad (8.25)$$

The distance r' that appears here can be written in terms of the distance r between the point of observation and the center of the dipole shown in

Figure 8–8 as

$$r' = [r^2 + z^2 - 2rz\cos\theta]^{1/2} \approx r - z\cos\theta \qquad (8.26)$$

We are allowed to make this approximation since the field distribution in the far field is to be determined, that is $r \gg z$. The difference in magnitude between $1/r'$ and $1/r$ is insignificant, since z is a small quantity and $\cos\theta$ varies between -1 and 1. Thus, the product of the two terms is also a small quantity and the difference between r' and r can be neglected. However, it is important that we incorporate this difference in the phase term $e^{-jkr'}$. Small changes in distance may be a reasonable fraction of a wavelength λ that could cause this term to change sign from $+$ to $-$. This has dramatic effects as shown below.

In order to compute the electromagnetic fields radiated from an antenna, we have to select the distribution for the current and perform an integration over the z coordinate of the antenna. This is done with the current distribution given in (8.24), for which we write that

$$E_\theta = Z_0 H_\phi = j\frac{Z_0 I_m k \sin\theta \, e^{-jkr}}{4\pi \, r} \int_{-\mathscr{L}/2}^{\mathscr{L}/2} \sin\left[k\left(\frac{\mathscr{L}}{2} - |z|\right)\right] e^{jkz\cos\theta} dz \qquad (8.27)$$

Before performing the integration required at this stage, let us look at the terms within the integrand. The term with a sine is an even function in the variable of integration z as shown in Figure 8–9b. The product of this term and

$$e^{jkz\cos\theta} = \cos(kz\cos\theta) + j\sin(kz\cos\theta)$$

will yield two terms—one an odd function and the other an even function—in the variable z. Since the limits of the integral are symmetric about the origin, only the integrand that results in an even function will yield a nonzero result. The integral (8.27) reduces to

$$E_\theta = Z_0 H_\phi = j2\frac{Z_0 I_m k \, e^{-jkr}}{4\pi \, r}\sin\theta \int_0^{\mathscr{L}/2} \sin\left[k\left(\frac{\mathscr{L}}{2} - z\right)\right]\cos(kz\cos\theta)dz \qquad (8.28)$$

After a lengthy integration, we finally obtain[1]

$$E_\theta = j60 I_m \frac{e^{-jkr}}{r} F(\theta) \qquad (8.29)$$

---

[1] S. Ramo, J. Whinnery, and T. van Duzer, *Fields and Waves in Communication Electronics*, 3rd edition (New York: Wiley, 1994).

8.3 Finite Electric Dipole Antenna

where the following explicit expression for the **radiation pattern** $F(\theta)$ is found.

$$F(\theta) \equiv F_1(\theta)F_a(\theta) = [\sin\theta] \times \left[\frac{\cos\left(\frac{k\mathscr{L}}{2}\cos\theta\right) - \cos\left(\frac{k\mathscr{L}}{2}\right)}{\sin^2\theta}\right] \quad (8.30)$$

$$= \frac{\cos\left(\frac{k\mathscr{L}}{2}\cos\theta\right) - \cos\left(\frac{k\mathscr{L}}{2}\right)}{\sin\theta}$$

The final solution is a product of two terms. The first term, $F_1(\theta)$, represents the radiation characteristics of one of the current elements that are used to make up the complete antenna. This term is usually called the element factor. In this case the basic element is an infinitesimal dipole located at $z = 0$. The second term, $F_a(\theta)$, is called the array factor or the space factor of the finite dipole antenna. This term represents the result of adding up all the radiation contributions of the various elements that make up the antenna array, as well as their interactions. In Section 8.6 of this chapter we will review these facts in more detail.

In Figure 8–10, we illustrate the *E*-plane radiation pattern for four different dipole lengths, where the arrays consist of a collection of uniformly

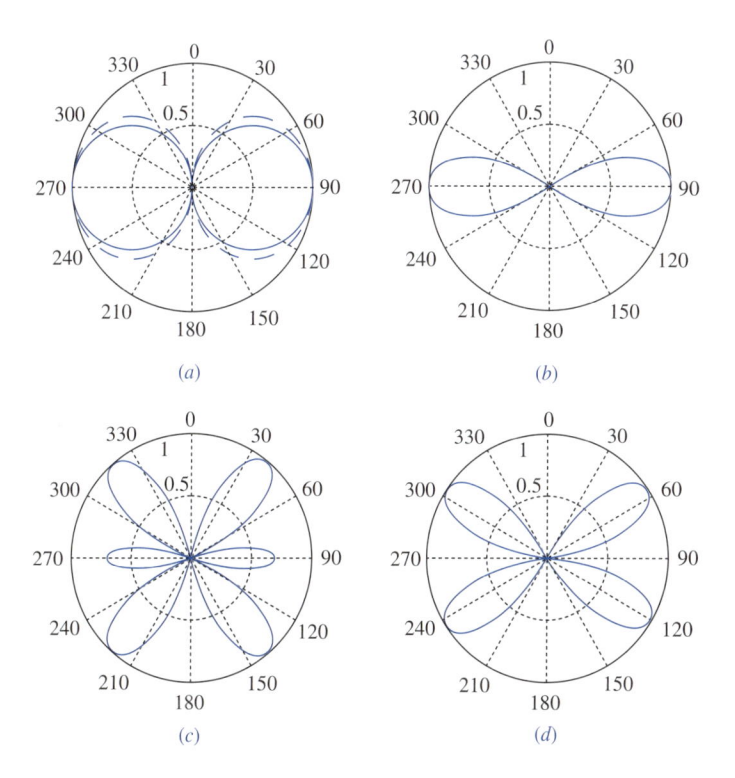

FIGURE 8–10

E-plane radiation patterns for center-fed dipole antennas of different lengths:
(*a*) $\mathscr{L} = \lambda/2$,
(*b*) $\mathscr{L} = \lambda$,
(*c*) $\mathscr{L} = 3\lambda/2$, and
(*d*) $\mathscr{L} = 2\lambda$.
The antenna with the dimension $\mathscr{L} = \lambda/2$ is called a half-wave dipole. In addition, the radiation pattern for an infinitesimal dipole, previously shown in Figure 8–7, in indicated in (*a*) with a dashed line for easy comparison.

excited infinitesimal dipoles. The radiation pattern for a half-wave dipole is shown in Figure 8–10a, with the pattern for an infinitesimal dipole superimposed as a dashed line for comparison. Looking at the various patterns, the total radiation pattern has the property that $F(0°) = 0$ for all four cases. With respect to the first two cases, the total pattern maximum occurs at $F(90°)$. As the length of the dipole exceeds one wavelength, the location of the maximum starts shifting. At a length of $3\lambda/2$, the maximum occurs at 45°, while the maximum shifts back toward 90° as the length increases to 2λ. We also notice a change in the number of lobes in the structure. Additionally, the H-plane radiation patterns are azimuthally symmetric circles, since $F(\theta)$ is independent of the angle ϕ. We note from Figure 8–10c that the maximum in the radiated power tends to shift away from $\theta = 90°$ as the length $\mathscr{L}$ is changed. If we set $\mathscr{L} = 2\lambda$ in (8.29) we find that the radiation at $\theta = 90°$ is equal to zero (Figure 8–10d).

8.4 Loop Antennas

We next examine another fundamental antenna type—the small loop antenna. This antenna consists of a small conductive loop with a current circulating around it. In Chapter 3—see Example 3.11—we discussed the fact that a current carrying loop can generate a magnetic dipole moment. Thus, we may consider a small loop antenna as equivalent to a magnetic dipole antenna, analogous to the electric dipole antenna.

Loop antennas are usually classified into two categories, electrically small and electrically large, based upon the size of the loop circumference. Electrically small loop antennas are those whose circumference is less than a tenth of a wavelength, or $C < \lambda/10$. If the circumference of the loop antenna is on the order of the size of a wavelength or larger, we consider the antenna to be electrically large. Generally, loop antennas are more commonly used in the HF through UHF bands or from about 3 MHz to 3 GHz. Another common usage of loop antennas is as magnetic field probes, and in this application, their utility extends well into the microwave frequencies.

The loop carries a harmonic current $i(t) = I \cos \omega t$ around its circumference. This antenna is depicted in Figure 8–11, where $ka \equiv (2\pi/\lambda) a \ll 1$ is assumed. The retarded vector potential resulting from this current loop is determined from (8.13). Since the current is confined to the loop, this integral becomes

$$\mathbf{A}(r) = \frac{\mu_0 \mathbf{I}}{4\pi} \oint_{\mathscr{L}} \frac{e^{-jkr'}}{r'} dl' \qquad (8.31)$$

8.4 Loop Antennas

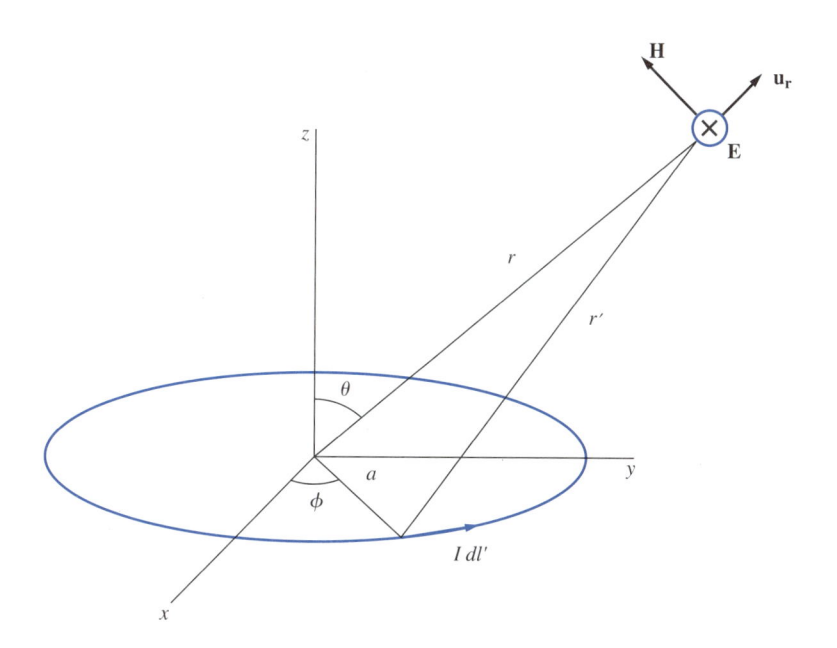

FIGURE 8-11

A magnetic loop antenna.

where $\mathbf{I} = I\mathbf{u}_\phi$ and the time-harmonic term $e^{j\omega t}$ is understood to be included. This integral is not easy to evaluate, since the terms within the integrand depend on the particular location where dl' is being evaluated. However, we can determine an approximate solution that illustrates the expected behavior by using the following procedure.[2] The exponential term can be written as

$$e^{-jkr'} = e^{-jkr}e^{-jk(r'-r)} \approx e^{-jkr}[1 - jk(r'-r)] \tag{8.32}$$

In expanding the second exponential term, we have made the approximation that the loop is small with respect to the distance r between the center of the loop and the point of observation ($a \ll r$). This is very similar to the approximation that we first encountered in describing an electric dipole. Hence, (8.31) can be written as

$$\mathbf{A}(r) = \frac{\mu_0 I}{4\pi} e^{-jkr} \left[(1 + jkr) \oint_\mathcal{L} \frac{dl'}{r'} - jk \oint_\mathcal{L} dl' \right] \mathbf{u}_\phi \tag{8.33}$$

---

[2] In order to obtain analytical solutions in electromagnetics, we have to resort to many approximations. The ingenuity of the practitioner is tested when it comes to making sure that the approximations are reasonable; the success of the practitioner is tested when it comes to deciding what "reasonable" means.

The second integral is equal to zero since this integral is akin to running around in a circle—we just return back to the original starting point and have progressed nowhere. The first integral is evaluated in Example 8.6.

EXAMPLE 8.6

Evaluate the integral $\oint_{\mathscr{L}} dl'/r'$.

Answer. Use the vector identity (see Appendix A)

$$\oint_{\mathscr{L}} b\, dl' = \int_{\Delta s} (\mathbf{u_n} \times \nabla b) \cdot \mathbf{ds}$$

to convert the closed line integral into a surface integral. The scalar quantity b is equal to $b = 1/r'$. With reference to Figure 8–11, we note that $\mathbf{u_n} = \mathbf{u_z}$ since the loop is in the x–y plane. Therefore

$$\oint_{\mathscr{L}} \frac{dl'}{r'} = \int_{\Delta s} \left(\mathbf{u_z} \times \nabla \frac{1}{r'} \right) \cdot \mathbf{u}_\phi \, ds = -\int_{\Delta s} \left(\mathbf{u_z} \times \frac{\mathbf{u}_{r'}}{(r')^2} \right) \cdot \mathbf{u}_\phi \, ds$$

where

$$\nabla \frac{1}{r'} = \frac{\mathbf{u}_r}{(r')^2}$$

For large distances from the current loop, we can replace the amplitude and the unit vector with $r' \approx r$ and $\mathbf{u}_{r'} \approx \mathbf{u}_r$. With these approximations, the integral becomes

$$\int_{\Delta s} \frac{(\mathbf{u}_z \times \mathbf{u}_r) \cdot \mathbf{u}_\phi}{r^2} ds = \frac{(\mathbf{u}_z \times \mathbf{u}_r) \cdot \mathbf{u}_\phi}{r^2} \int_{\Delta s} ds$$

The surface integral yields a factor of πa^2. Finally, we make use of the vector relation $\mathbf{u}_z = \cos\theta\, \mathbf{u}_r - \sin\theta\, \mathbf{u}_\theta$ to compute in spherical coordinates that

$$\mathbf{u}_z \times \mathbf{u}_r = \sin\theta\, \mathbf{u}_\phi$$

Hence, the final result of the integrations yields

$$\oint_{\mathscr{L}} \frac{dl'}{r'} = \frac{\pi a^2}{r^2} \sin\theta$$

We find from (8.33) that the final evaluation of this integral leads to writing the vector potential in the far field as

$$\mathbf{A}(\mathbf{r}) = \frac{\mu_0(I\pi a^2)(1+jkr)e^{-jkr}}{4\pi r^2}\sin\theta\, \mathbf{u}_\phi \approx j\frac{\mu_0(I\pi a^2)k e^{-jkr}}{4\pi r}\sin\theta\, \mathbf{u}_\phi \quad (8.34)$$

We recognize the term $I\pi a^2$ as being the magnitude $\mathbf{m}(r)$ of the *magnetic dipole moment* $\mathbf{m} = (I\pi a^2)\mathbf{u}_z$.

Having found the vector potential, we can evaluate the electromagnetic fields in the far field using (8.18) and (8.20)

$$H_\theta \approx -\frac{\omega\mu_0 mk}{4\pi Z_0}\frac{e^{-jkr}}{r}\sin\theta \tag{8.35}$$

$$E_\phi = -Z_0 H_\theta \approx \frac{\omega\mu_0 mk}{4\pi}\frac{e^{-jkr}}{r}\sin\theta \tag{8.36}$$

If we compare (8.35) through (8.36) with (8.18) and (8.21), we note a *similarity* in the field components of the short magnetic dipole and the short electric dipole. In the far field, the magnitude of the two fields each decay as r^{-1}, the ratio of the two fields is equal to the characteristic impedance Z_0 of free space, and the radiation pattern $F(\theta) = \sin\theta$ is the same and is shown in Figure 8–7.

8.5 Antenna Parameters

In addition to the radiation pattern for an antenna that was discussed in the previous sections, other parameters are used to characterize an antenna. If we connect the antenna to a transmission line, we could think of the antenna as being merely a load impedance. The radiation of electromagnetic power into the external environment removes the power from the circuit, so it acts like a resistor that just heats up. This is depicted in Figure 8–12.

8.5.1 Radiation Resistance

We consider the antenna shown in Figure 8–12c to be a load impedance Z_L that is connected to the generator with a transmission line of length $\mathcal{L}$, which has a characteristic impedance Z_c and a propagation constant. In order to compute the value of the load impedance Z_L, we have to return to the Poynting vector. Recall that this vector is a measure of the power density at a point in space that is calculated using the quantities in the electromagnetic wave. The *total power* radiated from the antenna can be computed by surrounding the antenna with a large imaginary sphere whose radius is r as shown in Figure 8–13. The radius r will be chosen so the surface of the sphere is in the far-field region. Then any power that is radiated from the antenna will have to pass through the sphere in order to propagate to distances greater than this

Radiation of Electromagnetic Waves

FIGURE 8-12

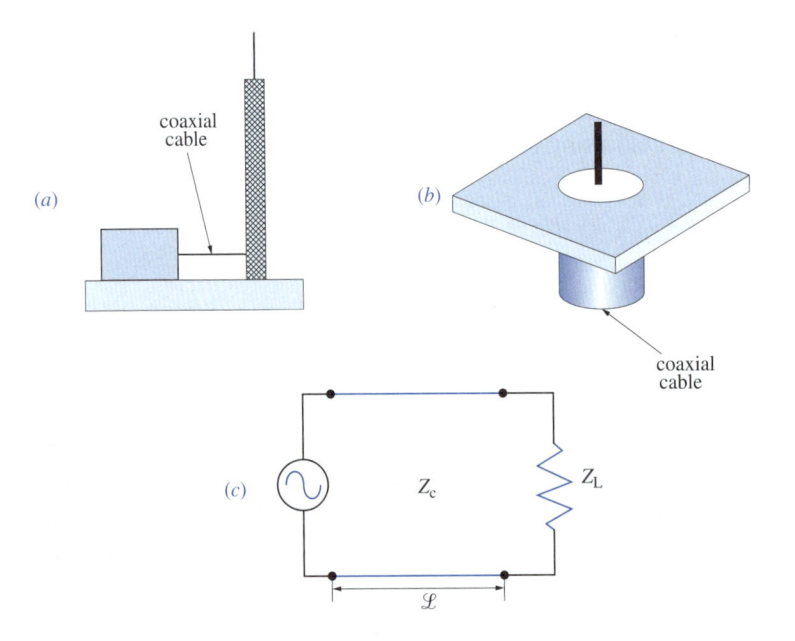

(*a*) An antenna that radiates electromagnetic energy is connected with a transmission line to a source of electromagnetic energy.
(*b*) Coaxial cable connected to a ground plane.
(*c*) Equivalent circuit of either structure.

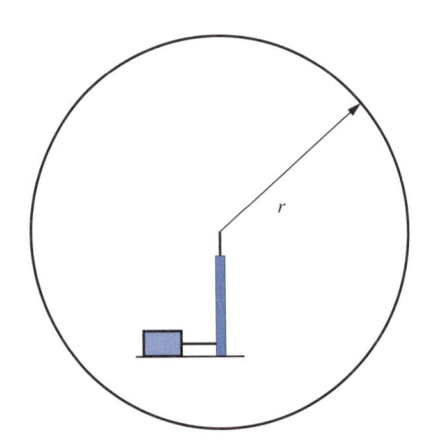

FIGURE 8-13

Electromagnetic power radiated from an antenna will pass through a sphere of radius *r*.

radius *r*. As shown in Example 8.13, this radius can be approximated in terms of antenna size $\mathscr{L}$ to be $r \approx \mathscr{L}^2/2\lambda$.

The total **radiated power** from the antenna is computed by integrating the time-average Poynting vector over this entire closed, spherical surface. From (8.1), this becomes

$$P_{\text{rad}} = \frac{1}{2}\text{Re}\left[\oint_{\Delta s} \mathbf{E}(\mathbf{r}) \times \mathbf{H}(\mathbf{r})^* \cdot \mathbf{ds}\right] = \frac{1}{2}\text{Re}\left[\int_0^{2\pi} d\phi \int_0^{\pi} r^2 \sin\theta (E_\theta H_\phi^*) d\theta\right] \quad (8.37)$$

8.5 Antenna Parameters

The factor of 1/2 arises since we are considering a time-average power over a temporal cycle of the oscillation. As shown in Chapter 5, the conjugated value of **H** must be employed. This average radiated power can be considered to be lost as far as the source is concerned and, therefore, the antenna is similar to a resistor in that it is dissipating the power from the source. This resistance is called the *radiation resistance* R_{rad}, and it is defined as

$$R_{rad} = \frac{2P_{rad}}{I_0^2} \quad (8.38)$$

where I_0 is the maximum amplitude of the current at the input terminals of the antenna. We will calculate the radiation resistance for several antennas below. Note that this resistance is different from the resistance associated with the conductor that makes up the antenna. The effect of having a conductor resistance will be a possible loss and will be discussed further when we discuss antenna gain below.

EXAMPLE 8.7

Find the radiation resistance of an infinitesimal dipole.

Answer. The radiated power P_{rad} from the Hertzian dipole is computed using (8.23). Substituting the terms for the Hertzian dipole into (8.23), using $Z_0 = 120\pi$ and $k = 2\pi/\lambda$ and employing the definition (8.38), we obtain

$$R_{rad} = 80\pi^2 \left(\frac{\mathscr{L}}{\lambda}\right)^2 \left(\frac{I_{av}}{I_0}\right)^2 = 80\pi^2 \left(\frac{\mathscr{L}}{\lambda}\right)^2$$

where a *uniform current distribution* is assumed, or $I_{av} = I_0$. If there were a triangular current distribution depicted in the next example, we would obtain $I_{av} = I_0/2$. In this case, the radiation resistance is 1/4 of the previous value.

$$R_{rad} = 20\pi^2 \left(\frac{\mathscr{L}}{\lambda}\right)^2$$

The radiation resistance for an infinitesimal dipole for $\mathscr{L}/\lambda < 0.15$ is shown in the figure below. The small values for the radiation resistance show that this antenna is not very efficient. This antenna will be discussed further in Example 8.11.

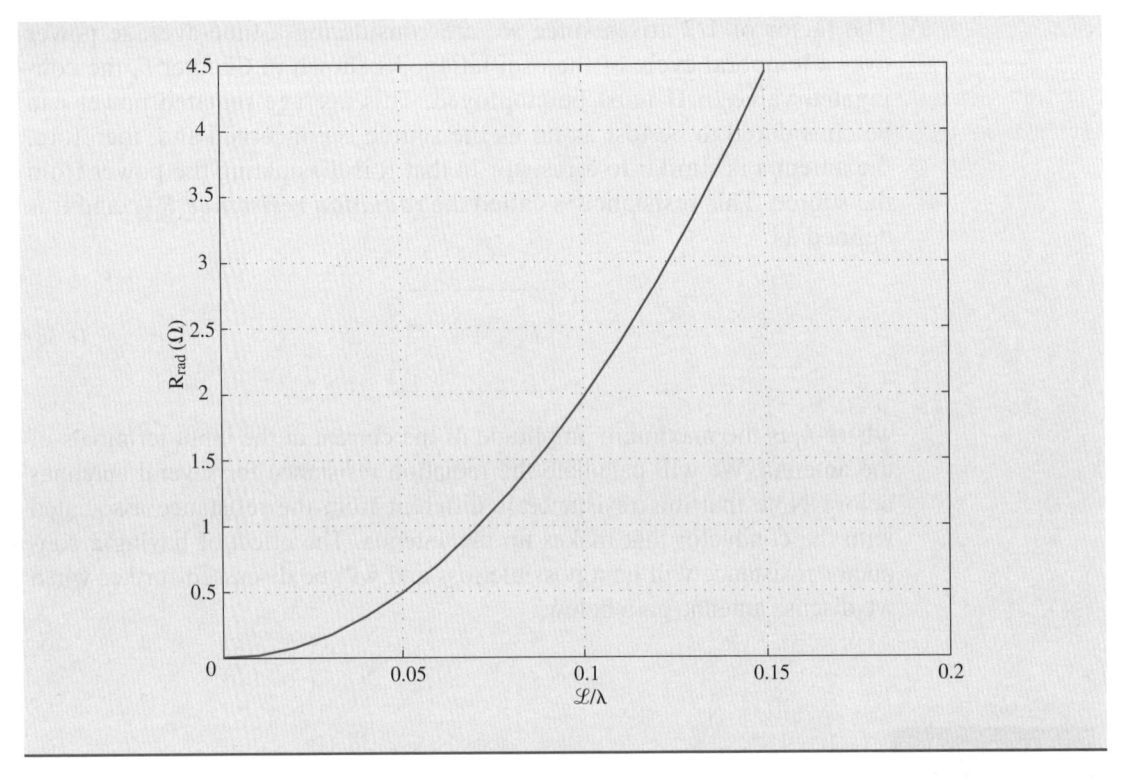

Taking another look at the loop antenna, we can calculate the antenna's radiation resistance using (8.38). Assuming a small loop with a uniform current distribution, we obtain

$$R_{rad} = 20\pi^2(ka)^4 \tag{8.39}$$

The radiation resistance is different than we found in Example 8.7 for a short electric dipole.

EXAMPLE 8.8

Find the current that is required to radiate 10 watts from a loop whose circumference is equal to $\lambda/5$.

Answer. Applying (8.39) for the case of $ka = 2\pi a/\lambda = 0.2$, we obtain

$$R_{rad} = 20 \times \pi^2 \times 0.2^4 = 0.316 \ \Omega$$

The radiated power can be found from (8.7) to be

$$P_{rad} = \frac{1}{2}R_{rad}I(\phi)^2$$

8.5 Antenna Parameters

which yields the current

$$I(\phi) = \sqrt{\frac{2P_{rad}}{R_{rad}}} = \sqrt{\frac{2 \times 10}{0.316}} = 7.95 \text{ A}$$

It should be apparent that the small antenna would not be very useful in radiating large amounts of power.

Next, we need to find a suitable expression for the radiation resistance of a large *loop antenna* where the radius a is comparable with the wavelength, or $ka \geqslant 1$. However, it is a difficult task to find the actual current distribution, as it requires the solution of an integral equation. Additionally, the current distribution strongly depends on the specific voltage excitation used to generate the current. For the special case of a sinusoidal current distribution, suitable equations and relevant design graphs are available.[3]

EXAMPLE 8.9

Find the radiation resistance of a short monopole antenna that is placed above a ground plane. The length of the antenna is $\lambda/8$.

Answer. The short electric monopole antenna near the ground is *equivalent* to a short dipole antenna with a length that is twice as long, as shown in the following figure. The additional contribution is from the "image antenna." In order to ensure that there is an excellent ground plane beneath the monopole antenna, conducting wires are typically implanted underneath the antenna, and they emanate radially away from the monopole for a distance that is approximately equal to the height of the antenna.

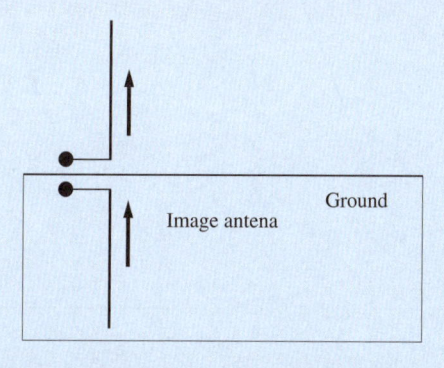

---

[3] S. V. Savov, "An Efficient Solution of a Class of Integrals Arising in Antenna Theory," *IEEE Antennas & Propag. Magazine* 44 (Oct. 2002): 98–101.

It is possible to estimate the validity of the approximation for a short dipole that is $\lambda/4$ in length. From the previous example, the radiation resistance is computed to be $R_{rad} \approx 12.3 \, \Omega$, while the actual value obtained from the theory of the linear antenna is $R_{rad} = 6.7 \, \Omega$. Hence, even for this small length, the short dipole approximation is not a good estimate.

EXAMPLE 8.10

Calculate the radiation resistance of a finite dipole antenna as a function of the physical length divided by the excitation wavelength.

Answer. The expression for the radiation resistance of a linear antenna is much more complicated than that derived in Example 8.7. It involves new special functions involving sine and cosine integrals, and it is usually considered in books dealing with advanced electromagnetics.[4] Because of this complexity, we will only provide the results here. For additional information, see the referenced text, where the equations are used to generate the graph below with MATLAB. One can see that for the popular case of a ***half-wave dipole***, the radiation resistance is found to be $R_{rad} = 73.1 \, \Omega$ as shown in Problem 8.4.7, which is very close to the measured value.

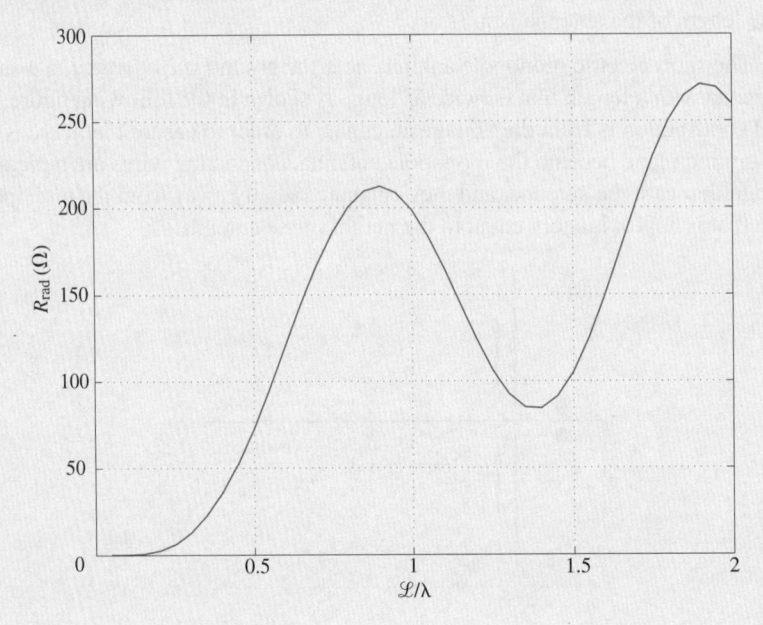

---

[4] C. Balanis, *"Antenna Theory, Analysis and Design,"* 3rd edition (New York: Wiley, 2005), pp. 173–178.

8.5.2 Directivity

The equation for the radiated power (8.37) can be written not only as a surface integral of the time-average radial component of the Poynting vector $S_r = 1/2 \, \text{Re}\{E_\theta H_\phi^*\}$, but also as an integral over a solid angle. For this purpose, we define the **radiation intensity** as $I(\theta, \phi) = r^2 S_r(\theta, \phi)$, which allows the radiated power to be written

$$P_{rad} = \oint_{4\pi} I(\theta, \phi) d\Omega \tag{8.40}$$

The ratio of $I(\theta, \phi)$ divided by its maximum value is the normalized **power radiation pattern**

$$I_n(\theta, \phi) = \frac{I(\theta, \phi)}{I(\theta, \phi)_{max}} \tag{8.41}$$

The **beam solid angle** of the antenna is defined as

$$\Omega_A = \oint_{4\pi} I_n(\theta, \phi) d\Omega \equiv \int_{\phi=0}^{2\pi} d\phi \int_{\theta=0}^{\pi} I_n(\theta, \phi) \sin\theta d\theta \tag{8.42}$$

In a complete sphere, there are 4π steradians. From the definition, it follows that, for an isotropic antenna, $I_n(\theta, \phi) \equiv 1$, and the beam solid angle is $\Omega_A = 4\pi$.

The next parameter that defines an antenna system is **directivity**. The directivity of an antenna is defined as the ratio of the maximum radiation intensity of a transmitting antenna divided by the average radiation intensity from an isotropic radiator with the same input power, given by

$$\boxed{D = \frac{I(\theta, \phi)_{max}}{\left(\frac{P_{rad}}{4\pi}\right)} = \frac{4\pi}{\oint_{4\pi} I_n(\theta, \phi) d\Omega} = \frac{4\pi}{\Omega_A}} \tag{8.43}$$

Because the denominator Ω_A is always less than 4π, the directivity $D > 1$.

EXAMPLE 8.11

Find the directivity of the Hertzian dipole.

Answer. Use the definition (8.43) for the directivity, including the normalized radiation intensity for a short dipole $I_n(\theta, \phi) = \sin^2\theta$, and obtain

$$D = \frac{4\pi}{2\pi \int_0^\pi \sin^2\theta \sin\theta d\theta} = \frac{2}{\int_0^\pi (\cos^2\theta - 1) d(\cos\theta)} = \frac{2}{-\frac{2}{3} + 2} = \frac{3}{2} = 1.5$$

For the short dipole, the directivity is $D = 1.5$ or $10 \log_{10}(1.5) = 1.76 \, \text{dB}$. A more complicated calculation performed for the half-wave dipole leads to a slightly higher value for the directivity $D = 1.64$ or $2.15 \, \text{dB}$ as given in Problem 8.4.6

8.5.3 Antenna Gain

The *gain* is a global characteristic of the antenna. Because of this, the *antenna efficiency* η is involved, and it is related to the directivity of the antenna. By definition the gain is

$$G = \eta D \qquad (8.44)$$

For the case of a lossless antenna, $\eta = 1$ and the gain is equal to the directivity. For a lossy antenna, $\eta < 1$ and the gain is less than the directivity. There are many types of loss that are associated with antennas, with the main ones being the losses due to energy dissipated in the dielectrics and conductors that make up the physical antenna, and reflection losses due to impedance mismatches between a transmission line and antenna. In an antenna course, we examine these and other losses in more detail, but will not discuss them further at this time.

8.5.4 Beamwidth

Beamwidth is a parameter associated with the lobes in an antenna pattern. It is defined as the angular separation between two identical points on opposite sides of the lobe maximum. There are several different types of beamwidth that are associated with an antenna pattern. The most common one is the *Half-Power Beamwidth* (HPBW). IEEE standard 149–1993 defines HPBW in this fashion: "In a radiation pattern cut containing the direction of the maximum of a lobe, the angle between the two directions in which the radiation intensity is one-half the maximum value." Thus, to find the HPBW location, we set the equation defining the radiation pattern equal to 0.5 and solve for the angle that gives that value. In dB values, this is equivalent to the points 3 dB down from the maximum value of the beam. Another useful beamwidth is the First-Null Beamwidth (FNBW), which is the angular separation between the first null on either side of the main beam.

Beamwidth is a very useful parameter and can be used as an antenna design parameter, as there is a trade-off between the beamwidth and the side lobe level. Beamwidth is also used to describe the resolution capabilities of an antenna when it is used to separate and identify two adjacent sources or the returns from two adjacent radar targets.

Because beamwidth is defined relative to the maximum value of the beam, it is common to specify an antenna's directivity with respect to the beamwidths of the antenna. For an antenna that generates a rotationally symmetric lobe, the half-power points of any two perpedicular planes are identical. Given this, we can approximate this antenna's directivity as

$$D \approx \frac{4\pi}{\theta_{HP} \phi_{HP}} \qquad (8.45)$$

8.5 Antenna Parameters

In the above expression, the values for the two beamwidths are given in radians. If the half-power beamwidths are given in degrees, then the approximate directivity is found by multiplying equation (8.45) by $(180/\pi)^2$. This yields the following expression

$$D \approx \frac{41,253}{(\theta°_{HP} - d)(\phi°_{HP})}$$

For additional information on gain and directivity, there is an excellent review article that appeared in 1998 in the IEEE Antennas and Propagation Magazine.[5]

EXAMPLE 8.12

Find the HPBW of an infinitesimal dipole.

Answer. The normalized radiation intensity for the electric field of an infinitesimal dipole is given by $I_n(\theta, \phi) = \sin^2\theta$, and it has a maximum value $I_n = 1$ for $\theta = \pi/2$. The value $I_n = 1/2$ is found at the angles $\theta = \pi/4$ and $3\pi/4$. Therefore, the HPBW for the electric field is $\theta_{HP} = \pi/2$. The normalized radiation intensity for the magnetic field of an infinitesimal dipole is $I_n(\theta, \phi) \equiv 1$. The HPBW for the magnetic field intensity is $\phi_{HP} = \pi/2$. From (8.45), we can approximate the directivity to be $D \approx 4/\pi = 1.27$. The value that was computed in Example 8.11 was $D = 1.5$.

8.5.5 Effective Aperture

Antennas are devices that exhibit the property of reciprocity. This means that the properties of an antenna are the same whether it is used as a transmitting antenna or a receiving antenna. For instance, the radiation pattern that is associated with an antenna is the same in either case. Although the antenna itself is a reciprocal device, reciprocity does not apply if a nonreciprocal device is added to the antenna. Thus, if a ferrite isolator is used with an antenna, the resultant combination will no longer exhibit reciprocity, since a ferrite isolator is a non-reciprocal device.

Despite the reciprocity of antennas, there are certain parameters that are more important for one use of an antenna instead of the other. For a receiving antenna, one of those parameters is the effective area A_e, which can be considered to be the reception cross-sectional area. The effective area of a

---

[5] "Estimating Directivity and Gain of Antennas," Warren L. Stutzman, IEEE Antennas and Propagation Magazine, Vol. 40, No. 4, August 1998, pp. 7–11.

receiving antenna is defined as follows: "In a given direction, the ratio of the available power at the terminals of a receiving antenna to the power flux density of a plane wave incident on the antenna from that direction, the wave being polarization matched to the antenna." The antenna being polarization matched means that the polarization of the incident wave is identical to that of the antenna. We can use the product of the effective area of the antenna and the incident power density, S_{av}, incident on the antenna to determine the power received by the antenna. The incident power density is given by:

$$S_{av} = \frac{E^2}{2Z_0} = \frac{E^2}{240\pi} \quad (8.46)$$

It is reasonable to assume that the plane wave is incident upon a receiving antenna which is terminated with a matched load impedance of Z_0 as shown in Figure 8–12c. Assume the incident wave arrives from a direction that matches the main beam of the antenna, or the direction of maximum power of the radiation pattern. Then

$$P_L = S_{av} A_e \quad (8.47)$$

As you may recall from circuit theory, maximum power can be delivered to a load impedance if it has a value that is equal to the complex conjugate of the antenna impedance, $Z_L = Z_A^*$. Replacing the antenna with an equivalent generator having the same voltage V and impedance Z_A, we determine a current at the terminals to be

$$I_0 = \frac{V}{Z_A + Z_L} \quad (8.48)$$

Therefore, the maximum power dissipated in the load is given by

$$P_L = \frac{1}{2} I_0^2 R_L = \frac{1}{2} \left(\frac{V}{Z_A + Z_A^*} \right)^2 R_L = \frac{V^2}{8 R_A} \quad (8.49)$$

where we have defined $Z_A + Z_A^* = 2 R_A$.

For the Hertzian dipole, the antenna resistance $R_A = R_{rad}$ was calculated in Example 8.7, and the maximum voltage in the direction $\theta = \pi/2$ was found to be $V = (E \sin \theta) \mathscr{L} = E \mathscr{L}$.

$$P_L = \frac{(E\mathscr{L})^2}{8 \times 80 \pi^2 \left(\frac{\mathscr{L}}{\lambda} \right)^2} = \frac{E^2 \lambda^2}{640 \pi^2} \quad (8.50)$$

For the Hertzian dipole, we find the effective area to be

$$A_e = \frac{\lambda^2}{4\pi} \left(\frac{3}{2} \right) = \frac{3 \lambda^2}{8\pi} \quad (8.51)$$

8.5 Antenna Parameters

using (8.47), (8.46), and (8.50). In general, the effective area A_e of any antenna is related to its directivity D by the following equation.

$$\boxed{\begin{aligned} A_e &= \frac{\lambda^2}{4\pi} D \\ A_e &= \frac{\lambda^2}{4\pi} G \end{aligned}} \quad (8.52)$$

We can also write the effective area in terms of gain. While both of these expressions are used, the expression in terms of G is more commonly used.

8.5.6 Friis Transmission Equation

In this section, we identify the relationship between an antenna used for transmission and another antenna used for reception. This is depicted in Figure 8–14.

Let antenna A in Figure 8–14 transmit to antenna B. Both antennas are in the far-field region. The gain of the transmitting antenna A in the direction of B is G_A, which in the lossless case is equal to D_A. Hence, the time-average power density at B is

$$S_{av} = \frac{P_t}{4\pi R^2} G_A \quad (8.53)$$

Writing (8.47) for the received power

$$P_r = S_{av} A_{e,r} = \left[\frac{P_t}{4\pi R^2} G_A\right] \frac{\lambda^2}{4\pi} G_r = \frac{P_t G_A G_r \lambda^2}{(4\pi R)^2} \quad (8.54)$$

$$\boxed{\frac{P_r}{P_t} = \frac{A_{e,t} A_{e,r}}{\lambda^2 R^2}} \quad (8.55)$$

Equation (8.55) is called the ***Friis transmission equation***.

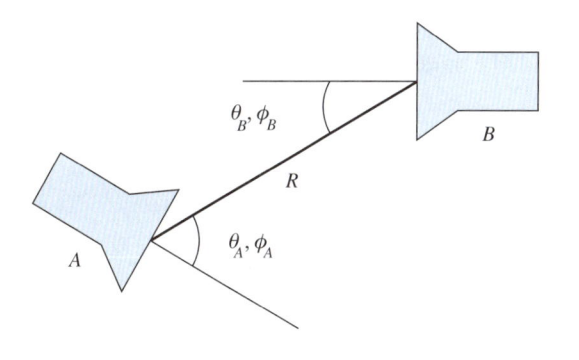

FIGURE 8–14

Two antennas separated by a distance R.

EXAMPLE 8.13

Find a criterion for the argument that a receiving antenna actually is in the far field of a transmitting antenna. Estimate the distance d of the far zone if $D = 10$ cm and $\lambda = 3$ cm. Assume that they have antenna gains $G_A = 1.5$ and $G_B = 1.64$, respectively, and find the ratio P_r / P_t at that distance.

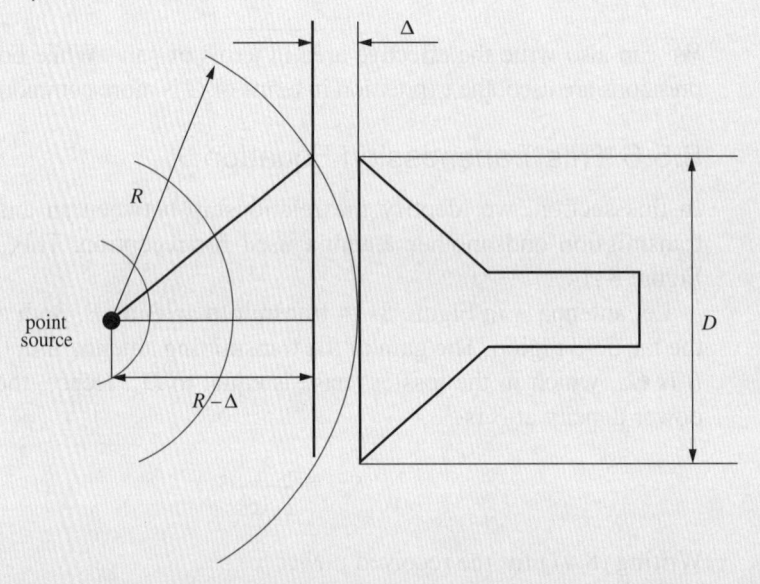

Answer. For the Friis transmission equation to be applicable, both antennas must be in the far field. The receiving antenna will be in the far field if the incident spherical wave deviates from an actual plane wave by less than a small fraction of a wavelength. Assume the largest dimension of the receiving antenna is D. By convention, we assume this deviation is approximately $\Delta \approx \lambda/16$, which means a phase difference of 22.5°. From the figure, we write

$$R^2 = (R - \Delta)^2 + \left(\frac{D}{2}\right)^2 \approx R^2 - 2R\Delta + \frac{D^2}{4}$$

This implies that the receiving antenna will be in the far field, or the **Fraunhofer zone**, if the second term is at least comparable with the third one or $R \approx D^2/8\Delta$. This means

$$R \approx \frac{2D^2}{\lambda}$$

For the particular values, we obtain

$$R = \frac{2 \times 0.1^2}{0.03} = 66.7 \text{ cm}$$

Beyond this distance, the receiving antenna will be in the far-field region of the transmitting antenna.

The Friis equation (8.55) gives the ratio

$$\frac{P_r}{P_t} = \frac{1.07 \times 1.17}{16.7^2 \times 3^2} = 5 \times 10^{-4}$$

The degradation of a received signal is approximately

$$-10 \log_{10}(5 \times 10^{-4}) = 33 \text{ dB}$$

8.6 Antenna Arrays

In many antenna applications, we will find that a single antenna will not be able to provide the desired radiation characteristics, Instead, it might take a combination of antennas to generate the required radiation characteristics. Because there are many ways to combine various types of antennas, we need a technique that will allow us to generate the required radiation pattern. Let us turn our attention to a technique that answers the need for such a radiation pattern through setting up an *antenna array*.

An antenna array is defined as a cluster of antennas that are arranged in a prescribed physical configuration—a straight line, rectangle, circle, etc. Each individual antenna is called an *element* of the array. We will initially assume that each element composing the array is physically identical. However, the amplitude and phase of excitation applied to each individual element may differ—a simplification that will prove useful. The far-field radiation from the array in a linear medium is computed from the vector addition of the components of the electromagnetic fields that are radiated from the individual elements. This is called the ***principle of superposition***.

There are several possible configurations for an antenna array. We will initially examine an antenna array where the elements are located in a straight line, also known as a *linear array*. To introduce the procedure, we will first examine an array that consists of two elements that are excited with the same amplitude signals but with phases that differ by an amount δ. This simple configuration illustrates the array concept.

A linear array consisting of two elements is shown in Figure 8–15. The individual element can be characterized by its *element pattern* $F_1(\theta, \phi)$, which generalizes our previous definition to the dependence on both angles in spherical coordinates.

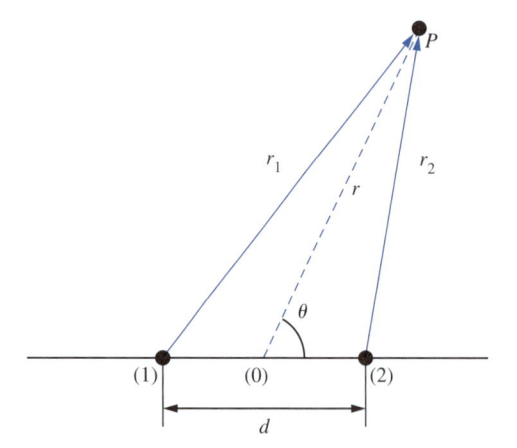

FIGURE 8-15

A two-element linear array.

At point P, the total far-field electric field component consists of the sum of the contributions of the two individual elements

$$E(\mathbf{r}) = E_1(\mathbf{r})e^{j\psi/2} + E_2(\mathbf{r})e^{-j\psi/2} \qquad (8.56)$$

where $E_1(\mathbf{r})$ is the electric field at distance r_1 due to source (1), $E_2(\mathbf{r})$ is the electric field at distance r_2 due to source (2), and the phase difference between the fields of the two sources due to the physical separation d and the different phase excitation δ is $\psi = kd\cos\theta + \delta$. The **phase center** is taken at the point (0)—the midpoint of the array. Since the elements are identical, we can assert that $E_2(\mathbf{r}) = E_1(\mathbf{r})$ and write

$$E(\mathbf{r}) = 2E_1(\mathbf{r})\left(\frac{e^{j\psi/2} + e^{-j\psi/2}}{2}\right) = 2E_1(\mathbf{r})\cos\left(\frac{\psi}{2}\right) \qquad (8.57)$$

It can easily be shown that relocating the phase center point (0) changes only the phase of the result but not its amplitude. The **total field pattern** $F(\theta, \phi)$ in (8.57) can be written as a *product* of the radiation pattern of an individual element $F_1(\theta, \phi)$ and the radiation pattern of the array $F_a(\theta, \phi)$. This term is called the **array factor**. This product solution is called the **multiplication of patterns**, and it will frequently be encountered in practice

$$\boxed{F(\theta, \phi) = F_1(\theta, \phi)F_a(\theta, \phi)} \qquad (8.58)$$

where

$$F_a(\theta, \phi) = \cos\left(\frac{kd\cos\theta + \delta}{2}\right) \qquad (8.59)$$

8.6 Antenna Arrays

where δ is the phase difference between the two antennas. Note that this latter term depends on the geometry of the array and the amplitude and phase of the individual excitation signals applied to each element. In (8.56), the amplitudes were set equal, but this need not be a general requirement.

In example 8.14, we will provide several examples that illustrate the far-field radiation pattern for two isotropic radiating elements placed along the x axis. In each case, the two elements are separated by a distance d, and the individual elements are excited with equal amplitude signals having a phase difference of delta, as shown in Figure 8-15.

EXAMPLE 8.14

Find and plot the array factor for three two-element antenna arrays, that differ only in the separation distance between the elements. Assume the two elements for each antenna array are isotropic radiators. The antennas are separated by 5, 10, and 20 cm, and each antenna is excited in phase. The frequency of the signal applied to each antenna is 1.5 GHz.

Answer. The separation between the elements is normalized by the wavelength via $\xi = kd/2 = \pi d/\lambda$. The wavelength in the free space is

$$\lambda = \frac{c}{f} = \frac{3 \times 10^8}{1.5 \times 10^9} = 20 \text{ cm}$$

or the normalized separation d is $\lambda/4$, $\lambda/2$, and λ, respectively. This yields the following values for the parameter ξ: (a) $\pi/4$, (b) $\pi/2$, and (c) π. The phase difference is zero ($\delta = 0$). The corresponding array factor $F_a(\theta)$ is found from equation (8.59), and it is plotted in the figure below for these three cases. Because the element pattern is uniform ($F_1(\theta) \equiv 1$), it follows from (8.58) that the total radiation pattern $F(\theta) = F_a(\theta)$. The number of the lobes that are found in the pattern increases linearly with an increase of the normalized length of the array, and it can be approximated by the nearest integer number to the real parameter $4d/\lambda$.

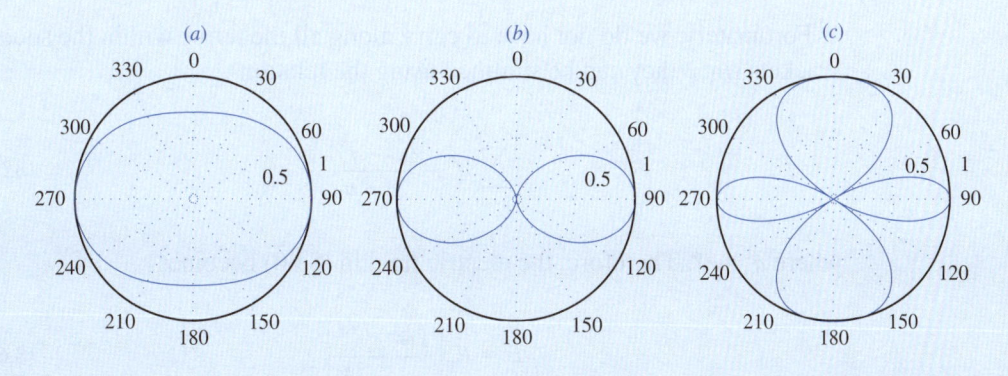

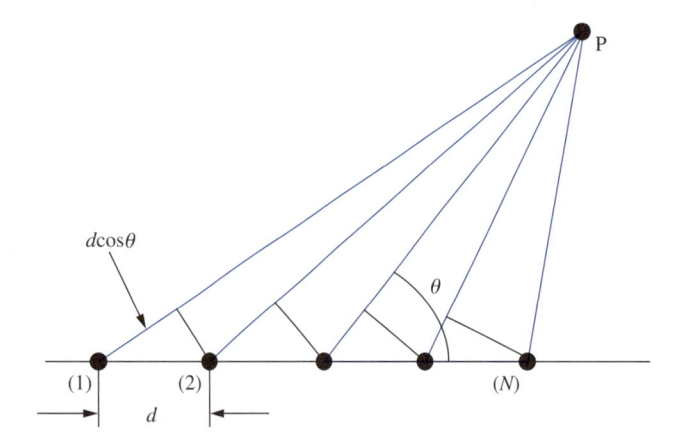

FIGURE 8-16

A uniform linear array. The current on the first element is $I(z)$, the current on the second element is $I(z)\, e^{j\Delta}$, the current on the third element is $I(z)\, e^{j2\Delta}$, etc.

Another method of altering the radiation pattern of the array is to electronically change the phase parameter δ of the applied signal—we can have the antenna "sweep" through certain regions of space. Such a structure is called a **phased-array antenna**. Antennas of this type are particularly important in large radar installations, where it would be mechanically impossible to rotate an antenna that may be the size of a football field.

We can extend our investigation into this type of antenna array in several ways—say, to consider more *identical* elements than two, as shown in Figure 8–16. In this linear array, there is a linearly progressive phase shift in the excitation signal that feeds the identical N elements. In this case, (8.56) generalizes to

$$E(r) = E_0(r)[1 + e^{j\psi} + e^{j2\psi} + \cdots + e^{j(N-1)\psi}] \tag{8.60}$$

Fortunately, we do not have to carry along all the terms within the square brackets since they can be summed using the relation

$$\sum_{n=0}^{N-1} q^n = \frac{1 - q^N}{1 - q} \tag{8.61}$$

where $q = e^{j\psi}$. Therefore, the electric field in (8.60) becomes

$$E = E_0 \left[\frac{1 - e^{jN\psi}}{1 - e^{j\psi}} \right] \tag{8.62}$$

8.6 Antenna Arrays

If we examine only the *magnitude* of the electric field |E|, we can simplify (8.62) with the relation

$$\left|1 - e^{j\xi}\right| = \left|2je^{j\xi/2}\sin\frac{\xi}{2}\right| = 2\sin\frac{\xi}{2}$$

Hence (8.62) identifies the magnitude of the electric field.

$$E(\theta) = E_0 \frac{\sin\left(\frac{N\psi}{2}\right)}{\sin\left(\frac{\psi}{2}\right)} \tag{8.63}$$

where $\psi(\theta) = kd\cos\theta + \delta$ and δ is the progressive phase difference between elements. The maximum value occurs when $\psi \to 0$, and it is

$$E_{\max} = NE_0 \tag{8.64}$$

In the direction of the maximum value, the condition $\psi = 0$ or $kd\cos\theta = -\delta$ is satisfied. Dividing (8.63) by (8.64) yields a **normalized array factor**

$$F_a(\theta) = \frac{\sin\left(\frac{N\psi}{2}\right)}{N\sin\left(\frac{\psi}{2}\right)} \tag{8.65}$$

The angles where the first null in the numerator of (8.62) occur will define the main beam in the radiation pattern of the linear array. This happens for $e^{jN\psi} = 1$, provided $e^{j\psi} \neq 1$ or for $\psi = \pm k2\pi/N$ (k is an integer value). Similarly, zeroes in the denominator will yield maxima in the pattern.

EXAMPLE 8.15

Find and plot the radiation pattern of two parallel thin half-wavelength electric dipoles separated by $d = \lambda/2$, λ, and $3\lambda/2$. The mutual coupling of the dipoles is neglected.

Answer. Here the radiation pattern of this linear array is calculated by applying the multiplication property (8.58), where the radiation pattern $F_1(\theta)$ of a single element (half-wavelength dipole with $\mathscr{L} = \lambda/2$) is given by (8.30) and presented by the first plot in Figure 8–10. The array factor for this case (in phase, excited with $\delta = 0$) is obtained from (8.65) for $N = 2$, which yields $F_a(\theta) = \cos((kd/2)\cos\theta)$. The results for the total radiation pattern $F(\theta)$ for these three cases are plotted in the figures below for *(a)* $d = \lambda/2$, *(b)* $d = \lambda$, and *(c)* $d = 3\lambda/2$.

460   Radiation of Electromagnetic Waves

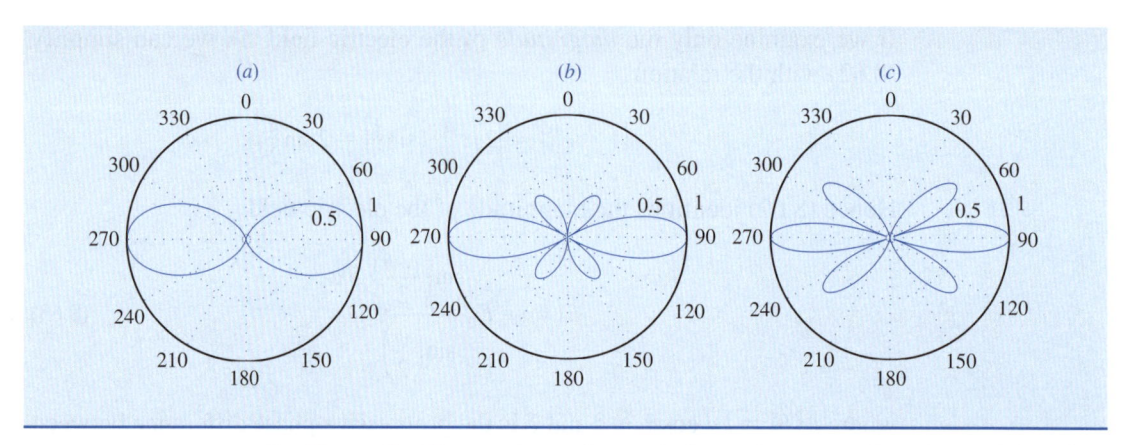

In Figure 8–17, we show the variation of $F(\theta)$ as the phase delay δ is changed in equal increments of $\pi/4$ for a four-element array ($N = 4$). The separation of the elements $d = \lambda/2$. Hence, we observe that the antenna radiation pattern can be altered by changing the *phase* even though the physical elements are not changed.

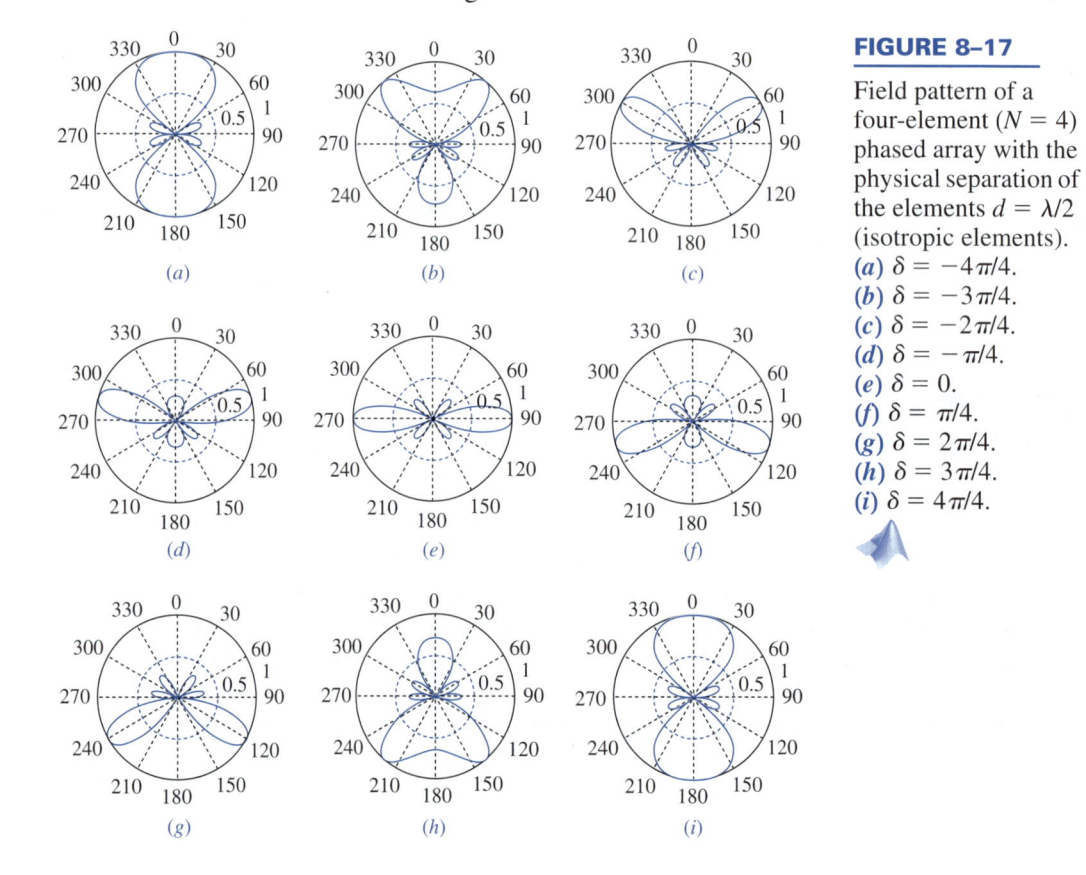

FIGURE 8–17

Field pattern of a four-element ($N = 4$) phased array with the physical separation of the elements $d = \lambda/2$ (isotropic elements).
(a) $\delta = -4\pi/4$.
(b) $\delta = -3\pi/4$.
(c) $\delta = -2\pi/4$.
(d) $\delta = -\pi/4$.
(e) $\delta = 0$.
(f) $\delta = \pi/4$.
(g) $\delta = 2\pi/4$.
(h) $\delta = 3\pi/4$.
(i) $\delta = 4\pi/4$.

8.6 Antenna Arrays

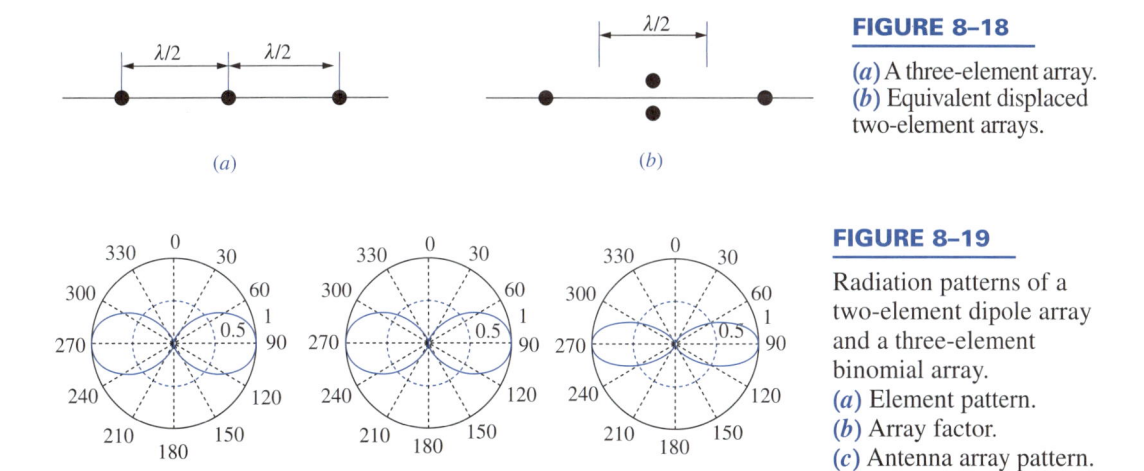

FIGURE 8-18

(*a*) A three-element array.
(*b*) Equivalent displaced two-element arrays.

FIGURE 8-19

Radiation patterns of a two-element dipole array and a three-element binomial array.
(*a*) Element pattern.
(*b*) Array factor.
(*c*) Antenna array pattern.

A second method would be to examine the expected behavior if there is a prescribed *nonuniform* excitation of the elements. For example, let us assume that we have a linear array that consists of three elements that are physically displaced by a distance $d = \lambda/2$, and each element is excited *in phase* ($\delta = 0$). The excitation of the center element is twice as large as that of the outer two elements, as shown in Figure 8–18*a*. The choice of this distribution of excitation amplitudes is based on the fact that 1:2:1 are the leading terms of a binomial series. The resulting array, which could be generalized to include more elements, is called a **binomial array**.

Because of the excitation at the center element being twice that of the outer two elements, we can consider that this three-element array is equivalent to two two-element arrays that are displaced by a distance $\lambda/2$ from each other. This allows us to make use of (8.65) for $N = 2$, where it is interpreted to be the radiation pattern of this new element. The result is

$$F(\theta) = \cos\left(\frac{\pi}{2}\cos\theta\right) \tag{8.66}$$

The array factor for these new elements is the same as the radiation pattern of one of the elements. Therefore, from (8.58) we write that the magnitude of the far-field radiated electric field from this structure is given by

$$F(\theta) = \cos^2\left(\frac{\pi}{2}\cos\theta\right) \tag{8.67}$$

The radiation pattern for this array is shown in Figure 8–19. It is contrasted with the two-element array, and we note that the radiation pattern of the three-element array with a nonuniform excitation is narrower. We note that in this binomial array there are *no side lobes* to absorb power. If more elements are included in the array, the beam width becomes narrower.

In drawing the composite figure for the antenna array that comprises two small dipoles separated by a half wavelength, we have *multiplied* the radiation pattern of the individual antenna by the array factor. In this case, the array factor is the same since this is a binomial array. The multiplication is best illustrated by working through an example.

EXAMPLE 8.16

Using the concept of the multiplication of patterns, find the radiation pattern of the array of four elements as shown in the figure.

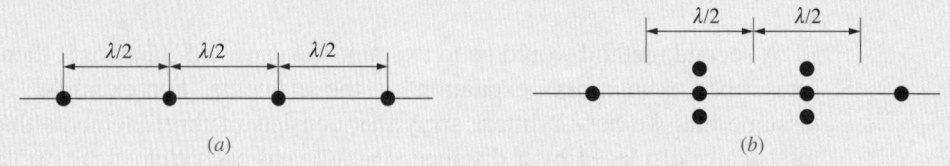

Answer. This array is to be replaced with an array of two elements containing three subelements (1:2:1) each. The new array will have the individual excitations (1:3:3:1). The result according to the multiplication property (8.58) will have a radiation pattern as follows:

$$F(\theta) = \cos\left(\frac{\pi}{2}\cos\theta\right)\cos^2\left(\frac{\pi}{2}\cos\theta\right) = \cos^3\left(\frac{\pi}{2}\cos\theta\right)$$

The results are shown in the figure below: (*a*) the element pattern, (*b*) the array factor, and (*c*) the antenna array pattern. We note that the final pattern is narrower and has no side lobes.

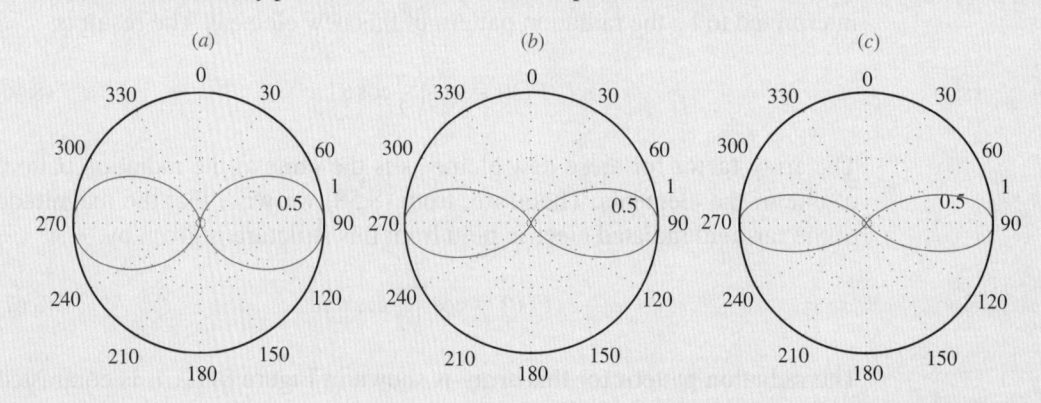

8.6 Antenna Arrays

Continuing the process, it is possible to obtain a pattern with arbitrarily high directivity and no side lobes if the amplitudes of the sources in the array correspond to the coefficients of binomial series. This means that the amplitude of the kth source in the binomial array of N elements has to be calculated with the following equation:

$$I_k = \frac{N!}{k!(N-k)!} \quad (k = 0, 1, \ldots, N) \tag{8.68}$$

By definition, it is clear that this array will be symmetrically excited or $I_{N-k} = I_k$. The resulting radiation pattern of the binomial array of N elements that are separated by a half wavelength is

$$F(\theta) = \cos^{(N-1)}\left(\frac{\pi}{2}\cos\theta\right) \tag{8.69}$$

In the analysis above, the **mutual coupling** between the elements of the antenna array was neglected. The simplest case is to consider an array of two elements—say, dipoles with lengths $\mathscr{L}_1$ and $\mathscr{L}_2$. The first dipole is driven by a voltage $V_1(\mathbf{r}')$ while the second one is passive (see Figure 8–20). Assume the currents in both terminals are I_1 and I_2, respectively, and the following circuit relations are fulfilled

$$\begin{aligned} Z_{11}I_1 + Z_{12}I_2 &= V_1 \\ Z_{21}I_1 + Z_{22}I_2 &= 0 \end{aligned} \tag{8.70}$$

where Z_{11} and Z_{22} are the self-impedances of elements (1) and (2) and $Z_{12} = Z_{21}$ are the mutual impedances between the elements. Suppose the dipoles are equal in length ($\mathscr{L}_1 = \mathscr{L}_2 = \mathscr{L}$); then the self-impedances are also equal. For

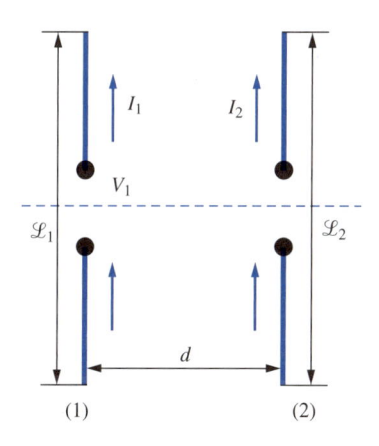

FIGURE 8–20

Array of two coupled parallel dipoles.

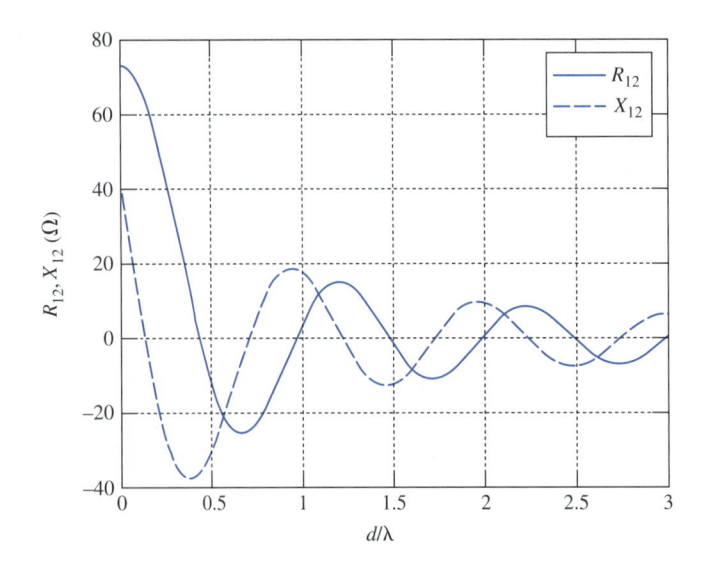

FIGURE 8–21

Mutual impedance between two parallel, thin, half-wavelength dipoles.

the particular case of thin half-wavelength dipoles as in Example 8.10, we have for the self-impedance the value $Z_{11} = 73.1 + j42.5 \; \Omega$. The dependence of the mutual impedance between two similar thin half-wavelength dipoles[6] as a function of the normalized distance d/λ is presented in Figure 8–21. Here R_{12} is shown with a solid line, while X_{12} is shown with a dashed line. In the limiting case of the separation distance $d \to 0$, the mutual impedance approaches the self-impedance, which is to be expected.

EXAMPLE 8.17

Find the current in one of the dipoles in an antenna array if there is

(a) No mutual coupling between the dipoles.
(b) A mutual coupling between the dipoles.

The driving voltage is $V_1 = 100$ V and the distance between the two half-wave dipoles is $d = \lambda/2$. Estimate the amplitude and the phase change.

---

[6] R. C. Hansen, ed., *Array Theory and Practice*, vol. 2, (San Diego, California: Academic Press, 1966).

8.6 Antenna Arrays

Answer.

(a) For the case with no mutual coupling, we can apply (8.70) with $Z_{12} = 0$. This results in

$$I_1 = \frac{V_1}{Z_{11}} = 1.183 e^{-j30.17°} \text{ A}$$

(b) For the case with mutual coupling, we obtain from Figure 8–21 that $Z_{12} = -12.5 - j29.9 \, \Omega$. The solution of (8.70) yields the current

$$I_1 = 1.218 e^{-j21.80°} \text{ A}$$

The amplitude of the input current changes slightly, while the change in phase is more significant.

One may wish to find the directivity of one antenna array. Applying (8.43) for the particular case of $N = 2M + 1$ identical elements separated by a distance $d = \lambda/2$, the following equation is obtained.

$$D = \frac{\left(\sum_{n=-M}^{M} I_n\right)^2}{\sum_{n=-M}^{M} I_n^2} \tag{8.71}$$

which means that the directivity measures the degree of the coherence of the total electric field.

EXAMPLE 8.18

Compare the directivities of two arrays consisting of three identical elements separated by a half wavelength for the following two cases:

(a) Uniform array: $(I_{-1} = I_0 = I_1 = 1 \text{ A})$.
(b) Binomial array: $(I_{-1} = I_1 = 1 \text{ A}; I_0 = 2 \text{ A})$.

Answer. From (8.71), we compute

(a) Uniform array $D = \dfrac{(1 + 1 + 1)^2}{1 + 1 + 1} = 3 \Rightarrow D = 4.77$ dB.

(b) Binomial array $D = \dfrac{(1 + 2 + 1)^2}{1 + 4 + 1} = \dfrac{16}{6} = 2.667 \Rightarrow D = 4.26$ dB.

The directivity of a uniform array is greater than that of a binomial array.

The uniform array and the binomial array are just two examples of linear antenna arrays. The first one has a narrow HPBW, but a relatively high **Side-Lobe Level** (SLL), while the second one has a relatively wide HPBW but a very low SLL. The designer of real antenna arrays has to make a trade-off between these two important characteristics: (1) narrow HPBW and (2) low SLL. There are two different approaches for solving this **optimization problem**, such as (1) choosing special excitation (i.e., Chebyshev array) or (2) choosing nonuniform distance between the elements. However, these methods are topics for more advanced investigations.

8.7 Conclusion

We have reviewed the fundamental radiation characteristics of electromagnetic waves and their relationship to accelerating and decelerating charges. Using these concepts, the small Hertzian dipole radiator was described, and we identified a radiation pattern for such an antenna.

A formal procedure using the concept of the vector potential was introduced and applied to several antennas. We then directed our attention to the far-field properties of antennas. If the medium in which the waves are propagating is linear, the principle of superposition applies. Constructive and destructive interference between fields radiated by displaced antenna elements with differing phases in the applied currents led to different radiation or reception characteristics of an antenna. Terms such as beamwidth, main and side lobes, radiation resistance, gain, directivity, and effective area were defined and applied.

The solution that we obtained for the radiation patterns was predicated on assuming a valid approximation for the current distribution on the antenna. Several distributions were analyzed. The method of moments introduced in Chapter 4 can be equally well applied to antenna calculations. In this case, the field distribution is known from the experimental measurements, and the current distribution becomes the unknown term that must be ascertained.

We then introduced the subject of antenna arrays consisting of several identical antennas. By controlling either the phase or the amplitude of the signal applied to each individual antenna element or its spatial separation, we saw that the resulting radiation pattern could be changed. We found that predictions of the radiation pattern could be made by multiplying the radiation pattern of an individual antenna times an array factor in order to find the radiation pattern of the entire antenna array.

8.8 Problems

8.1.1. Perform the integration of the integral

$$\int_0^\infty \frac{\rho^3 \, d\rho}{(\rho^2 + x^2)^3}$$

which arises in Example 8.1.

8.1.2. Using dimensional arguments, show that the term

$$\frac{Q^2}{32\pi\varepsilon_0 x_0} \; \mathrm{J} \; (x_0 > 0)$$

corresponds to the electrostatic energy stored in the region $x > x_0$ in Example 8.1.

8.2.1. A short electric dipole with a length $\mathscr{L}\,(\mathscr{L} \ll \lambda)$ is located above a ground plane a distance $h = \lambda/4$ $(h \gg \mathscr{L})$. The dipole is perpendicular to the ground plane. Find the directivity of this antenna. How does the ground plane change the directivity?

Hint: Use the **method of images** and the array concept, then apply a numerical integration.

8.4.1. Find the directivity of two short electric dipoles that are excited out of phase and are physically separated by one half-wavelength.

8.4.2. A short magnetic dipole with a radius a ($a \ll \lambda$) is located above a ground plane a distance $h < \lambda/4$ ($h \gg a$). The dipole is parallel to the ground plane. Find the directivity of this antenna. How does the ground plane change the directivity?

8.4.3. Find the directivity of two short magnetic dipoles that are excited in phase and are physically separated by one half-wavelength.

8.4.4. Determine the effects that a ground plane has on the radiation resistance of a short electric dipole with a length $\mathscr{L}\,(\mathscr{L} \ll \lambda)$ that is located a distance $h < \lambda/4$ above the ground plane.

8.4.5. Determine the effects that a ground plane has on the radiation resistance of a small magnetic dipole with a radius a ($a \ll 1$) that is located a distance $h < \lambda/4$ above the ground plane.

8.4.6. Find the directivity D of a half-wavelength electric dipole.

8.4.7. Find the radiation resistance R_r of a half-wavelength electric dipole.

8.4.8. A thin, vertical, quarter-wavelength monopole is located above a horizontal conducting ground. Find the radiation resistance of the antenna.

8.4.9. Find the current required to radiate a power of 100 W at 1 GHz from a 15 cm dipole.

8.4.10. A Hertzian dipole of length $L = 0.2$ m operates at 10 MHz. Find the *radiation efficiency* η_r if the copper conductor has the following parameters: $\sigma = 5.8 \times 10^7$ S/m, $\mu_r = 1$, and radius $a = 1$ mm. By definition, $\eta_r = R_r/(R_r + R_L)$ where R_r is the radiation resistance and R_L is the ohmic resistance.

8.4.11. Calculate the effective aperture A_e of the Hertzian dipole.

8.4.12. Calculate the effective aperture A_e of the half-wavelength dipole.

8.4.13. Find the HPBW and the SLL of an electric dipole with a length $L = 1.5\lambda$.

8.4.14. A TV broadcasting station (T_x) radiates a power of 500 W from an antenna on a 100 m tower above a perfectly conducting ground. The antenna is omnidirectional in the horizontal plane, but has a HPBW = 20° in the vertical plane. If the wavelength is 1 m, what is the optimum height for this antenna at a location of 2 km from the antenna for (a) vertical polarization, and (b) horizontal polarization? What is the received power in the receiver (R_x) for both cases if the receiving antenna is a half-wave dipole?

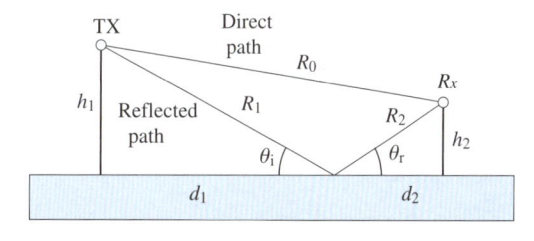

8.4.15. Find an expression for the *radar cross section* σ_s of a target. By definition, $\sigma_s = P_s/S_i$ where P_s is the backscattered power and S_i is the incident time-average power density.

8.4.16. Derive the *radar equation*, expressing the power received by a monostatic radar P_r in terms of the transmitted power from the same antenna P_t. Here, the antenna gain is G and the operating wavelength is λ. The distance between the antenna and the target is R and the radar cross section of the target is σ_s.

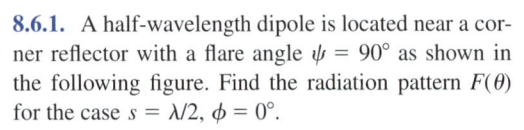

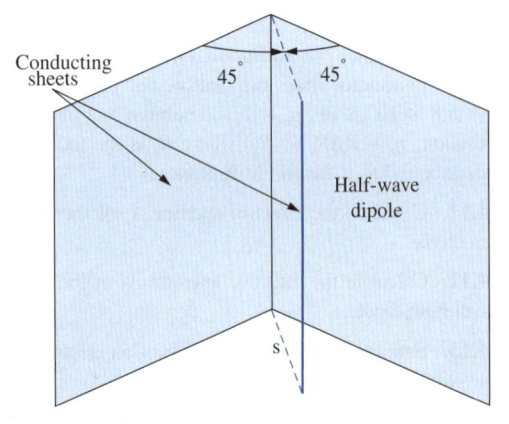

8.6.1. A half-wavelength dipole is located near a corner reflector with a flare angle $\psi = 90°$ as shown in the following figure. Find the radiation pattern $F(\theta)$ for the case $s = \lambda/2$, $\phi = 0°$.

8.6.2. Consider an antenna array that is excited in-phase ($\delta = 0°$) with N elements ($N \gg 1$). Each element of the array is separated by a distance d. Show that the direction of the main maximum is $\theta_0 = 90°$ (**broadside array**). Find an approximate expression for the HPBW.

8.6.3. Consider an antenna array consisting of N elements with each of the elements excited with a phase delay ($\delta \neq 0°$). Each element of the array is separated by a distance d. Show that it is possible to have the direction on the main maximum be $\theta_0 = 0°$. This is called an *end-fire array*. Find an approximate expression for the HPBW.

8.6.4. Plot the radiation pattern $F(\phi)$ of two parallel half-wave dipoles that are separated by a distance $d = \lambda/4$ and are excited by currents that have a phase difference $\delta = -90°$.

8.6.5. Calculate the input impedance for an array of two half-wave parallel dipoles that are separated by a distance $d = \lambda/2$ (see Figure 8–20). The currents are $I_2 = -I_1$.

8.6.6. Calculate the input impedance for an array of two parallel half-wave dipoles that are separated by a distance $d = \lambda/2$. The currents are $I_2 = +I_1$.

8.6.7. A half-wave dipole with a terminal current I_0 is placed a distance $s = 0.1\lambda$ from a perfectly conducting x–y plane as shown in the figure.

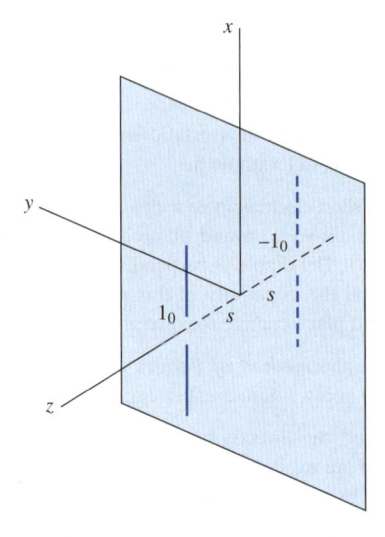

Neglecting ohmic losses, compare the terminal currents with and without the reflector if the radiated power is 2 W.

8.6.8. Assume that three identical antennas are separated by a distance $d = \lambda/4$ along a straight line. Each antenna is fed with the same current, but there is a uniform progressive phase shift δ along the line. Find and plot the array factor in three cases: (a) $\delta = 0$, (b) $\delta = \pi/2$, and (c) $\delta = \pi$.

8.6.9. Repeat Problem 8.6.8 with a larger separation $d = \lambda/2$ for the same phase differences.

APPENDIX A

Mathematical Formulas

A.1 Vector Identities

A, **B**, and **C** are vectors and a and b are scalars.

$$(\mathbf{A} \times \mathbf{B}) \cdot \mathbf{C} = (\mathbf{B} \times \mathbf{C}) \cdot \mathbf{A} = (\mathbf{C} \times \mathbf{A}) \cdot \mathbf{B} \tag{A.1}$$

$$\mathbf{A} \times (\mathbf{B} \times \mathbf{C}) = \mathbf{B}(\mathbf{A} \cdot \mathbf{C}) - \mathbf{C}(\mathbf{A} \cdot \mathbf{B}) \tag{A.2}$$

$$\nabla \cdot (\mathbf{A} + \mathbf{B}) = \nabla \cdot \mathbf{A} + \nabla \cdot \mathbf{B} \tag{A.3}$$

$$\nabla(a + b) = \nabla a + \nabla b \tag{A.4}$$

$$\nabla \times (\mathbf{A} + \mathbf{B}) = \nabla \times \mathbf{A} + \nabla \times \mathbf{B} \tag{A.5}$$

$$\nabla \cdot (a\mathbf{B}) = \mathbf{B} \cdot \nabla a + a \nabla \cdot \mathbf{B} \tag{A.6}$$

$$\nabla(ab) = a\nabla b + b\nabla a \tag{A.7}$$

$$\nabla \times (a\mathbf{B}) = \nabla a \times \mathbf{B} + a \nabla \times \mathbf{B} \tag{A.8}$$

$$\nabla \cdot (\mathbf{A} \times \mathbf{B}) = \mathbf{B} \cdot \nabla \times \mathbf{A} - \mathbf{A} \cdot \nabla \times \mathbf{B} \tag{A.9}$$

$$\nabla(\mathbf{A} \cdot \mathbf{B}) = (\mathbf{A} \cdot \nabla)\mathbf{B} + (\mathbf{B} \cdot \nabla)\mathbf{A} + \mathbf{A} \times (\nabla \times \mathbf{B}) + \mathbf{B} \times (\nabla \times \mathbf{A}) \tag{A.10}$$

$$\nabla \times (\mathbf{A} \times \mathbf{B}) = \mathbf{A}\nabla \cdot \mathbf{B} - \mathbf{B}\nabla \cdot \mathbf{A} + (\mathbf{B} \cdot \nabla)\mathbf{A} - (\mathbf{A} \cdot \nabla)\mathbf{B} \tag{A.11}$$

$$\nabla \cdot \nabla a = \nabla^2 a \tag{A.12}$$

$$\nabla \cdot \nabla \times \mathbf{A} = 0 \tag{A.13}$$

$$\nabla \times \nabla a = 0 \tag{A.14}$$

$$\nabla \times \nabla \times \mathbf{A} = \nabla(\nabla \cdot \mathbf{A}) - \nabla^2 \mathbf{A} \tag{A.15}$$

A.2 Vector Operations in the Three Coordinate Systems

Cartesian

$$\nabla a = \frac{\partial a}{\partial x}\mathbf{u}_x + \frac{\partial a}{\partial y}\mathbf{u}_y + \frac{\partial a}{\partial z}\mathbf{u}_z \tag{A.16}$$

$$\nabla \cdot \mathbf{A} = \frac{\partial A_x}{\partial x} + \frac{\partial A_y}{\partial y} + \frac{\partial A_z}{\partial z} \tag{A.17}$$

$$\nabla \times \mathbf{A} = \left(\frac{\partial A_z}{\partial y} - \frac{\partial A_y}{\partial z}\right)\mathbf{u}_x + \left(\frac{\partial A_x}{\partial z} - \frac{\partial A_z}{\partial x}\right)\mathbf{u}_y + \left(\frac{\partial A_y}{\partial x} - \frac{\partial A_x}{\partial y}\right)\mathbf{u}_z \tag{A.18}$$

$$\nabla^2 a = \frac{\partial^2 a}{\partial x^2} + \frac{\partial^2 a}{\partial y^2} + \frac{\partial^2 a}{\partial z^2} \tag{A.19}$$

Cylindrical

$$\nabla a = \frac{\partial a}{\partial \rho}\mathbf{u}_\rho + \frac{1}{\rho}\frac{\partial a}{\partial \phi}\mathbf{u}_\phi + \frac{\partial a}{\partial z}\mathbf{u}_z \tag{A.20}$$

$$\nabla \cdot \mathbf{A} = \frac{1}{\rho}\frac{\partial(\rho A_\rho)}{\partial \rho} + \frac{1}{\rho}\frac{\partial A_\phi}{\partial \phi} + \frac{\partial A_z}{\partial z} \tag{A.21}$$

$$\nabla \times \mathbf{A} = \left(\frac{1}{\rho}\frac{\partial A_z}{\partial \phi} - \frac{\partial A_\phi}{\partial z}\right)\mathbf{u}_\rho + \left(\frac{\partial A_\rho}{\partial z} - \frac{\partial A_z}{\partial \rho}\right)\mathbf{u}_\phi + \frac{1}{\rho}\left(\frac{\partial(\rho A_\phi)}{\partial \rho} - \frac{\partial A_\rho}{\partial \phi}\right)\mathbf{u}_z \tag{A.22}$$

$$\nabla^2 a = \frac{1}{\rho}\frac{\partial\left(\rho\frac{\partial a}{\partial \rho}\right)}{\partial \rho} + \frac{1}{\rho^2}\frac{\partial^2 a}{\partial \phi^2} + \frac{\partial^2 a}{\partial z^2} \tag{A.23}$$

Spherical

$$\nabla a = \frac{\partial a}{\partial r}\mathbf{u}_r + \frac{1}{r}\frac{\partial a}{\partial \theta}\mathbf{u}_\theta + \frac{1}{r\sin\theta}\frac{\partial a}{\partial \phi}\mathbf{u}_\phi \tag{A.24}$$

$$\nabla \cdot \mathbf{A} = \frac{1}{r^2}\frac{\partial(r^2 A_r)}{\partial r} + \frac{1}{r\sin\theta}\frac{\partial(A_\theta \sin\theta)}{\partial \theta} + \frac{1}{r\sin\theta}\frac{\partial A_\phi}{\partial \phi} \tag{A.25}$$

$$\nabla \times \mathbf{A} = \frac{1}{r\sin\theta}\left(\frac{\partial(\sin\theta A_\phi)}{\partial \theta} - \frac{\partial A_\theta}{\partial \phi}\right)\mathbf{u}_r$$
$$+ \frac{1}{r}\left(\frac{1}{\sin\theta}\frac{\partial A_r}{\partial \phi} - \frac{\partial(rA_\phi)}{\partial r}\right)\mathbf{u}_\theta + \frac{1}{r}\left(\frac{\partial(rA_\theta)}{\partial r} - \frac{\partial A_r}{\partial \theta}\right)\mathbf{u}_\phi \tag{A.26}$$

A.3 Summary of the Transformations Between Coordinate Systems

$$\nabla^2 a = \frac{1}{r^2}\frac{\partial\left(r^2 \frac{\partial a}{\partial r}\right)}{\partial r} + \frac{1}{r^2 \sin\theta}\frac{\partial\left(\sin\theta \frac{\partial a}{\partial \theta}\right)}{\partial \theta} + \frac{1}{r^2 \sin^2\theta}\frac{\partial^2 a}{\partial \phi^2} \quad (A.27)$$

A.3 Summary of the Transformations Between Coordinate Systems

Cartesian–cylindrical

$$\begin{cases} x = \rho\cos\phi \\ y = \rho\sin\phi \\ z = z \end{cases} \quad \begin{cases} \rho = \sqrt{x^2 + y^2} \\ \phi = \tan^{-1}\left(\frac{y}{x}\right) \\ z = z \end{cases} \quad (A.28)$$

	$\mathbf{u}_\rho$	$\mathbf{u}_\phi$	$\mathbf{u}_z$
$\mathbf{u}_x$	$\cos\phi$	$-\sin\phi$	0
$\mathbf{u}_y$	$\sin\phi$	$\cos\phi$	0
$\mathbf{u}_z$	0	0	1

(A.29)

Cartesian–spherical

$$\begin{cases} x = r\sin\theta\cos\phi \\ y = r\sin\theta\sin\phi \\ z = r\cos\theta \end{cases} \quad \begin{cases} r = \sqrt{x^2 + y^2 + z^2} \\ \theta = \tan^{-1}\left(\frac{\sqrt{x^2 + y^2}}{z}\right) \\ \phi = \tan^{-1}\left(\frac{y}{x}\right) \end{cases} \quad (A.30)$$

	$\mathbf{u}_r$	$\mathbf{u}_\theta$	$\mathbf{u}_\phi$
$\mathbf{u}_x$	$\sin\theta\cos\phi$	$\cos\theta\cos\phi$	$-\sin\phi$
$\mathbf{u}_y$	$\sin\theta\sin\phi$	$\cos\theta\sin\phi$	$\cos\phi$
$\mathbf{u}_z$	$\cos\theta$	$-\sin\theta$	0

(A.31)

Distance $R = |r_2 - r_1|$ in the three coordinate systems

Cartesian

$$R = [(x_2 - x_1)^2 + (y_2 - y_1)^2 + (z_2 - z_1)^2]^{1/2} \quad (A.32)$$

Cylindrical

$$R = [\rho_2^2 + \rho_1^2 - 2\rho_2\rho_1 \cos(\phi_2 - \phi_1) + (z_2 - z_1)^2]^{1/2} \quad (A.33)$$

Spherical

$$R = \{r_2^2 + r_1^2 - 2r_2 r_1 [\cos\theta_2 \cos\theta_1 + \sin\theta_2 \sin\theta_1 \cos(\phi_2 - \phi_1)]\}^{1/2} \quad (A.34)$$

A.4 Integral Relations

Divergence theorem

$$\int_{\Delta v} \nabla \cdot \mathbf{A}\, dv = \oint \mathbf{A} \cdot \mathbf{ds} \quad (A.35)$$

Stokes's theorem

$$\int_{\Delta s} \nabla \times \mathbf{A} \cdot \mathbf{ds} = \oint \mathbf{A} \cdot \mathbf{dl} \quad (A.36)$$

$$\int_{\Delta v} \nabla \times \mathbf{A}\, dv = -\oint \mathbf{A} \times \mathbf{ds} \quad (A.37)$$

$$\int_{\Delta v} \nabla a\, dv = \oint a\mathbf{ds} \quad (A.38)$$

$$\int_{\Delta s} \nabla a \times \mathbf{ds} = -\oint a\, \mathbf{dl} \quad (A.39)$$

A.4 Integral Relations

Coordinate system	Cartesian (x, y, z)	Cylindrical (ρ, ϕ, z)	Spherical (r, θ, ϕ)
Unit vectors	$\mathbf{u_x}\,\mathbf{u_y}\,\mathbf{u_z}$	$\mathbf{u_\rho}\,\mathbf{u_\phi}\,\mathbf{u_z}$	$\mathbf{u_r}\,\mathbf{u_\theta}\,\mathbf{u_\phi}$
Differential length **dl**	$dx\,\mathbf{u_x}$ $+dy\,\mathbf{u_y}$ $+dz\,\mathbf{u_z}$	$d\rho\,\mathbf{u_\rho}$ $+\rho d\phi\,\mathbf{u_\phi}$ $+dz\,\mathbf{u_z}$	$dr\,\mathbf{u_r}$ $+r\,d\theta\,\mathbf{u_\theta}$ $+r\sin\theta\,d\phi\,\mathbf{u_\phi}$
Differential surface area **ds**	$dy\,dz\,\mathbf{u_x}$ $dx\,dz\,\mathbf{u_y}$ $dx\,dy\,\mathbf{u_z}$	$\rho d\phi\,dz\,\mathbf{u_\rho}$ $d\rho\,dz\,\mathbf{u_\phi}$ $\rho d\rho\,d\phi\,\mathbf{u_z}$	$r^2\sin\theta\,d\theta\,d\phi\,\mathbf{u_r}$ $r\sin\theta\,dr\,d\phi\,\mathbf{u_\theta}$ $r\,dr\,d\theta\,\mathbf{u_\phi}$
Differential volume dv	$dx\,dy\,dz$	$\rho\,d\rho\,d\phi\,dz$	$r^2\sin\theta\,dr\,d\theta\,d\phi$

APPENDIX B

Material Parameters

Conductors	Conductivity σ (S/m)
Aluminum	3.5×10^7
Brass	1.6×10^7
Carbon	3.0×10^4
Copper	5.8×10^7
Germanium	2.3
Gold	4.1×10^7
Graphite	10^5
Iron	10^7
Mercury	10^6
Seawater	4
Silver	6.2×10^7
Tungsten	1.8×10^7

Material Parameters

Insulators	Conductivity σ (S/m)
Bakelite	10^{-9}
Distilled water	10^{-4}
Dry earth	10^{-5}
Glass	10^{-12}
Mica	10^{-15}
Porcelain	2×10^{-13}
Quartz	10^{-17}
Rubber	10^{-15}
Silicon	3.9×10^{-4}
Transformer oil	10^{-11}
Wax	10^{-17}
Wet earth	10^{-3}

Dielectrics	Relative dielectric constant ε_r
Air	1
Barium titanate	1200
Glass	6
Mica	6
Oil	2.3
Paper	3
Paraffin	2
Polystyrene	2.6
Porcelain	7
Quartz (fused)	4
Rubber	2.3–4.0
Teflon	2.1
Water (distilled)	80

Materials	Relative permeability μ_r
Ferromagnetic	
Cobalt	600
Commercial iron	250–9000
Nickel	250
Permalloy	8×10^3–10^5
Purified iron	10^4–2×10^5
Superpermalloy	10^5–10^6
Paramagnetic	
Aluminum	1.000021
Magnesium	1.000012
Palladium	1.00082
Titanium	1.00018
Diamagnetic	
Bismuth	0.99983
Gold	0.99996
Silver	0.99998
Copper	0.99999

APPENDIX C

Mathematical Foundation of the Finite Element Method

C.1 Minimization of Energy Result

Let us assume that $V(x, y)$ is the *true solution* of Laplace's equation. In addition, let us assume that $U(x, y)$ is another function that can be differentiated and is equal to *zero* on the boundary L_1 of the region s. Then, the sum of the two solutions which we will call the variation is given by $V(x, y) + \alpha U(x, y)$ where α is a small real parameter. The variation will have the same value on the boundary L_1 as $V(x, y)$. The electrostatic energy of the summation of the two terms is obtained from the expression (2.67)

$$W(V + \alpha U) = \int_A \frac{\varepsilon}{2} |\nabla(V + \alpha U)|^2 ds \tag{C.1}$$

This functional-energy $W(V + \alpha U)$ is expanded in powers of the small parameter α

$$W(V + \alpha U) = W(V) + \alpha \varepsilon \int_A \nabla V \cdot \nabla U \, ds + \frac{\alpha^2}{2} \varepsilon \int_A |\nabla U|^2 ds \tag{C.2}$$

The third term on the right hand side can be identified from (2.67) as being the energy of the additional functions U. The second term on the right hand side can be transformed using the vector identity (A.6)

$$\nabla \cdot (U \nabla V) = \nabla V \cdot \nabla U + U \nabla \cdot \nabla V \tag{C.3}$$

where U is a scalar and ∇V is a vector. The last term in (C.3) can be written as $U \nabla^2 V$. Therefore, we finally write (C.2) as

$$W(V + \alpha U) = W(V) + \alpha^2 W(U) - \alpha \varepsilon \int_A U \nabla^2 V \, ds + \alpha \varepsilon \oint_L U \nabla V \cdot \mathbf{u_n} \, dl \tag{C.4}$$

The term $\nabla V \cdot \mathbf{u_n} = \partial V / \partial n$. From Figure 4–17a, we have $U = 0$ on L_1 and $\partial V / \partial n = 0$ on L_2. This means that the fourth term in (C.4), which is a line integral, *vanishes* on the entire boundary L. The third term in (C.4), which is a surface integral on A, also vanishes since Laplace's equation $\nabla^2 V = 0$ must

477

be satisfied. Therefore, we finally obtain

$$W(V + \alpha U) = W(V) + \alpha^2 W(U) \tag{C.5}$$

Since $\alpha^2 > 0$ and $W(V) > 0$, the term on the left hand side is greater than $W(V)$ which proves that the energy has a *minimum* when V is a solution of Laplace's equation.

C.2 Interpolation Conditions

We calculate $\alpha_1(x_1, y_1)$ from (4.82)

$$\begin{aligned}\alpha_1(x_1, y_1) &= \frac{1}{2A_e}[(x_2 y_3 - x_3 y_2) + (y_2 - y_3)x_1 + (x_3 - x_2)y_1] \\ &= \frac{1}{2A_e}[(x_2 y_3 - x_3 y_2) + (x_3 y_1 - x_1 y_3) + (x_1 y_2 - x_2 y_1)] = 1\end{aligned} \tag{C.6}$$

because the expression within the brackets is equal to twice the area of the triangle A_e.

Let us calculate the same term at a different point, say $\alpha_1(x_2, y_2)$ using (4.82)

$$\begin{aligned}\alpha_1(x_2, y_2) &= \frac{1}{2A_e}[(x_2 y_3 - x_3 y_2) + (y_2 - y_3)x_2 + (x_3 - x_2)y_2] \\ &= 0\end{aligned} \tag{C.7}$$

These results are summarized in (4.83).

C.3 S-Matrix Elements

Using the explicit expression for the α-functions (4.82), we find that the gradient of these functions are

$$\begin{aligned}\nabla \alpha_1 &= \frac{1}{2A_e}[(y_2 - y_3)\mathbf{u_x} - (x_2 - x_3)\mathbf{u_y}] \\ \nabla \alpha_2 &= \frac{1}{2A_e}[(y_3 - y_1)\mathbf{u_x} - (x_3 - x_1)\mathbf{u_y}] \\ \nabla \alpha_3 &= \frac{1}{2A_e}[(y_1 - y_2)\mathbf{u_x} - (x_1 - x_2)\mathbf{u_y}]\end{aligned} \tag{C.8}$$

Applying the definition (4.89) after taking the scalar product and performing the integration over the element, the following elements of the $S^{(e)}$-matrix

C.4 Decoupled and Coupled Node Potentials

are obtained:

$$S_{1,1} = \frac{\varepsilon}{4A_e}[(y_2 - y_3)^2 + (x_2 - x_3)^2];$$

$$S_{1,2} = S_{2,1} = \frac{\varepsilon}{4A_e}[(y_1 - y_3)(y_3 - y_2) + (x_1 - x_3)(x_3 - x_2)];$$

$$S_{1,3} = S_{3,1} = \frac{\varepsilon}{4A_e}[(y_1 - y_2)(y_2 - y_3) + (x_1 - x_2)(x_2 - x_3)];$$

$$S_{2,2} = \frac{\varepsilon}{4A_e}[(y_3 - y_1)^2 + (x_3 - x_1)^2];$$

$$S_{2,3} = S_{3,2} = \frac{\varepsilon}{4A_e}[(y_2 - y_1)(y_1 - y_3) + (x_2 - x_1)(x_1 - x_3)];$$

$$S_{3,3} = \frac{\varepsilon}{4A_e}[(y_1 - y_2)^2 + (x_1 - x_2)^2].$$

(C.9)

They are used in Example 4.19.

C.4 Decoupled and Coupled Node Potentials

The *coupling matrix* [C] has a dimension $(N_d \times N)$ where N_d is the total number of the decoupled nodes and N is the total number of the coupled nodes after assembling. In the case of two elements, the number of the decoupled nodes is $N_d = 6$ (see Figure 4–19a), while the number of the coupled nodes is $N = 4$ (see Figure 4–19b). The decoupled nodes are numbered with a subscript $k = 1, 2, ..., N_d$, while the coupled nodes are numbered with a subscript $m = 1, 2, ..., N$. The elements of the coupling matrix are defined by the following *rule*: $C_{k,m} = 1$ when the k-th decoupled node corresponds to the m-th coupled node and $C_{k,m} = 0$ otherwise. There is only one "1" on every row, but there can be one or two "1" on every column (the last case reflects the boundary conditions). For the particular problem considered in Figure 4–19, the corresponding C-matrix is

$$[C] = \begin{bmatrix} 1 & 0 & 0 & 0 \\ 0 & 1 & 0 & 0 \\ 0 & 0 & 1 & 0 \\ 0 & 0 & 0 & 1 \\ 0 & 1 & 0 & 0 \\ 0 & 0 & 1 & 0 \end{bmatrix}$$

(C.10)

which coincides with (4.85). The following relation between the column matrices of the decoupled potentials $[V]_d$ and coupled potentials $[V]$ is obtained

$$[V]_d = [C][V] \qquad (C.11)$$

where the coupling matrix $[C]$ is involved. This equation matches with (4.84).

After the determination of the energy of the element from (4.86) via the potentials of the decoupled nodes and via the potentials of the coupled nodes as

$$W = \frac{1}{2}[V]^T[S][V] \qquad (C.12)$$

we obtain for the *global* S-matrix of the coupled system using the following relation

$$[S] = [C]^T[S]_d[C] \qquad (C.13)$$

We can show that the last equation yields (4.90) for the global S-matrix in its explicit form. For convenience we can write this equation in two stages

$$\begin{aligned}[B] &= [S]_d[C], \\ [S] &= [C]^T[B]\end{aligned} \qquad (C.14)$$

From (C.10) for the decoupled matrix $[S]_d$ of the system of two triangles, the following expression is obtained

$$[S]_d = \begin{bmatrix} S^{(1)}_{1,1} & S^{(1)}_{1,2} & S^{(1)}_{1,3} & 0 & 0 & 0 \\ S^{(1)}_{2,1} & S^{(1)}_{2,2} & S^{(1)}_{2,3} & 0 & 0 & 0 \\ S^{(1)}_{3,1} & S^{(1)}_{3,1} & S^{(1)}_{3,1} & 0 & 0 & 0 \\ 0 & 0 & 0 & S^{(2)}_{4,4} & S^{(2)}_{4,5} & S^{(2)}_{4,6} \\ 0 & 0 & 0 & S^{(2)}_{5,4} & S^{(2)}_{5,5} & S^{(2)}_{5,6} \\ 0 & 0 & 0 & S^{(2)}_{6,4} & S^{(2)}_{6,5} & S^{(2)}_{6,6} \end{bmatrix} \qquad (C.15)$$

C.5 The Matrix Equation for the Unknown Potentials

Now after a simple matrix multiplication, we get for the intermediate matrix $[B]$ the first equation of (C.14) yields

$$[B] = \begin{bmatrix} S_{1,1}^{(1)} & S_{1,2}^{(1)} & S_{1,3}^{(1)} & 0 \\ S_{2,1}^{(1)} & S_{2,2}^{(1)} & S_{2,3}^{(1)} & 0 \\ S_{3,1}^{(1)} & S_{3,2}^{(1)} & S_{3,3}^{(1)} & 0 \\ 0 & S_{4,5}^{(2)} & S_{4,6}^{(2)} & S_{4,4}^{(2)} \\ 0 & S_{5,5}^{(2)} & S_{5,6}^{(2)} & S_{5,4}^{(2)} \\ 0 & S_{6,5}^{(2)} & S_{6,6}^{(2)} & S_{6,4}^{(2)} \end{bmatrix}$$

Then from the second equation of (C.14) after a matrix multiplication, we obtain the result presented by (4.93):

$$[S] = \begin{bmatrix} S_{1,1}^{(1)} & S_{1,2}^{(1)} & S_{1,3}^{(1)} & 0 \\ S_{2,1}^{(1)} & S_{2,2}^{(1)} + S_{5,5}^{(2)} & S_{2,3}^{(1)} + S_{5,6}^{(2)} & S_{5,4}^{(2)} \\ S_{3,1}^{(1)} & S_{3,2}^{(1)} + S_{6,5}^{(2)} & S_{3,3}^{(1)} + S_{6,6}^{(2)} & S_{6,4}^{(2)} \\ 0 & S_{4,5}^{(2)} & S_{4,6}^{(2)} & S_{4,4}^{(2)} \end{bmatrix} \quad (C.16)$$

The Matrix Equation for the Unknown Potentials

The necessary condition that the function (C.12) has a minimum is that the appropriate derivatives must be equal to zero

$$\frac{\partial W}{\partial [V_j]_u} = 0 \quad (j = 1, 2, ..., N) \quad (C.17)$$

First, the potential matrix-column is split into two parts: known $[V]_k$ and unknown $[V]_u$ as shown in (4.94):

$$[V] = \begin{bmatrix} [V]_k \\ [V]_u \end{bmatrix}$$

and the same is done with the S-matrix, as shown in (4.95):

$$[S] = \begin{bmatrix} [S]_{k,k} & [S]_{k,u} \\ [S]_{u,k} & [S]_{u,u} \end{bmatrix}$$

The following explicit expression is found from (C.12) for the energy:

$$W = \frac{1}{2}[V]_k^T[S]_{k,k}[V]_k + [V]_u^T[S]_{u,k}[V]_k + \frac{1}{2}[V]_u^T[S]_{u,u}[V]_u \qquad (C.18)$$

where the symmetry of the S-matrix is used.

From (C.17), we write

$$[S]_{u,u}[V]_u + [S]_{u,k}[V]_k = 0 \qquad (C.19)$$

which leads to the final equation (4.97):

$$[V]_u = -[S]_{u,u}^{-1}[S]_{u,k}[V]_k \qquad (C.20)$$

APPENDIX D

Transmission Line Parameters of Two Parallel Wires

We calculate the capacitance of two parallel wires that are shown in Figure D–1. The radius of each wire is a.

The two wires can be replaced with two line charges $+\rho_\ell$ and $-\rho_\ell$. The precise location of these equivalent line charges is determined from the requirement that the surfaces of the metal wires be equipotential surfaces. This implies that the tangential electric fields will always be equal to zero on these surfaces. The potential $V(x, y)$ at the point P is given by

$$V(x, y) = \frac{\rho_\ell}{2\pi\varepsilon}\ln\left(\frac{s}{r_2}\right) + \frac{-\rho_\ell}{2\pi\varepsilon}\ln\left(\frac{s}{r_1}\right) = \frac{\rho_\ell}{2\pi\varepsilon}\ln\left(\frac{r_1}{r_2}\right) \tag{D.1}$$

The plane at the midpoint between the two wires is an equipotential surface that is equal to zero potential. Other equipotential contours are found by setting

$$\frac{r_1}{r_2} = \frac{\sqrt{(s+x)^2 + y^2}}{\sqrt{(s-x)^2 + y^2}} = k = \text{constant} \tag{D.2}$$

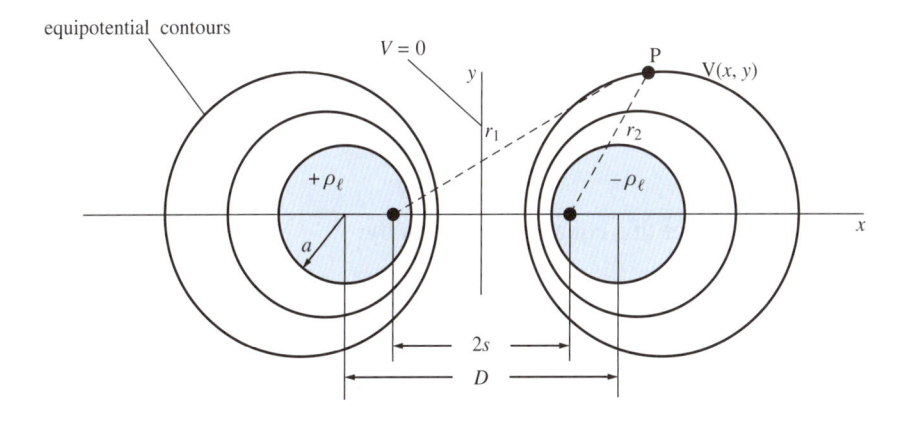

FIGURE D–1

Equipotential contours surrounding two line charges $+\rho_\ell$ and $-\rho_\ell$. The surfaces at $r = a$ are equipotential surfaces, therefore the electric field will always be normal to the metal surfaces.

This can be written as

$$(s + x)^2 + y^2 = k^2[(s - x)^2 + y^2] \qquad (D.3)$$

or

$$\left(x - s\frac{k^2 + 1}{k^2 - 1}\right)^2 + y^2 = \left(\frac{2ks}{k^2 - 1}\right)^2 \qquad (D.4)$$

The common factor

$$s^2\left(\frac{k^2 + 1}{k^2 - 1}\right)^2$$

has been added to both sides of (D.3) in order to complete the squares. Equation (D.4) is an equation for a family of circles that have radii

$$r_0 = \frac{2ks}{k^2 - 1} \qquad (D.5)$$

and are centered at the points

$$(h, 0) = \left[s\frac{k^2 + 1}{k^2 - 1}, 0\right] \qquad (D.6)$$

where

$$h = s\frac{k^2 + 1}{k^2 - 1}$$

Eliminating the term s between (D.5) and (D.6), we obtain

$$k^2 - \frac{2kh}{r_0} + 1 = 0 \qquad (D.7)$$

The two solutions for this equation are given by

$$k = \frac{h}{r_0} \pm \sqrt{\left(\frac{h}{r_0}\right)^2 - 1} \qquad (D.8)$$

The root k with the $+$ sign will give the equipotential contours in the region $x > 0$, and the root $\hat{k}$ with the $-$ sign will give the equipotential contours in the region $x < 0$. We will have particular interest in the equipotential contour at the surface of the wire at $r_0 = a$. The spacing h must also be greater than this radius a of the wire, and we will set it equal to $D/2$.

Transmission Line Parameters of Two Parallel Wires

The potential difference between the wires is given by

$$\Delta V = \frac{\rho_\ell}{2\pi\varepsilon}\ln k - \frac{\rho_\ell}{2\pi\varepsilon}\ln \hat{k} = \frac{\rho_\ell}{\pi\varepsilon}\ln k = \frac{\rho_\ell}{\pi\varepsilon}\ln\left[\frac{D}{2a} \pm \sqrt{\left(\frac{D}{2a}\right)^2 - 1}\right] \quad (D.9)$$

Making use of the identity

$$\ln[\xi + \sqrt{\xi^2 - 1}] = \cosh^{-1}\xi \quad (D.10)$$

we write the capacitance as

$$C = \frac{\rho_\ell \Delta z}{\Delta V} = \frac{\Delta z}{\dfrac{1}{\pi\varepsilon}\cosh^{-1}\left(\dfrac{D}{2a}\right)} = \frac{\pi\varepsilon\Delta z}{\cosh^{-1}\left(\dfrac{D}{2a}\right)} \quad (D.11)$$

In order to calculate the inductance per unit length, we make use of the relation $(L/\Delta z)(C/\Delta z) = \mu\varepsilon$ and write

$$L = \frac{\mu\Delta z}{\pi}\cosh^{-1}\left(\frac{D}{2a}\right) \quad (D.12)$$

There are alternative derivations that can be used to approximate the equivalent circuit parameters. For example, if we define a point between the two cylinders as $\hat{x}$ and place the left cylinder at $x = 0$, then the magnetic flux density is as shown in Figure D–2.

If the current is into the paper in the wire centered at $x = 0$ and out of the paper in the wire centered at $x = D$, the magnetic flux densities will add in the center. From Ampere's circuital law, we write the magnetic flux density as

$$B = \frac{\mu I}{2}\left[\frac{1}{\hat{x}} + \frac{1}{D-\hat{x}}\right] \quad (D.13)$$

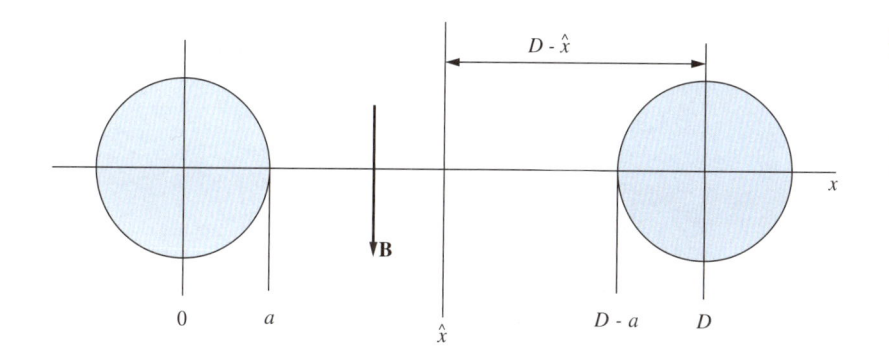

FIGURE D–2

Alternative cross section of a twin-lead transmission line.

The total magnetic flux that passes between the two wires is found from

$$\frac{\Psi_m}{\Delta z} = \int_a^{D-a} B\,d\hat{x} = \frac{\mu I}{2\pi}[\ln\hat{x} - \ln(D-\hat{x})]_a^{D-a} = \frac{\mu I}{\pi}\ln\left(\frac{D-a}{a}\right) \quad (D.14)$$

The inductance per unit length is given by

$$\boxed{\frac{L}{\Delta z} \equiv \frac{\left(\frac{\Psi_m}{\Delta z}\right)}{I} = \frac{\mu}{\pi}\ln\left(\frac{D-a}{a}\right) \approx \frac{\mu}{\pi}\ln\left(\frac{D}{a}\right)} \quad (D.15)$$

which is in agreement with (D.12) in the limit of $D \gg a$.

The electric field at $\hat{x}$ is obtained using Gauss's law by assuming that each wire has a charge uniformly distributed on the wire. If the charges are of the opposite sign, we obtain linear charge densities of $+\rho_\ell$ and $-\rho_\ell$. This results in

$$E = \frac{\rho_\ell \Delta z}{2\pi\varepsilon}\left[\frac{1}{\hat{x}} + \frac{1}{D-\hat{x}}\right] \quad (D.16)$$

The potential difference ΔV between the two wires is obtained by integrating the electric field between the two wires to yield

$$\Delta V = \int_a^{D-a} E\,d\hat{x} = \frac{\rho_\ell \Delta z}{\pi\varepsilon}\ln\left(\frac{D-a}{a}\right) \quad (D.17)$$

The capacitance per unit length is given by

$$\boxed{\frac{C}{\Delta z} \equiv \frac{\rho_\ell}{\Delta V} = \frac{\pi\varepsilon}{\ln\left(\frac{D-a}{a}\right)} \approx \frac{\pi\varepsilon}{\ln\left(\frac{D}{a}\right)}} \quad (D.18)$$

This is in agreement with (D.11) in the limit of $D \gg a$.

We present both methods to obtain the same results since you may encounter them in different books. The twin lead is commonly used in practice. In neither calculation did we include the self-inductance of each wire that arises when $D \approx a$. This will cause an additional small constant term to appear in the final results.

APPENDIX E

Plasma Evolution Adjacent to a Metallic Surface

The temporal and spatial evolution of the plasma adjacent to a metallic electrode whose voltage is suddenly decreased from zero to a large negative value has certain implications. A gaseous plasma consists of negatively charged electrons and positively charged ions whose mass is significantly greater than the mass of the electrons. In Example 6–3, we determined the potential profile at the time $t = 0^+$ just after the switch connecting the negative potential source was closed at a time $t = 0$. During this initial time interval, the electrons were expelled from the region adjacent to the metallic plate but the ions had not yet started to move. The temporal and spatial evolution of these ions toward the negatively biased metallic electrode and the expansion of the ion density rarefaction into the plasma requires a numerical computation.[1] The plasma is modeled with a dimensionless-fluid-model description.

The ion density perturbation n_i and the ion velocity perturbation v_i are described with the equation of continuity

$$\frac{\partial n_i}{\partial t} + \frac{\partial (n_i v_i)}{\partial z} = 0 \tag{E.1}$$

and the equation of motion

$$\frac{\partial v_i}{\partial t} + v_i \frac{\partial v_i}{\partial z} = -\frac{\partial V}{\partial z} \tag{E.2}$$

Since the mass of the electron is so much smaller than the mass of the ion, the electron density perturbation n_e can be approximated with a Maxwell-Boltzmann distribution.

$$n_e = e^V \tag{E.3}$$

---

[1] M. Widner, I. Alexeff, W. D. Jones, and K. E. Lonngren, "Ion-Acoustic Wave Excitation and Ion Sheath Evolution," *Physics of Fluids*, Vol. 13, (October 1970): 2532–2540.

In order to reflect the non-neutrality of the density perturbations, there will be an electric field $E = -dV/dz$ that is governed by Poisson's equation

$$\frac{\partial^2 V}{\partial z^2} = n_e - n_i \qquad (5.4)$$

These equations have been written in dimensionless units that are defined by

$$n_i = \frac{n_i'}{n_0}; \quad n_e = \frac{n_e'}{n_0}; \quad v_i = \frac{v_i'}{v_s}; \quad V = \frac{eV'}{\kappa_B T_e}; \quad z = \frac{z'}{\lambda_D}; \quad t = \omega_{pi} t' \qquad (5.5)$$

where n_0 is the equilibrium electron and ion density, the ' indicates the laboratory variables, κ_B is Boltzmann's constant $\kappa_B = 1.38 \times 10^{-23}$ J/°K, and

$$v_s = \sqrt{\frac{\kappa_B T_e}{m_i}} \quad \text{— the ion acoustic velocity} \qquad (E.6)$$

$$\lambda_D = \sqrt{\frac{\kappa_B T_e}{\left(\frac{n_0 e^2}{\varepsilon_0}\right)}} \quad \text{— the electron Debye length} \qquad (E.7)$$

$$\omega_{pi} = \sqrt{\frac{n_0 e^2}{m_i \varepsilon_0}} \quad \text{— the ion plasma frequency} \qquad (E.8)$$

This model is valid if the electronic temperature T_e is much greater than the ion temperature T_i. This is a valid approximation in a gaseous plasma that one normally encounters in the laboratory. For example, a typical laboratory argon plasma will have an approximate ion acoustic velocity of 10^3 m/s and an ion plasma frequency of $2\pi \times 10^6$ radians/s.

The calculated results of the evolution of the density perturbations are shown in Figure E–1a. It is noted that there is an instantaneous decrease of the electron density adjacent to the electrode, which is the "transient sheath" or the "ion matrix sheath" that was described in Example 6–3. As time increases, this electron density rarefaction expands into the plasma. Initially, this expansion is faster than the final asymptotic value, which is the ion acoustic velocity. In addition, the heavy positive ions are attracted to and move toward the electrode. This results in a flux of the ions that impinges upon the electrode. The temporal evolution of this flux is shown in Figure E–1b. These ions can implant themselves into the electrode and change the surface characteristics of the electrode.

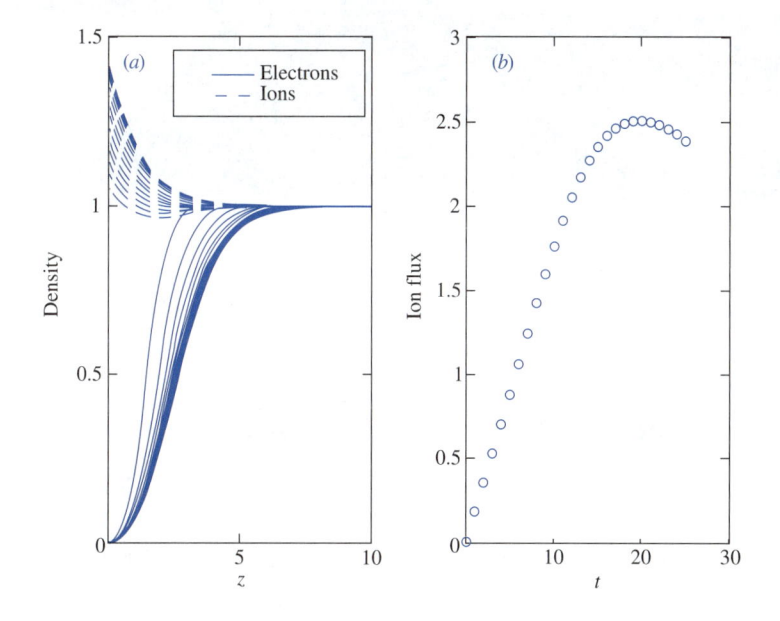

Plasma Evolution Adjacent to a Metallic Surface

FIGURE E–1

(*a*) Evolution of the normalized electron density and the normalized ion density in space at equal increments in time after the application of a negative potential to a metallic electrode that had been inserted into a plasma.
(*b*) Temporal evolution of the ion flux impinging upon the electrode.

APPENDIX F

Bibliography

*Most of these books are available at a discount and for immediate shipment on SciTech's website: www.scitechpub.com

Best Student References – Short and Economical

Edminister, S. *Schaum's Easy Outline of Electromagnetics: A Crash Course*, New York: McGraw-Hill, 2003.

Edminister, S. *Schaum's Outline of Electromagnetics*. New York: McGraw-Hill, 1994.

Maxum, B. Field Mathematics for Electromagnetics, Photonics, and Materials Science. Bellingham, WA: SPIE Press, 2005.

Nasar, S. *2000 Solved Problems in Electromagnetics*. Raleigh, NC: SciTech Publishing, 2008.

Schmidt, R. *Electromagnetics Explained*. San Diego, CA: Elsevier, 2002.

Schey, H.M. *Div, Grad, Curl and All That: An Informal Text on Vector Calculus*. 4th Ed. New York: WW. Norton, 2005.

Basic Electromagnetic Theory: Textbooks

Cheng, D. K. *Field and Wave Electromagnetics*. 2nd ed. Upper Saddle River, New Jersey: Prentice Hall, 1989.

Griffiths, D. J. *Introduction to Electrodynamics*. 3rd Ed., Upper Saddle River, New Jersey: Benjamin Cummings, 1999.

Hayt, W. H. and J. A. Buck *Engineering Electromagnetics*. 7th ed. New York: McGraw-Hill, 2006.

Hirose, A. and K. E. Lonngren. *Introduction to Wave Phenomena*. Melbourne, Florida: Krieger, 2002. Kraus, J. D. and D. Fleisch, *Electromagnetics*. 5th ed. New York: McGraw-Hill, 1999.

Ramo, S., J. R. Whinnery, and T. Van Duzer. *Fields and Waves in Communication Electronics*. 3rd ed. New York: Wiley, 1994.

Rao, N. N. *Elements of Engineering Electromagnetics*. 6th ed. Upper Saddle River, New Jersey: Prentice Hall, 2004.

Sadiku, M. N. O. *Elements of Electromagnetics*. 4th ed., New York: Oxford University Press, 2007.

Ulaby, F. T. *Fundamentals of Applied Electromagnetics*. 5th ed. Upper Saddle River, New Jersey: Prentice Hall, 2006.

MATLAB Guides for Engineers

Chapman, S. J. *MATLAB Programming for Engineers*. Cincinnati, Ohio: Brooks/Cole-Thomson Learning, 2002.

Daku B. *MATLAB Tutorial CD: Learning MATLAB SuperFast*, Hoboken, NJ: John Wiley & Sons, 2006.

Etter, D. et al *Introduction to MATLAB 7*, Upper Saddle River, NJ: Prentice Hall, 2005.

Davis, T. A. and Sigmon K. *MATLAB Primer*. 7th ed. Boca Raton, FL: CRC Press, 2005.

Gilat, A. *MATLAB: An Introduction with Applications* 2nd Ed., Hoboken, New Jersey: Wiley, 2003.

Hahn, B.D. *Essential MATLAB for Scientists and Engineers*. San Diego, California: Elsevier, 2002.

Hanselman, D. and B. Littlefield. *Mastering MATLAB 7: A Comprehensive Tutorial and Reference*. Upper Saddle River, New Jersey: Prentice Hall, 2005.

Magrab, E.B. et al *An Engineer's Guide to MATLAB 2nd Ed.* Upper Saddle River, NJ: Prentice Hall 2005.

Moler, C. *Numerical Computing with MATLAB*. Philadelphia, Pennsylvania: SIAM, 2004.

Moore, H. *MATLAB for Engineers*, Upper Saddle River, NJ: Prentice Hall, 2007.

Palm III, W. J. *Introduction to MATLAB 7 for Engineers*. New York: McGraw-Hill, 2005.

Advanced Topics in Electromagnetic Theory

Balanis, C. A. *Advanced Engineering Electromagnetics*. New York: Wiley, 1989.

Balanis, C. A. *Antenna Theory: Analysis and Design*. 3rd ed. New York: Wiley, 2005.

Bansal, R. *Handbook of Engineering Electromagnetics*. Boca Raton, FL: CRC Press, 2005.

Christopoulis, C. *The Transmission Line Method TLM*. Piscataway, New Jersey: IEEE Press, 1995.

Chew, W. C. *Wave and Fields in Inhomogeneous Media*. Piscataway, New Jersey: IEEE Press, 1995.

Clemmow, P. C. *The Plane Wave Spectrum Representation of Electromagnetic Fields*. Piscataway, New Jersey: IEEE Press, 1996.

Cohen, D. L. *Demystifying Electromagnetic Equations*. Bellingham, Washington: SPIE, 2001.

Collin, R. E. *Field Theory of Guided Waves*. 2nd ed. Piscataway, New Jersey: IEEE Press, 1991.

Collin, R. E. *Foundations for Microwave Engineering*. 2nd ed. Piscataway, New Jersey: IEEE Press, 2001.

Dudley, D. G. *Mathematical Foundations for Electromagnetic Theory*. Piscataway, New Jersey: IEEE Press, 1994.

Elliott, R. S. *Antenna Theory and Design*. revised ed. Piscataway, New Jersey: IEEE Press, 2003.

Elliott, R. E. *Electromagnetics, History, Theory, and Applications*. Piscataway, New Jersey: IEEE Press, 1993.

Felsen, L. B. and N. Marcuvitz. *Radiation and Scattering of Waves*. Piscataway, New Jersey: IEEE Press, 1994.

Hansen, T. B. and A.D. Yaghjian. *Plane Wave Theory of Time-Domain Fields*. Piscataway, New Jersey: IEEE Press, 1999.

Harrington, R. F. *Field Computation by Moment Methods*. Piscataway, New Jersey: IEEE Press, 1993.

Harrington, R. F. *Time-Harmonic Electromagnetic Fields*. Piscataway, New Jersey: IEEE Press, 2001.

Ishimaru, A. *Wave Propagation and Scattering in Random Media*. Piscataway, New Jersey: IEEE Press, 1997.

Jackson, J. D. *Classical Electrodynamics*. 3rd ed. New York: Wiley, 1998.

Jin, J. *The Finite Element Method in Electromagnetics*. 2nd ed. New York: Wiley, 2002.

Jones, D. S. *Methods in Electromagnetic Wave Propagation*. 2nd ed. Piscataway, New Jersey: IEEE Press, 1994.

Lindell, I. V. *Differential Forms in Electromagnetics*. Piscataway, New Jersey: IEEE Press, 2004.

Lindell, I. V. *Methods for Electromagnetic Field Analysis*. Piscataway, New Jersey: IEEE Press, 1995.

Makarov, S. *Antenna and Electromagnetic Modeling with MATLAB*. Hoboken, NJ: John Wiley & Sons, 2002.

Peterson, A. F., S. L. Ray, and R. Mittra *Computational Methods for Electromagnetics*. Piscataway, New Jersey: IEEE Press, 1997.

Sadiku, M. N. O. *Numerical Techniques in Electromagnetics*. 4th ed. Boca Raton, Florida: CRC Press, 2007.

Silvester, P. P. and R. L. Ferrari. *Finite Elements for Electrical Engineers*. Cambridge: Cambridge University Press, 1996.

Stratton, J. A. *Electromagnetic Theory*. New York: IEEE Press/Wiley, 1941/2007.

Tai, C.-T. *Dyadic Green Functions in Electromagnetic Theory*. 2nd ed. Piscataway, New Jersey: IEEE Press, 1994.

Tai, C.-T. *Generalized Vector and Dyadic Analysis*. 2nd ed. Piscataway, New Jersey: IEEE Press, 1997.

Van Bladel, J. *Singular Electromagnetic Fields and Sources*. 2nd Ed. Piscataway, New Jersey: IEEE Press, 2007.

Volakis, J. L. A. Chatterjee, and L. C. Kempel. *Finite Element Method for Electromagnetics*. Piscataway, New Jersey: IEEE Press, 1998.

Wait, J. R. *Electromagnetic Waves in Stratified Media*. Piscataway, New Jersey: IEEE Press, 1996.

Werner, D. H. and R. Mittra *Frontiers in Electromagnetics*. New York: Wiley, 2000.

APPENDIX G

Selected Answers

Chapter 1

1.1.1. $\mathbf{C} = \mathbf{A} + \mathbf{B} = -2\mathbf{u}_x + 8\mathbf{u}_y + 2\mathbf{u}_z;\quad \mathbf{D} = \mathbf{A} - \mathbf{B} = 8\mathbf{u}_x + 0\mathbf{u}_y + 8\mathbf{u}_z$

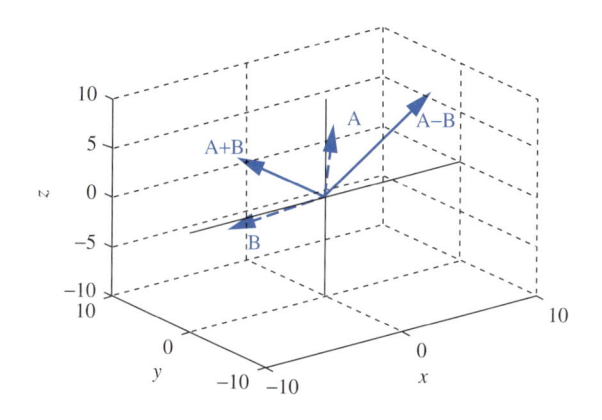

1.1.2. $\mathbf{A} \cdot \mathbf{B} = -14;\quad \mathbf{A} \times \mathbf{B} = -32\mathbf{u}_x - 16\mathbf{u}_y + 32\mathbf{u}_z$

1.1.5.

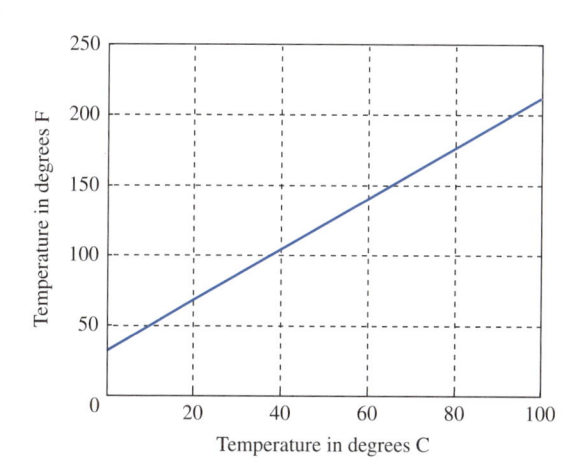

493

1.1.8.

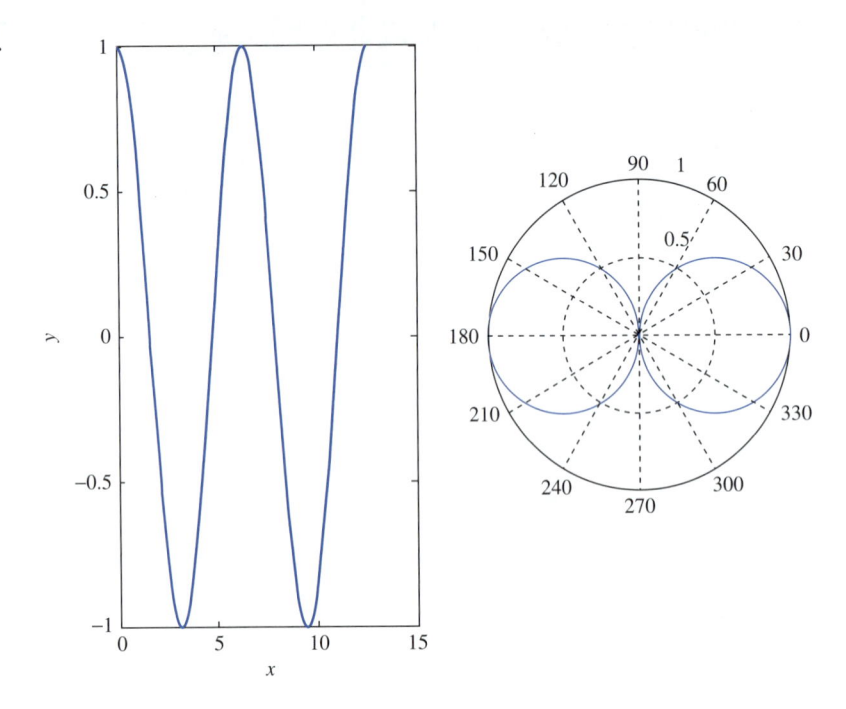

1.1.9. $\mathbf{A}(3, 2) = 4\mathbf{u_x} - 3\mathbf{u_y}$, $|\mathbf{A}(3, 2)| = \sqrt{4^2 + (-3)^2} = 5$

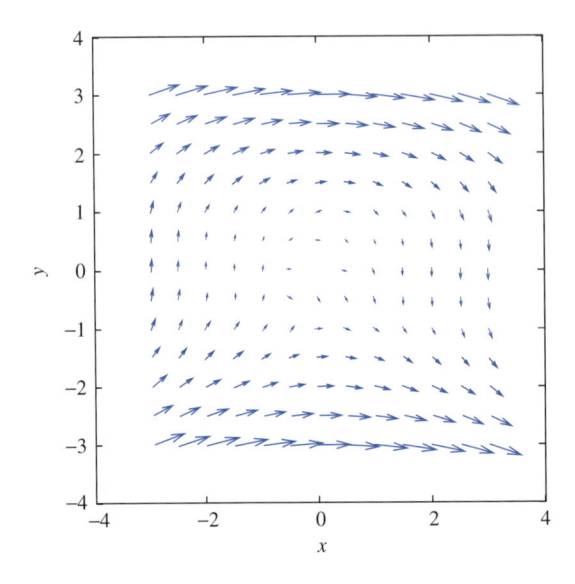

Chapter 1

1.2.1. $\mathbf{u}_A = \dfrac{\mathbf{u}_x + \mathbf{u}_y + \mathbf{u}_z}{\sqrt{3}}$

1.2.3. $\mathbf{A} \cdot \mathbf{B} = -14;\qquad \theta = 106.26°$

1.2.5. $\mathbf{A} \cdot \mathbf{B} = 22;\qquad \theta = 23.33°$

1.2.7. $\mathbf{A} \times \mathbf{B} = 2\mathbf{u}_x - 4\mathbf{u}_y + 2\mathbf{u}_z;\qquad \mathbf{B} \times \mathbf{A} = -2\mathbf{u}_x + 4\mathbf{u}_y - 2\mathbf{u}_z$
$\Rightarrow \mathbf{A} \times \mathbf{B} = -\mathbf{B} \times \mathbf{A}$

1.2.8. $\mathbf{A} = 5\mathbf{u}_\rho + 0.927\mathbf{u}_\phi + 5\mathbf{u}_z$

1.2.10. $\mathbf{A} = 7.071\mathbf{u}_r + 0.785\mathbf{u}_\theta + 0.927\mathbf{u}_\phi$

1.2.13. $\mathbf{A} \times (\mathbf{B} \times \mathbf{C}) = -52\mathbf{u}_x - 16\mathbf{u}_y + 20\mathbf{u}_z$
$\mathbf{B}(\mathbf{A} \cdot \mathbf{C}) - \mathbf{C}(\mathbf{A} \cdot \mathbf{B}) = -52\mathbf{u}_x - 16\mathbf{u}_y + 20\mathbf{u}_z$

1.2.14. Area $= 24$

1.2.16. $(\mathbf{A} \times \mathbf{B}) \cdot \mathbf{C} = -30;\qquad \mathbf{A} \cdot (\mathbf{B} \times \mathbf{C}) = -30$

1.3.1. $\Delta W = 53$

1.3.2. $W_{abc} = 24\ \ W_{adc} = 24$

1.3.4. $W = 2a^2\pi^2$

1.3.5. $\oint_{\Delta s} \mathbf{A} \cdot \mathbf{ds} = 16;\qquad \int_{\Delta v} \nabla \cdot \mathbf{A}\, dv = 16$

1.3.7. $\oint \mathbf{A} \cdot \mathbf{ds} = 12\pi a^2$

1.4.1. $\nabla H = -2x\mathbf{u}_x - 6y\mathbf{u}_y$

1.4.3. $\nabla\left(\dfrac{1}{r}\right) = -\dfrac{1}{r^2}\nabla r$

$= -\dfrac{2}{r^3}[(x-x')\mathbf{u_x} + (y-y')\mathbf{u_y} + (z-z')\mathbf{u_z}]$

$= -\nabla'\left(\dfrac{1}{r}\right)$

1.4.5. $\nabla \cdot \mathbf{A} = 1$; $\oint \mathbf{A} \cdot d\mathbf{s} = 8a^3$; $\int \nabla \cdot \mathbf{A}\, dv = 8a^3$

1.4.7. $\nabla \times \mathbf{v}\left(x = \dfrac{a}{4}, z = 1\right) = -\dfrac{3a^2}{8}\mathbf{u_x} + \dfrac{a}{2}\mathbf{u_z} \Rightarrow (\nabla \times \mathbf{v}) \cdot \mathbf{u_z} > 0$

$\nabla \times \mathbf{v}\left(x = \dfrac{3a}{4}, z = 1\right) = -\dfrac{3a^2}{8}\mathbf{u_x} - \dfrac{a}{2}\mathbf{u_z} \Rightarrow (\nabla \times \mathbf{v}) \cdot \mathbf{u_z} < 0$

The paddle wheel will rotate if the axis is parallel to the x axis since there is $(\nabla \times \mathbf{v}) \cdot \mathbf{u_x} \neq 0$.
The paddle wheel will not rotate if the axis is parallel to the y axis since $(\nabla \times \mathbf{v}) \cdot \mathbf{u_y} = 0$.

1.4.8. $\int \nabla \times \mathbf{A} \cdot d\mathbf{s} = \dfrac{1}{2}$; $\oint \mathbf{A} \cdot d\mathbf{l} = \dfrac{1}{2}$

1.4.11. $\nabla \cdot \nabla \times \mathbf{A} = 0$

1.4.12. $\nabla \times \nabla a = 0$

1.4.13. $\nabla \times (a\mathbf{A}) = (-2x^2yz + 4x^2z^2 - 3x^2y^2)\mathbf{u_x}$
$\qquad + (3x^2y^2 + 8x^2yz - 6xy^2z - 8xyz^2)\mathbf{u_y}$
$\qquad + (6xy^2z + 8xyz^2 - 6x^2yz - 4x^2z^2)\mathbf{u_z}$

$(\nabla a) \times \mathbf{A} = -x^2yz\mathbf{u_x} + (4x^2yz - 3xy^2z - 4xyz^2)\mathbf{u_y}$
$\qquad + (3xy^2z + 4xyz^2 - 3x^2yz)\mathbf{u_z}$

Chapter 1

$$a(\nabla \times \mathbf{A}) = (4x^2z^2 - 3x^2y^2 - x^2yz)\mathbf{u_x}$$
$$+ (3x^2y^2 - 3xy^2z + 4x^2yz - 4xyz^2)\mathbf{u_y}$$
$$+ (3xy^2z - 3x^2yz + 4xyz^2 - 4x^2z^2)\mathbf{u_z}$$

which verifies the identity $\nabla \times (a\mathbf{A}) = (\nabla a) \times \mathbf{A} + a(\nabla \times \mathbf{A})$.

1.4.15. From (A.27)

$$\nabla^2\left(\frac{1}{r}\right) = \frac{1}{r^2}\frac{\partial}{\partial r}\left(r^2 \frac{\partial}{\partial r}\left(\frac{1}{r}\right)\right) = 0$$

1.5.1. $v(t) = \operatorname{Re}\{Ve^{j120\pi t}\}$ V where $V = \dfrac{100}{\sqrt{2}}(1-j)$ V

1.5.3. $V = 0.518 e^{j15°}$ V

1.5.5. $i(t) = 6.669\cos(120\pi - 3.17°)$ A

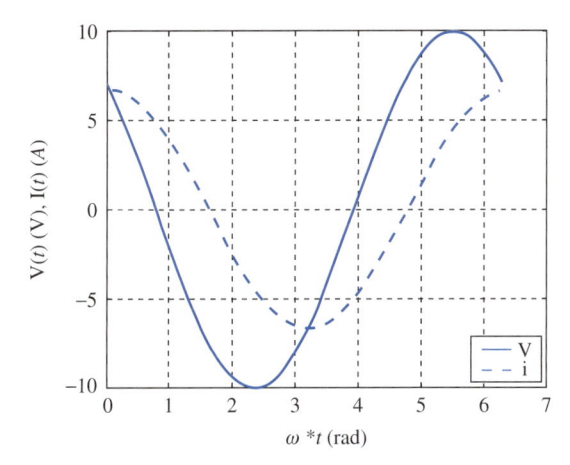

Chapter 2

2.1.1. $\varepsilon_0 = \dfrac{Q^2 T^2}{ML^3}$ F/m

2.1.3. $\mathbf{F}_2 = \dfrac{1}{4\pi\varepsilon_0}[-0.750\mathbf{u_x} - 2.222\mathbf{u_y}]$ N

2.1.5. $d = \sqrt{\dfrac{Q^2}{4\pi\varepsilon_0 Mg \sin\dfrac{\pi}{4}}}$ m

2.2.1. $E = \dfrac{ML}{QT^2}$ V/m

2.2.3. $\mathbf{E} = \dfrac{1}{4\pi\varepsilon_0}[-0.186\mathbf{u_x} - 0.048\mathbf{u_y}]$ V/m

2.2.5. $E_y(y) = \dfrac{Q}{4\pi\varepsilon_0}\left(\dfrac{(2(y-2))}{[(y-2)^2+4]^{3/2}}\right) = \dfrac{4(y-2)}{[(y-2)^2+4]^{3/2}}$ V/m

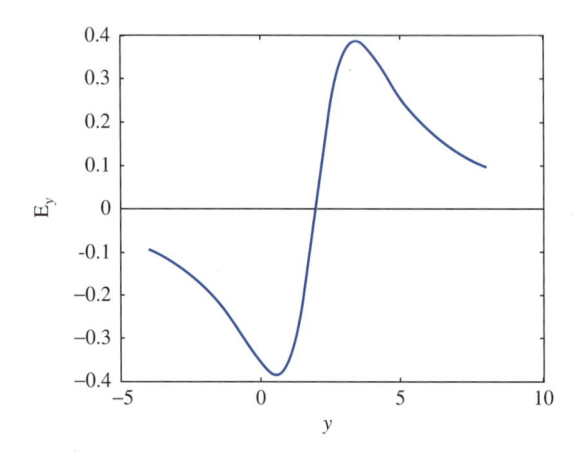

2.3.3. $E_z = \dfrac{\rho_s z}{2\varepsilon_0}\left[\dfrac{1}{\sqrt{a^2+z^2}} - \dfrac{1}{\sqrt{b^2+z^2}}\right]$ V/m

Chapter 2

2.3.5. $E_z = \dfrac{\rho_s}{2\varepsilon_0}$ V/m

2.4.1. $0 < \rho < a$, $\quad E_\rho = \dfrac{\rho^2 \rho_0}{4a\epsilon_0}$

$0 > a$, $\quad E_\rho = \dfrac{a^3 \rho_0}{4\epsilon_0 \rho^2}$

2.4.2. $0 < \rho < a \Rightarrow E_\rho = \dfrac{\rho_\ell \rho}{2\pi\varepsilon_0 a^2}$ V/m; $\quad a < \rho < b \Rightarrow E_\rho = \dfrac{\rho_\ell}{2\pi\varepsilon_0 \rho}$ V/m

$b < \rho < c \Rightarrow E_\rho = \dfrac{\rho_\ell}{2\pi\rho\varepsilon_0}\left[1 - \dfrac{\rho^2 - b^2}{c^2 - b^2}\right]$ V/m; $\quad \rho > c \Rightarrow E_\rho = 0$

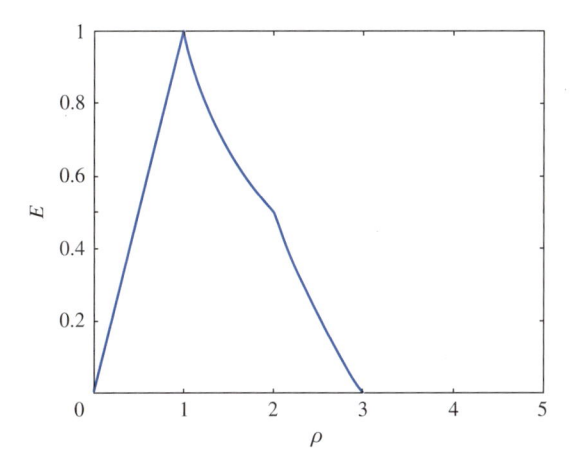

2.4.5. $0 < r < a \Rightarrow E_r = 0$; $\quad a < r < b \Rightarrow E_r = \dfrac{2}{4\pi\varepsilon_0 r^2}$ V/m;

$b < r < c \Rightarrow E_r = \dfrac{-2}{4\pi\varepsilon_0 r^2}$ V/m; $\quad r > c \Rightarrow E_\rho = \dfrac{2}{4\pi\varepsilon_0 r^2}$ V/m

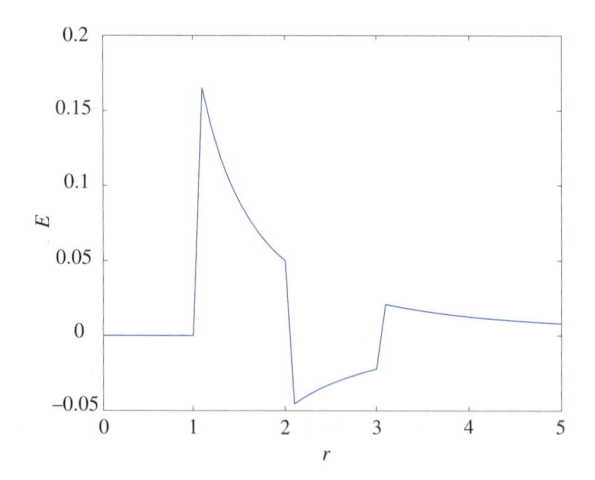

2.5.1. $V(x, y) = \dfrac{2}{[(x-2)^2 + (y-2)^2]^{1/2}} - \dfrac{4}{[(x-4)^2 + (y-5)^2]^{1/2}}$ V

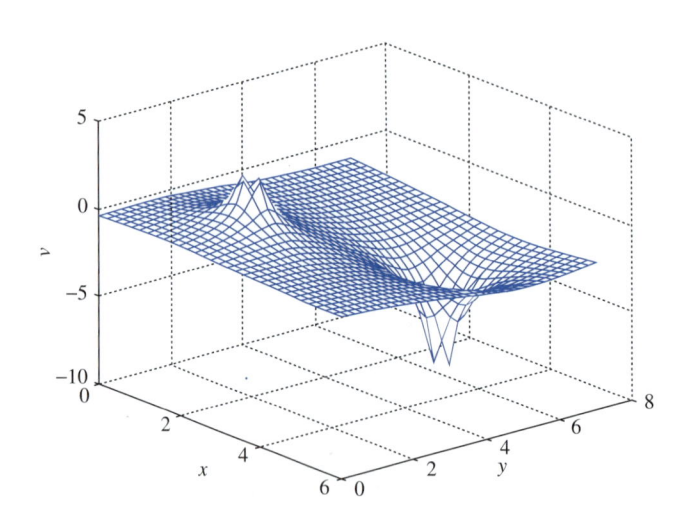

Chapter 2

2.5.2.

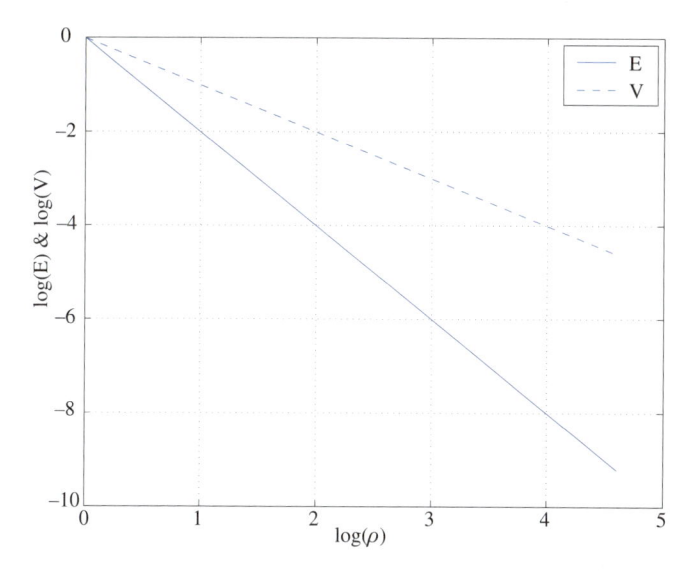

2.5.4. (a) $V_{AB} = 4$ V; (b) $V_{AB} = 4$ V

2.5.8.
$$V_{\text{monopole}} = \frac{Q}{4\pi\varepsilon_0 r} \text{ V}; \qquad V_{\text{dipole}} \approx \frac{Qd\cos\theta}{4\pi\varepsilon_0 r^2} \text{ V};$$

$$V_{\text{quadripole}} \approx \frac{Qd\cos\theta\cos\phi}{4\pi\varepsilon_0 r^3} \text{ V}$$

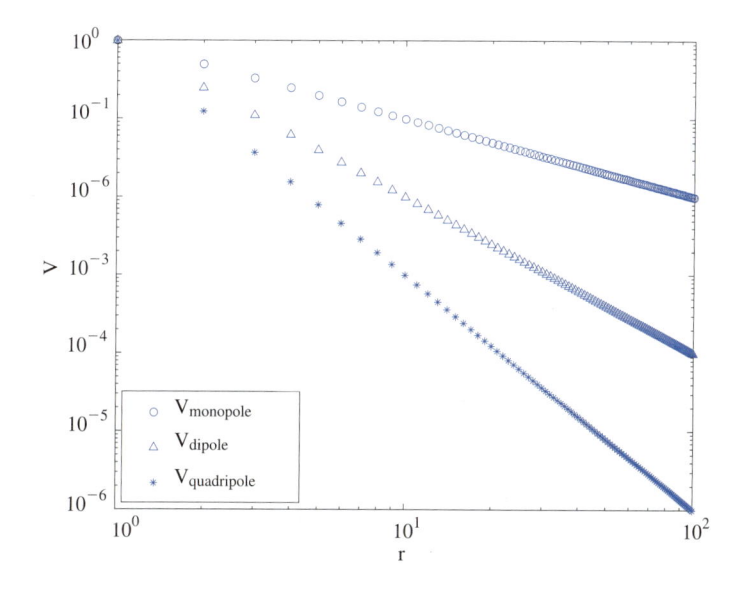

Chapter 3

3.1.1. $I = 3$ A

3.1.3. $I = \dfrac{\pi a^2}{2} J_0$ A

3.2.1. $\mu_0 = \dfrac{ML}{Q^2}$ H/m

3.2.3.
$0 < \rho < a \Rightarrow B_\phi = 0; \qquad a < \rho < b \Rightarrow B_\phi = -\dfrac{\mu_0 I}{2\pi\rho}$ T;

$\rho > b \Rightarrow B_\phi = \dfrac{\mu_0 I}{2\pi\rho}$ T

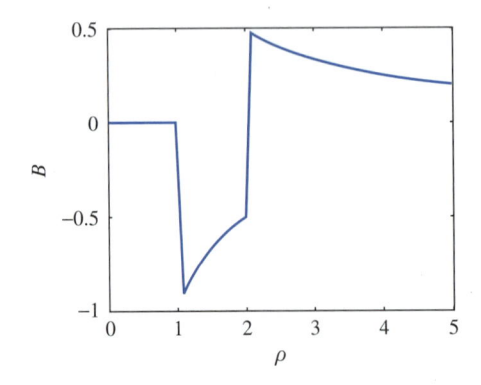

3.2.5. $0 < \rho < a \Rightarrow B_\phi = \mu_0 J_0 \left(\dfrac{\rho}{2} - \dfrac{\rho^3}{4a^2} \right)$ T; $\qquad \rho > a \Rightarrow B_\phi = \dfrac{\mu_0 a^2 J_0}{4\rho}$ T

3.2.7. $\mathscr{L} = \dfrac{\mu_0 NI}{B_\phi} = 251$ m

3.3.1. $\nabla\left(\dfrac{1}{R}\right) = -\dfrac{\mathbf{R}}{R^3} = -\dfrac{\mathbf{u_R}}{R^2}$

3.3.3. $\mathbf{A}(b) = \dfrac{\mu_0}{2\pi} \dfrac{I\Delta L}{(b^2 + a^2)^{1/2}} \mathbf{u}_z$ T-m; $\qquad \mathbf{B}(b) = -\dfrac{\mu_0}{2\pi} \dfrac{I\Delta L b}{(b^2 + a^2)^{3/2}} \mathbf{u}_x$ T

3.3.5. $B_\phi = \dfrac{\mu_0 I}{4\pi\rho} (\cos\theta_1 - \cos\theta_2)$ T; as $L \rightarrow \infty$, $B_\phi \rightarrow \dfrac{\mu_0 I}{2\pi\rho}$ T

3.3.7. $B_\phi = \dfrac{\mu_0 I}{2\pi}\left[\dfrac{L}{\rho_1\sqrt{\rho_1^2 + L^2}} - \dfrac{L}{\rho_2\sqrt{\rho_2^2 + L^2}}\right]$ T

3.3.9. $\mathbf{B} = -\dfrac{2\mu_0 I}{\pi W}\mathbf{u_z}$ T

3.4.1. $\dfrac{\mathbf{F}}{\Delta L} = 100\mathbf{u_z}$ N/m

3.4.2. $\Delta V = Ed = v_0 B_0 d$ V

3.4.4. The charged particle will drift with a velocity $\mathbf{v} = \dfrac{\mathbf{E} \times \mathbf{B}}{B^2}$ m/s.

3.4.5. $\dfrac{\mathbf{F}_1}{L} = \dfrac{\mu_0 I^2}{4\pi}(\mathbf{u_x} + \mathbf{u_y})$ N/m

3.4.7. $F = I\mathscr{L}B = \dfrac{\mu_0 I^2}{b}\mathscr{L}$ N

3.5.1. $H_x = \begin{cases} \dfrac{0.1}{2\mu_0} \text{ A/m} & |x| \leq 1 \\ \dfrac{0.1}{\mu_0} \text{ A/m} & |x| > 1 \end{cases}$, $M_x = \begin{cases} \dfrac{0.1}{2\mu_0} \text{ A/m} & |x| \leq 1 \\ 0 & |x| > 1 \end{cases}$

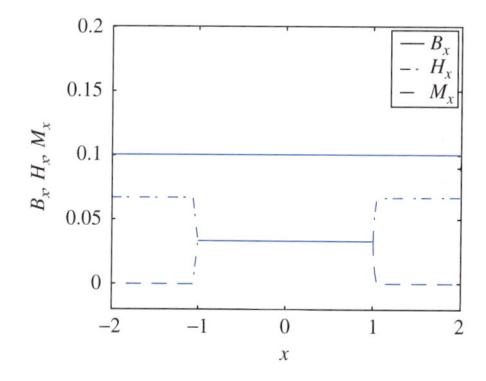

3.6.1. $B_g = \dfrac{\psi_1 + \psi_2}{A}$ T

Chapter 4

4.1.1. $\mathbf{E}_2 = 1\mathbf{u}_x + 2\mathbf{u}_y$ V/m

4.1.3. $\mathbf{E}_2 = \frac{3}{4}\mathbf{u}_x + 4\mathbf{u}_y$ V/m

4.1.5. $\mathbf{B}_2 = 4\mathbf{u}_x + 8\mathbf{u}_y$ T

4.1.7. No

4.3.1. $V(r) = \dfrac{V_0}{\frac{1}{a} - \frac{1}{b}}\left(\dfrac{1}{a} - \dfrac{1}{r}\right)$ V

4.3.3. $V(\theta) = \dfrac{V_1}{2\ln\left[\tan\left(\frac{\theta_1}{2}\right)\right]} \ln\left\{\left[\tan\left(\frac{\theta_1}{2}\right)\right]\left[\tan\left(\frac{\theta}{2}\right)\right]\right\}$ V

4.3.5. $V(x) = 4(2 - x)$ V

4.3.7. $E_x(x) = (5 - 4x)$ V/m

4.3.9. $V(P) = \dfrac{Q}{4\pi\varepsilon_0 L} \ln\left(\dfrac{L}{r} + \sqrt{\left(\dfrac{L}{r}\right)^2 + 1}\right) = \dfrac{1}{4\pi\varepsilon_0}(0.881)$ V

4.3.11. $E_r(P) = \dfrac{1}{4\pi\varepsilon_0} \dfrac{1}{+\sqrt{2}} = \dfrac{1}{4\pi\varepsilon_0}(0.707)$ V/m

4.3.13. $E_z(z) = \dfrac{Q}{4\pi\varepsilon_0} \dfrac{z}{(a^2 + z^2)^{3/2}}$ V/m

when $a \to 0$, $E_z(z) \to \dfrac{Q}{4\pi\varepsilon_0 z^2}$ V/m

4.3.15. $E_z(z) = \dfrac{\rho_s}{2\varepsilon_0}\left[1 - \dfrac{z}{\sqrt{a^2 + z^2}}\right]$ V/m; $a \rightarrow \infty,\, E_z(z) \rightarrow \dfrac{\rho_s}{2\varepsilon_0}$ V/m

4.5.1. $V(x, y) = \dfrac{aE_0}{\pi} e^{-\left(\frac{\pi y}{a}\right)} \sin\left(\dfrac{\pi x}{a}\right)$ V

4.5.3. $V(x, y) = \dfrac{4V_0}{\pi} \displaystyle\sum_{n=1,3,5\ldots} \dfrac{e^{-\left(\frac{2n\pi y}{a}\right)}}{n} \sin\left(\dfrac{2n\pi x}{a}\right)$ V

4.5.5. $V(x) = \dfrac{2V_0}{\pi} \displaystyle\sum_{n=1}^{\infty} \dfrac{(-1)^{n+1} \sin\left(\dfrac{n\pi x}{a}\right)}{n}$ V

4.5.7. $V(x, y) = \dfrac{4V_0}{\pi} \displaystyle\sum_{m=0}^{\infty} \dfrac{\sin\left(\dfrac{(2m+1)\pi x}{a}\right) \cosh\left[\dfrac{(2m+1)\pi}{a}\left(y - \dfrac{a}{2}\right)\right]}{(2m+1)\cosh\left(\dfrac{(2m+1)\pi}{2}\right)}$

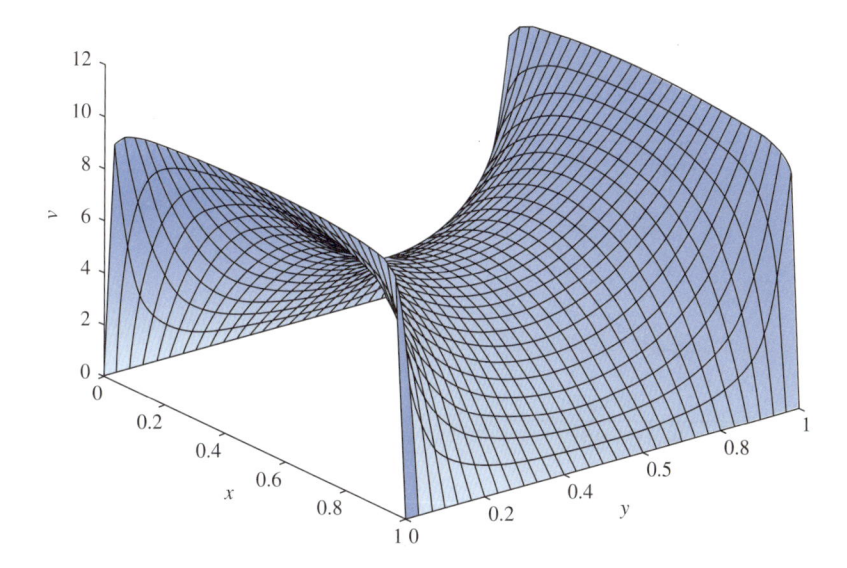

4.5.8. Bessel functions:

$$J_n(s) = \sum_{m=0}^{\infty} \frac{(-1)^m}{m!(n+m)!}\left(\frac{s}{2}\right)^{n+2m}; \qquad (s = \lambda r)$$

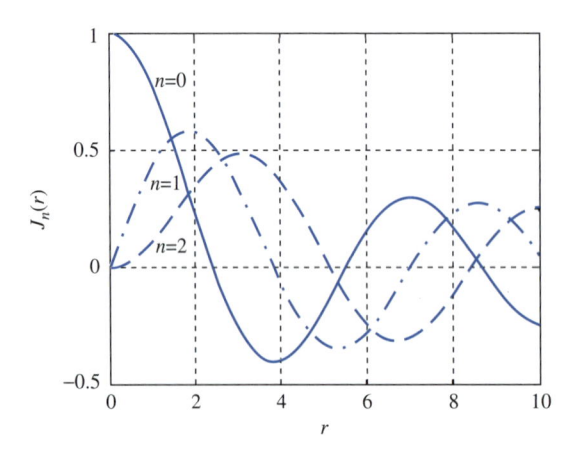

4.5.10. Legendre polynomials:

$$P_n(\xi) = \frac{1}{2^n n!}\left(\frac{d}{d\xi}\right)^n (\xi^2 - 1)^n \quad \text{or}$$

$$P_0(\xi) = 1, \qquad P_1(\xi) = \xi, \qquad P_2(\xi) = \tfrac{1}{2}(3\xi^2 - 1), \text{ etc.}$$

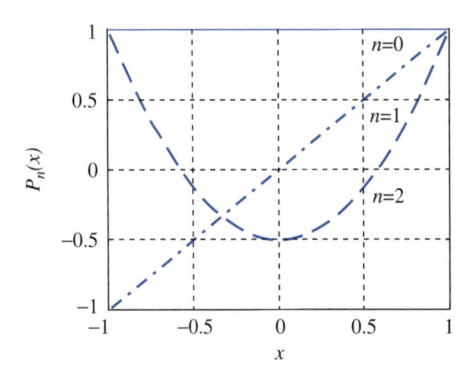

Chapter 4

4.7.1. FEM-application (see Example 4–23)—with 16 triangular elements, 15 nodes.

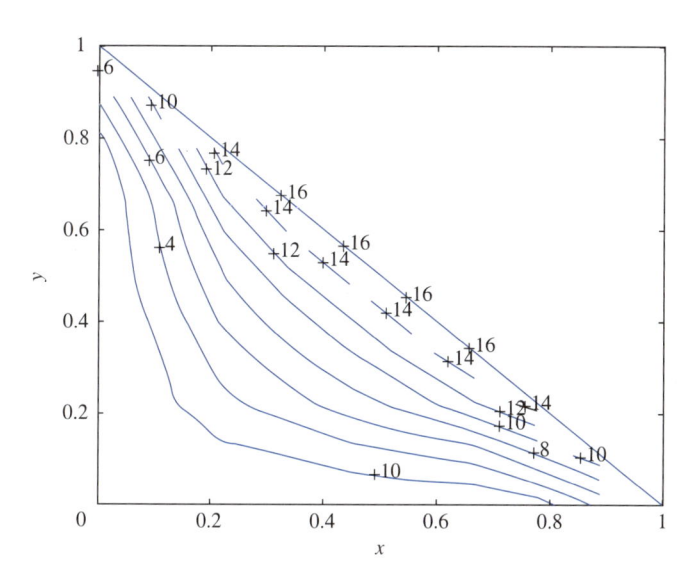

4.8.1. MoM-application (see Example 4–26) with $9 \times 9 = 81$ nodes in the grid.

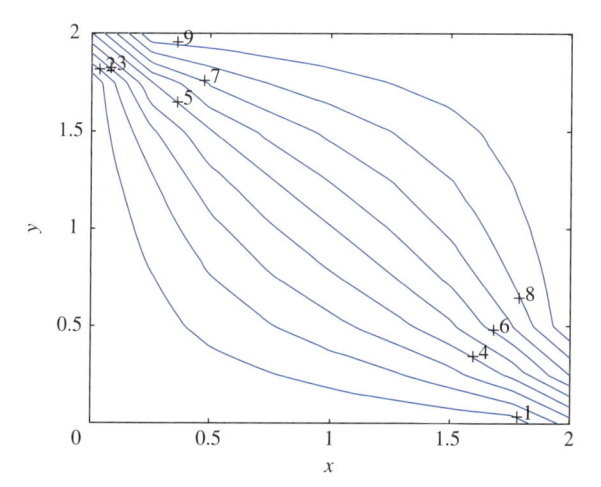

Chapter 5

5.1.1. The three components of the vector $\nabla \times \mathbf{B}$ are

$$(\nabla \times \mathbf{B})_x = \frac{\partial B_z}{\partial y} - \frac{\partial B_y}{\partial z} = -1; \quad (\nabla \times \mathbf{B})_y = \frac{\partial B_x}{\partial z} - \frac{\partial B_z}{\partial x} = -1;$$

$$(\nabla \times \mathbf{B})_z = \frac{\partial B_y}{\partial x} - \frac{\partial B_x}{\partial y} = 0$$

$\Rightarrow$ the time derivative of the vector $\mathbf{E} \neq 0$ or this vector varies with time.

5.1.3. $V = \oint_L (\mathbf{v}_0 \times \mathbf{B}) \cdot \mathbf{dl} = -0.2$ V

5.1.5. $V = -6$ V

5.1.7. $\Delta V = 1180$ V

5.1.9. $I(t) = 2[\sin(1000t) - \sin(500t)]$ A

5.1.11. $V(t) = \dfrac{\mu_0 I(2ab)}{2\pi} \dfrac{1}{v_0 t(v_0 t + 2b)}$ V

5.2.1. $\left.\dfrac{\partial \rho_v}{\partial t}\right|_{x=1} = \pi$ C/m^{3}s

5.3.1. $r \equiv \dfrac{J_c}{J_d} = 18 \times 10^9 \dfrac{\sigma}{\varepsilon_r f};$

	60 Hz	1 MHz	1 GHz
Copper	$r = 17.4 \times 10^{15}$	$r = 10.4 \times 10^{11}$	$r = 10.4 \times 10^{8}$
Sea Water	$r = 14.8 \times 10^{6}$	$r = 889$	$r = 0.889$
Earth	$r = 3 \times 10^{4}$	$r = 1.8$	$r = 0.0018$

5.3.2. $J_d = \dfrac{\partial D}{\partial t} = 72.3 \cos(2\pi 10^9 t)$ A/m^2

Chapter 5   509

5.4.1. $k = \pm \dfrac{\omega}{c}$ rad/m

5.4.3. $(\nabla \times \mathbf{B})_y \neq \dfrac{1}{c^2} \dfrac{\partial E_y}{\partial t}$

5.4.5. $J_d(\rho, t) = -\dfrac{\varepsilon \omega V_0}{\rho \ln\left(\dfrac{b}{a}\right)} \sin \omega t$ A/m^2

5.4.7. $\dfrac{\partial \rho_v}{\partial t} = -\nabla \cdot \mathbf{J}$ A/m^3

5.4.8. Maxwell's equations in Cartesian coordinates are

$$(5.29) \Rightarrow \begin{cases} \dfrac{\partial E_z}{\partial y} - \dfrac{\partial E_y}{\partial z} = -\dfrac{\partial B_x}{\partial t} \\ \dfrac{\partial E_x}{\partial z} - \dfrac{\partial E_z}{\partial x} = -\dfrac{\partial B_y}{\partial t} \\ \dfrac{\partial E_y}{\partial x} - \dfrac{\partial E_x}{\partial y} = -\dfrac{\partial B_z}{\partial t} \end{cases}$$

$$(5.30) \Rightarrow \begin{cases} \dfrac{\partial H_z}{\partial y} - \dfrac{\partial H_y}{\partial z} = \dfrac{\partial D_x}{\partial t} + J_x \\ \dfrac{\partial H_x}{\partial z} - \dfrac{\partial H_z}{\partial x} = \dfrac{\partial D_y}{\partial t} + J_y \\ \dfrac{\partial H_y}{\partial x} - \dfrac{\partial H_x}{\partial y} = \dfrac{\partial D_z}{\partial t} + J_z. \end{cases}$$

$$(5.31) \Rightarrow \dfrac{\partial D_x}{\partial x} + \dfrac{\partial D_y}{\partial y} + \dfrac{\partial D_z}{\partial z} = \rho_v.$$

$$(5.32) \Rightarrow \dfrac{\partial B_x}{\partial x} + \dfrac{\partial B_y}{\partial y} + \dfrac{\partial B_z}{\partial z} = 0.$$

5.5.1. $\mathbf{S}_{av} = \dfrac{1}{2}\operatorname{Re}(\mathbf{E} \times \mathbf{H}^*) = \dfrac{|E_0|^2}{2} \operatorname{Re}\left\{\dfrac{1}{Z_c^*}\right\} \mathbf{u_z}$ W/m^2

5.5.3. (a) $W_{e,a} = 4.42 \times 10^{-4}$ J
       (b) $W_{e,b} = 81\, W_{e,a}$

5.6.1. $\mathbf{E} = 120\pi e^{-j10z}\mathbf{u_x}$ V/m; $\mathbf{H} = 1e^{-j10z}\mathbf{u_y}$ A/m

5.6.3. (a) Copper at $f = 1$ MHz, $r = 10.4 \times 10^{11} \gg 1$

(b) Quartz at $f = 1$ MHz, $r = 4.5 \times 10^{-14} \ll 1$

Therefore, copper is a good conductor at this frequency, and quartz is a good insulator at this frequency.

5.6.5. $f_c = 1.04 \times 10^{18}$ Hz

Chapter 6

6.1.1. $\left(\dfrac{F}{m} \cdot \dfrac{H}{m}\right)^{-1/2} = \dfrac{m}{s}$

6.1.3. $\mu_r = 3.183, \varepsilon_r = 1.257$

6.1.5. $\beta = 2085.6$ °/m, $\alpha = 130.8$ dB/m

6.2.1.

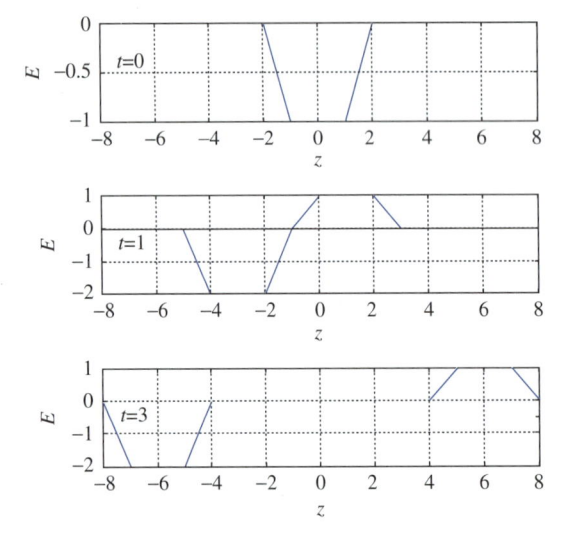

6.2.3. $\dfrac{\varepsilon_2}{\varepsilon_1} = \dfrac{1}{4}$

Chapter 6

6.2.5.

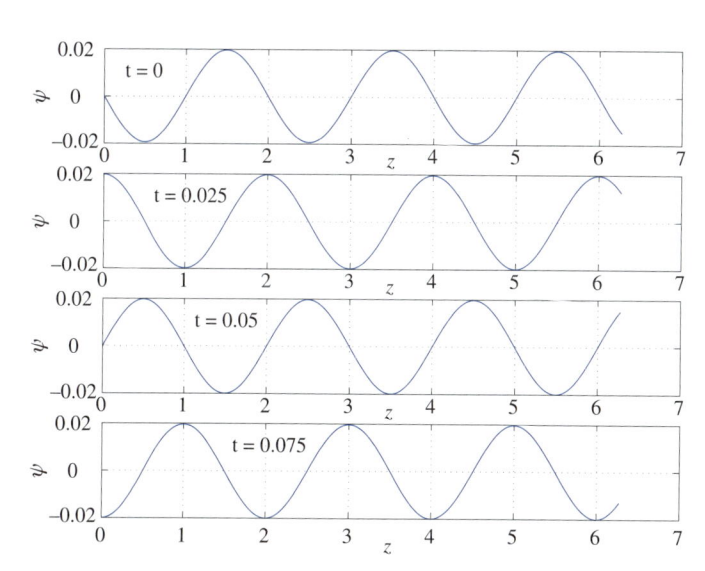

6.2.7. With $\dfrac{h}{c\tau} = \dfrac{1}{2}$, unstable

6.2.12. $c = 1$

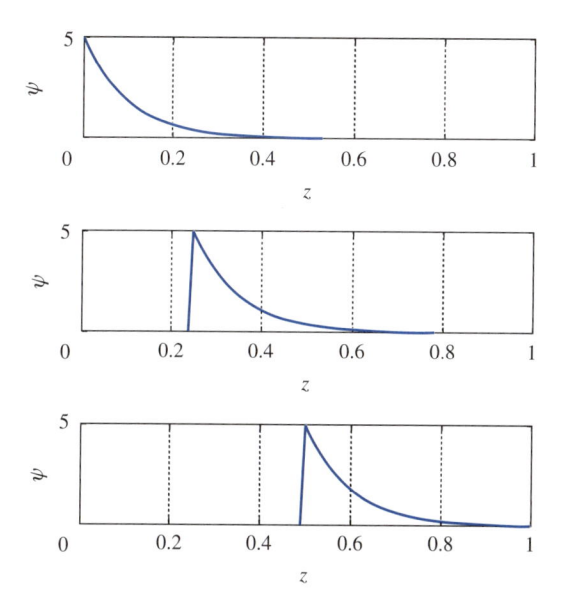

6.2.14. $c = 0.5$

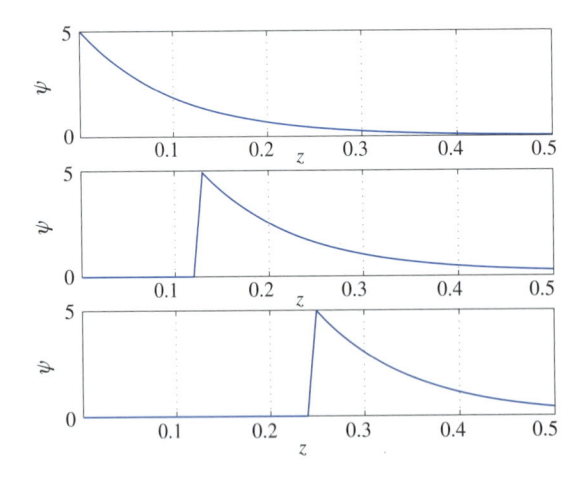

6.3.1. $E(z, t) = \sin\left[2\pi \times 75 \times 10^6 t - \dfrac{2\pi}{4} z\right]$ V/m

6.3.4. Circular polarization: $a = 1$, $\delta = \pi/2$; Linear polarization: $\delta = 0$

6.3.5. $\mathbf{S_{av}} = \frac{1}{2}\text{Re}\{\mathbf{E} \times \mathbf{H}^*\} = S_{av}\mathbf{u_z}$, $S_{av} = \frac{1}{2}Z_0 H_0^2$ W/m^2

6.4.1. $v_z = 1.5 \times 10^8$ m/s, $k_z = 0.0419$ 1/m, $\omega = 6.28 \times 10^6$ rad/s, $Z_c = 60\pi$ Ω, $H_x = -0.0069\cos(\omega t - k_z z)$ A/m

6.4.6. $\alpha \approx 0.333$ Np/m $= 2.89$ dB/m, $Z_c \approx 133$ Ω

6.4.8. (1) copper $\delta \approx 1.3$ μm, (2) graphite $\delta = 32$ μm, (3) seawater $\delta \approx 5.1$ mm

6.5.2.

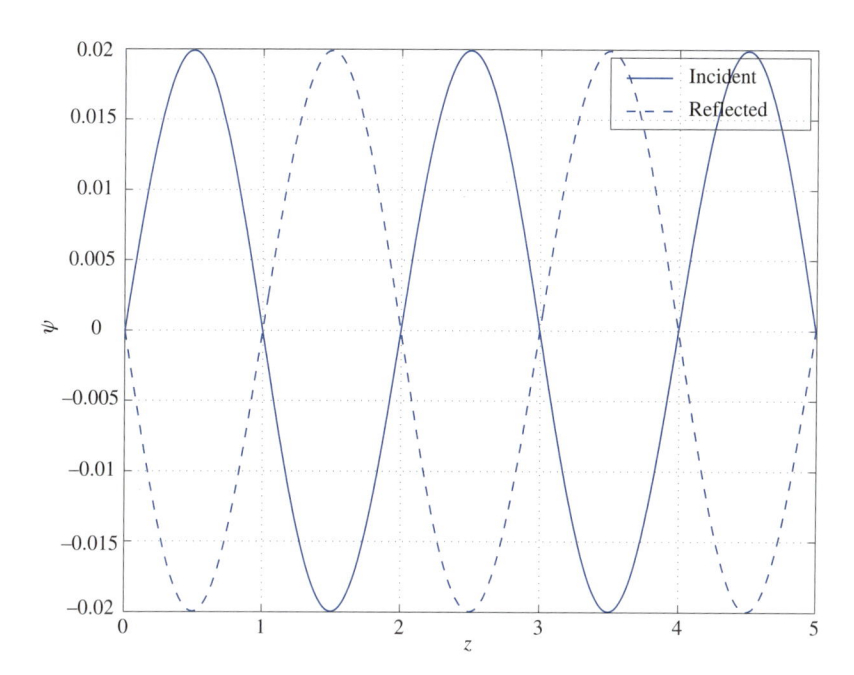

6.5.4. $H_i = 2.65$ mA/m, $H_r = 0.884$ mA/m, $H_t = 3.536$ mA/m

6.5.5. $E_r = \frac{1}{3}$ V/m, $E_t = \frac{4}{3}$ V/m, $p_i = 2.65$ mW, $p_r = \frac{1}{9}p_i$, $p_t = \frac{8}{9}p_i$

6.5.6. $\varepsilon_r = 1.5625$

6.5.8. $\Gamma = \dfrac{\Gamma_1(1 - e^{j2k_2 d})}{1 - \Gamma_1^2 e^{j2k_2 d}} = 0$ where $d = \dfrac{n\lambda}{2\sqrt{\varepsilon_r}}$ (n is an integer)

6.5.10. $\varepsilon_{r2} = \sqrt{\varepsilon_{r1}\varepsilon_{r3}}$

6.5.12. $\lambda = 0.6328\ \mu m$, $\quad f = 474.8\ \text{THz}$, $\quad T = 0.0021\ \text{ns}$, $\quad k = 9.93 \times 10^6\ 1/m$

6.5.13. $N \approx 1580000$

6.5.15. $\varepsilon_{rav} \approx 1.01$, $\quad p = 98\%$

Chapter 7

7.1.1. $\hat{L}\hat{C} = \mu\varepsilon$

7.2.2. $\dfrac{1}{\hat{R}\hat{C}} = \dfrac{L^2}{T} \Rightarrow D$

7.2.3. $V(\psi) = V(z + vt)$ is a solution of the transmission line equation.

7.3.1. $Z_c = \sqrt{\dfrac{\mu_0}{\varepsilon}\dfrac{d}{w}}$

7.3.3. $Z_c = 133.3\ \Omega$, $\quad v = 2.1 \times 10^8\ m/s$

7.3.5. $\hat{C} = 27.8\ \text{pF/m}$, $\quad Z_c = 207.4\ \Omega$

7.4.1. $Z_L < Z_c \Rightarrow \Gamma = -\tfrac{1}{3}$, $\quad Z_L > Z_c \Rightarrow \Gamma = +\tfrac{1}{3}$

7.4.3.

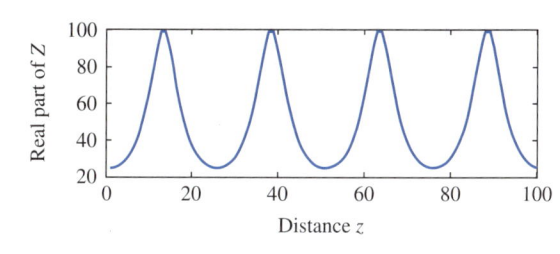

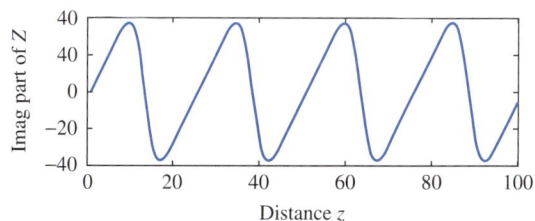

7.5.1. Periodic since $\tan\left[\frac{2\pi}{\lambda}\left(L + \frac{\lambda}{2}\right)\right] = \tan\left(\frac{2\pi L}{\lambda}\right)$

7.5.3. $Z_L = \infty$ & $\frac{\lambda}{4} < L < \frac{\lambda}{2} \Rightarrow$ inductive

7.5.4. $Z_{in} = Z_L = (25 + j25)\;\Omega$

7.5.6. Center of circles: $\left(\frac{r}{r+1}, 0\right)$ & $\left(1, \frac{1}{x}\right)$

Respective radii of circles: $\frac{1}{r+1}$ & $\frac{1}{|x|}$

7.6.1. $Z_{in} = (20 - j10)\;\Omega$

7.6.3. Two solutions: $+j1 \Rightarrow L_1 = \frac{\lambda}{8}$ & $-j1 \Rightarrow L_2 = \frac{3\lambda}{8}$

7.7.1. Reflection coefficient $\Gamma = -1$ at both ends

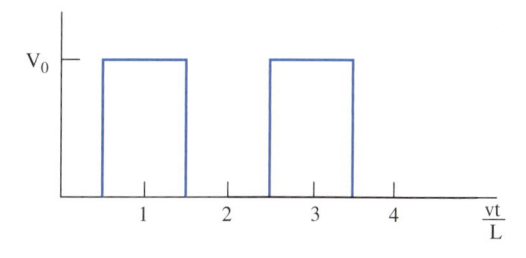

7.7.3. Reflection coefficient $\Gamma = +1$ at the load and $\Gamma = -1$ at the source

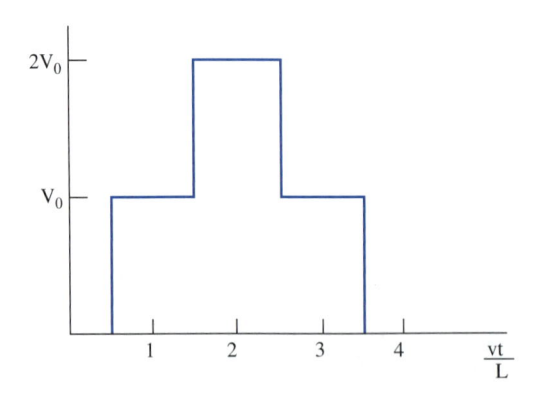

7.7.5. $V_T = \dfrac{Z_{c2}}{R_L + Z_{c2}}(1 + \Gamma)\, V_{inc}$

7.7.7. Reflection coefficient $\Gamma = -1$ at the load and $\Gamma = +1$ at the source

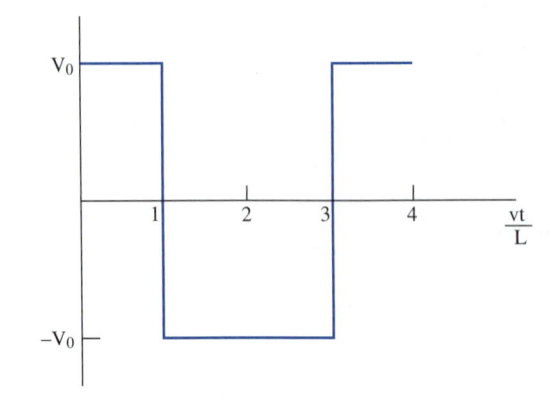

7.7.8. $T = 20$ ns, $V_{inc} = \frac{5}{8}$ V, $\Gamma_L = -\frac{3}{7}$, $\Gamma_g = -\frac{1}{4}$,
At the middle point:

t(ns)	10	30	50	70	90
amp(V)	$+\dfrac{5}{8}$	$-\dfrac{15}{56}$	$+\dfrac{15}{224}$	$-\dfrac{45}{1568}$	$+\dfrac{45}{6272}$

7.8.1. $\delta t < \dfrac{L}{v}$

Chapter 7

7.9.1.

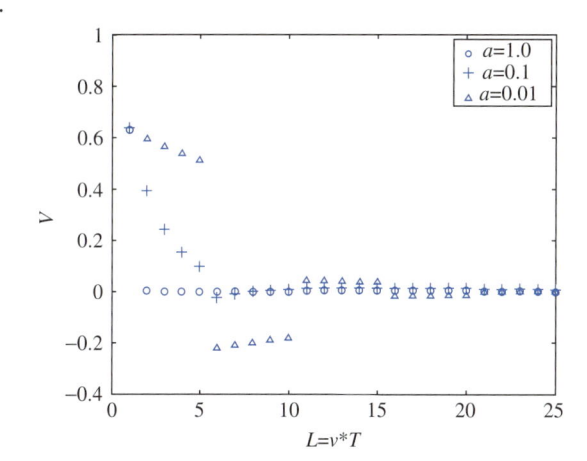

[Plot: V vs L=v*T, showing data for a=1.0 (circles), a=0.1 (plus), a=0.01 (triangles)]

7.9.2. $\gamma = \alpha + j\beta = \sqrt{\dfrac{\hat{C}_1}{\hat{C}_2} - \omega^2 \hat{L}_1 \hat{C}_1};$ with $\dfrac{\hat{C}_1}{\hat{C}_2} = \hat{L}_1 \hat{C}_1 = 1$

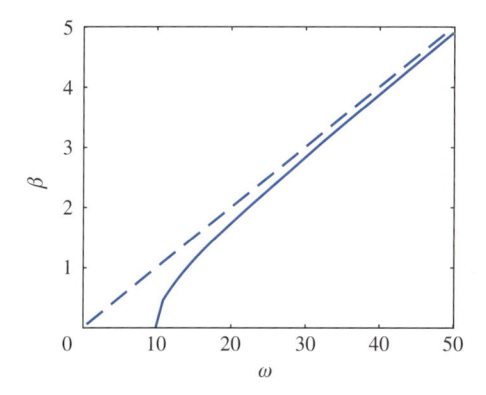

[Plot: β vs ω]

7.10.1. $\gamma = \alpha + j\beta = \sqrt{\dfrac{\hat{C}_1}{\hat{C}_2} - \omega^2 \hat{L}_1 \hat{C}_1 + j\omega \hat{R}_s \hat{C}_1},$ with

$\dfrac{\hat{C}_1}{\hat{C}_2} = \hat{L}_1 \hat{C}_1 = 1$ & $\hat{R}_s \hat{C}_1 = 2,$ $\gamma = 1 + j\omega,$ there is no dispersion

Chapter 8

8.1.1. $\dfrac{1}{4x^2}$

8.2.1. $D = \dfrac{1}{\displaystyle\int_0^{\pi/2} \sin^3\theta \cos^2\left(\dfrac{\pi}{2}\cos\theta\right)d\theta} = 3.62 \text{ dB}$

8.4.1. $D = \dfrac{1}{\displaystyle\int_0^{\pi/2} \sin^3\theta \sin^2\left(\dfrac{\pi}{2}\cos\theta\right)d\theta} = 4.31 \text{ dB}$

8.4.3. $D = \dfrac{1}{\displaystyle\int_0^{\pi/2} \sin^3\theta \cos^2\left(\dfrac{\pi}{2}\cos\theta\right)d\theta} = 3.62 \text{ dB}$

8.4.5. The radiation resistance decreases by a factor of $K = 2.88$.

8.4.7. $R_r = 120\displaystyle\int_0^{\pi/2} \dfrac{\cos^2\left(\dfrac{\pi}{2}\cos\theta\right)}{\sin\theta}d\theta = 73.1 \ \Omega$

8.4.10. Assume that the current is confined within one skin depth of the surface.

This yields a resistance of $R_L = \dfrac{L}{\sigma(2\pi a)\delta} = 0.026 \ \Omega$.

The radiation resistance of the antenna is $R_r = 80\pi^2(L/\lambda)^2 = 0.035 \ \Omega$.

Then the radiation efficiency is $\eta_r = \dfrac{0.035}{0.035 + 0.026} = 57.4\%$.

8.4.11. $A_e = \dfrac{\lambda^2}{4\pi}D = \dfrac{1.5}{4\pi}\lambda^2 = 0.119\lambda^2$

Chapter 8   519

8.4.14. (a) $R_1 + R_2 - R_0 = \dfrac{m\lambda}{2}$, m is an integer

(b) $R_1 + R_2 - R_0 = m\lambda$, m is an integer

The path difference is $R_1 + R_2 - R_0 = \dfrac{2h_1 h_2}{d}$, where $h_2 = \dfrac{d\lambda}{2h_1}$.

The power that is received by the antenna is
$P_t = 500$ W, $R = d = 2000$ m, $\lambda = 1$ m, $p_r = 29.3$ μW

8.4.15. $\sigma_s \equiv \dfrac{4\pi r^2 \left(\dfrac{|E_s|^2}{2Z_0}\right)}{\left(\dfrac{|E_i|^2}{2Z_0}\right)} = 4\pi r^2 \left|\dfrac{E_s}{E_i}\right|^2$ m^2

8.6.1. Use of method of images and superposition.

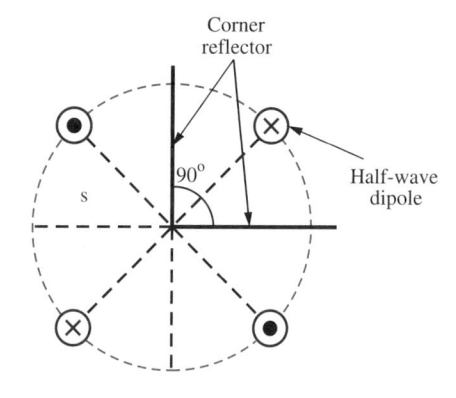

$F(\theta, \phi) = \dfrac{\cos\left(\dfrac{\pi}{2}\cos\theta\right)}{\sin\theta} \dfrac{1}{4} \sum_{n=0}^{3} (-1)^n e^{jks\sin\theta\cos(n90° - \phi)}$. Choosing $s = \dfrac{\lambda}{2}$, $\phi = 0°$, we obtain the array pattern below.

8.6.3. $\text{HPBW} \approx \dfrac{108°}{\sqrt{L/\lambda}}$

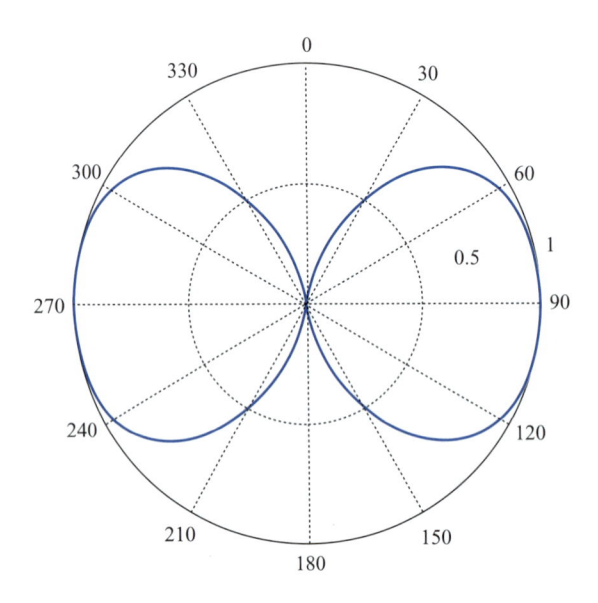

8.6.4.

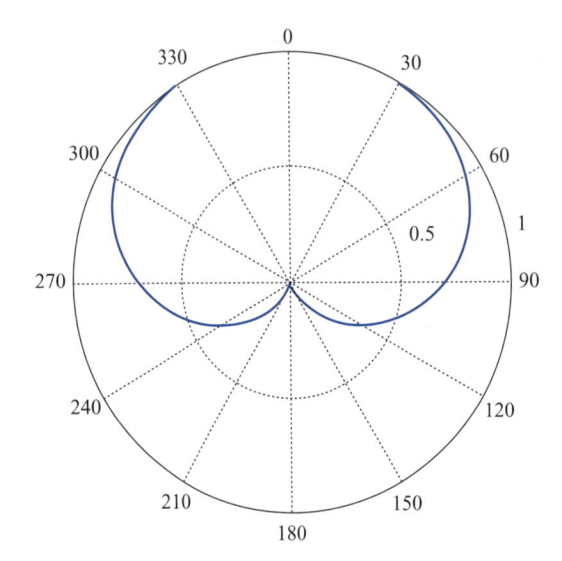

8.6.5. $Z_1 = 85.6 + j72.4 \ \Omega$

Chapter 8

8.6.7. (a) without a reflector: $R_r = 73.1\ \Omega$, $\quad I_0 = 0.234$ A
(b) with a reflector: $R_r = 21.7\ \Omega$, $\quad I_0 = 0.429$ A

8.6.9. The plots for (a) $\delta = 0$, (b) $\delta = \dfrac{\pi}{4}$, (c) $\delta = \dfrac{\pi}{2}$ are shown.

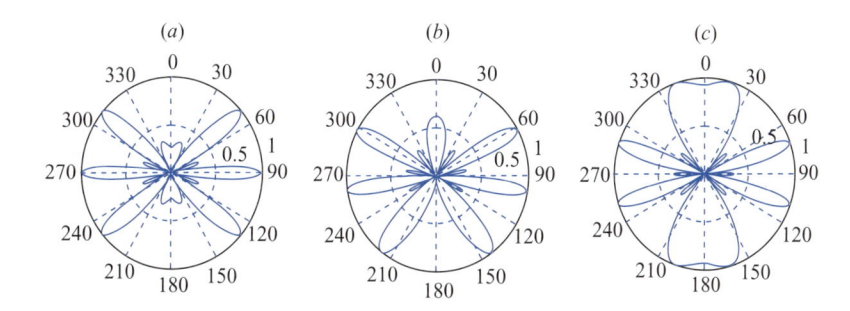

APPENDIX H

Greek Alphabet

Upper Case	Lower Case	Name
A	α	Alpha
B	β	Beta
Γ	γ	Gamma
Δ	δ	Delta
E	ε	Epsilon
Z	ζ	Zeta
H	η	Eta
Θ	θ	Theta
I	ι	Iota
K	κ	Kappa
Λ	λ	Lambda
M	μ	Mu
N	ν	Nu
Ξ	ξ	Xi
O	o	Omicron
Π	π	Pi
P	ρ	Rho
Σ	σ	Sigma
T	τ	Tau
Υ	υ	Upsilon
Φ	ϕ	Phi
X	χ	Chi
Ψ	ψ	Psi
Ω	ω	Omega

Index

A

absolute potential, 86, 89
acceleration, 425
advection equation, 312–314
amber (élektron), 61, 62
Ampere's circuital law, 160, 163–64, 181
    magnetic flux density, 130–32, 137–38
Ampere's Law, 131, 137–39, 275
ampere-turns, 164
Angstrom, 96
antenna
    beam solid angle, 449
    beam width, 450
    directivity, 450
    Friis transmission equation, 453
    gain, 450
antenna array (s), definition of, 455–56
array factor, 456
array size, 207, 223
attenuation constant, 329

B

beam solid angle, 449
beam width, 450–51
Bessel function, 506
binomial array, 461–63
Biot-Savart law, 143, 154, 422
Blumlein transmission line, 417
Boltzmann's constant, 488
bounce diagram, 392
boundary conditions, 178–185

C

capacitance, 114–18
    of metallic objects, 242
    of parallel plate capacitors, 116, 247

Cartesian coordinates, 18–21
central difference method, 202–04, 314
characteristic impedance, 323–24
    of the conductor, 329
    for lossy transmission lines, 403
    of matching transmission line, 376–79
    propagation characteristics, 337
    of the transmission line, 362, 365–67, 369
charge density distribution, 247–49
closed line integral, 30, 32, 45, 49, 51, 100
closed path, 32, 99
closed surface integral, 35–36, 41, 78, 80, 129, 272, 287, 293
    units of, 287
conduction current, 124
conservative fields, 32, 99
constitutive relations, 280, 298
coordinate systems, 17–29
Coulomb gauge, 139, 189
Coulomb, Charles-Augustin de, 64
Coulomb's Law 61–67
coupled potentials, 230, 234–35
curl operation, 46–51
current density, 125–26
cylinder command, 25
cylindrical coordinates, 22–25
    potential at the surface, 245

D

dblquad command, 107
Debye length, 199–200
Debye's sheath, 195
del operator, 41, 186, 187
depletion layer, 195
diamagnetic material, 158
dielectric interface potential, 225
dielectric materials, 96, 98–99, 113–14

Index

differential electric field, 75
differential form of the energy, 288
differential surface area, 28
diffusion equation, 281–82, 361
dipole antenna
    Hertzian, 430
    radiation pattern, 432, 436–439,
Dirac's delta function, 361
direct integration method, 191–95
directivity, 449, 465
Dirichlet boundary condition; 226
Dirichlet's matrix, 232
dispersion relation, 409–11
dispersion, 356, 408–411, 413
dispersive medium, 328–29, 413
displacement current, 276, 278–79
displacement flux density, 112, 178–84
distortionless line, 406
distributed charge densities, 72–73
divergence theorem, 45–46, 79, 472
divergence, 40–46
domain wall, 159

E

effective aperture, 451–53
eigen-function, 216
eigenvalues, 216
Einstein, Albert, 257, 319, 426
electric dipole moment, 98
electric field for a wave derivation, 318–21
electric field, 67–71, 75–77
electric flux density, 112
electric force, 62
electric potential difference, 89–92
electric potentials
    absolute potential, 86, 90
    electric dipole, 100–101
    total electrostatic energy, 85–89, 96
    voltage difference, 92
electric susceptibility, 113
electrical currents, 123–27
electrical engineering, 62
electromagnetic energy density, 287
electromagnetic fields, 40–41, 287, 290–93
electromagnetic wave propagation
    for perpendicular waves, 335–42
    for time-harmonic signal, 343–47
    in varying dielectric medium (s), 325–33
electromagnetics terminology, 5–6

electrostatic field, 67, 98–99, 119, 129
electrostatic force (Coulomb force), 63, 64, 66
electrostatic stored energy, 87–89
elliptical polarization, 299
enclosed charge, 79, 82
energy of triangular elements. *See* finite element method (FEM) using MATLAB
E-plane, 439
equation of continuity, 271–74
equipotential contours, 483–84
equipotential surface, 38, 91
Euler's identity, 319

F

Fabry-Perot Resonator, 346–47
Faraday disc generator, 267
Faraday's Law of Induction, 257–70
Farads, 115
FDM using MATLAB, 220–26
FEMLAB, 237
ferromagnetic material, 159
finite difference in time domain (FDTD) method, 314–15
finite difference method (FDM), 201–08
finite dipole (linear antenna), 434–40
finite element method (FEM) using MATLAB, 226–41
forward difference method, 314
Fourier coefficient, 218
Fourier series expansion method, 211–15
Fourier sine series, 218
fplot command, 13
Franklin, Benjamin, 62, 124
Fraunhofer zone, 454
Friis transmission equation, 453
fringing field, 116

G

gain, 450
Gauss's Law (Gaussian surface), 77–85
Gauss's theorem, 45
Gaussian pulse traveling wave, 311
Gaussian surface, 78
Gedanken experiment, 85–88, 275–77
Gibb's phenomenon, 220
gradient of scalar quantities, 37–40
ground potential, 90–91, 192
group velocity, 329, 407–12
gyroradius, 148

H

H field, 160
half-power beamwidth (HPBW), 450, 466
half-wave dipole, 439
half wavelength, 414
Hall voltage, 174
heat equation. *See* diffusion equation
heat flow, 282
Heaviside equations, 358
Helmholtz equation, 318
Henries, 166, 258
Hertzian dipole, 438, 445–46, 449
hold command, 13
H-plane, 440
hyperbolic equations, 313
hysteresis curve, 162

I

image charge, 184
incident wave, 370, 373, 452
index of refraction, 327
individual elements, 455–56
inductance, 166–71
infinite transmission line termination, 367–73
infinitesimal electric dipole, 427–34
inline command, 109
input impedance of transmission line, 374–75
integrals, 29–37
integrated circuit, 110
interpolation conditions, 478
intrinsic wave impedance, 323
ion matrix sheath, 488
irrotational fields, 99

J

Joule heating, 127, 287–88

K

Kirchhoff's laws, 163, 165, 272, 358

L

Lagrangian mass variables, 84, 137
Laplace's equation, 94, 178, 186–91, 208, 212
laplacian operator, 52, 187, 187–89, 215
Larmor radius, 148
Lax method, 316–17
Leibnitz rule, 83
Lenz's law, 158, 259
line integral, 29–34
linear array, 455–56, 458, 461
linear charge density, 73–74, 76
linear polarization, 299–300
lobe(s), 432
lodestones, 128
longitudinal wave, 305
loop antenna, 440
Lorentz force, 147
Lorentz gauge, 429
lossless transmission line, 357–59, 367–69, 378, 381–82
lossy dielectric medium, 327
lossy transmission lines, 403–406

M

magnetic circuits, 162–66
magnetic dipole moment, 157–59, 443
magnetic dipoles, 145, 157–58, 171
magnetic energy, 170–71
magnetic field component of plane wave, 322–24
magnetic field fundamentals, 128–38, 142–43, 148
magnetic field intensity, 160–61, 322–25, 426
  calculation through vector potential, 431
  electric field components, 336
  magnitude of, 420
  relation to magnetic flux density, 164
  tangential component of, 181–82
magnetic flux density, 128–29, 259, 485. *See also* magnetic vector potential
  dependence on magnetic field intensity, 161–62
  normal component (s) of, 180
  self inductance of a solenoid, 166–68
  within the wire (s), 132–33, 154
magnetic flux linkage, 166–67
magnetic flux, 128, 164, 258–59, 486
magnetic forces, 146–56
magnetic materials, 157–62
magnetic monopoles, 128, 171, 180,
magnetic susceptibility, 158, 161
magnetic vector potential, 138–45, 188. *See also* magnetic flux density
magnetization, 159–160
magnetohydrodynamics, 174
magnetomotive force (mmf), 164
mass spectrometer, 148

Index

mathematical foundation of FEM, 477–82
MATLAB applications
    numerical integration, 101–109
    plotting, 97–98, 185
MATLAB commands, 7, 10–13, 19, 21, 25
MATLAB, 3–5
    3D graphical representations, 14–16
    graphical plotting routines, 13
matrix equation, 481–82
Maxwell's equations, 280–85
method of moments (MoM) using MATLAB, 241–51
    capacitance of metallic objects, 242
    one-dimensional charge distribution, 249–51
    potential at the perimeter of the circle, 247
minimization of energy result, 477–78
molecules, 110
multiplication of patterns, 456
mutual capacitance, 115
mutual coupling, 246
mutual inductance, 166, 265

N

negative dispersion, 411
nepers, 329
Neumann boundary condition, 226
Newton, Isaac, 257
node potentials, 227, 479–81
nonconservative fields, 32
nonlinear varactor diode, 409, 413–14
norm command, 7
normalized array factor, 459
num2str command, 109

O

Oersted, Hans Christian, 128, 130
Ohm's law, 124–25
one-dimensional charge distribution, 249–51
one-dimensional wave equation, 302–17
ordinary differential equations, 213, 215, 298, 403
orthogonal coordinate systems, 18, 28, 30, 188
orthogonal, 6, 17–19, 383, 427
oscillating current, 427

P

paddle wheel, 46
parabolic equation, 313
parallel current-carrying wires, 154–57
parallel plate capacitor, 115, 116, 118, 278
paramagnetic material, 158
periodic boundary conditions, 314, 316
permeability of free space, 64–65, 130
phase velocity, 319–20, 327–28, 412–13
phased-array antenna, 458
phasor notation (s), 54–56, 292, 344
phasor(s), 52–56
pillbox, 178–79
plasma adjacent to metallic electrode, 487–89
Poisson's equation, 94, 186–91, 193
polarization charge, 111–12
polarization field, 111–13
polarization, 75, 112–14, 299–301, 324, 452
potential distribution, 52, 211–12
    in a bounded space, 215–16
power density, 127, 452, 453
power loss density, 287
power radiation pattern, 449
Poynting vector, 287, 291, 324, 426
Poynting's theorem, 286–90, 420
Priestley, Joseph, 62
propagation constant, 328, 331, 404–05, 409, 411
pulse propagation, 397

Q

Q_{enc}, enclosed charge, 78–79, 82–84
quad command, 102
quadrupole, 185
quarter-wave transformer, 376
quiver command, 40, 44

R

radial electric field, 76, 78, 80
radiation intensity, 449
radiation of electromagnetic waves, 419–26
radiation patterns, 438–40,
    through antenna array(s), 455–62
radiation resistance, 443–48
rail gun, 154
reactive energy, 293
real pulse, 341–42
reflected wave, 307, 373, 394, 398
reflection coefficient, 337, 516–17
    of lossless transmission lines, 381-82, 385–87
    of transmission line (s), 368–70
    with a battery impedance, 391–94
relative dielectric constant, 180, 348, 399, 475
relative permeability, 161–65

reluctance, 163–164
repeated vector operation application, 269–70
repeated vector operations, 51–52

S

scalar product, 8–12, 21, 285.
   See also vector product
scalars, 6, 16
self capacitance, 115
self inductance, 166–68
separation of variables, 211–19
side-lobe level (SLL), 466
Simpson's rule, 101
single-stub matching, 380-81, 390
sinusoidal waves, 362–63, 365–67
skin depth, 332–334
skin effect, 333
small loop antenna, 440–43
S-matrix elements, 478–79
Smith Chart, 381–90
spherical coordinates, 25
standing wave, 343, 370–73
static electric field, 186, 422
Stokes theorem, 49–51
strip line, 353, 354
subplot command, 13
superposition principles, 69–76
surf command, 14
surface charge density, 72, 113, 114, 179–80, 225
surface integral, 34, 131–32, 270
   over two surfaces, 41–42
symmetrical components, 322

T

Taylor series expansion, 202, 203
telegraphers' equations, 353, 357, 403
TEM mode of propagation, 355
Tesla, 128
Thales of Miletus, 61
three-dimensional vector wave equation, 299
time derivatives, 286
time domain reflectometry, 400
time-harmonic electromagnetic fields, 291–293
total electric field, 69, 70–71, 74, 343, 425
total field pattern, 456
total magnetic flux, 128, 164, 166–68, 486
total magnetization, 159–60
total radiated power, 421, 444–45
transformation of variables, 28–29

transient characteristics of step voltage, 391–97
transient sheath, 196, 488
transmission coefficient, 337, 398–99
transmission line equations, 357–62
   velocity of propagation, 359
transmission line matching, 388–89
transmission line parameters, 483–86
transmission lines, 401–402
   types of, 354–55
transverse electromagnetic propagation mode, 355–57
transverse waves, 305
trapezoidal command, 102
traveling wave(s), 307–310
trombone line, 380
twin lead, 354

U

unit vectors, 17–22
unknown charge distribution, 245

V

Van Allen belt, 151
vector potential, 139–41, 146, 188, 428–30, 431
vector product, 10–12, 21, 143, 146, 152.
   See also scalar product
vector wave equation, 299
vectors, 6–12
velocity of light, 65, 131
velocity of propagation, 304–07, 349, 359, 366
virtual pulse, 341
voltage divider rule, 391
voltage pulse along transmission line, 397–400
voltage standing wave ratio, 372
voltage standing wave ratio, 372
volume charge density, 36, 72, 78–79
volume charge density, 72, 78
volume integral, 36
volume integral, 36–37

W

wave equation, 297–302, 353–54, 364
   and numerical solution, 312–17
wave experiments, 302–305
wave number, 318–20, 326, 346, 363, 410, 430
wave propagation, 308–09
wave vector, 320
wavemaker, 302
webers, 128

Useful Integrals

$$\int u\,dv = uv - \int v\,du$$

$$\int \frac{dx}{x} = \ln x$$

$$\int e^x\,dx = e^x$$

$$\int \sin ax\,dx = -\frac{1}{a}\cos ax$$

$$\int \cos ax\,dx = \frac{1}{a}\sin ax$$

$$\int \frac{dx}{\sqrt{x^2 + a^2}} = \ln(x + \sqrt{x^2 + a^2})$$

$$\int \frac{dx}{(x^2 + a^2)^{3/2}} = \frac{x}{a^2\sqrt{x^2 + a^2}}$$

$$\int \frac{dx}{x^2 + a^2} = \frac{1}{a}\tan^{-1}\frac{x}{a}$$

$$\int e^{ax}\cos bx\,dx = \frac{e^{ax}}{a^2 + b^2}(a\cos bx + b\sin bx)$$

$$\int e^{ax}\cos(c + bx)\,dx =$$
$$\frac{e^{ax}(a\cos(bx + c) + b\sin(bx + c))}{a^2 + b^2}$$

Trigonometric Relations

$$\sin(x \pm y) = \sin x \cos y \pm \cos x \sin y$$
$$\cos(x \pm y) = \cos x \cos y \mp \sin x \sin y$$
$$2\sin x \sin y = \cos(x - y) - \cos(x + y)$$
$$2\sin x \cos y = \sin(x + y) + \sin(x - y)$$
$$2\cos x \cos y = \cos(x + y) + \cos(x - y)$$
$$\sin 2x = 2\sin x \cos x$$
$$\cos 2x = 1 - 2\sin^2 x$$
$$\sin x + \sin y = 2\sin\left(\frac{x + y}{2}\right)\cos\left(\frac{x - y}{2}\right)$$
$$\sin x - \sin y = 2\cos\left(\frac{x + y}{2}\right)\sin\left(\frac{x - y}{2}\right)$$
$$\cos x + \cos y = 2\cos\left(\frac{x + y}{2}\right)\cos\left(\frac{x - y}{2}\right)$$
$$\cos x - \cos y = -2\sin\left(\frac{x + y}{2}\right)\sin\left(\frac{x - y}{2}\right)$$
$$\cos(x \pm 90°) = \mp \sin x$$
$$\cos(-x) = \cos x$$
$$\sin(x \pm 90°) = \pm \cos x$$
$$\sin(-x) = -\sin x$$
$$e^{jx} = \cos x + j\sin x$$
$$\sin x = \frac{e^{jx} - e^{-jx}}{2j}$$
$$\cos x = \frac{e^{jx} + e^{-jx}}{2}$$

Approximations for Small Quantities

For $|x| \ll 1$,

$$(1 \pm x)^n \simeq 1 \pm nx$$
$$(1 \pm x)^2 \simeq 1 \pm 2x$$
$$\sqrt{1 \pm x} \simeq 1 \pm \frac{x}{2}$$
$$\frac{1}{\sqrt{1 \pm x}} \simeq 1 \mp \frac{x}{2}$$
$$e^x = 1 + x + \frac{x^2}{2!} + \cdots \simeq 1 + x$$
$$\ln(1 + x) \simeq x$$
$$\sin x = x - \frac{x^3}{3!} + \frac{x^5}{5!} + \cdots \simeq x$$
$$\cos x = 1 - \frac{x^2}{2!} + \frac{x^4}{4!} + \cdots \simeq 1 - \frac{x^2}{2}$$
$$\lim_{x \to 0} \frac{\sin x}{x} = 1$$

Half-Angle Formulas

$$\sin^2 \theta = \tfrac{1}{2}(1 - \cos 2\theta)$$
$$\cos^2 \theta = \tfrac{1}{2}(1 + \cos 2\theta)$$

Hyperbolic Functions

$$\sinh x = \tfrac{1}{2}(e^x - e^{-x})$$
$$\cosh x = \tfrac{1}{2}(e^x + e^{-x})$$
$$\tanh x = \frac{\sinh x}{\cosh x}$$